Engineering Tenders, Sales and Contracts

STANDARD FORMS AND PROCEDURES

Engineering Tenders, Sales and Contracts

STANDARD FORMS AND PROCEDURES

ANDREW PIKE
LL.B. (Birmingham)
Solicitor of the Supreme Court and Notary Public

LONDON NEW YORK
E. & F.N. SPON · SWEET & MAXWELL

To Anne, Arabella and Alexandra

First published 1982 by E. & F.N. Spon Ltd
and Sweet & Maxwell Ltd
11 New Fetter Lane, London EC4P 4EE
Published in the USA by E. & F.N. Spon
and Sweet & Maxwell
733 Third Avenue, New York NY 10017

Printed in Great Britain at
the University Press, Cambridge

ISBN 0 419 12530 2

British Library Cataloguing in Publication Data

Pike, Andrew

Engineering tenders, sales and contracts:
standard forms and procedures.
1. Engineering—Contracts and specifications—
England
I. Title
338.7′620 KD1641

ISBN 0-419-12530-2

Library of Congress Cataloging in Publication Data

Pike, Andrew

Engineering tenders, sales, and contracts.
Bibliography: p.
Includes index.
1. Engineering – Contracts and specifications – Great Britain. 2. Engineering – Contracts and specifications. I. Title.
KD1641.P53 1982 343.41′07862 82-10735

ISBN 0-419-12530-2 344.1037862

Contents

Part Two Standard tenders

(*Note:* For explanation of terms such as 'FOB', please refer to Section 7 of the Preliminary Text on Trade Terms)

Based on Conditions of Contract published by The Institution of Mechanical Engineers (IMechE), The Institution of Electrical Engineers (IEE) and The Association of Consulting Engineers (ACE)

Form number

Based on Conditions of Contract published by Fédération Internationale des Ingénieurs – Conseils (FIDIC)

Based on Conditions of Sale and Contract published by The British Electrical and Allied Manufacturers' Association Limited (BEAMA)

Based on Conditions of Contract published by Organisme de Liaison des Industries Métalliques Européennes (ORGALIME)

Based on Sub-Contract published by The Federation of Civil Engineering Contractors

Based on Conditions of Contract published by The Institution of Chemical Engineers (IChemE)

Part Three Additional Forms

Part Four Appendices

Appendix number

Preface

The purpose of this book is to help engineering suppliers and contractors with their sales, tenders and contracts, ranging from simple sales of stock articles in the United Kingdom to major projects on overseas sites. The book attempts to achieve that purpose through a Preliminary Text giving general guidance, and by a numerous and varied collection of Standard Tenders and Additional Forms for the reader to use. The book is directed to those engaged in mechanical, electrical and chemical engineering, not building or civil engineering.

In the hectic rushes so typical of the selling and tendering process, mistakes can easily be made, even by the most experienced. It is also a very time-consuming chore to draft the commercial/contractual terms of sales and tenders on a 'one-off' basis, when time is at a premium. This book is designed to enable readers to respond to enquiries quickly and efficiently, with the minimum of 'one-off' drafting.

Enquiries reach engineering suppliers and contractors in all shapes and forms. The enquiry (large or small) may be by letter, telex or telephone, with no proposed terms of payment, or any other commercial terms. If the prospective customer does provide such terms, the company may, if market conditions permit, make its own counter-proposals. The enquiry may be from a main contractor wishing to pass on formidable main contract risks, which the company is not prepared to accept. The possible circumstances are infinitely variable, but in all cases one must be able to decide correctly, respond speedily and state clearly what is required.

In a number of cases, usually major projects, the enquiry will be accompanied by voluminous tender documents, often prepared by consulting engineers. The company will be invited to insert its prices in the documents, and sign a standard formal tender. If the tender documents as offered are acceptable, all well and good, but examination will frequently disclose vital points which should, again if market conditions permit, be qualified or clarified in the tender. In particular, the proposed terms of payment may fail to secure the tenderer's reasonable interests. It is vital to be able to recognize and counter these problems. The information contained in this book will, I hope, contribute to that end.

ANDREW PIKE
31 March 1982
Whitchurch
Hampshire

Note for overseas readers

This book is primarily intended for a UK readership, but it should also have considerable use for overseas readers. This is especially true for those, in Commonwealth countries and the United States of America, whose legal systems have the same roots as the English legal system. For overseas readers, much of the Preliminary Text will be relevant, and so will the Forms relating to documents published by the international associations referred to. Overseas readers may also find assistance in the Forms relating to documents published by the British institutions, as they may find that there is no published document in their own countries corresponding to the relevant British document.

Acknowledgements

My particular thanks are due to the following friends and colleagues who have read, and commented upon, portions of the manuscript: Mr J. N. Miller a Director of Permutit-Boby Limited, a member of the Portals Group; the late Mr B. E. Smith who was also, until his most untimely death, a Director of Permutit-Boby Limited; Mr L. E. Day, of Commercial & Political Risk Consultants Limited; and Mr R. E. Booth, of J. H. Minet & Co. Limited.

I should also like to thank all those who have, over the years, enabled me to accumulate the knowledge necessary to write this book, and Mrs Debbie Collier for typing the manuscript.

I gratefully acknowledge the help of The Institution of Electrical Engineers and of The Institution of Mechanical Engineers, who kindly gave permission for the reproduction of their Model Forms in this book. These forms are obtainable as stated on their title pages.

I also gratefully acknowledge the help of The British Electrical and Allied Manufacturers' Association Limited, who kindly gave permission for the reproduction of their Conditions and Contract Price Adjustment Clauses in this book. These forms are obtainable from BEAMA at 8 Leicester Street, Leicester Square, London WC2H 7BN.

Notice

PART ONE

Preliminary Text

1: Introduction

Description of the book

Part One, the Preliminary Text of this book, is designed to give general information and guidance to companies involved in engineering sales, tenders and contracts in the mechanical, electrical and chemical engineering fields.

Sections 2 to 8 of the Preliminary Text deal in order with the seven fundamental elements of the Standard Tenders which form Part Two of this book, namely Prices, Programme, Terms of Payment, Conditions of Contract, Contract Price Adjustment, Trade Terms and Validity. Basically, the Standard Tenders have been drafted in the tenderer's favour. They are terms which most tenderers might prefer to have, but cannot always obtain. Obviously, each reader will need to adapt the suggested forms to his own particular field of operations and market conditions.

The remaining sections of the Preliminary Text, numbered 9 to 15, deal with allied subjects, such as Bonds and Financial Guarantees.

Some sections of the Preliminary Text deal with subjects already well covered by other publications, such as the publications of the Export Credits Guarantee Department (ECGD) on export credit insurance and their other services. In such cases, the reader is referred to those publications, but a basic outline is given in this book, which should be sufficient to guide the tenderer along the right lines.

The other publications referred to in this book are gathered together in a Bibliography, for ease of reference.

Part Two, the Standard Tenders, provides ready-made forms of tender or quotation for a very wide range of projects and sales. The Table of Contents shows that there is a suitable tender for almost every purpose which is likely to be required by a UK tenderer. The forms are also readily adaptable and variable, without undue cross-referencing between different sections or paragraphs of the forms. For ease of use, each form is self-contained. The reader is not asked to flip to and fro in the book in order to find standard paragraphs on such matters as VAT. Appendix I, the Table of Suitability of Conditions of Contract, indicates which conditions of contract are suitable for which tenders and quotations.

Part Three, the Additional Forms, provides variations and supplements to the Standard Tenders. For example, nearly all the Standard Tenders ask for a Confirmed Irrevocable Letter of Credit. This will by no means be always realistic. Therefore, Part Three contains a wide variety of substitute Terms of Payment, which can readily be used without consequential amendment to other sections of the Standard Tender selected for use.

Part Four, the Appendices, besides the Table of Suitability of Conditions of Contract, contains the full text of the Model Forms published by The Institution of Electrical Engineers (IEE) in association with The Institution of Mechanical Engineers (IMechE) and The Association of Consulting Engineers (ACE) (commonly called the IMechE/IEE Model Forms), of The British Electrical and Allied Manufacturers' Association Limited (BEAMA) Conditions of Contract, and of the BEAMA Contract Price Adjustment Clauses referred to in the Standard Tenders. Clearly, it would have been impossible to include all the conditions of contract referred to in this book. Those selected for reproduction in the Appendices are, it is hoped, those most likely to be useful to UK readers.

General law of contract

In this book 'tender' refers to any form of offer or quotation to provide goods and/or services, and 'contract' means any legally binding agreement to provide goods and/or services.

Under English law, contracts are formed by offer and acceptance. A tender or quotation is an offer, and a letter of award or order is an acceptance. A qualified or conditional response from the customer, such as a letter of award requiring the customer's standard conditions of purchase to apply, or delivery within a lesser time than that quoted, is not an acceptance, but a counter-offer, which may then be accepted by the tenderer. Acceptance may be express or implied, e.g. by proceeding to deliver the goods.

In practice, many communications may pass to and fro before a contract is formed, and it is very often difficult to determine exactly when the contract has been formed. If one or both of the parties have put forward their own standard terms of contract, the question of whose terms have been accepted is also frequently obscure. Disputes and litigation regularly occur on these points. The way to avoid this problem is for the parties to agree clearly, when making their contract, whose conditions of contract shall apply.

The parties should also agree exactly what documents are to form part of the contract. Frequently, it will be essential for the tenderer to see that his tender is incorporated in the contract documents, because his tender may vary or qualify the terms put forward by the customer in other documents. If he omits to see that his tender is incorporated in the contract documents, he will generally lose the right to rely upon the terms of his tender in any dispute with the customer.

If more than one language is involved, the ruling language and the status of any translations should be specified in the contract.

Lack of care in identifying which party's conditions of contract shall apply, which documents form part of the contract, and which language shall prevail in case of translation difficulties, frequently leads to a duplication of disputes, because a dispute about the actual performance of the contract is joined by a dispute about what the terms of the contract are.

There is a vast body of statute and case-law in the UK on the law of contract, in particular contracts for the sale of goods and the provision of services. However, it is impracticable and unnecessary for tenderers to have any detailed knowledge of the general law of contract, because the conditions of contract adopted by the parties will, to all practical intents and purposes, replace the general law of contract in so far as it affects the business transaction concerned. Readers who wish to learn in detail about the law relating to engineering contracts should refer to *Hudson's Building and Engineering Contracts* by I.N. Duncan Wallace.

Tenderers should be wary of entering into contracts which do not incorporate appropriate conditions of contract. For example, as pointed out in Section 5, Conditions of Contract, interest on late payments and consequential loss need to be expressly covered. Under English law, a creditor has no effective implied right to interest on late payments, and a supplier/contractor has very wide prospective liability for consequential loss, however out of proportion such loss may be to the contract price.

Tenderers should also be aware that, once a contract has been formed, it cannot be unilaterally altered. If conditions of contract have not been referred to before the relevant offer is accepted, and the contract formed, neither the tenderer nor the customer may introduce them afterwards, except by mutual agreement. For example, if the contract is informally made by telex or letter, conditions of contract incorporated in the customer's subsequent formal order, or in the tenderer's subsequent invoice, are not binding unless accepted, expressly or by implication, by the other party.

Letters of intent

The customer may be favourable to a particular tender, and ready to pay for work to be commenced immediately. However, he may not be prepared to award a contract for the entire work pending further negotiations with the tenderer, which may take a considerable time. In such cases, the customer may issue a 'letter of intent'. Much confusion surrounds these documents. In fact, the essential feature of a 'letter of intent' is that it should undertake to pay the tenderer for the work he is asked to do, whether or not the parties subsequently enter into a contract for the entire work, at ascertained rates. An example of a 'letter of intent' is given in this book (see p. 295). When accepted by the tenderer, such a 'letter of intent' becomes contractually binding. It is effectively a preliminary contract for the execution of the work, usually preliminary design work, covered by its terms.

2: Prices

The Prices sections of all Standard Tenders contain the following information:

(a) A description of the basic obligations the tenderer is prepared to undertake, whether it be supply, delivery to and erection on Site, export FOB, repairs at the tenderer's works, or any other possible combination of obligations. If a trade term such as FOB is used, it is always defined, in a subsequent section of the Standard Tender, by reference to Incoterms. If site work is offered, the geographical location of the Site is given.
(b) A description of the goods offered, by reference

to an accompanying Specification. Of course, in suitable cases, the goods may just as well be listed in the Prices section. In cases where site services are offered, the tender and accompanying documents should identify, as closely as possible, what is being offered, e.g. how many engineers or technicians on Site, at what rates per hour or day, and for how many hours per week.

Many of the Standard Tenders involving site work provide for payment on a time basis. Tenderers will frequently be reluctant to offer site services, especially on a foreign Site, for a lump sum.

(c) The quoted price, which will usually be a total of various prices shown in the Specification or other documents. The total should be shown in the Prices section, without leaving it to the customer to add up prices shown on accompanying documents. If an approximate price is being quoted, it is expressly stated. It is preferable not to use the term 'budget price', as this is open to misunderstanding, especially by foreign customers.

(d) The currency of payment, which is sterling in all Standard Tenders. The term 'pounds' is used in all Standard Tenders to UK customers, and 'pounds sterling' to export customers. Examples are also given of tenders involving foreign currencies. Contracts providing for payment in foreign currency impose on the exporter a risk of fluctuation of exchange rates, which is considered elsewhere in this book. An example is also given of a clause providing for tender comparison with foreign currency.

3: Programme

All the Programme sections begin with the words 'Our best estimate is that'. An example is also given of a firm commitment to a programme. Obviously, many customers will demand the latter.

Use of the phrase about 'best estimate' may well avoid any legal commitment to a specific completion date at all. In the absence of a fixed contractual date, English law implies an obligation to complete within a reasonable time which, of course, will almost always be very difficult to determine in the case of a contractual dispute.

Clause 26 (Delay in Completion) of IMechE Model Form A/1976/1978 reads:

If the Contractor fail to complete the Works in accordance with the Contract (except the maintenance thereof as provided in Clause 30 (Defects after Taking Over) and such tests as are to be made in accordance with Clause 27 (Tests on Completion)) within the time fixed by the Contract for the completion of the Works or any extension of such time, or if no time be fixed, within a reasonable time, and the Purchaser shall have suffered any loss from such failure, there shall be deducted from the Contract Price the percentage named in the Appendix of the Contract Value of such portion or portions only of the Works as cannot in consequence of the said failure be put to the use intended for each week between the time for completion of the Works as aforesaid and the actual date of completion, but the amount so deducted shall not in any case exceed the maximum percentage named in the Appendix of the Contract Value of such portion or portion of the Works, and such deduction shall be in full satisfaction of the Contractor's liability for the said failure.

The Contractor's obligation is to 'complete . . . within the time fixed by the Contract . . . or any extension of such time, or if no time be fixed, within a reasonable time. . . .' A 'best estimate' included in the tender may not amount to a 'time fixed by the Contract' at all, even if the tenderer has also, as is done in the Standard Tenders, quoted a certain rate of liquidated damages for delay.

This clause, and equivalent clauses in certain other conditions of contract quoted in this book, require that the Purchaser shall have suffered some loss (however small) from the failure to complete on time. This is by no means universal in contractual clauses providing for liquidated damages for delay. In cases where the customer is, for example, a non-profit-making body, such as a hospital authority, it may be impossible for the customer to prove any loss at all, and therefore liquidated damages under the above clause would not be recoverable. Such matters as lack of, or delay in providing, an amenity to the public would not be a loss to the non-profit-making body.

The requirement in the clause that the liquidated damages shall be 'deducted' leaves doubt as to whether they may be recovered in any other way, e.g. by suing the Contractor, should the liquidated damages for delay exceed the sum, if any, owing from the Purchaser to the Contractor under the Contract.

The closing words of the clause state the usual position, i.e. that where liquidated damages are payable, the customer is entitled to such damages however small his actual loss may have been, and is entitled to no more, however great his loss may have been.

Where a tender offers site services as well as goods, especially in cases where there will be a gap in time between arrival of the goods on Site and the commencement of site work, it is wise to show in a suitable document, such as a bar chart or critical path network, how the programme is calculated. This document should show what is required to be done on the Site before the site services commence.

4: Terms of Payment

Place and currency of payment

All export Standard Tenders specify that all prices are payable in pounds sterling in the UK. Alternatives are given for other terms of payment, such as payment in foreign currency. The place of payment (whether in the UK or overseas) is material. If the exporter has contracted to accept payment abroad, in whatever currency, the customer has fulfilled his obligation at the moment of making such payment. The risk and responsibility of remitting the funds from the foreign country to the UK is imposed on the exporter. This is called 'extra-contractual transfer risk' and is insurable with ECGD or other insurers.

Advance payments

All appropriate Standard Tenders also require an advance payment, which is very common in export business, and increasingly common in the UK. The question of bonds or financial guarantees in respect of advance payments is covered elsewhere in this book.

United Kingdom

All Standard Tenders to UK customers provide for the balance of the price to be paid in accordance with the conditions of contract referred to in the tender. In the case of IMechE Model Form A/1976/1978, this means 95% against interim certificates, 97½% after the taking-over certificate, and the balance after the final certificate, which is issued after the defects liability period of twelve months.

Export

In export business, payment by the customer on 'open account', i.e. without any security for payment, is much less common than in the UK, because the UK exporter will frequently have small means of checking the credit-worthiness of the foreign customer, and because of the difficulty and expense of making claims for payment in foreign jurisdictions. The exporter will generally wish to achieve some security for payment, such as a letter of credit, or documentary drafts or Bills of Exchange, as described below.

United Kingdom exporters will also need to be especially concerned about retentions of part of the price by foreign customers for prolonged periods. All export Standard Tenders therefore provide for payment 'in full' on presentation of documents under the relevant letter of credit.

Letters of credit

Letters of credit (strictly called Irrevocable Documentary Credits) provide a method by which the exporter is enabled to claim payment at a bank in his own country upon presenting to that bank invoices, shipping documents such as Bills of Lading, or whatever other documents are required by the terms of the letter of credit.

Exporters should be familiar with the following useful publications on letters of credit and associated topics:

Guide to Documentary Credit Operations, ICC Publication No 305.
Standard Forms for Issuing Documentary Credits, ICC Publication No 323.
The Problem of Clean Bills of Lading, ICC Publication No 283.
Uniform Rules for a Combined Transport Document, ICC Publication No 298.

Croner's Reference Book for Exporters also contains a most useful section on letters of credit.

All export Standard Tenders provide for a Confirmed Irrevocable Letter of Credit (CILC), which guarantees that the UK exporter will be paid by a UK bank, if he complies with the terms of the letter of credit.

Confirmation of a letter of credit means that the bank in the UK advising the exporter of the establishment of the credit in his favour undertakes to the exporter that the credit will be honoured. Therefore, the exporter will be relying primarily on the creditworthiness of the bank in the UK, not on the creditworthiness of the foreign bank, or of the customer. The Standard Tenders require confirmation by an 'acceptable bank', so that the exporter may reject unsuitable banks. However, a foreign bank with an office in the UK may very well be 'acceptable'. The exporter may prefer to nominate his own UK bank, or, for example, list a few banks, such as the main UK clearing banks, which are 'acceptable' to him. If the exporter is based away from London, he may prefer to specify another city where it will be more convenient for him to present documents at the confirming bank.

An exporter who accepts an unconfirmed Irrevocable Letter of Credit (ILC) from a foreign bank relies upon the creditworthiness of that bank. The credit will be 'advised' to him by a bank in the UK, where he will present documents, without any engagement on the part of that bank. In the event of non-payment upon presentation of documents, the exporter has no grounds for complaint against the advising bank. Of course, exporters need have no hesitation in accepting ILC's from, for example, first-class American or Western European banks, but in all cases the question should be carefully considered. It is by no means unknown for foreign banks to default on ILCs because, for example, the foreign country concerned has a foreign exchange crisis.

Both CILCs and ILCs are 'irrevocable'. They cannot be amended or cancelled without the consent of the beneficiary. Revocable Letters of Credit, which are only rarely encountered, can be amended or cancelled at any time, and therefore provide the exporter with no security for payment.

The export Standard Tenders also give full details of the CILC required, in a form following as closely as possible the *Standard Procedure for the Issuing of Documentary Credits by Cable, Telegram or Telex* (App VII *Standard Forms for Issuing Documentary Credits*), ICC Publication No 323. If these details are shown to the foreign bank by the customer, the bank should be able to understand them readily, and formulate a clear message to the UK confirming or advising bank. This should help to avoid the confusion which very often surrounds the establishment of letters of credit.

The procedure for the establishment of a letter of credit is shown in Fig. 4.1. When terms have been agreed, the customer instructs his bank to establish the letter of credit, by informing the bank of the terms of payment contained in the contract. Naturally, as the bank is undertaking obligations on behalf of the customer, it may require security from him, such as a deposit of part or the whole of the cash required. Official permission to establish the letter of credit may also be required under the foreign law.

The foreign bank will then communicate with the UK bank, either by mail, or by cable, telegram or telex. If the means used is other than mail, the foreign bank will quote its 'test number' which is a code number known only to other banks. This is because non-mail communications will not be manually signed by the foreign bank's officials, and will not be on the foreign bank's official stationery. Non-mail communications may therefore be more easily forged.

The ICC standard forms for mailing of letters of credit are efficiently designed, so that typing only needs to take place once. The foreign bank will usually send to the UK bank one copy called 'Advice for the Beneficiary' and another called 'Advice for

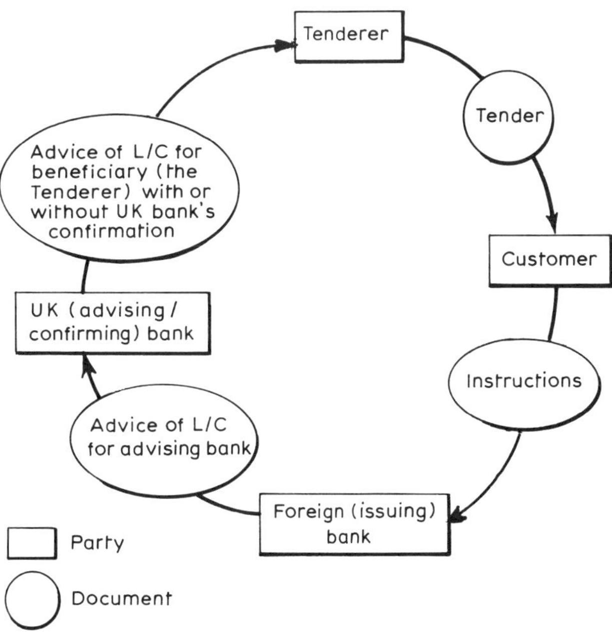

Fig. 4.1 Procedure for establishment of letter of credit

the Advising Bank'. The latter will indicate whether or not the UK bank is requested or authorized to add its confirmation. If the UK bank is only 'authorized' to add its confirmation, it may or may not do so. If it is 'requested' to add its confirmation, it will do so. It should be added here that UK banks will sometimes confirm ILCs from foreign banks at the request of the UK exporter, if he will pay their charges for so doing.

The UK bank will only be willing to advise or confirm the letter of credit for the foreign bank if the terms for doing business between those banks are met. For example, the UK bank may not be prepared to confirm unless the foreign bank has deposited with the UK bank part or the whole of the cash required.

The UK bank will then issue the 'Advice for the Beneficiary' to the exporter, with a notification indicating either (in the case of an unconfirmed ILC) that the Advice is sent to the exporter without engagement on the part of the UK bank, or (in the case of a CILC) that it is confirmed by the UK bank. Such notification may be endorsed on the Advice, or separate.

This involved procedure shows how important it is that the instructions to the foreign bank should be clear in the first place. The best thing that the tenderer can do towards this end is to make the terms of payment clear in his tender.

If the contract provides for increases in the contract price by means of contract price adjustment (CPA), the letter of credit amount cannot definitely be known until the CPA has taken place. Therefore, the export Standard Tenders indicate that the letter of credit is 'to be increased according to any contractual increases in the price'.

The time required for completion of the contract may be unexpectedly extended, and therefore the export Standard Tenders indicate that the letter of credit is 'to be extended according to any increase in the period required for execution of our respective obligations under the Contract'.

Letters of credit may be available by payment, acceptance or negotiation. All the export Standard Tenders stipulate for a 'Credit available with confirming bank by payment'. Under this, the simplest method, the exporter presents the documents required at the confirming bank and is paid, without the need for any further documents. If a letter of credit is 'available with advising bank by payment', in the case of an unconfirmed ILC, the same procedure applies.

Letters of credit available by acceptance or negotiation involve the use of a 'draft' or 'Bill of Exchange'. The exporter has no reason normally to ask for letters of credit to be available otherwise than by payment, but the procedures in the case of letters of credit available by acceptance or negotiation are also explained here, as they are quite often encountered in practice.

Section 3 of the United Kingdom Bills of Exchange Act 1882 defines a Bill of Exchange as:

> an unconditional order in writing, addressed by one person to another, signed by the person giving it, requiring the person to whom it is addressed to pay on demand or at a fixed or determinable future time a sum certain in money to or to the order of a specified person, or to bearer.

Suppose an exporter receives a CILC which is expressed to be available with the confirming bank by acceptance of his draft 'at sight drawn on' the confirming bank, accompanied by the requisite shipping documents. He will present, with the shipping documents, a draft in something like the following form:

(Exporter's address and date)

£________

At sight pay this Bill of Exchange to our order the sum of pounds sterling value received. Drawn under Credit No dated of (name of foreign issuing bank) and Credit No of (name of UK confirming bank).

(Exporter's signature)

To: (Name and address of confirming bank)

In modern practice, 'value received' is often replaced or expanded by adding e.g. the name of the vessel, a description of the contract, and details of the goods.

The draft being 'at sight' means that it is payable immediately. It is 'drawn on' the confirming bank by being addressed to it.

If all the documents, including the draft, are in order, the confirming bank will, as the payment is due immediately, pay the exporter. However, if the CILC is expressed to be available with the confirming bank by acceptance of the exporter's draft at say '90 days after sight' drawn on the confirming bank, the draft will be in something like the following form:

(Exporter's address and date)

£________

At ninety days after sight pay this Bill of Exchange to our order the sum of pounds sterling value received.

Drawn under Credit No dated of (name of foreign issuing bank) and Credit No of (name of U K confirming bank)

(Exporter's signature)

To: (Name and address of confirming bank)

This is commonly called a 'usance' draft, as opposed to a 'sight draft'.

If all the documents, including the draft, are in order, the confirming bank will accept the draft by signing it, and return it to the exporter. The confirming bank, by acceptng the draft, accepts legal liability to pay the holder of the Bill when it falls due ninety days later.

The exporter may hold the Bill and present it to the confirming bank for payment ninety days later, when the confirming bank will pay him. However, if he wishes to receive cash immediately, he may sell the Bill to a person or institution in business for the purpose. Naturally, the purchaser will pay the exporter rather less than the face amount of the Bill, and the difference is called a 'discount'. The process of thus selling a Bill is therefore called 'discounting' it and the institutions specializing in purchasing such Bills are called 'discount houses'. The exporter may well use his own bank to discount Bills.

The Bill is transferred to the purchaser by endorsing it, i.e. by the exporter signing his name on the back of the Bill. The purchaser then presents the Bill to the confirming bank at the due time, and is paid. Transfer may take place more than once. The transfer of Bills of Exchange is usually called 'negotiation'. Bills of Exchange are one of the classes of document known as 'negotiable instruments'.

A letter of credit available by negotiation will state that it is available, e.g. 'by negotiation against presentation of the documents detailed herein and of your drafts at 180 days after sight drawn on (name of the customer)'. The exporter's draft will be in something like the following form:

(Exporter's address and date)

£________

At one hundred and eighty days after sight pay this Bill of Exchange to our order the sum of pounds sterling value received. Drawn under Credit No dated of (name of foreign issuing bank) and Credit No of (name of U K confirming bank)

(Exporter's signature)

To: (Name and address of customer)

If all the documents, including the draft, are in order, the U K bank will negotiate the draft and pay the exporter the agreed sum. This will be the face amount of the draft, less the bank's discount or interest and commission. The U K bank will then forward the draft to the foreign issuing bank, with the other documents presented under the letter of credit, and will be paid by that bank.

If the U K bank has confirmed the letter of credit, i.e. it is a CILC, the exporter is not concerned with the question of whether the U K bank really is paid. The U K bank is 'without recourse' to the exporter.

If, on the other hand, the U K bank is only an 'advising' bank and has not confirmed the letter of credit, i.e. it is an ILC, the U K bank has 'recourse' to the exporter if it is not paid. That is to say, the exporter will be liable to the U K bank for the face amount of the draft. This important distinction is based on Article 3, *Uniform Customs and Practice for Documentary Credits,* ICC Publication No 290 (UCP).

All the export Standard Tenders request a CILC with partial shipments and transhipment allowed. Both these are allowed, unless specifically prohibited by the terms of the letter of credit (see Articles 21(a) and 35(a), UCP). However, as the customer will have to instruct his bank on these matters, it is as well to cover them expressly in the tender.

Bills of Lading stating that the goods are loaded on deck will be refused, unless specifically authorized by the letter of credit (see Article 22(a), UCP). Therefore, this should be referred to in the tender, if the goods are suitable for shipment 'on deck', and the export Standard Tenders so provide.

Banks will, unless otherwise instructed by the terms of the letter of credit, refuse documents presented to them later than twenty-one days after the date of issuance of the Bills of Lading or other shipping documents (see Article 41, UCP). The export Standard Tenders leave a blank for the exporter to request the period of time allowed. Usually, twenty-one days is enough. In the case of air freight, a lesser time will be reasonable. The purpose of imposing a time limit is to allow the customer to be in possession of the documents of title to the goods in good time for their arrival at the destination. Without such documents, he will not generally be able to collect the goods, because the carrier will not release them to him. However, in such cases, the carrier will often in practice release the goods, in return for an indemnity from the customer, if he is considered sufficiently trustworthy or can give the carrier security, such as a bank guarantee.

The export Standard Tenders stipulate that the letter of credit shall be subject to the UCP, which is a voluntary code almost universally accepted by banks.

The documents required to be presented under a letter of credit require little comment. Notes referring to documents required in various export Standard Tenders give certain specific points of information. It is not in the tenderer's interest to go into great detail in describing the documents required. The UCP at various places, especially Article 33, allows banks to accept documents as tendered, where there is lack of detail in the definition of such documents given in the letter of credit. The terms of a letter of credit must be strictly complied with by the exporter. It is no good presenting documents which are 'just as good as' those required by the letter of credit, or presenting them even one day late. The bank is required to carry out the terms of the letter of credit rigorously. If it does not, it may itself incur legal liability to the issuing bank or the ultimate customer, or fail to obtain re-imbursement. The bank to which documents are presented cannot become involved in the details of the dealings between the exporter and the customer.

Sooner or later, exporters are likely to find themselves unable to comply with the terms of a letter of credit, e.g. because documents are delayed, and are presented more than twenty-one days from the date of issuance of shipping documents. In that case, it may still be possible to persuade the bank to pay, provided it is given an acceptable indemnity by the exporter. This is, of course, undesirable from the exporter's point of view, because he cannot really be sure that the money is safe until the indemnity is released, usually a long time afterwards. However, it is better than being denied payment altogether.

Letters of credit by instalments

In the case of major projects over a long period, where the customer will normally not be prepared to open a letter of credit for the whole contract price valid for the whole period, the contract may provide for a letter of credit by instalments. An example of such Terms of Payment is given in this book. In view of bank charges, which are often charged quarterly or half-yearly as a percentage of the face value of the letter of credit, the customer's objections are understandable. He may also have to deposit cash with the issuing bank in order to open the letter of credit, related to its face value, and such cash will be tied up for the whole period of validity of the letter of credit. However, if an instalment is not duly established in accordance with the contract, that is a matter contractually between the exporter and the customer. The issuing and advising banks are not affected by the terms of the contract between the exporter and the customer. The exporter is only entitled to look to the issuing and advising banks to fulfil their obligations precisely as stated in the letter of credit itself. If a new instalment is due, the customer must instruct the issuing bank accordingly, and that bank will arrange for the advising bank in the UK to issue an amendment increasing the face amount of the letter of credit.

Revolving letters of credit

This expression means a letter of credit under which the amount is renewed or reinstated without specific amendment. A credit revolving by time would be, for example, a credit available for £10 000 per month for twelve months. A credit revolving by value would be, for example, a credit whose amount was automatically reinstated upon utilization, up to a given overall maximum.

However, in practice the term is also sometimes taken to refer to letters of credit by instalments. The distinction is fundamental. Under a revolving letter of credit, as correctly defined, the bank's obligation to pay future instalments is stated on the face of the letter of credit. The obligation is therefore not only that of the customer but of the issuing bank and, if the letter of credit is confirmed, of the confirming bank also. An example of Terms of Payment calling for a true revolving letter of credit is given in this book.

Bills of Exchange

These documents have been introduced as ancillary documents used in relation to letters of credit. However, in cases where no letter of credit is provided for in the terms of payment, payment may be provided for by Bills of Exchange alone, either 'at sight' or at an 'usance', i.e. a period of credit.

Croner's Reference Book for Exporters contains useful information on Bills of Exchange and exporters should also be familiar with *Uniform Rules for Collections*, ICC Publication No 322, which sets out the relevant banking procedures.

The procedure for payment by Bill of Exchange 'at sight' is known as 'Documents against payment' or 'D/P', because the customer is able to collect the documents of title to the goods from a bank in his local area in return for payment. A sight draft for this purpose might read:

(Exporter's address and date)

£________

At sight pay this First Bill of Exchange (Second of this tenor and date unpaid) to our order the sum of pounds sterling value received. Documents against payment. All charges for account of drawee.

(Exporter's signature)

To: (Name and address of customer)

The 'drawee' is the customer. This draft should be drawn and signed in duplicate. The duplicate will be in the same terms, except that it will say 'pay this Second Bill of Exchange (First of this tenor and date unpaid)'.

The exporter should endorse both drafts by signing them on the back. They are then handed to a UK bank, with the relevant shipping documents, with appropriate instructions. The UK bank forwards the documents to the nominated bank ('the collecting bank, with the relevant shipping documents and draft to the customer. He accepts it by signing it and, as it is immediately payable, pays the collecting bank. The shipping documents are then released to him.

This procedure gives fairly good security to the exporter, but if the customer does not pay, the exporter has the goods back on his hands, with freight charges, possible demurrage, bank charges etc. Possibly the goods will be unsaleable, or saleable only at a loss.

Payment also takes place in the foreign country, which places the onus of remitting from that country to the UK on the exporter. Sometimes, banks will 'negotiate' or purchase the draft from the exporter, so giving him immediate cash before payment by the customer. In such cases, banks will, unless special terms to the contrary have been agreed, have 'recourse' to the exporter if the customer defaults.

The procedure for payment by Bills of Exchange at an 'usance' is known as 'Documents against acceptance' or 'D/A', because the customer is able to collect the documents of title to the goods from a bank in his local area in return for his accepting the exporter's draft. Obviously, this entails trusting the customer to pay on time.

An 'usance' draft for this purpose might read:

(Exporter's address and date)

£________

At one hundred and eight days after sight pay this First Bill of Exchange (Second of this tenor and date unpaid) to our order the sum of pounds sterling value received. Documents against acceptance. All charges for account of drawee.

(Exporter's signature)

To: (Name and address of customer)

The same procedure is followed as in 'Documents against payment', except that the collecting bank releases the shipping documents upon the customer's acceptance of the draft, and presents it to him for payment when due.

In the case of D/P or D/A, the law of most countries requires 'protesting' of the Bill, either for non-acceptance or non-payment, which involves the re-presentation of the Bill to the customer by a Notary Public as a preliminary to legal proceedings.

Payment for site work

As payment for site work usually requires a document from the customer, e.g. an Interim Certificate from the customer's Engineer, or the customer's approval evidenced by his counter-signature of the exporter's invoice, payment for such work may be much more difficult to secure than for goods shipped. The exporter may, however, protect himself to a certain extent by framing the price for goods shipped so that it includes most or all of the profit on the project, and by suspending site work unless he is promptly paid for it. Suspension of site work may or may not be a credible threat, depending on whether the customer is holding 'on demand' bank guarantees, or is capable of denying the exporter's employees exit from the foreign country concerned, and on all other circumstances of the case.

Instalment sales

Where any sale provides for payment by instalments, default in payment for more than a short time of any principal or interest should result in the entire price falling immediately due, and a term should be included to this effect.

5: Conditions of Contract

General

The Standard Tenders are based on the conditions of contract published by various well-known institutions. Tenderers are well advised, where they have the opportunity to nominate conditions of contract, to select a published form, rather than the tenderer's own 'house' form. Published conditions of contract are much more likely to gain acceptance, because the customer will believe, usually quite rightly, that the tenderer's 'house' form is slanted in the tenderer's favour. Appendix I, the Table of Suitability of Conditions of Contract, indicates which conditions of contract are suitable for which tenders and quotations.

When the customer nominates conditions of contract, the tenderer should compare them with published conditions of contract designed for use in conjunction with projects of a similar nature. Very often, it will be possible to detect that, although the customer's nominated conditions of contract appear to be his 'house' form, or specially drafted for a particular project, they are really copied in large part from published conditions of contract. This facilitates the tenderer's examination of the customer's conditions of contract, because clearly the important points to note will usually be those the customer has changed from the published form.

British tenderers have little reason to depart from The Institution of Mechanical Engineers (IMechE) and British Electrical and Allied Manufacturers Association Limited (BEAMA) forms, which are strongly recommended to readers, except where the customer will probably have a preference for internationally drafted forms, such as Fédération Internationale des Ingénieurs-Conseils (FIDIC) or United Nations Economic Commission for Europe (UNECE) forms.

IMechE forms

These forms are in fact published by three institutions – The Institution of Mechanical Engineers, The Institution of Electrical Engineers, and the Association of Consulting Engineers. All three institutions recommend Model Forms A, B1, B2 and B3, covering respectively home contracts with erection, export contracts for the supply of plant and machinery, export contracts with delivery FOB, CIF or FOR with supervision of erection, and export contracts including delivery to and erection on site. Model Form C, covering home contracts without erection, is recommended by the IMechE and IEE. Model Form E, covering home cable contracts with installation, is recommended by the IEE. Model Forms A, B3 and E are very similar, as they all cover contracts involving erection on site.

Within their scope, these forms are suitable for projects in the UK or overseas, however large. For some applications the IMechE forms are too complex, being orientated towards projects of medium to high value, rather than to simple or lower-value sales. In such cases the equivalent BEAMA forms are appropriate. Both IMechE and BEAMA forms protect the tenderer's interests very well.

This book does not attempt detailed commentary on published conditions of contract. For further information on the IMechE forms, see *Further Building and Engineering Standard Forms* by I. N. Duncan Wallace, which deals with the now superseded 1966 edition of Model Form A, but which is nevertheless still extremely useful. See also *Electrical and Mechanical Engineering Contracts* by K. F. A. Johnston, which deals with delay in completion, defects after completion, and liability for accidents and damage under Model Forms A, B1, B2, B3 and C.

FIDIC forms

The FIDIC Conditions of Contract (International) for Electrical and Mechanical Works (Including Erection on Site) (often called 'FIDIC M&E') are suitable for large overseas projects only. FIDIC also publishes conditions of contract for civil engineering works, with which this book is not concerned.

The only reasons why a British tenderer should nominate FIDIC M&E are that the customer has expressed a preference for them, or it is otherwise thought unsuitable to nominate a British form. IMechE Model Form B3 is as good or better from the contractor's point of view, whatever the size of the project, and is less complex.

Very generally, Model Form B3 is more likely to be considered favourably in Commonwealth and ex-Commonwealth countries than elsewhere. In the Middle East, it is usually more suitable to nominate FIDIC M&E, if the tenderer is given the opportunity to nominate conditions of contract.

BEAMA forms

These numerous forms of conditions of sale and contract are generally suitable for less complex or smaller sales or projects. They also cover many situations for which the IMechE forms are not intended, e.g. the sale of stock and catalogue articles.

UNECE forms

These forms are rather exceptional because they appear in two series, 188 and 574, the latter being designed for East–West trade. For all practical purposes, 574 are much the same as 188. One major drafting difference is that industrial disputes are not specifically stated to be 'cases of relief', entitling the parties to relaxation of their contractual obligations concerning timely performance and payment, under 574, which only defines cases of relief in general terms as: 'Any circumstances beyond the control of the parties intervening after the formation of the contract and impeding its reasonable performance.' (See Clause 10.1, UNECE Conditions No 574.) A list of matters, such as industrial disputes and fire is given in 188, as well as describing cases of relief in general terms as 'other circumstances . . . beyond the control of the parties'. It is submitted that the legal position under the 188 and 574 series in this respect is practically indistinguishable, at least under English law.

Another major drafting difference is that no specific arbitration procedure is laid down in the 574 series, whereas ICC arbitration is specified in the 188 series. This is not mentioned in the Standard Tenders because, unless otherwise agreed, the contract will be governed by the law of the vendor's country, i.e. English, Scots or Northern Ireland law, which lay down general arbitration procedures in the absence of specific procedures agreed by the parties.

Unlike other UNECE Conditions, UNECE Conditions No 730 are not divided into series for use respectively in East–West trade and other trade.

K. F. A. Johnston's book, *Electrical and Mechanical Engineering Contracts*, provides a commentary on the UNECE forms in relation to delay in completion, defects after completion, and liability for accidents and damage. Several short commentaries on the UNECE forms are also published by BEAMA.

ORGALIME forms

These are several forms designed for somewhat diverse purposes. Organisme de Liaison des Industries Métalliques Européenes (ORGALIME) Conditions for the Provision of Technical Personnel Abroad are in two parts. Both should be used where the contract is for the provision of technical personnel abroad, independent of other transactions. Part 1 only should be used in cases where the provision of technical personnel abroad is ancillary to another transaction entered into on the basis of UNECE conditions of contract.

ORGALIME General Conditions for the Import and Export of Semi-Processed Goods and Components for Incorporation in other Goods, is a simplified form of UNECE Conditions No 730.

The purposes of the ORGALIME Model Forms of Maintenance and Processing Contracts is self-evident.

FCEC Sub-Contract ('the ICE Sub-Contract')

This form is dealt with in Section 12, Sub-Contracts. It is impossible to generalize about the form which should be adopted when tendering as a sub-contractor, because this will depend on the terms of the main contract, and on the requirements of the main contractor. If tenderers are not compelled to offer a 'back-to-back' sub-contract (the meaning of which is explained in Section 12) they should select, and tender on the basis of, the IMechE, BEAMA or other conditions of contract which suit them best. It is the main contractor's problem if there is a shortfall between the sub-contractor's obligations to the main contractor under the sub-contract, and the main contractor's obligations to the employer under the main contract, in relation to the sub-contractor's work.

I. N. Duncan Wallace's book, *Further Building and Engineering Standard Forms*, contains a detailed commentary on the ICE Sub-Contract.

IChemE forms

These are two specialized and complex forms for UK chemical process plants. One is lump-sum, the other re-imbursable.

Interest

In order to act as an incentive for the customer to pay, it is very important to obtain interest on delayed payments. Clause 34(iii) of IMechE Model Form A/1976/1978 reads:

If the payment of any sum payable under Sub-Clause (i) of this clause shall be improperly delayed by the Purchaser or the Engineer interest at the rate of two per cent per annum over the Bank of England minimum lending rate from time to time in force on the amount of the delayed payment for the period of the delay shall be added to the Contract Price.

However, references in published conditions of contract to the Bank of England minimum lending rate are now obsolete, because on 5 August 1981 the Bank of England announced that, from 20 August 1981, it would no longer continuously publish a minimum lending rate. Accordingly, provisions in the Standard Tenders about interest refer to commercial bank base lending rates. Provisions for interest are included in all appropriate Standard Tenders.

Under the UNECE forms, the Vendor must give notice to the Purchaser to obtain interest. See for example Clause 8.7 UNECE General Conditions No 188, which reads in part:

. . . if the Purchaser delays in making any payment, the Vendor shall on giving to the Purchaser within a reasonable time notice in writing be entitled to the payment of interest on the sum due at the rate fixed in paragraph E of the Appendix from the date on which such sum became due.

In some countries, customers, especially public authorities, will not pay interest for legal and/or religious reasons.

'Consequential loss'

Clause 22 of IMechE Model Form A/1976/1978 provides:

Subject as provided in Clause 26 (Delay in Completion) for the deduction of liquidated damages for delay, the Contractor shall not be liable to the Purchaser by way of indemnity or by reason of any breach of the Contract for loss of use (whether complete or partial) of the Works or of profit or of any contract that may be suffered by the Purchaser.

It is very important from the tenderer's point of view to have such a clause incorporated in any resulting contract. Otherwise, the tenderer may, for example, become liable for loss of production of a large industrial plant, through late supply of a part costing a few pounds. The published conditions of contract referred to in this book, where appropriate, contain an equivalent clause.

Tenderers should avoid drafting such clauses of their own, without legal advice. In particular, it is wise to avoid the term 'consequential loss', in contracts, as its meaning in law is not sufficiently certain.

Technical guarantees of plant performance

Suppliers/contractors are frequently required to guarantee that the plant will achieve a certain technical performance. Typically, the guarantee will be that, given a certain input of raw material, a certain output of product will be achieved. Sometimes, the guarantees will extend to the amount of power, consumables and other things required for the operation of the plant, and the quality of the effluent. The contract will often provide that, if the guaranteed performance is not achieved, the customer may reject the plant. The right of rejection is also often made conditional upon the plant falling more than a certain small percentage below the guaranteed figures. Within that small percentage, the supplier/contractor may be required to pay liquidated damages.

The clear disadvantage of contractual guarantee figures from the supplier/contractor's point of view is that they leave precious little room for argument in the case of marginal performance. If the plant must produce X tonnes per day, with the customer having the right of rejection if it falls below X minus 5%, the customer is entitled to reject if the plant is only capable of X minus 6%. If the customer has made up his mind to reject, it is no good arguing that the 1% does not really matter, or can be compensated for by a slight reduction in price. The supplier/contractor is caught by the express terms of the contract.

During the progress of the contract, it may be that the guarantees become inapplicable, because circumstances change. This will most frequently be an unforeseen change in the quality of the raw material going into the plant. In the interests of the customer, the guarantees should state precisely what the guarantee figures are going to be, given a range of raw material quality. However, the customer by no means always does this. Therefore, the change invalidates the guarantee.

In such cases, the supplier/contractor is often asked to agree to fresh guarantee figures, but he is ill-advised to agree to this, and is under no obligation to do so. In the absence of contractual guarantee figures, the customer is entitled to whatever is a reasonable plant performance, in all the circumstances, and will only be able to reject the plant in practice if its performance is demonstrably bad.

UK Value Added Tax

A tenderer should always make clear, in a tender involving taxable supplies in the UK, whether his tender price is inclusive or exclusive of Value Added Tax (VAT). The better course is to quote a price exclusive of VAT, and to charge VAT extra, at the ruling rate at the appropriate time.

In all the Standard Tenders in this book involving taxable supplies in the UK, either the published conditions of contract referred to in the tender state that the price is exclusive of VAT, or a clause so stating has been included in the Standard Tender.

Exports are zero-rated, but generally the exporter must provide proof of export to the Customs & Excise, such as a copy of a Bill of Lading, a Certificate of Shipment from the shipping company, or a copy of an air waybill. The detailed requirements are set out in Customs & Excise Notices. Provided the exporter will be able to provide the Customs & Excise with proof of export as required, he need not deal with VAT in his tender.

Croner's Reference Book for Exporters provides very useful information on VAT for exporters.

Reservation of title

By means of a reservation of title clause, a supplier/contractor may stipulate that title to goods sold will not pass to the customer until payment in full, and that, if the customer sells the goods before payment in full takes place, the proceeds of sale will be held by the customer in trust for the supplier/contractor. Then, if the customer becomes insolvent, the supplier/contractor will have a prior right, over all other creditors, to repossess the goods or to receive the proceeds of sale. However, these clauses are fraught with difficulties, and their effect cannot be said to be clear in English law. If it is a foreign customer who becomes insolvent, the law of his jurisdiction, which is where the goods or proceeds of sale will be, would also be relevant.

The drafting of such clauses is extremely difficult, and even appears to be affected in some degree by the nature of the goods involved. If suppliers/contractors are interested in such clauses, specific legal advice should be taken.

Retention of title clauses may be useful in the occasional case of a customer's insolvency, but they are no substitutes for credit control, credit insurance, secure means of payment, and the other methods of protection normally used.

Arbitration

Settlement of disputes by arbitration, rather than litigation in the Courts, is usually in the interests of the supplier/contractor, because arbitration is private. The other advantages sometimes claimed for arbitration, such as speed or lesser expense than litigation, often prove to be illusory. Even privacy is sacrificed if, during or after the arbitration, the dispute reaches the Courts. All the published conditions of contract referred to in the Standard Tenders provide for arbitration.

6: Contract Price Adjustment

All suitable Standard Tenders provide for contract price adjustment (CPA) by reference to Clauses and Formulae published by the British Electrical and Allied Manufacturers' Association Limited (BEAMA). These are well-tried contractual provisions.

The Standard Tenders concerned refer to the general CPA Clauses and Formulae for electrical machinery published by BEAMA, one of which is for home contracts, and the other of which is for exports. These are reproduced in the Appendices to this book. The equivalents for mechanical plant are the BEAMA *Contract Price Adjustment Clause and Formula for use with Home Contracts – Mechanical Plant: (for which there is no other specific Formulae),* edition January 1979, corrected edition and the BEAMA *Contract Price Adjustment Clause and Formula for use with Export Contracts – Mechanical Plant: (for which there is no other specific Formulae),* edition January 1979, corrected edition. These are identical in substance to the equivalent electrical machinery Clauses and Formulae, except that paragraph (b) of each refers to 'Mechanical Engineering Industry', rather than 'Electrical Machinery Industry'. BEAMA also publishes many more specialized CPA Clauses and Formulae.

Of course, there are many published CPA Clauses and Formulae, other than those published by BEAMA, e.g. those associated with the Joint Contracts Tribunal's Standard Forms of Building Contract, but BEAMA CPA Clauses and Formulae are the best known in the British engineering industry.

BEAMA also operates a CPA Advisory Service, available by subscription, which circulates all relevant index figures referred to in the various BEAMA CPA Clauses and Formulae.

The following points on the BEAMA CPA Clauses and Formulae referred to in the Standard Tenders may be useful to readers:

(a) CPA is only available if the Contractor's costs are increased or reduced by reason of the specified causes, namely 'any rise or fall in labour costs or in the cost of material or transport'. Other causes of increased costs, e.g. increased cost of insurance, appear to give no right to CPA. Such elements of cost are fixed price.

(b) CPA by reference to the algebraic formulae only applies to variations in the cost of materials and labour. In the case of other permitted causes of CPA, e.g. increases in the cost of transport, CPA still appears to be obtainable, but the Contractor will have to prove the actual difference between the cost ruling at the date of tender and the cost in fact incurred.

When the algebraic formulae do apply, the increases in the Contract Price resulting from their application are payable regardless of the actual labour and material costs incurred by the Contractor.

(c) The 'Home' Clause and Formulae refer to a 'Contract Period', whereas the 'Export' Clause and Formula refers to a 'Delivery Period'. The former is suitable for contracts excluding or including erection, whereas the latter are not suitable for contracts including work on a foreign site, unless amended. Where export Standard Tenders do involve site work, an adapted BEAMA Clause and Formula is used.

(d) All Standard Tenders omit the CPA Clauses, incorporated in certain published conditions of contract, which do not refer to a formula. The calculation of CPA without the use of a formula is a very difficult process, involving mountains of paperwork and endless disputes, especially in export sales and projects.

If a tender is to be fixed price, the CPA section of any Standard Tender may simply be omitted.

7: Trade Terms

The trade terms offered in the tender should always, if possible, be defined by reference to a standard definition. International Chamber of Commerce (ICC) Incoterms are by far the most widely used definitions of trade terms. The ICC, besides publishing Incoterms themselves, also publish a larger booklet, *Guide to Incoterms,* ICC Publication No 354 (1980 edition). The current fourteen Incoterms, listed in increasing order of responsibility on the supplier, are:

Term in full	*Usual abbreviation*	*ICC official abbreviation*
Ex Works	—	EXW
Free Carrier (named point)	—	FRC
Free on Rail/Free on Truck (named departure point)	FOR/FOT	FOR
Free on Board Airport (named airport of departure)	FOB Airport	FOA

Term in full	*Usual abbreviation*	*ICC official abbreviation*
Free Alongside Ship (named port of shipment)	FAS	FAS
Free on Board (named port of shipment)	FOB	FOB
Cost and Freight (named port of destination)	C & F	CFR
Cost, Insurance and Freight (named port of destination)	CIF	CIF
Freight or Carriage paid to (named point of destination)	—	DCP
Freight or Carriage and Insurance paid to (named point of destination)	—	CIP
Ex Ship (named port of destination)	—	EXS
Ex Quay (Duty Paid) (named port of destination)	—	EXQ
or		
Ex Quay (Duties on Buyer's Account) (named port of destination)		
Delivered at Frontier (named place of delivery at frontier)	—	DAF
Delivered Duty Paid (named place of destination in the country of importation)	—	DDP

Perhaps the trade terms most commonly met with in export contracts from the UK are FOB, C&F, CIF, and FOB Airport. The Standard Tenders give FOB, C&F and CIF alternatives, and an example of FOB Airport is given in the Additional Forms.

Guide to Incoterms gives excellent advice and information on the use of the above trade terms, and little can usefully be added in this book. However, it should be emphasized that the risk really does pass to the buyer at the crucial point named in Incoterms. If, for example, a sale is CIF, the seller is under no contractual duty to see that the goods arrive at their destination. He discharges his obligation by tendering the shipping and marine insurance documents. If the goods have perished in the interval, whether the parties know of it or not, this makes no difference. The seller is still entitled to be paid the price, upon tendering the documents. Frequently, of course, the UK exporter will be tendering the documents to a bank under the terms of a letter of credit, and he is perfectly entitled so to do, regardless of what may have happened to the goods after they have passed the ship's rail at the port of shipment.

Under the FOB Incoterm, the seller does not necessarily obtain the Bills of Lading. His obligation is to:

> A.7 Provide at his own expense the customary clean document in proof of delivery of the goods on board the named vessel. . . .
>
> A.9 Render the buyer, at the latter's request, risk and expense, every assistance in obtaining a bill of lading. . . .

The 'clean document' will usually be a Mate's Receipt, which is generally used to obtain the Bill of Lading. However, the seller frequently in practice makes all necessary arrangements, including procuring the Bills of Lading. He must obtain the Bills of Lading, if they are required under the terms of the relevant letter of credit. The FOB Standard Tenders all require the seller to present Bills of Lading under the letter of credit, and therefore make it clear that it will be the seller's right and duty to obtain them. An alternative is given for Mate's Receipts in Part Three, Additional Forms.

Clearance through the foreign part, which is required by the term 'Delivered Duty Paid' and by all contracts which require delivery to a foreign Site, can be a very onerous task in some countries.

8: Validity

All Standard Tenders include a self-explanatory validity clause. Nevertheless, under English law, unlike some foreign legal systems, tenders may be withdrawn or amended at any time before communication of acceptance by the customer, despite such a validity clause. Obviously, a reputable tenderer would only take such action for weighty reasons.

In cases where an 'on demand' or other tender bond or guarantee has been given, unilateral withdrawal or amendment of the tender by the tenderer may cause forfeiture to the customer of the amount of the bond or guarantee. If the tenderer does not wish to adhere to his tender in such a case because, for example, he has made some miscalculation in the tender, his best course may be to remain silent and hope that another tenderer will be successful. If the customer calls him for negotiations, he may attempt to lose the order by means of his negotiating tactics. In tender bond cases, the project will usually be large and complex, and the order is unlikely to be given without detailed negotiations.

9: Bonds and Financial Guarantees

General

The term 'guarantees', as used in this section, does not refer to guarantees given by the supplier/contractor relating to the technical performance of a finished plant or product, but refers only to financial guarantees given by a third party in respect of the supplier/contractor's obligations under the contract or tender.

There is much use of confusing nomenclature in this field. Bonds and financial guarantees (collectively called 'bonds' in this section) given by suppliers/contractors to customers fall into two main types, namely, those which are 'on demand', and those which are not.

The Form of Guarantee incorporated in the IMechE Model Form A/1976/1978 reads as follows:

Whereas by an Agreement dated .... and made between .... (hereinafter called 'the Purchaser') and .... (hereinafter called 'the Contractor') the parties thereto entered into a contract as therein stated: Now we hereby jointly and severally guarantee to the Purchaser punctual true and faithful performance and observance by the Contractor of the covenant on his part contained in the said Agreement and undertake to be responsible to the Purchaser his legal personal representatives successors or assigns as Sureties for the Contractor for the payment by him of all sums of money losses damages costs charges and expenses that may become due or payable to the Purchaser his legal personal representatives successors or assigns by or from the Contractor by reason or in consequence of the default of the Contractor in the performance or observance of his said covenant but so nevertheless that the total amount to be demanded or recovered by the Purchaser his legal personal representatives successors or assigns of or from us as Sureties shall not exceed 15 per cent of the Contract Price.

This Guarantee shall not be revocable by notice or by reason of the death of us or either of us and our liability as Sureties hereunder shall not be impaired or discharged by any extensions of time or variations or alterations made given conceded or agreed (with or without our knowledge or consent) under the General Conditions referred to in the said Agreement or (where the Purchaser or the Contractor is a firm) by any change in the constitution of the Purchaser's or the Contractor's respective firms.

This is an example of a bond which is not 'on demand'. Such bonds are variously described as 'conditional', 'surety' or 'traditional'.

Under the above document, a Purchaser alleging breach of contract by the Contractor must, unless the Contractor and/or Sureties admit liability, or the Purchaser has withheld sufficient funds otherwise owing to the Contractor, pursue the Contractor for damages through litigation or arbitration, perhaps for years. If the Purchaser is successful, he will recover damages for breach of contract from the Contractor. Only if the Contractor is insolvent will the Sureties have to pay the damages to the Purchaser, up to the stated percentage limit of the Contract Price.

Therefore, the Purchaser only benefits from documents such as the above if he successfully establishes a financial claim under the Contract against the Contractor, who is then incapable of paying it. This type of bond is generally given by surety companies or insurance companies, not banks.

'On demand' bonds oblige the institutions providing them to pay the customer the stipulated sum on his demand, without any proof of loss, or that the supplier/contractor has defaulted in his obligations, or of anything else. Their terms vary widely, and they are generally given by banks. They are obviously very dangerous documents as far as the supplier/contractor is concerned. He will, of course, always be obliged to indemnify the institution concerned, if the bond is 'called'. The use of such bonds has rapidly increased in recent years, especially in Middle East projects. They are seldom, if ever, used in UK projects. They are sometimes also called 'unconditional bonds' or 'bank guarantees' and the banks sometimes refer to them as 'terminable indemnities'.

Suppliers/contractors may be required to provide customers with bonds of either main type in respect of any or all of the following:

(a) The tender ('tender bonds');
(b) The supplier/contractor's general performance of his obligations under the contract ('performance bonds');
(c) Advance payments made by the customer under the contract ('advance payment bonds'); and
(d) The early release of retention money under the contract ('retention bonds').

Tender bonds

These are also sometimes called 'bid bonds' and are practically invariably 'on demand'. Sometimes, an actual cash deposit is required as a condition of tendering, rather than a tender bond. Such bonds or deposits are intended to be forfeited if the tenderer withdraws his tender within the stipulated period of validity, or declines to accept the contract, if awarded to him. The customer's expenses in setting up the tender procedure, and evaluating tenders, may be very substantial, and the use of tender bonds or deposits is designed to safeguard him against frivolous tenders, which might cause such expenses to be wasted.

Performance bonds

The purpose of these is self-evident, and is explained above. In cases where a tender bond has been used, the tenderer is usually required to deliver his performance bond and sign the contract, in return for the release of the tender bond.

Advance payment bonds

Many contracts, especially for export projects, provide for advance payments (sometimes called 'mobilization fees' or 'mobilization payments') to the supplier/contractor. The customer is obviously concerned to see that the payee cannot depart with the money scot-free. He therefore often requires a bond in return for the advance payment. Such bonds are often reduced by stages, e.g. if the contract is for £10 million with a 20% (£2 million) advance payment, and the payee has executed work to the value of £3 million, he will be paid 80% of £3 million through interim or progress payments, i.e. £2.4 million, and the customer will release £600 000 of the £2 million advance payment bond. In this way, the customer has always either received the work for which he has paid, or holds an advance payment bond in respect of advance payments for which the work has not yet been executed.

If possible, it is desirable for the advance payment bond to stipulate that it shall not come into force until the advance payment is actually made. Otherwise, the customer may keep the bond, or even call it, without making the advance payment. In order to avoid such a possibility, advance payments are sometimes made by means of letters of credit in the U K, providing for payment upon presentation of the requisite advance payment bond.

Retention bonds

Many contracts provide for a percentage of the price, e.g. 5%, to be withheld until the end of a period called 'defects liability period' or 'period of guarantee' or 'maintenance period', or by other names. This period is often twelve months from completion of the work. Suppliers/contractors may often persuade customers to part with this retention money, in return for a bond. This often makes financial sense for the supplier/contractor, as the use of the retention money throughout the relevant period usually outweighs the bank or other charges for the bond, and the premiums for insuring it against 'unfair calling', if it is an 'on demand' bond.

Parent-company guarantees

Subsidiaries of substantial parent companies may sometimes be able to avoid giving bonds by giving a parent-company guarantee instead, thus saving charges which would otherwise have to be paid to institutions issuing bonds.

Procedure for establishment of bonds

The procedure for the establishment of bonds, like that for the establishment of letters of credit, is often attended by confusion, because of the involvement of one or more third parties between the tenderer and the customer.

The procedures involved for the establishment of, respectively, conditional bonds, 'on demand' bank guarantees involving a foreign bank, and 'on demand' bank guarantees not involving a foreign bank are shown in Figs 9.1 to 9.3.

In each case, the legal document binding the tenderer to pay is the indemnity or counter-indemnity given by him to the bank or other surety with whom he deals directly. This document will often be in general terms, to apply to all bonds or other obligations undertaken by the bank or surety at the

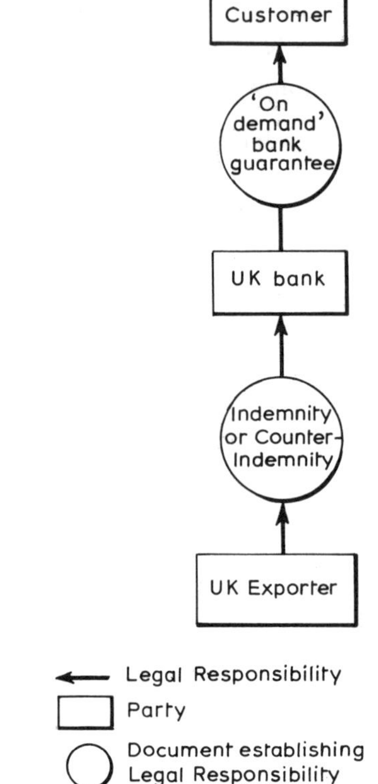

Fig. 9.3 'On demand' bank guarantees not involving foreign bank

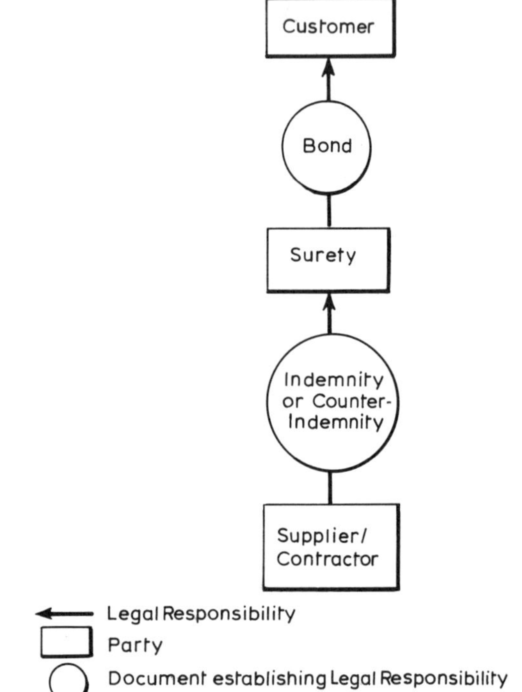

Fig. 9.1 Conditional bond

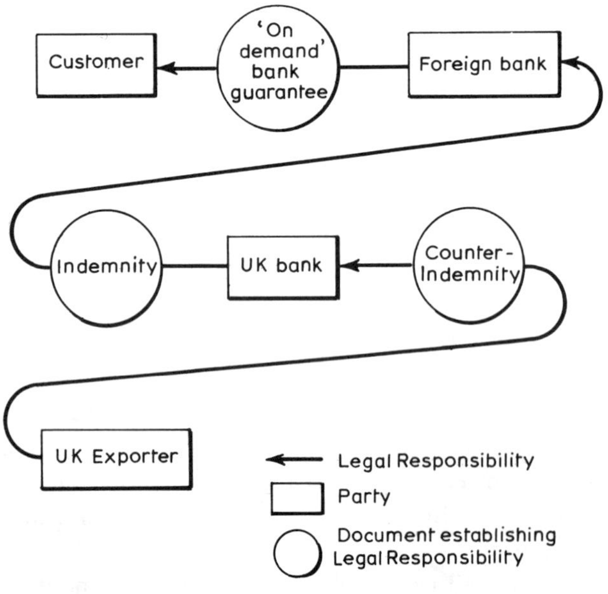

Fig. 9.2 'On demand' bank guarantees involving foreign bank

tenderer's request, rather than referring to a particular transaction or transactions.

It is important for the tenderer to be sure, in each particular case, that the bank or other surety with whom he deals will be prepared to undertake obligations in the terms requested by the customer. Some banks or sureties will, for example, be reluctant to issue bonds or indemnities with long expiry dates, or no expiry dates.

The procedure for the establishment of 'on demand' bank guarantees involving a foreign bank is required when the customer insists on a bond from a bank in his own country. In such a case, the foreign bank acts on the instructions of the UK bank, not of the exporter, and relies upon the UK bank for an indemnity, if the bond is called. This procedure is prevalent in the Middle East.

A foreign bank need not be involved, if the customer will trust to a UK bank's bond directly to the customer. This saves the exporter the foreign bank's charges.

If an 'on demand' bond is called, it is no good the exporter protesting to the UK or the foreign bank that the call is unjustified. That question is strictly between the exporter and the customer. The UK bank will indemnify the foreign bank on demand, and will promptly debit the exporter's account. It is up to the exporter to protect himself by means of 'unfair calling' insurance, as described below.

Procedure for release of bonds

Conditional bonds are very often formal documents executed under seal by the surety. When they are due for release, they should be physically recovered from the customer, if possible, and returned to the surety. Alternatively, a release in writing addressed to the surety should be obtained from the customer.

'On demand' bonds are, more often than not, established by telex or cable. In such cases, a release in writing addressed to the bank which has given the bond should be obtained from the customer. If this is a foreign bank, it should be ensured that the foreign bank then notifies the UK bank of the release. Until duly released, the surety or bank will generally consider itself still bound, and will continue to make periodic charges for keeping the bond in force.

In relation to tender, performance and retention bonds, release generally only happens once, but advance payment bonds are frequently released in stages, as previously described. Obtaining numerous staged releases can be a laborious process, but it is indispensable, if the exporter is not to remain at risk for longer, and for larger amounts, than is necessary.

Time validity of bonds

Conditional bonds, like the IMechE Model Form A Form of Guarantee, often have no expiry date.

'On demand' bonds usually provide that the demand must be made by the customer by a certain date. However, this is by no means a real protection. It has been known for customers to make a demand near the expiry date, in terms that the bank concerned must either pay immediately, or extend the time validity of the bond for a stated period.

'On demand' bonds: bond support and unfair calling insurance

'On demand' bonds would clearly place many exporters in an impossibly risky position, if they were not able to protect themselves by means of insurance. Further, as banks often treat 'on demand' bonds as the equivalent of overdraft, such bonds may effectively reduce the exporter's borrowing limit.

The Export Credits Guarantee Department (ECGD) bond-support facility solves both these problems. This facility is only usually available in respect of 'on demand' bonds given to public sector buyers. Subject to certain eligibility criteria, ECGD will give an unconditional indemnity to the UK bank in respect of the liability under the 'on demand' bond. The bank is then relying on the creditworthiness of ECGD, rather than that of the exporter. If the bond is called, and ECGD indemnifies the UK bank, ECGD will in turn claim an indemnity from the UK exporter, unless it is established that the exporter is not in default under the terms of the relevant contract, or that his failure to comply is due to specified causes outside his control, in which case the exporter will not be required to indemnify ECGD at all. In order to make use of the ECGD bond-support facility, exporters should apply to ECGD as early as possible, before submitting a tender.

ECGD also offers insurance against unfair calling of 'on demand' bonds, independently of the bond-support facility, subject to certain eligibility criteria. The cover is for 100% of the sum called either unfairly or for the above specified causes outside the exporter's control.

Unfair calling insurance is also available in the private market, particularly from Lloyd's syndicates. Such insurance is generally only for 90% or less of the amount lost, rather than for 100%, as is the case with ECGD.

For further information, exporters should refer to ECGD publications, particularly *ECGD Services*, the Confederation of British Industry's *Contract Bonds & Guarantees* (September 1981) and *Uniform Rules for Contract Guarantees*, ICC Publication No 325.

10: Contract and Product Liability Insurance

Contract insurance

This sub-section deals with insurance cover which the supplier/contractor may be required to effect under the terms of the contract. Generally, these insurances are likely to be for marine and inland transit, for the safety of Works on Site, for the supplier/contractor's liability as an employer of labour, for automotive liability, and for other third party liabilities.

The first step to be taken is to study any tender documents issued by the customer, to see if they require any contract insurances, whether expressly or by implication. If they do, the likely cost should be ascertained by obtaining quotations through the company's insurance brokers. If there is no guidance in tender documents, the tenderer will have to decide what terms, if any, regarding contract insurances he is going to offer. Usually, this will be covered by insurance clauses in the standard conditions of contract referred to in the tender. Again, quotations should be obtained through the brokers.

It should be noted that brokers' services are usually free to the supplier/contractor, because they are paid for by the insurers through a commission on the eventual premium paid. It is wise to transfer as much responsibility as possible to the brokers. They should not be casually asked to quote for cover described vaguely. They should be given full and exact details in writing of the cover required. Preferably, they should actually be sent copies of all the relevant clauses from the conditions of contract being used. Then, if the resulting quotation or cover does not accord with the requirements of the contract, the brokers will usually be liable to the contractor for negligence. This could be important, for example, if there is an uninsured loss, or the supplier/contractor has to pay much more than the premium quoted because the brokers had mistakenly obtained quotations for cover not in compliance with the contract.

It will sometimes be suggested by foreign customers that insurers in their own countries should be used. There is every objection to this, unless the relevant insurers are of the international standing upon which the supplier/contractor may totally rely. It must be remembered that, under very many contracts, the supplier/contractor must reinstate the Works after a disaster, whether or not he is reimbursed by insurers, if the disaster occurs before a specified point in time, which is usually take-over by the customer.

In some cases, the foreign customer is sufficiently powerful to insist that insurers in his country shall be used, or the foreign law may so insist. If the insurers are not deemed reliable, this may place the supplier/contractor at risk. He may protect himself by taking out further insurance with insurers selected by him, in order to safeguard against non-payment of legitimate claims by the foreign insurers. If this cost is anticipated, it should be covered in the tender, which means considering the question at an early stage.

The form of the foreign insurer's policy may also bristle with exclusions and small print to the insured's disadvantage. Again, he can protect himself by taking out further insurance to cover the exclusions and pitfalls. This is sometimes called 'difference in conditions' or 'DIC' cover. The currency in which the insurance is denominated should preferably correspond with the currency which will have to be paid out, if an insured risk occurs. For example, a supplier/contractor sending American equipment to Saudi Arabia, which he will have to pay for in US dollars, will generally wish to insure that equipment in US dollars, because that is the currency he will have to pay for replacements. However, if he is using vehicles in Saudi Arabia, he will generally wish to have third party motor accident cover in Saudi Riyals, because that is the currency which he would have to pay out in compensation for such accidents.

Insurance must be monitored regularly throughout the project, and adjustments made for such matters as extensions of the contract period and increases in value arising from contract price adjustment, inflation, or additional orders placed under the contract. If there is an important change in the contract, so that the insured risks are altered, the insurers must be informed, and asked for their consent.

Sometimes, the supplier/contractor will not be

required to take out any contract insurances, with the customer undertaking to insure. In such cases it is vital to find out exactly what cover the customer has or proposes to effect. If necessary, the relevant copy policy should be translated. It should then be ascertained, generally with the broker's help, whether the supplier/contractor is protected. There are grave pitfalls in such a situation. For example, if the supplier/contractor is not specifically included in the cover as an insured party, he may not be able to recover from the insurers if the Works are damaged by a fire or other accident caused by his employees' negligence. This would apply to damage suffered either by the supplier/contractor's own portion of the Works, or other portions. The supplier/contractor would have to pay for the entire damage, or reinstate it free of charge.

Naturally, accidents on Site caused by employees' negligence are one of the main risks covered by the usual forms of 'Contractors' All Risks' or 'CAR' policies. If examination shows that the supplier/contractor is not sufficiently protected by the customer's insurance cover, the supplier/contractor, unless he can persuade the customer to put the matter right, will have to consider whether to protect himself by taking out additional cover required not by the contract, but by his own commercial interests.

Excesses may also be an important consideration. For large contracts, where cover is arranged by the customer, the policy may be subject to a substantial excess, and for overseas contracts a figure of £250 000 is not unheard of. It is therefore up to the supplier/contractor to decide whether to run the risk of a large excess, or insure himself against it.

It is also advisable for the supplier/contractor to ensure that any sub-contractors he may employ have their own employer's liability and public/product liability insurance for adequate limits of indemnity. As a general rule, it is advisable for a supplier/contractor to check that sub-contractors have insurances to protect their own liabilities.

Insurance for projects undertaken by joint venture partners, as joint suppliers/contractors, raise special problems which are referred to in Section 13, Joint Ventures for Projects.

Product liability insurance

Suppliers/contractors must always have sufficient product liability cover, especially for exports to the US. This cover is for personal injuries or property damage which may be suffered by the customer or third parties as a result of product defects for which the supplier/contractor is legally liable. The cover usually also includes consequential losses suffered by a customer, resulting from property damage for which the supplier/contractor is legally liable. In very many countries, the Courts are imposing progressively stricter and higher standards of product safety, and awarding higher and higher damages against suppliers/contractors. Because liability is generally stricter and damages are higher in the US than elsewhere, cover for this market is especially costly in premiums, and sometimes hard to obtain at all.

11: Credit Insurance and Export Finance

General

The comments about insurance brokers made in the sub-section on contract insurance are also applicable to this section. There are insurance brokers who handle vast amounts of credit insurance business, and are of immense help to traders and exporters.

Where no credit insurance is effected, reports of credit reference agencies or banks may be relied upon. If the customer gives his bank as a reference, the correct procedure is for the supplier/contractor to ask his own bank to contact the customer's bank for a report. This will be passed back through the supplier/contractor's bank.

Credit insurance within the UK is offered by well-established commercial insurers. The great bulk of credit insurance for exports is provided by the Export Credits Guarantee Department (ECGD), a British Government Department.

ECGD services

The services provided by ECGD are very well described in ECGD's publications, particularly the booklet *ECGD Services*. The Department also publishes booklets and pamphlets on its individual services. The main services are also most helpfully explained in *Schmitthoff's Export Trade*.

ECGD documents are exceptionally complex. Exporters should, immediately something peculiar or unusual happens or is proposed in relation to an export sale, consider whether it affects their ECGD cover. As the ECGD policy will be difficult for many exporters to understand in detail, it may be best to put the matter to the broker for his advice. If in doubt, the wise course is to inform the Department of the unusual situation.

Failure to bear in mind ECGD's involvement may easily lead to an exporter losing his cover. For example, the standard form of ECGD Specific Guarantee forbids alteration of the contract by mutual agreement between the exporter and the customer, without the Department's consent. As a further example, the standard form of Comprehensive Short Term Guarantee forbids the exporter to consent to any extension of a due date of payment, with some exceptions for extensions not exceeding ninety days, unless ECGD gives consent.

The exporter should always err on the side of keeping the Department informed of anything which could conceivably affect the cover. Just like all other insurers, ECGD will not be pleased to be handed a claim relating to stale matters, about which it has not previously been informed, and may not then accept liability. The Department publishes a booklet entitled *Cutting Your Losses*, which should be referred to when problems arise.

Department services are constantly being varied and improved, and there is a wide number of permutations which may be used, depending on the exporter's requirements. The credit insurance broker is indispensable for advice in this field.

The description of ECGD services given below is merely a brief outline of the subject, so that readers will be generally aware of what kind of services are in existence.

ECGD credit insurance

Credit insurance covers the exporter for 90% (in the case of some risks 95%) of his losses. Some of the major credit insurance policies offered by ECGD are:

(a) Comprehensive Short Term Guarantees, which generally cover the exporter's whole export turnover on credit terms of up to six months. This is by far the most important ECGD policy.

(b) Supplemental Extended Terms Guarantees, which are available to holders of Comprehensive Short Term Guarantees, and which generally cover contracts where the delivery period does not exceed two years, and the credit period five years.

(c) Supplemental Stocks Guarantees, which are available to holders of Comprehensive Short Term Guarantees, and which cover exporters' stocks held in foreign countries against certain risks such as war between the foreign country concerned and the UK, requisition, confiscation or prevention of re-export.

(d) External Trade Guarantees, which cover trade by UK parties in goods which are sent direct from one foreign country to another, without ever coming into the UK, on terms generally similar to the Comprehensive Short Term Guarantee.

(e) Specific Guarantees for major individual capital-goods contracts.

(f) Constructional Works Guarantees for major individual overseas construction projects.

(g) Services policies for technical and professional earnings, as opposed to the export of goods.

The following are the insured 'causes of loss' in a standard Comprehensive Short Term Guarantee. The 'Insurer' is the Secretary of State acting by ECGD:

1. The Insolvency of the buyer;
2. The failure of the buyer to pay to the Insured within 6 months after the Due Date of Payment the amount owing in connection with goods delivered to and accepted by the buyer;
3. The failure or refusal of the buyer to accept goods despatched, where such failure or refusal does not arise from any breach of contract on the part of the Insured.
   Provided that this cause of loss shall not apply unless the Insurer has stated in writing that he is satisfied that no useful purpose would be served by the institution or continuation of legal proceedings against the buyer;
4. A general moratorium decreed by the government of the buyer's country or by that of a third country through which payment must be effected;
5. Any other measure or decision of the government of a foreign country which in whole or in part prevents performance of the contract;

6. Political events, or economic difficulties arising outside the United Kingdom or legislative or administrative measures taken outside the United Kingdom, being events, difficulties or measures which prevent or delay the transfer of payments or deposits made in respect of the contract;
7. The operation of a law (including an order, decree or regulation having the force of law) in the buyer's country which has the effect of giving the buyer a valid discharge of the debt under that law (not being a valid discharge under the proper law of the contract) for payment made notwithstanding that, as a result of fluctuations in exchange rates, such payments, when converted into the currency of the contract, are less than the amount of the debt at the date of transfer;
8. The occurrence outside the United Kingdom of war (including civil war, hostilities, rebellion and insurrection), revolution or riot, cyclone, flood, earthquake, volcanic eruption or tidal wave which in whole or in part prevents performance of the contract: Provided that no liability shall arise under this cause of loss in respect of any risk which is normally insured with commercial insurers;
9. (i) The cancellation or non-renewal of an export licence; or
   (ii) the operation, after the Date of Contract, of any law in the United Kingdom which prohibits or restricts the export of the goods to the buyer's country, other than the refusal to grant an export licence in relation to goods which on the said Date of Contract were subject to licence; or
10. Where in respect of any contract the Insurer has stated in writing that –
    (i) the buyer under that contract is a Public Buyer; or
    (ii) he is satisfied, in the case of a guarantee of payment and indemnity for breach of that contract, that the giver of the guarantee and indemnity is a national government authority, and the Insurer has confirmed that this cause shall apply (subject only to such conditions as the Insurer may think fit), the failure or refusal on the part of the buyer to fulfil any of the terms of that contract.

The detailed cover varies from one ECGD credit insurance policy to another, and from one insured 'cause of loss' to another. Examination of the exact terms of each individual cover is essential, as there may be exceptions to the cover which are very important to the exporter. For example, in the case of the third cause of loss listed above – non-acceptance of goods by the buyer – the exporter has to bear a 'first loss' of 20% of the full original price, and ECGD bears 90% of the balance. As a further example, under Specific Guarantees, no cover is given against the failure of private buyers to take up exported goods.

In appropriate cases 'pre-credit risk' cover is available, for cases where the loss materializes before the goods are shipped, e.g. where the customer becomes insolvent. In such cases, the exporter may have invested heavily in the manufacture of the goods, and they may not be readily saleable to other customers.

ECGD guarantees for supplier credit financing

'Supplier credit' occurs when the exporter extends credit to his customer. As far as the customer is concerned, it is the exporter who is his creditor. The exporter will usually have to finance this by borrowing from his own bank. As part of his borrowing arrangements with his bank, he may assign to his bank his rights under the ECGD credit insurance he will have effected, so giving extra security to the bank. The Department will not normally object to such an assignment.

For a further premium, ECGD will issue a direct guarantee to the bank for the repayment of the exporter's borrowing. The bank is then completely safe, because it can rely on the credit of the government, rather than that of the exporter. This may induce banks to lend to exporters when they otherwise would not. Banks will often not count ECGD guaranteed borrowing against the exporter's borrowing facilities. Should the exporter become insolvent, the bank will recover 100% of the amount guaranteed from ECGD.

The major ECGD guarantees for supplier credit financing are:

(a) Comprehensive Bills Guarantees, where the credit period is less than two years, and the customer gives the exporter a Promissory Note or Bill of Exchange.
(b) Comprehensive Open Account Guarantees, where the business is done on open account, unsupported by Bills or Notes, and where the credit period is less that six months.
(c) Specific Guarantees to Banks, where the credit period is two years or more.

Comprehensive Bills Guarantees and Comprehensive Open Account Guarantees are 'global' guarantees, applying to a specified volume of business of the relevant class. Specific Guarantees to Banks are considered by ECGD on a case-by-case basis, like Specific Guarantees issued for credit-insurance purposes.

Finance granted to exporters by banks under Comprehensive Bills Guarantees and Comprehensive Open Account Guarantees are granted at fine rates of interest, as the lending is guaranteed by the government.

Finance granted to exporters by banks under Specific Guarantees to Banks are granted at fixed, especially low, rates of interest, except where the export is to another EEC member State.

ECGD guarantees for buyer credit financing

'Buyer credit' occurs when a UK bank lends directly to the export customer, the exporter being paid on cash terms. As far as the customer is concerned, it is the UK bank which is his creditor. The exporter will, in fact, generally have arranged the loan for the benefit of the customer. ECGD Buyer Credit Guarantees are available to banks making such finance available in respect of contracts worth £1 million or more. Exporters should approach their banks and ECGD at an early stage about possible buyer credit financing. Unlike supplier credit, the customer must deal with two UK parties – the exporter and the bank. Therefore, negotiations with the customer may be more complex than over supplier credit.

Where a UK bank extends credit to a foreign borrower for a wide range of purchases of UK capital goods, with the repayment of the loans being guaranteed to the bank by ECGD on a buyer credit basis, this is known as a 'line of credit'. A line of credit will often be allocated to a particular large project. Alternatively, a 'general purpose' line of credit may be granted to a foreign bank, who will allocate the line of credit among its own customers who want to buy UK capital goods. Unlike Buyer Credit Guarantees, the UK exporter may benefit under line of credit arrangements even if his export order is quite small.

Interest rates on buyer credits and lines of credit are analogous to rates applied under Specific Guarantees to Banks for supplier-credit financing.

ECGD cost escalation cover

In relation to certain UK cost increases on non-EEC contracts worth over £5 million, with a manufacturing period of at least two years, a limited cost escalation cover is available. A percentage of the exporter's UK costs (excluding fixed price subcontracts) is designated as 'eligible'. The percentage is generally 70% for cash business, 75% for credit. ECGD and the exporter agree a 'threshold' up to which the exporter has to bear increases in eligible UK costs. After the 'threshold' is reached, ECGD will compensate the exporter up to the agreed 'ceiling' of the cover. After the 'ceiling' is reached, the exporter again is on his own.

ECGD bond support and cover against unfair calling of bonds

These ECGD services are described in Section 9, Bonds and Financial Guarantees.

ECGD project participants insolvency cover

This facility will insure UK contractors against 90% of the losses they may incur in relation to an export project of £20 million or more because of the insolvency of a sub-contractor or another consortium member.

ECGD joint and several cover

This cover is available in relation to export projects of more than £50 million, where they are judged to be of exceptional national interest. The facility is available to main contractors in relation to UK subcontracts amounting to 5% or more of the total project value. It can also be adapted to cover UK members of consortia or joint ventures. ECGD indemnifies the insured against cost over-runs outside the insured's control in connection with subcontracts. The two causes of loss covered are the default of a sub-contractor necessitating his replacement by another sub-contractor, and additional costs caused by a sub-contractor but not recoverable from him because of the terms of the relevant sub-contract. Cover is for 80% of losses, with a maximum of 20% of the total UK value of the project contract.

ECGD insurance for overseas investments

This facility offers the UK investor insurance against expropriation, war damage, restriction on remittances, and allied political risks. Commercial risks are not covered. Normal cover is for fifteen years against 90% of losses.

ECGD tender to contract cover

The main ECGD services have allied to them ancillary services to deal with contracts involving foreign currency, and the extra complication and risks to the exporter that may arise as a result of being paid in foreign currency. Exporters with contracts providing for foreign currency payments in the future may protect themselves also by using the forward exchange market. For example, an exporter due to receive US dollars in six months may 'sell them forward' in return for a known sterling sum. However, a tenderer expressing his prices in foreign currency cannot do this before he has been awarded the contract, and the 'tender to contract' period may be long. During that time, forward exchange rates may move to his disadvantage. ECGD will protect the exporter against this risk by providing some protection against such forward exchange rate movements if they alter beyond a small margin.

The exporter who wins the order may thus be sure of the amount of sterling he will receive, regardless of movements in forward exchange rates since the tender date.

Non-ECGD insurance

In some cases, export business may not be eligible for ECGD services at all. ECGD may have decided, following bad claims experience, to write no further credit insurance for exports to a particular country, until it mends its ways. Alternatively, the UK content of the tender may be insufficient for the grant of ECGD cover. This might happen in a case where a UK contractor acts as main contractor for an overseas project, but only the supervising engineers are provided from the UK. The labour and equipment might originate from the country in which the project is situated, or a third country. As a further example, a political decision might have been taken in the UK not to provide British Government encouragement to certain business with a foreign country.

In these circumstances, the absence of ECGD services is not necessarily fatal. Commercial insurers will also provide export credit insurance, even against certain political risks. They will also provide cover against unfair calling of 'on demand' bonds. However, premiums are likely to be steep, the cover less wide than that granted by ECGD, and claims may be examined more rigorously than by ECGD. Nevertheless, it is perfectly possible to win export business without ECGD services, and to obtain reasonable protection through the medium of commercial insurers.

Lending institutions

Tender documents for major projects will sometimes state that the funds for the project are being provided by a loan to the prospective customer from a particular lending institution, such as the European Investment Bank (EIB) (an European Community institution) or the International Bank for Reconstruction and Development (IBRD or the World Bank). It is important to understand that this gives no right to the supplier/contractor to look directly to the lending institution for payment. If the loan is never made, or withdrawn by the lending institution, or for any other reason does not materialize, the customer may be left with insufficient funds to complete the project. The supplier/contractor may, of course, protect himself by ECGD or other credit insurance, but this is only partial protection, because the credit insurance will still leave him with an uninsured percentage of loss, and the loss must be within one of the insured causes of loss to be claimable at all.

There is, nevertheless, considerable practical protection for the supplier/contractor, provided the loan funds are actually advanced by the lending institution to the customer, because the lending institution will generally see to it that the funds are used for the particular project. If a customer uses funds advanced for one particular project for another purpose, he will be in considerable trouble with the lending institution, which will usually be so powerful that the customer will be very loath to offend it. However, this practical protection is subject to withdrawal if the lending institution and the customer do actually fall out, with the result that the loan funds are not forthcoming.

If the supplier/contractor is able to negotiate letter of credit terms for the contract, this may eliminate the risk. However, the project in question is likely to be a major one, extending over several years, and it may very well be impossible to induce the customer to open a letter of credit for the whole price, valid for the whole anticipated term of the contract. If letter of credit terms are forthcoming at all in such a case, tranches of the price will often have to be established through the letter of credit from time to time. Therefore, the risk will remain that future tranches will not be established.

An alternative to watertight letter of credit terms is for the customer to arrange that the lending institution shall enter into a special commitment direct to the supplier/contractor to pay the price, notwithstanding any subsequent suspension or cancellation of the loan arrangements between the lending institution and the customer. This will usually mean that the customer will have to pay the lending institution extra charges. Such special commitments are expressely provided for in the publications of certain lending institutions, including the World Bank.

Suppliers/contractors are well advised to be familiar with the publications of the lending institutions with which they may come into contact, in order to understand how they operate.

12: Sub-Contracts

A main contractor, in the absence of particular terms in the main contract relieving him in some measure of the responsibility, is responsible to the employer for all sub-contractors' work, just as if such work was the main contractor's own. Default or insolvency of a sub-contractor will normally be no excuse in favour of the main contractor in the face of claims against him by the employer under the main contract for delay in completion, defective work, or anything else, and neither will such default or insolvency give rise to any valid claim by the main contractor against the employer.

This statement may be subject to exceptions, under certain published conditions of contract, in respect of defaulting or insolvent sub-contractors who have been nominatd by the employer. However, such possible exceptions do not alter the fact that main contractors always have the greatest interest in ensuring that all sub-contractors are technically competent, financially sound, and prepared to enter into suitable sub-contracts protecting the main contractor's position.

A well-drafted form of sub-contract in common use is the 'Form of Sub-Contract (as amended 30 March 1973) designed for use in conjunction with the ICE General Conditions of Contract', published by the Federation of Civil Engineering Contractors. It has no official short name, although it is commonly called 'the ICE Sub-Contract'. This form of sub-contract, although designed for use in connection with civil engineering projects, can quite easily be adapted for use in connection with mechanical, electrical and chemical engineering projects.

It is an example of a 'back-to-back' sub-contract. That is, it gives to the main contractor a complete indemnity against loss or liability which he may suffer as a result of any breach of the sub-contract by the sub-contractor, however great that liability may be. This is achieved by Clause 3 which is now considered in detail, as it illuminates the whole sub-contracting process. Sub-Clause 3(1) reads:

> The Sub-Contractor shall be deemed to have full knowledge of the provisions of the Main Contract (other than the details of the Contractor's prices thereunder as stated in the bills of quantities or schedules of rates and prices as the case may be), and the Contractor shall, if so requested by the Sub-Contractor, provide the Sub-Contractor with a true copy of the Main Contract (less such details), at the Sub-Contractor's expense.

The 'Contractor', as defined in this form of sub-contract, is the main contractor. This sub-clause records that the sub-contractor has notice of the provisions of the main contract, including any provisions in the main contract imposing liquidated damages for delay in completion of the main works which may be imposed on the main contractor by the employer, and including any other provisions imposing losses or liabilities on the main contractor for breach of the main contract. The full significance of this appears from the later sub-clauses.

Sub-Clause 3(2) reads in part:

> Save where the provisions of the Sub-Contract otherwise require, the Sub-Contractor shall so execute, complete and maintain the Sub-Contract Works that no act or omission of his in relation thereto shall constitute, cause or contribute to any breach by the Contractor of any of his obligations under the Main Contract and the Sub-Contractor shall, save as aforesaid, assume and perform hereunder all the obligations and liabilities of the Contractor under the Main Contract in relation to the Sub-Contract Works.

This establishes that the sub-contractor must execute, complete and maintain the sub-contract works so that they comply with the requirements of the main contract, unless the sub-contractor is relieved of this obligation by the sub-contract itself.

The significance of this may be shown by two examples, the first concerning delay in completion, and the second concerning defects in the sub-contract works.

With regard to delay, if the main contract is for £50 million, with liquidated damages for delay running against the main contractor at a rate of ½% per week, without a maximum limit, a sub-contractor who is contractually responsible to the main contractor for such delay must indemnify the main contractor in respect of such liquidated damages, without limit. It makes no difference whether the sub-contract is for £50 000 or £20 million, although, of course, a sub-contractor whose price is only a small proportion of the main contract price should be very reluctant to enter into a sub-contract on such terms. There is no doctrine under English law limiting the sub-contractor's liability to the amount of the sub-contract price, or in any other way.

With regard to defects in the sub-contract works, the main works may be rejected by the employer as a result of defects in the sub-contract works, and the main contractor will, under nearly all main contracts, be contractually responsible to the employer to remedy the defects, whatever the cost may be, be it £1 million or £100 million, and whatever the main contract price may be. In those circumstances, a sub-contractor who is contractually responsible to the main contractor for the defects in the sub-contract works must again indemnify the main contractor against such costs, without limit. Usually, this will mean that the sub-contractor will himself have to carry out the remedial works, without charge to the main contractor.

Naturally, where there are severe defects in important sub-contract works, the delay caused by the remedial works will cause liquidated damages for delay to be levied against the main contractor under the main contract in many cases, and in that event the main contractor will also be able to claim an indemnity from the sub-contractor in respect of such liquidated damages.

The last sentence of Sub-Clause 3(2) reads:

Nothing herein shall be construed as creating any privity of contract between the Sub-Contractor and the Employer.

This is a statement of the general principle of English law that a contract only binds and benefits the parties to that contract. No third party, however closely involved in fact, may normally enforce a contract to which he is not a contracting party. This is known as the doctrine of 'privity of contract'. In the field of sub-contracting, it has the following important results.

First, the sub-contractor may only recover payment from the main contractor. He cannot sue the employer for the sub-contract price. If the main contractor is or becomes insolvent, the sub-contractor will normally be only an unsecured creditor of the main contractor. The fact that the main contractor may have received in full from the employer a relevant payment in respect of the sub-contract works does not necessarily mean that the sub-contractor has any prior right to that payment. For example, if the main contractor becomes insolvent after receiving from the employer £50 000 in respect of sub-contract works, and the winding-up of the main contractor's affairs only yields a payment of 5p in £1 for unsecured creditors, the sub-contractor will usually only receive £2 500, and that after considerable delay. Some published forms of main contract do contain powers for the employer to pay sub-contractors direct, but this is a power, not a duty. The sub-contractor has no legal right to compel the employer to exercise that power, or indeed any contractual legal rights against the employer at all.

A sub-contractor should therefore always consider his position in the event of the insolvency of the main contractor, whether the main contractor is British or foreign. If necessary, he may be able to protect his position in that event, by commercial credit insurance if it is a British main contractor, or via ECGD if it is a foreign main contractor.

Second, only the main contractor, not the employer, may enforce the sub-contract by, for example, recovering damages from the sub-contractor for delay or for defects in the sub-contract works. However, it is not uncommon for sub-contractors to be required, by the terms of their sub-contracts, to give direct 'guarantees' or 'warranties' to the employer in respect of the time for completion and the quality of their sub-contract works. Such documents constitute separate contracts between the employer and sub-contractor, and therefore the employer may enforce them directly on the sub-contractor.

Third, no terms in a sub-contract, nor any default of a sub-contractor, will normally relieve the main contractor of any responsibility to the employer which he has assumed under the main contract. For example, if the main contractor is responsible to the employer for delays in completion of the main works, as a result of delays by a sub-contractor, or if the main contractor is responsible to the employer for defects in the sub-contract works as a result of bad work by a sub-contractor, the sub-contractor will, under the ICE Sub-Contract, be liable to indemnify the main contractor. However, it will not alter the main contractor's liability to the employer at all if the sub-contractor has disappeared, or has become insolvent, or is, by reason of some exclusion clause in the sub-contract, not liable to indemnify the main contractor.

Sub-Clause 3(3) reads:

> The Sub-Contractor shall indemnify the Contractor against every liability which the Contractor may incur to any other person whatsoever and against all claims, demands, proceedings, damages, costs and expenses made against or incurred by the Contractor by reason of any breach by the Sub-Contractor of the Sub-Contract.

This makes explicit the implied indemnity to the main contractor, in respect of claims by the employer against the main contractor, which is given by Sub-Clause 3(2). Further, the main contractor is given an indemnity against claims by other parties resulting from the sub-contractor's breach of the sub-contract. In a complex project, a single sub-contractor's delay or defective work may delay and disrupt the work of many other sub-contractors employed by the main contractor, who will be entitled to claim in respect of such delay and disruption against the main contractor under the terms of their sub-contracts. Under Sub-Clause 3(3), the main contractor is entitled to an indemnity from the defaulting sub-contractor in respect of such claims, as well as for the main contractor's own losses.

Sub-Clause 3(4) reads:

> The Sub-Contractor hereby acknowledges that any breach by him of the Sub-Contract may result in the Contractor's committing breaches of and becoming liable in damages under the Main Contract and other contracts made by him in connection with the Main Works and may occasion further loss or expense to the Contractor in connection with the Main Works and all such damages loss and expense are hereby agreed to be within the contemplation of the parties as being probable results of any such breaches by the Sub-Contractor.

This Sub-Clause emphasizes the main purpose of the whole of Clause 3, which is to impose 'back-to-back' liability on the sub-contractor.

Because it is 'back-to-back', engineering sub-contractors will not generally wish to tender by reference to the ICE Sub-Contract, if they have any choice. Nevertheless, a Standard Tender based on the ICE Sub-Contract is included in this book.

It is notable that the ICE Sub-Contract contains no provisions imposing liquidated damages on the sub-contractor for delay, although most published forms of main contract contain provisions imposing liquidated damages for delay on main contractors. This is because the existence of a clause imposing liquidated damages for delay on a sub-contractor is basically inconsistent with the nature of a 'back-to-back' sub-contract.

It may very well not be in the interests of a main contractor to provide for liquidated damages in sub-contracts. For example, if the main contract is for £10 million, with liquidated damages for delay running against the main contractor at a rate of ½% per week, rising to a maximum of 5%, after ten weeks, the main contractor's liability for liquidated damages for delay is, of course, £500 000. If the main contractor awards a sub-contract for £1.5 million, incorporating equivalent percentages for liquidated damages for delay, the sub-contractor's maximum liability for liquidated damages for delay is, of course, £75 000. Therefore, if the main contractor is contractually responsible to the employer for ten weeks' delay in completion, and if the sub-contractor is contractually responsible to the main contractor for such delay, the result will be that the main contractor will suffer an irrecoverable loss of £425 000.

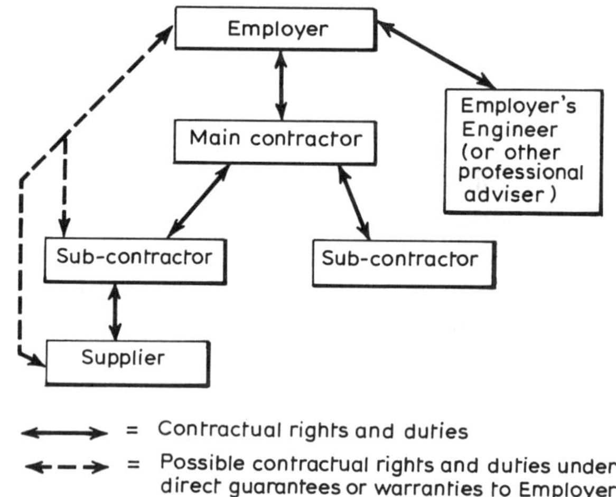

Fig. 12.1 Rights and duties under main and sub-contracts

If the sub-contractor is a subsidiary company within a group of companies, it will usually be in the main contractor's interests to obtain a guarantee from the sub-contractor's ultimate parent company that the sub-contractor will duly perform and observe the sub-contract. Otherwise, it will not be possible for the main contractor to claim successfully against the parent company in the event of the sub-contractor's breach of the sub-contract and insolvency. A parent company is not liable for its subsidiaries' debts and contractual obligations, in the absence of such a guarantee. That is why it may be desirable for a parent company to take risky business through a subsidiary possessing the minimum of assets.

While the main contractor may have to provide, under the terms of the main contract, bonds and guarantees in favour of the employer from banks or sureties, it does not automatically follow that equivalent bonds and guarantees should be required by the main contractor in respect of the sub-contractor. Such equivalent bonds and guarantees will increase the sub-contract price, as the sub-contractor will naturally have to provide for this considerable cost in his tender to the main contractor.

The main contractor should consider in each case whether it is really necessary for his security to require bonds and guarantees in respect of the sub-contractor. If the sub-contractor is financially sound, and legally accessible for claims, it is reasonable to omit them, whatever the bonds and guarantees required under the main contract.

For example, a main contractor might win a £50 million contract from a foreign Government, which might require a 5% 'on demand' bank guarantee for performance, and a 20% 'on demand' bank guarantee in respect of the advance payment under the main contract – a total of £12.5 million in 'on demand' bank guarantees. If the main contractor awards a substantial sub-contract to another British company which is considered to be financially sound, with a parent company guarantee if the sub-contractor is a subsidiary, it is reasonable not to require equivalent bonds and guarantees in respect of the sub-contractor. If the sub-contractor, or its parent company, is not considered to be financially sound, wherever situated, it is almost certainly better not to award the sub-contract to that sub-contractor, regardless of whether that sub-contractor can provide bonds and guarantees. If the sub-contractor is little known and/or situated in a country where it is not practicable for the main contractor to bring legal proceedings, bonds and guarantees are certainly indicated. It should be noted that, in very many countries, it is much more difficult than in the UK to ascertain the financial position of another company, often because the foreign law may not require nearly as much financial information to be publicly available about the company as would be required in the UK.

Many main contracts provide for the employer to nominate sub-contractors whom the main contractor must accept, on certain conditions. For example Clause 39.2 (Objections to Nominations) of the FIDIC Conditions of Contract (International) for Electrical and Mechanical Works, Second Edition 1980, provides:

> The Contractor shall not be required by the Employer or the Engineer or be deemed to be under any obligation to employ any Nominated Sub-Contractor whose performance warranties are not acceptable to the Contractor or against whom the Contractor may raise reasonable objection, or who shall decline to enter into a sub-contract with the Contractor containing provisions:
>
> (a) that in respect of the work, goods, materials or services the subject of the Sub-Contract the Nominated Sub-Contractor will undertake towards the Contractor the like obligations and liabilities as are imposed on the Contractor towards the Employer by the terms of the Contract and will save harmless and indemnify the Contractor from and against the same and from all claims, proceedings, damages, costs, charges and expenses whatsoever arising out of or in connection therewith, or arising out of or in connection with any failure to perform such obligations or to fulfil such liabilities. Provided that nothing in this paragraph shall entitle the Contractor to object to a Nominated Sub-Contractor who requires that the liabilities and indemnities to the Contractor shall be limited pro rata to the Sub-Contract price;
>
> (b) that the Nominated Sub-Contractor will save harmless and indemnify the Contractor from and against any negligence by the Nominated Sub-Contractor, his servants or agents, and from and against any misuse by him or them of any Contractor's Equipment provided by the Contractor for the purposes of the Contract and from all claims as aforesaid.

If a main contractor is entitled to object to a nominated sub-contractor, he should certainly do so. If, for example, a main contractor enters into a sub-contract with a nominated sub-contractor who will only accept a twelve-month defects liability period in respect of the sub-contract works, while the main contract imposes a twenty-four-month defects liability period on the main contractor in respect of the whole works, the cost of remedying all defects in the sub-contract works during the second twelve-month period will fall on the main contractor, and he will have no claim against the nominated sub-contractor or the employer.

It sometimes happens that a nominated sub-contractor is in a strong enough position to insist on his own contractual terms, while the employer still insists on the nomination. In such a case, the main contractor should stipulate to the employer that, if the main contractor is compelled to enter into a sub-contract with the nominated sub-contractor on terms less favourable to the main contractor than those to which he is entitled, under the main contract, to require from nominated sub-contractors, then the main contractor's responsibility in respect of the sub-contract works shall be correspondingly reduced. In the above example, the main contractor's defects liability period in respect of the sub-contract works should be reduced from twenty-four to twelve months. Further, the employer should be required to indemnify the main contractor against loss or expense or damage suffered arising out of the refusal of the nominated sub-contractor to enter into a 'back-to-back' sub-contract.

The proviso to Clause 39.2(a) of the FIDIC Conditions quoted above highlights a point which often arises in practice, and makes it clear that, under those particular conditions, a main contractor cannot require a nominated sub-contractor to enter into a truly 'back-to-back' sub-contract.

Sub-contracts frequently contain terms making payments from the main contractor to the sub-contractor contingent upon such payments first being received, under the main contract, by the main contractor from the employer. Under such terms, the main contractor is only liable to pay the sub-contractor 'as and when' he is paid. If the sub-contractor is prepared to accept 'as and when' terms of payment under the sub-contract, it should be appreciated that this will mean that, if the main contractor is never paid in respect of the sub-contract works because of the employer's default or insolvency, the sub-contractor will never be paid. In such a case, therefore, the creditworthiness of the employer, as well as that of the main contractor, is of great importance to the sub-contractor.

Sub-contracts often contain terms of payment which are 'as and when', but with an ultimate date by which the main contractor must pay the sub-contractor, regardless of whether the employer has paid the main contractor. For example, Clause 15 (Payment) of the ICE Sub-Contract provides:

(1) From time to time during the execution of the Sub-Contract Works the Sub-Contractor shall within 7 days of the Contractor's request so to do, submit to the Contractor a written statement of the value of all work properly done under the Sub-Contract and of all materials delivered to the Site for incorporation in the Sub-Contract Works at such date as may be specified in the Contractor's request. Such written statement shall be in such form and contain such details as the Contractor may reasonably require and the value of work done shall be calculated in accordance with the rates and prices, if any, specified in the Sub-Contract, or if there are no such rates or prices, then by reference to the Price.

(2) The Contractor shall from time to time make prompt applications for payment under and in accordance with the Main Contract and subject to the Sub-Contractor's having complied with the preceding sub-clause, shall include in such applications claims for work done and, if allowable under the Main Contract, for materials delivered to the Site by the Sub-Contractor and the Contractor shall use his best endeavours to obtain prompt payment of all sums due to him in respect of the Sub-Contract Works under the Main Contract.

(3) Within 7 days of his receiving from the Employer on account of the Main Works any payment which includes a sum in respect of the Sub-Contract Works, the Contractor shall pay to the Sub-Contractor in respect of the work done or materials provided by the Sub-Contractor and allowed for in such payment under the Main Contract, a sum calculated in accordance with the rates and prices specified in this Sub-Contract, or by reference to the Price, as the case may require, but subject to a deduction of retention monies at the rate specified in the Third Schedule hereto until such time as the limit of retention, (if any), therein specified has been reached.

(4) Within 7 days of the Contractor's receipt of any payment under the Main Contract which is by way of release either of the first or second half of the retention monies for the Main Works of where under the Main Contract the Main Works are to be completed by sections, then for the last of such sections in which the Sub-Contract Works are comprised, the Contractor shall pay to the Sub-Contractor the first or second half as appropriate of the retention monies held under this Sub-Contract.

(5) Within three months after the Sub-Contractor has finally performed his obligations under Clause 14 (Maintenance and Defects), or within 14 days after the Contractor has recovered full payment under the Main Contract in respect of the Sub-Contract Works, whichever is the sooner, and provided that one month has expired since the submission by the Sub-Contractor of his final account to the Contractor, the Contractor shall pay to the Sub-Contractor the Price together with any other sums that may have become due to the Sub-Contractor under the Sub-Contract, less such sums as have already been received by the Sub-Contractor on account of the Price or of such other sums.

Provided always that if the Contractor shall have been required by the Main Contract to give to the Employer or to procure the Sub-Contractor to give to the Employer any undertaking as to the completion or maintenance of the Sub-Contract Works, the Sub-

Contractor shall not be entitled to payment under this Sub-Contract until he has given a like undertaking to the Contractor, or has given the required undertaking to the Employer, as the case may be.

(6) The Contractor shall not be liable to the Sub-Contractor for any matter or thing arising out of or in connection with this Sub-Contract or the execution of the Sub-Contract Works unless the Sub-Contractor has made a written claim in respect thereof to the Contractor before the Engineer issues the Maintenance Certificate in respect of the Main Works, or, where under the Main Contract the Main Works are to be completed by sections, the Maintenance Certificate in respect of the last of such sections in which the Sub-Contract Works are comprised.

Under Sub-Clauses (3) and (4), interim payments during the progress of the sub-contract works, and payments of retention money, are 'as and when'. However, under Sub-Clause (5) the main contractor must pay the price, regardless of whether he has been paid by the employer in respect of the sub-contract works.

A UK contractor winning an export sale or project may require UK sub-contractors to accept purely 'as and when' terms. He will often have ECGD or other credit insurance cover against not being paid under the main contract by the foreign employer.

The sub-contractor may protect himself in such cases by arranging with the main contractor that, in return for accepting 'as and when' payment terms under the sub-contract, the sub-contractor will be included as an insured party under the ECGD or other credit insurance held or to be arranged by the main contractor. The sub-contractor and the main contractor will then be equally protected by the credit insurance. An example of an appropriate clause is given in this book.

If this is not done, the main contractor has no absolute need to hold credit insurance in respect of the sub-contract price. For example, if the main contractor wins a £50 million foreign project, and a sub-contractor is prepared to accept purely 'as and when' payment terms in respect of a £10 million sub-contract, the main contractor need not necessarily hold any credit insurance in respect of the £10 million. In such a case, obviously a sub-contractor is at grave risk because, if the employer does not pay the main contractor, the sub-contractor will never be entitled to payment from the main contractor, nor will the sub-contractor have any credit insurance cover in respect of the loss.

13: Joint Ventures for Projects

This section refers to joint ventures between one or more contractors to tender for and, if successful, execute a project for an employer. Joint ventures for trading purposes on a permanent or semi-permanent basis are not covered. Joint ventures are sometimes referred to as consortia – the two terms are not distinct.

Joint ventures are adopted for a wide variety of reasons. They may be required by an employer, as a condition of tendering for a project, because the employer requires each and all of the joint venture partners to be directly responsible to him for the whole of the works. Or the bonding and financial requirements of the contract may be beyond the financial capacity of the individual joint-venture partners. Or the project risk may be greater than one tenderer could accept alone. These are good reasons, and it is essential only to enter into joint ventures for such reasons, because joint ventures involve serious risks and problems, especially for minority partners.

It is possible for the contractors to incorporate a special company, owned jointly by them, to tender for and execute the project. The interests of the contractors will then be as shareholders of the joint company and, unless the contractors are required to give direct guarantees to the employer that the joint company will duly perform the contract, the contractors' liability will be limited to the amount of the share capital. This follows from the general principle of English and most other systems of law that a shareholder is not responsible for the liabilities of a limited liability company in which he holds shares. His risk is limited to the amount of cash he has paid the company for his shares, plus any amount which he has been allowed to retain until the company calls for the balance, e.g. the shareholder may have paid 50p for a £1 share in a company, with the remaining 50p to be paid on the 'call' of the board of directors.

However, unless the project is exceptionally large or special, tendering through a joint company is not often adopted in practice. The most usual reason for this is that there is simply not enough time. The employer will also, if he is well advised, always require the direct guarantees of the several contractors as described above, so nullifying the main advantage of tendering through a joint company.

Therefore, the typical joint venture for a project involves a joint tender by the parties, signed by or on behalf of both or all of them. If the tender is successful, the contract will also be signed by or on behalf of both or all of them. They will each and all be responsible to the employer, as joint main contractors, for the performance of the whole of the main contract, and for the execution of the entire works required by the main contract. If one of them becomes insolvent, the surviving joint-venture partner or partners will not be relieved in any way. They must still perform the whole of the main contract. To add to their responsibilities, the insolvency of one of them will, under the conditions of many main contracts, entitle the employer to terminate the main contract, even if the survivors are willing to go on. The employer will, after such termination, generally be entitled to employ another main contractor to finish the works, and to sue the survivors for damages if he thereby suffers costs he would not have suffered if the original main contract had been duly performed. The calling by the employer of 'on demand' bonds in the event of the insolvency and default of one of the joint-venture partners may, depending on the precise circumstances, not be 'unfair' calling within the terms of ECGD insurance against unfair calling of bonds.

Contractors should, therefore, look at prospective joint-venture partners for major projects with great care. Only those with unimpeachable characteristics should be considered. Prospective partners with an insufficient or dubious financial record or standing are perilous. The joint venture will often be for a major project lasting for years. All the partners should be capable of weathering a serious storm.

Joint ventures present problems of management and control, which can only be solved by painstaking negotiation and detailed agreement between the parties. As this will generally have to take place immediately prior to tender, when time is almost invariably short, the time consumed by such negotiations is a big disadvantage in itself. The time consumed is likely to be substantial because, while contractors' directors and senior employees will be generally familiar with the procedures for main contractor/sub-contractor relationships, they will not generally be so familiar with the procedures for two or more contractors acting as joint venturers, as joint ventures are necessarily only occasional *ad hoc* arrangements. There is also the fact that, while there are many published forms of sub-contract widely accepted in practice, there are no equivalent benchmarks for joint ventures. Each one is unique, and has to be negotiated from scratch.

It very often happens that the relevant project will involve two or more contractors from different industries. Typically, this may be a building or civil engineering contractor, and a contractor for mechanical and electrical works. The proportion of the tender price attributable to one may be very much more than for the other, e.g. the civil works may account for 80%, and the mechanical and electrical works for 20%. However, the share attributable to the contractor with the lesser share may be absolutely vital, as the civil works may be practically useless without the mechanical and electrical equipment. However, in such a case, the position of the major contractor can be perfectly well protected by taking the main contract with the employer himself, and by entering into an appropriate 'back-to-back' sub-contract with the minor contractor in the way described in Section 12, Sub-Contracts. A joint venture, under which the major and minor contractors both accept liability to the employer as joint main contractors, can protect the major contractor's position no better, while it opens up the formidable risk for the minor contractor that he may have to complete the whole project alone, if the major contractor becomes insolvent. In that case, the surviving contractor will often face a crippling loss because of the disruption of the project, the higher prices demanded by substitute contractors, and other factors.

The prices required by substitute contractors will frequently be very much higher, owing to mobilization costs (which may be incurred twice, if the insolvent contractor has already incurred such costs once) and the lesser volume of work to be done by the substitute contractor, the insolvent contractor having completed part. A lesser volume will almost inevitably lead to higher unit prices for the unfulfilled work of the insolvent contractor.

Therefore, the importance of the minor contractor's work to the project as a whole is no justification, by itself, for a joint venture, and joint ventures are not necessary for that reason alone. This is illustrated by the fact that, in the UK building industry, it is common practice for main contractors to sub-contract piling for foundations and structural steelwork, and one can hardly imagine work which is more important to the construction project as a whole.

The minor contractor sometimes feels that he somehow has more control or influence over the project if he is a joint-venture partner. This is usually quite incorrect. Whatever form of management and control for the joint venture is adopted, the major contractor will be most unlikely to concede any real control to the minor contractor – indeed, the major contractor would be ill-advised to do so, because he needs to maintain all the control which a main contractor normally would have over a sub-contractor. While the advantages to the minor contractor of a joint venture are often illusory, his added contractual responsibilities to the employer as a joint main contractor are always substantial and real.

It is sometimes asserted that joint venture tenders lead to a cheaper tender price. Naturally, the existence of a joint venture, rather than a main contractor/sub-contractor relationship, in no way affects the volume of goods and services which are to be provided, and paid for, under the contract with the employer. However, the answer may lie in the financial policies of some contractors, which set a target or budget profit as a percentage of turnover. If a sub-contract is awarded, the sub-contract price will be counted as turnover, and therefore the main contractor will add a percentage to the sub-contract price in the make-up of his own tender price for the main contract. However, if a joint venture is entered into, the main contractor may not count the joint venture partner's price as part of his own turnover, and will not therefore add the percentage. The fact is that there is no logical and inevitable reason why a joint venture tender price should be any less than the price of a main contractor who proposes to enter into a major 'back-to-back' sub-contract with a sub-contractor.

With regard to bonds and financial guarantees, there may possibly be some saving in bonding costs in joint ventures, because it is only necessary to have one set of bonds, i.e. those from the joint-venture partners to the employer required by the main contract. There will be no doubling up of bonds, i.e. bonds under the main contract in favour of the employer, plus bonds under the sub-contract in favour of the main contractor. However, the saved cost will often not be really substantial when considered against the risk to the minor contractor of undertaking joint-venture responsibilities. Further, it is often possible for the main contractor safely to dispense with bonds from the sub-contractor, for reasons explained in Section 12, Sub-Contracts.

A joint venture does not necessarily mean that the partners share profits or losses. The Joint Venture Agreement may well provide that each partner is to receive the proportion of the main contract price attributable to his part of the works. In that and similar cases, it is quite possible for one partner to make a profit, but others a loss.

Joint ventures may create special insurance problems, basically because the joint venturers will be joint insured parties under the policies taken out for the purposes of the project. Special care should be taken to see that all insurance cover taken out is effective and appropriate. For example, it is no good one contractor insuring his part of the works, in his name, while the other contractor insures his part of the works in his name. If that is done, and, for example, the employees of one contractor negligently damage the part of the works executed by the other, the contractor who executed that part of the works will be entitled to an indemnity from his insurers, but such insurers may well be able to recover the money from the contractor whose employees caused the negligent damage, under the doctrine of 'subrogation'. Because the first contractor's employees were negligent, the first contractor is liable for the loss to the second contractor. The insurers, who have indemnified the second contractor against the loss, have all the first contractor's rights against the second contractor. That is, they are 'subrogated' to the first contractor's rights. Both contractors should have been insured parties in respect of the whole works.

As described in the sub-section on Contract Insurance (p. 22), as much responsibility as possible should be put on the insurance brokers.

The complications may also extend to ECGD or other credit insurance. The position under joint ventures will often be that the employer pays all money jointly to the joint-venture partners. One partner may be foreign, and so not entitled to the benefit of ECGD. The possible difficulties can be serious, unless the problem is fully disclosed to competent insurance brokers at the proper stage, which is pre-tender. They will then usually be able to make satisfactory arrangements, with ECGD or the other credit insurers concerned, which will be ready to go into operation if the joint-venture tender is successful.

Any contractor entering into a joint venture with a company which is a subsidiary member of a group should seek to secure parent-company guarantees that the subsidiary will duly perform the joint venture, for the same reasons that such guarantees should be sought when awarding a major sub-contract to such a subsidiary.

Contractors considering entering into joint ventures for projects should take legal advice at the earliest possible time. It is no good negotiating a deal with a joint venture partner or partners and then going to legal advisers afterwards. The legal advisers will then probably raise so many points the negotiators have not covered that re-negotiation becomes necessary, with much waste of time and loss of goodwill. Legal advisers should be involved in the negotiations, in order to make their contribution at that time.

It is not the purpose of this section to give any detailed guidance on the drawing up of Joint Venture Agreements. There is something of a dearth of generally available information on joint ventures for projects, and Joint Venture Agreements must be negotiated *ad hoc* in each case. However, readers should find helpful The Export Group for the Constructional Industries' *Check List: Establishing a Joint Venture* and the ORGALIME *Guide for Drawing up an International Consortium Agreement.*

14: Foreign Taxes

The first rule to follow is to avoid involvement with foreign taxes altogether, if possible, by supplying on terms that do not attract foreign taxation. It is important for the exporter to determine whether he is trading 'with' or 'within' the foreign country. Whether the proceeds of an export contract are liable to foreign taxes or not depends on the tax laws of the foreign country concerned. Generally speaking, foreign tax laws will only affect a UK contractor if the contractor undertakes work in the foreign country. It is possible, through very slight involvement in the foreign country, such as delivering the goods to Site, to make the whole contract price subject to foreign taxes. This is also possible, in certain countries, if the UK contractor is permanently represented in the foreign country in such a way that all the contractor's business in that country will be subject to that country's taxes.

An exporter who exports goods against secure means of payment, such as a letter of credit payable in the UK, will usually not have to worry about foreign taxes. Regardless of the foreign tax laws, he will normally obtain payment in full in the UK, and the foreign tax authorities will have small means of collecting any theoretical foreign tax liability from him.

If the contractor must undertake obligations in the foreign country, it is sometimes possible to improve the position by splitting the project into two contracts, e.g. one for export FOB, and another for supervision of erection on Site. This is done by submitting separate tenders, and receiving separate orders, and obviously this procedure requires the customer's active collaboration. Only the contract for supervision of erection on Site will then, it is hoped, give rise to foreign tax liability. Whether this works depends on the tax laws and practice of the foreign country.

While avoidance is the best course, taking an indemnity from the customer is the next best, but it is a very poor second. The contractor becomes involved in filing tax returns, local accountancy, paying the taxes, and recovering them from the customer. Having to pay the contractor's taxes might enable the customer to ascertain the contractor's profits or costs on the contract which, to say the least, is unwelcome. Recovery of the taxes from the customer will often be impossible to secure by such means as a letter of credit. Further, the taxes may have to be paid in foreign currency. If the contract is not payable in that foreign currency, this may expose the contractor to currency fluctuation risks.

The Standard Tender based on the FIDIC M&E Conditions contain a tax indemnity clause and such a clause is also given in the Additional Forms.

Rather rarely, a contract which would otherwise be subject to foreign taxes is exempted by special legal measures in the country concerned, usually because the project is of national importance. The tenderer will need to be absolutely certain of his ground in relation to this, and will also need to consider the possibility that such special legal measures may be revoked. It is probably wise, nevertheless, in any such case, to provide in the tender that the customer shall indemnify the contractor against all foreign taxes, even though the customer or others state that the contract is exempt. If it really is exempt, the customer will have to pay nothing to the contractor under the tax indemnity clause. If it is not exempt, or the exemption is revoked, the tenderer would have to bear the foreign taxes himself, without any indemnity from the customer, unless a tax indemnity clause had been duly incorporated in the contract.

If a contract is subject to foreign taxes, because the contractor is working in the foreign country, failure to file tax returns, accounts or other documents required by foreign tax laws may have serious consequences. Large financial penalties, or even personal difficulties for the contractor's employees in the foreign country, may result.

Contractors should be aware that some foreign tax authorities have power to take such measures as seizing money directly from the customer, without warning, because of alleged tax claims against the contractor. It may then prove very difficult in practice to recover such taxes.

If an exporter is going to become involved with foreign taxes, he will need to learn everything possible about the foreign tax system, preferably before tendering. For example, he may find that the contract, when and if awarded to him, will be liable to a heavy stamp duty: or he may find that supplies obtained locally may be subject to a heavy VAT, which he may not be able to recover unless he is entitled to register as a trader with the foreign VAT authorities. If possible, the exporter should visit the country concerned, and discuss with more than one financial and/or legal expert the local tax, customs duties and import licence regulations. It should be remembered that tax and commercial laws change very frequently and abruptly in certain countries.

Double taxation would arise if the proceeds of the contract were to be liable both to foreign taxes and to corporation tax in the UK. However, double taxation in the UK has been substantially reduced by unilateral relief and by a number of double tax agreements between the UK and foreign countries.

For example, if a UK company executes a £1 million contract on site in a foreign country, with which the UK has no double tax agreement, the UK company may be assessed by the foreign tax authorities to have made a profit of say £200 000 in the foreign country, and to be subject to a 30% foreign profits tax, i.e. £60 000. The profit is nevertheless subject to UK corporation tax at say 52%, because the company is resident for tax purposes in the UK, and therefore liable to corporation tax on its worldwide income. Under UK legislation providing for unilateral relief, the foreign tax is credited against the UK corporation tax. The result therefore is:

	£	£
Profit		200 000
Less foreign profits tax		60 000
Corporation tax at 52% on £200 000	104 000	
Less foreign profits tax relieved	60 000	
		44 000
Retained by Contractor		96 000

The result is therefore the same as if the contract had been in the UK. However, the effective rate of UK corporation tax paid by many UK companies is less than 52%, and may well be less than the rate of tax paid in the foreign country. If this is the case, the UK tax relief is restricted. In addition, certain foreign taxes are not allowed against UK tax. These foreign taxes are then allowed as a cost of the contract, and the UK tax relief is consequently less than the foreign tax suffered.

In the case of relief for foreign taxes as a result of a double tax agreement between the UK and the foreign country (usually called 'treaty relief') the position always depends on the precise wording of the agreement. However, the usual rule adopted in such agreements is that a UK resident company's business profits will only be taxed in the foreign country if it has a permanent establishment there, and then only on the permanent establishment's profits. Certain costs may be disallowed by the foreign tax authorities for the purpose of calculating such profits, e.g. management charges from the UK. The agreement definition of 'permanent establishment' will often include a building site or construction or installation project lasting more than twelve months.

It is necessary that the company subject to UK corporation tax should have sufficient taxable profits to cancel out the foreign tax. If the UK company has many different contracts, it may find itself in the position that, although it is subject to foreign tax on a particular contract, it still has no liability to pay UK corporation tax, because its overall trading does not result in UK taxable profits. This problem may be overcome by taking the relevant contract through a subsidiary company which has no other significant trading, e.g. a dormant company, or a company obtained ready-made for the purpose.

There are many complexities and pitfalls in double taxation relief including:

Foreign taxes not corresponding to UK income tax, corporation tax, or capital gains tax.
Foreign tax rates exceeding UK tax rates.
'Tax sparing', under which credit will be given for foreign taxes against UK corporation tax, even if such foreign taxes have not been paid because of a special foreign tax exemption.

Therefore, tenderers should always obtain expert tax advice on tenders giving rise to double taxation questions. *Double Taxation Relief* by Deloitte Haskins & Sells is a very useful book to consult on this subject.

Employees, such as project managers, will usually be liable to foreign income and other taxes if they reside in the foreign country, and this must be taken into account when settling their terms of employment. They may be effectively compensated by ceasing to be liable to UK income tax, if the foreign tour is of sufficient length.

15: Notarial Requirements

Notarization is often required by foreign customers in respect of documents such as powers of attorney given to the tenderer's employees or agents empowering them to sign tenders or contracts or to act as project managers in the foreign country.

Some tenders must also be accompanied by documents such as notarially certified copies of the constituent documents of the tenderer, e.g. its Certificate of Incorporation or its Memorandum and Articles of Association. Sometimes, notarially certified translations of documents are also required. It is usually the case that the most onerous requirements in this regard are imposed by foreign public authorities.

The purpose of notarization is to provide independent and authoritative verification of the legal effect and accuracy of the relevant documents. The foreign customer may know little of the tenderer, and nothing of the individuals who have purported to sign documents on the tenderer's behalf. An example of a Notarial Certificate is given in this book.

In order to give the Notarial Certificate, the Notary Public should actually witness the signing and (if applicable) sealing of the document. In many countries, particularly Commonwealth countries, no further formal steps are required. However, some countries require the Notary's signature and seal on the Notarial Certificate to be authenticated by the Foreign and Commonwealth Office in London. This is done by a Foreign Office official signing a document called an 'Apostille', which also bears the Foreign Office seal, and which is usually attached to the back of the Notarial Certificate. The Foreign and Commonwealth Office will only do this if they hold an example of the Notary's signature and seal.

Yet other countries require that the 'Apostille' shall itself be authenticated by their own Embassy or Consulate in London, to the effect that the Foreign Office official who signed the 'Apostille' is duly recognized.

A further variation is that some Embassies and Consulates will authenticate the Notary's signature and seal, without an 'Apostille' from the Foreign and Commonwealth Office, but, like the Foreign and Commonwealth Office, they will only do this if they hold an example of the Notary's signature and seal.

The full process may thus involve attendance before a Notary, with the document subsequently having to be sent to the Foreign and Commonwealth Office, and then to the foreign Embassy or Consulate. Fees have to be paid at each stage, but this is the least of the tenderer's problems, because most tenderers will be under pressure of time to submit their tenders by a closing date.

The instructions to tenderers should make the customer's requirements clear. If the instructions to tenderers require notarization, but do not make clear whether further formalities are required, the Notary will usually be able to advise whether, in relation to the country concerned, reference to the Foreign and Commonwealth Office, and/or to the foreign Embassy or Consulate, is required.

Notaries are also often able to arrange for translation. In London at least, some Notaries have very close contacts with translators.

Exporters should take care to recognize notarial requirements in tender documents, to understand them, and to comply with them in a timely manner. It is quite possible for major tenders to be disqualified because of lack of these formalities. When in doubt, it should be remembered that a tender will never be disqualified because of an excess of formalities. For example, if there is doubt about whether or not to obtain authentication from the foreign Embassy or Consulate, it should be obtained, if time permits.

PART TWO

Standard Tenders

FORM 1

Tender based on IMechE Model Form A/1976/1978 Home Contracts – with erection

Dear Sirs

(heading)

Prices

We offer to supply, deliver to and erect on Site at

the items described in the accompanying Specification for:
£
(pounds)

Programme

Our best estimate is that completion of our work will be achieved within of receiving your order. (We enclose * showing the basis of this estimate.)†

*Appropriate document, e.g. bar chart or critical path network.
†Delete if inappropriate.

Terms of Payment

We propose that payment shall be made as follows:
(a) %(per cent) in advance.
(b) The balance in accordance with the Conditions of Contract referred to in this tender.

Conditions of Contract

Where consistent with this tender, we propose that 'Model Form A/1976/1978, Home – with erection', a copy of which is enclosed, shall apply, subject to the following:

The Appendix shall be completed as follows:
Clause
26. (Delay in Completion)
 (a) Percentage of Contract Value to be deducted as damages.
 (b) Maximum percentage of Contract Value which the deductions may not exceed.
32. Percentage on Prime Cost items.
37. The Institution of Mechanical Engineers.*
 The Institution of Electrical Engineers.*

*Delete one line.

Clause 34 (Terms of Payment) – the rate of interest shall apply to all delayed payments from the Purchaser, and shall be 2% (two per cent) per annum over the * Bank base lending rate from time to time in force on the amount of the delayed payment for the period of the delay.

*Insert name of bank.

Clause 39 (Variation in Costs) shall not apply.

Contract Price Adjustment

We propose that contract price adjustment shall apply in accordance with the BEAMA *Contract Price Adjustment Clause and Formula for use with Home Contracts – Electrical Machinery: (for which there is no other specific formula)*, edition January 1979, corrected edition,* a copy of which is enclosed.

The provisions of the Clause and Formulae applicable to contracts including erection shall apply.

The BEAMA Labour Cost Index referred to shall be the Labour Cost Index for Electrical Engineering.*

References to the *Trade and Industry Journal* shall be read as references to *British Business*, which is the new name of the journal.

*This is only one of several BEAMA CPA Clauses and Formulae. If another is adopted, this form may need consequential amendment. In particular, reference to the BEAMA CPA Clause applicable to Mechanical Plant, and to the Labour Cost Index for Mechanical Engineering, may be appropriate. See p. 16.

Validity

This tender is valid for acceptance until and including *, but shall, after that date, be null and void unless extended by us in writing.

*Date in full, e.g. 'Monday, 19 January 1981'.

Yours faithfully

(Signature)

FORM 2

FOB Tender based on IMechE Model Form B1/1981 Export Contracts for Supply of Plant and Machinery

Dear Sirs

(heading)

Prices

We offer to supply FOB * the items described in the accompanying Specification for:

£

(pounds sterling)

* Port of shipment.

Programme

Our best estimate is that delivery FOB will be achieved within of receiving your order.

Terms of Payment

All prices are payable in pounds sterling in the United Kingdom.

We propose that payment shall be made as follows:

(a) %(per cent) in advance.

(b) The balance through a Confirmed Irrevocable Letter of Credit providing for payment in full upon presentation of documents, to be increased according to any contractual increases in the price, to be extended according to any increase in the period required for execution of our respective obligations under the Contract, and otherwise complying with the following description:

(i) Confirmed by an acceptable bank in London.
(ii) Irrevocably issued, confirmed and advised to us by .[a]
(iii) Date and place of expiry – [a] in London.
(iv) Beneficiary – ourselves.
(v) Amount – the balance of the price.
(vi) Credit available with confirming bank by payment.
(vii) Partial shipments allowed.
(viii) Transhipment allowed.
(ix) Shipment from .[b]
(x) For transportation to .[c]
(xi) Shipment of .[d]
(xii) Documents required:

Commercial Invoice

Full set of clean on board marine Bills of Lading (which shall be acceptable although to the effect that the goods have been shipped on deck or unprotected*).

*Delete if inappropriate.

(xiii) Period after issuance of shipping documents for presentation – days.

(xiv) Subject otherwise to the *Uniform Customs and Practice for Documentary Credits*, 1974, ICC Publication No 290.[e]

[a] Date in full, e.g. 'Monday, 19 January 1981'.
[b] Port of shipment, or e.g. 'any United Kingdom port'.
[c] Port of destination.
[d] Short description of goods.
[e] This description follows as closely as possible the *Standard Procedure for the Issuing o Documentary Credits by Cable, Telegram or Telex* (App VII *Standard Forms for Issuing Documentary Credits*, ICC Publication No 323).

Conditions of Contract

Where consistent with this tender, we propose that 'Model Form B1/1981', a copy of which is enclosed, shall apply, subject to the following:

The Appendix shall be completed as follows:

Clause

17. (Delay in Completion)
    - (a) Percentage to be deducted as damages.
    - (b) Maximum percentage which the deductions may not exceed.

20. Percentage on Prime Cost items.

24. The Institution of Mechanical Engineers.*
    The Institution of Electrical Engineers.*

*Delete one line.

Clause 22 (Terms of Payment) – the rate of interest shall apply to all delayed payments from the Purchaser, and shall be 2% (two per cent) per annum over the * Bank base lending rate from time to time in force on the amount of the delayed payment for the period of the delay. Clause 22(i) (a) shall not apply.

*Insert name of bank.

Clause 26 (Variation in Costs) shall not apply.

Contract Price Adjustment

We propose that contract price adjustment shall apply in accordance with the BEAMA *Contract Price Adjustment Clause and Formula for use with Export Contracts – Electrical Machinery: (for which there is no other specific formula)*, edition January 1979, corrected edition,* a copy of which is enclosed.

The BEAMA Labour Cost Index referred to shall be the Labour Cost Index for Electrical Engineering.*

References to the *Trade and Industry Journal* shall be read as references to *British Business*, which is the new name of the journal.

*This is only one of several BEAMA CPA Clauses and Formulae. If another is adopted, this form may need consequential amendment. In particular, reference to the BEAMA CPA Clause applicable to Mechanical Plant, and to the Labour Cost Index for Mechanical Engineering, may be appropriate. See p. 16.

Trade Terms

The trade term 'FOB' shall be defined by the current edition of Incoterms, where consistent with this tender. A copy of the definition is enclosed.

Validity

This tender is valid for acceptance until and including   *, but shall, after that date, be null and void unless extended by us in writing.

*Date in full, e.g. 'Monday, 19 January 1981'.

Yours faithfully

(Signature)

FORM 3

C & F Tender based on IMechE Model Form B1/1981 Export Contracts for Supply of Plant and Machinery

Dear Sirs

(heading)

Prices
We offer to supply C & F * the items described in the accompanying Specification for:
£
(pounds sterling)
*Port of destination.

Programme
Our best estimate is that delivery C & F will be achieved within of receiving your order.

Terms of Payment
All prices are payable in pounds sterling in the United Kingdom.

We propose that payment shall be made as follows:

(a) % (per cent) in advance.

(b) The balance through a Confirmed Irrevocable Letter of Credit providing for payment in full upon presentation of documents, to be increased according to any contractual increases in the price, to be extended according to any increase in the period required for execution of our respective obligations under the Contract, and otherwise complying with the following description:

(i) Confirmed by an acceptable bank in London.
(ii) Irrevocably issued, confirmed and advised to us by .[a]
(iii) Date and place of expiry – [a] in London.
(iv) Beneficiary – ourselves.
(v) Amount – the balance of the price.
(vi) Credit available with confirming bank by payment.
(vii) Partial shipments allowed.
(viii) Transhipment allowed.
(ix) Shipment from .[b]
(x) For transportation to .[c]
(xi) Shipment of .[d]
(xii) Documents required:

Commercial Invoice

Full set of clean on board marine Bills of Lading (which shall be acceptable although to the effect that the goods have been shipped on deck or unprotected*).

*Delete if inappropriate.

(xiii) Period after issuance of shipping documents for presentation – days.

(xiv) Subject otherwise to the *Uniform Customs and Practice for Documentary Credits,* 1974, ICC Publication No 290.[e]

[a] Date in full, e.g. 'Monday, 19 January 1981'.
[b] Port of shipment, or e.g. 'any United Kingdom port'.
[c] Port of destination.
[d] Short description of goods.
[e] This description follows as closely as possible the *Standard Procedure for the Issuing of Documentary Credits by Cable, Telegram or Telex* (App VII *Standard Forms for Issuing Documentary Credits,* ICC Publication No 323).

Conditions of Contract

Where consistent with this tender, we propose that 'Model Form B1/1981', a copy of which is enclosed, shall apply, subject to the following:

The Appendix shall be completed as follows:
Clause
17. (Delay in Completion)
(a) Percentage to be deducted as damages.
(b) Maximum percentage which the deductions may not exceed.
20. Percentage on Prime Cost items.
24. The Institution of Mechanical Engineers.*
The Institution of Electrical Engineers.*

*Delete one line.

Clause 22 (Terms of Payment) – the rate of interest shall apply to all delayed payments from the Purchaser, and shall be 2% (two per cent) per annum over the * Bank base lending rate from time to time in force on the amount of the delayed payment for the period of the delay. Clause 22(i) (a) shall not apply.

*Insert name of bank.

Clause 26 (Variation in Costs) shall not apply.

Contract Price Adjustment

We propose that contract price adjustment shall apply in accordance with the BEAMA *Contract Price Adjustment Clause and Formula for use with Export Contracts – Electrical Machinery: (for which there is no other specific formula),* edition January 1979, corrected edition,* a copy of which is enclosed.

The BEAMA Labour Cost Index referred to shall be the Labour Cost Index for Electrical Engineering.*

References to the *Trade and Industry Journal* shall be read as references to *British Business,* which is the new name of the journal.

*This is only one of several BEAMA CPA Clauses and Formulae. If another is adopted, this form may need consequential amendment. In particular, reference to the BEAMA CPA Clause applicable to Mechanical Plant, and to the Labour Cost Index for Mechanical Engineering, may be appropriate. See p. 16.

Trade Terms

The trade term 'C & F' shall be defined by the current edition of Incoterms, where consistent with this tender. A copy of the definition is enclosed.

Validity

This tender is valid for acceptance until and including *, but shall, after that date, be null and void unless extended by us in writing.

*Date in full, e.g. 'Monday, 19 January 1981'.

Yours faithfully

(Signature)

FORM 4

CIF Tender based on IMechE Model Form B1/1981 Export Contracts for Supply of Plant and Machinery

Dear Sirs

(heading)

Prices

We offer to supply CIF * the items described in the accompanying Specification for:

£

(pounds sterling)

*Port of destination.

Programme

Our best estimate is that delivery CIF will be achieved within of receiving your order.

Terms of Payment

All prices are payable in pounds sterling in the United Kingdom.

We propose that payment shall be made as follows:

(a) %(per cent) in advance.

(b) The balance through a Confirmed Irrevocable Letter of Credit providing for payment in full upon presentation of documents, to be increased according to any contractual increases in the price, to be extended according to any increase in the period required for execution of our respective obligations under the Contract, and otherwise complying with the following description:

(i) Confirmed by an acceptable bank in London.

(ii) Irrevocably issued, confirmed and advised to us by .[a]

(iii) Date and place of expiry – [a] in London.

(iv) Beneficiary – ourselves.

(v) Amount – the balance of the price.

(vi) Credit available with confirming bank by payment.

(vii) Partial shipments allowed.

(viii) Transhipment allowed.

(ix) Shipment from .[b]

(x) For transportation to .[c]

(xi) Shipment of .[d]

(xii) Documents required:

Commercial Invoice

Full set of clean on board marine Bills of Lading (which shall be acceptable although to the effect that the goods have been shipped on deck or unprotected*).

Insurance Certificate.

*Delete if inappropriate.

(xiii) Period after issuance of shipping documents for presentation – days.

(xiv) Subject otherwise to the *Uniform Customs and Practice for Documentary Credits,* 1974, ICC Publication No 290.[e]

[a] Date in full, e.g. 'Monday, 19 January 1981'.

[b] Port of shipment, or e.g. 'any United Kingdom port'.

[c] Port of destination.

[d] Short description of goods.

[e] This description follows as closely as possible the *Standard Procedure for the Issuing of Documentary Credits by Cable, Telegram or Telex* (App VII *Standard Forms for Issuing Documentary Credits,* ICC Publication No 323).

Conditions of Contract

Where consistent with this tender, we propose that 'Model Form B1/1981', a copy of which is enclosed, shall apply, subject to the following:

The Appendix shall be completed as follows:

Clause

17. (Delay in Completion)
    - (a) Percentage to be deducted as damages.
    - (b) Maximum percentage which the deductions may not exceed.

20. Percentage on Prime Cost items.

24. The Institution of Mechanical Engineers.*
    The Institution of Electrical Engineers.*

*Delete one line.

Clause 22 (Terms of Payment) – the rate of interest shall apply to all delayed payments from the Purchaser, and shall be 2% (two per cent) per annum over the * Bank base lending rate from time to time in force on the amount of the delayed payment for the period of the delay. Clause 22(i) (a) shall not apply.

*Insert name of bank.

Clause 26 (Variation in Costs) shall not apply.

Contract Price Adjustment

We propose that contract price adjustment shall apply in accordance with the BEAMA *Contract Price Adjustment Clause and Formula for use with Export Contracts – Electrical Machinery: (for which there is no other specific formula),* edition January 1979, corrected edition,* a copy of which is enclosed.

The BEAMA Labour Cost Index referred to shall be the Labour Cost Index for Electrical Engineering.*

References to the *Trade and Industry Journal* shall be read as references to *British Business,* which is the new name of the journal.

*This is only one of several BEAMA CPA Clauses and Formulae. If another is adopted, this form may need consequential amendment. In particular, reference to the BEAMA CPA Clause applicable to Mechanical Plant, and to the Labour Cost Index for Mechanical Engineering, may be appropriate. See p. 16.

Trade Terms

The trade term 'CIF' shall be defined by the current edition of Incoterms, where consistent with this tender. A copy of the definition is enclosed.

Validity

This tender is valid for acceptance until and including __________ *, but shall, after that date, be null and void unless extended by us in writing.

*Date in full, e.g. 'Monday, 19 January 1981'.

Yours faithfully

(Signature)

FORM 5

FOB Tender based on IMechE Model Form B2/1981 Export Contracts, Delivery FOB, CIF or FOR, With Supervision of Erection

Dear Sirs

(heading)

Prices

We offer to supply FOB * and supervise the erection on Site at

of the items described in the accompanying Specification for:

£

(pounds sterling)

*Port of shipment.

Programme

Our best estimate is that delivery FOB will be achieved within of receiving your order, and that we could complete supervision of erection within a further of receiving your order, making a total of .

We enclose * showing the basis of this estimate.

*Appropriate document, e.g. bar chart or critical path network.

Terms of Payment

All prices are payable in pounds sterling in the United Kingdom.

We propose that payment shall be made as follows:

(a) %(per cent) in advance.

(b) The balance through a Confirmed Irrevocable Letter of Credit providing for payment in full upon presentation of documents, to be increased according to any contractual increases in the price, to be extended according to any increase in the period required for execution of our respective obligations under the Contract, and otherwise complying with the following description:

(i) Confirmed by an acceptable bank in London.

(ii) Irrevocably issued, confirmed and advised to us by .[a]

(iii) Date and place of expiry – [a] in London.

(iv) Beneficiary – ourselves.

(v) Amount – the balance of the price.

(vi) Credit available with confirming bank by payment.

(vii) Partial shipments allowed.

(viii) Transhipment allowed.

(ix) Shipment from .[b]

(x) For transportation to .[c]

(xi) Shipment of .[d]

(xii) Documents required:

Commercial Invoice
In respect of goods, full set of clean on board marine Bills of Lading (which shall be acceptable although to the effect that the goods have been shipped on deck or unprotected*).

In respect of other work, your or your Engineer's Certificate.†

*Delete if inappropriate.
†If this sentence is omitted, payment for such other work should be made out of the letter of credit on presentation only of commercial invoice.

(xiii) Period after issuance of shipping documents for presentation – days.

(xiv) Subject otherwise to the *Uniform Customs and Practice for Documentary Credits*, 1974, ICC Publication No 290.[e]

[a] Date in full, e.g. 'Monday, 19 January 1981'.
[b] Port of shipment, or e.g. 'any United Kingdom port'.
[c] Port of destination.
[d] Short description of goods and site work.
[e] This description follows as closely as possible the *Standard Procedure for the Issuing of Documentary Credits by Cable, Telegram or Telex* (App VII *Standard Forms for Issuing Documentary Credits*, ICC Publication No 323).

Conditions of Contract

Where consistent with this tender, we propose that 'Model Form B2/1981', a copy of which is enclosed, shall apply, subject to the following:

The Appendix shall be completed as follows:
Clause
21. (Delay in Completion)
(a) Percentage to be deducted as damages.
(b) Maximum percentage which the deductions may not exceed.
25. Percentage on Prime Cost items.
30. The Institution of Mechanical Engineers.*
The Institution of Electrical Engineers.*

*Delete one line.

Clause 27 (Terms of Payment) – the rate of interest shall apply to all delayed payments from the Purchaser, and shall be 2% (two per cent) per annum over the * Bank base lending rate from time to time in force on the amount of the delayed payment for the period of the delay. Clause 27(i) (a) shall not apply.

*Insert name of bank.

Clause 32 (Variation in Costs) shall not apply.

Contract Price Adjustment

We propose that contract price adjustment in respect of the price for goods shall apply in accordance with the BEAMA *Contract Price Adjustment Clause and Formula for use with Export Contracts – Electrical Machinery: (for which there is no other specific formula)*, edition January 1979, corrected edition,* a copy of which is enclosed.

We further propose that contract price adjustment in respect of the price for site work shall apply in accordance with the same BEAMA Clause and Formula subject to the amendment that adjustment shall take place at the rate of 0.95 per cent of such price per 1.0 per cent difference between the Labour Cost Index published for the month in which the tender date falls and the average of the Index figures published for the last two-thirds of the period during which the Contractor is carrying out site work, this difference being expressed as a percentage of the former Index figure.

The BEAMA Labour Cost Index referred to shall be the Labour Cost Index for Electrical Engineering.*

References to the *Trade and Industry Journal* shall be read as references to *British Business*, which is the new name of the journal.†

*This is only one of several BEAMA CPA Clauses and Formulae. If another is adopted, this form may need consequential amendment. In particular, reference to the BEAMA CPA Clause applicable to Mechanical Plant, and to the Labour Cost Index for Mechanical Engineering, may be appropriate. See p. 16.

†The above form requires that the separate prices to be adjusted shall be identifiable from the contract documents.

Trade Terms

The trade term 'FOB' shall be defined by the current edition of Incoterms, where consistent with this tender. A copy of the definition is enclosed.

Validity

This tender is valid for acceptance until and including *, but shall, after that date, be null and void unless extended by us in writing.

*Date in full, e.g. 'Monday, 19 January 1981'.

Yours faithfully

(Signature)

FORM 6

C & F Tender based on IMechE Model Form B2/1981 Export Contracts, Delivery FOB, CIF or FOR, With Supervision of Erection

Dear Sirs

(heading)

Prices

We offer to supply C & F ,* and supervise the erection on Site at

of the items described in the accompanying Specification for:

£

(pounds sterling)

*Port of destination.

Programme

Our best estimate is that delivery C & F will be achieved within of receiving your order, and that we could complete supervision of erection within a further of receiving your order, making a total of

We enclose * showing the basis of this estimate.

*Appropriate document, e.g. bar chart or critical path network.

Terms of Payment

All prices are payable in pounds sterling in the United Kingdom.

We propose that payment shall be made as follows:

(a) % (per cent) in advance.

(b) The balance through a Confirmed Irrevocable Letter of Credit providing for payment in full upon presentation of documents, to be increased according to any contractual increases in the price, to be extended according to any increase in the period required for execution of our respective obligations under the Contract, and otherwise complying with the following description:

(i) Confirmed by an acceptable bank in London.

(ii) Irrevocably issued, confirmed and advised to us by .[a]

(iii) Date and place of expiry – [a] in London.

(iv) Beneficiary – ourselves.

(v) Amount – the balance of the price.

(vi) Credit available with confirming bank by payment.

(vii) Partial shipments allowed.

(viii) Transhipment allowed.

(ix) Shipment from .[b]

(x) For transportation to .[c]

(xi) Shipment of .[d]

(xii) Documents required:

Commercial Invoice

In respect of goods, full set of clean on board marine Bills of Lading (which shall be acceptable although to the effect that the goods have been shipped on deck or unprotected*).

In respect of other work, your or your Engineer's Certificate.†

*Delete if inappropriate.

†If this sentence is omitted, payment for such other work should be made out of the letter of credit on presentation only of commercial invoice.

(xiii) Period after issuance of shipping documents for presentation – days.

(xiv) Subject otherwise to the *Uniform Customs and Practice for Documentary Credits*, 1974, ICC Publication No 290.[e]

[a]Date in full, e.g. 'Monday, 19 January 1981'.

[b]Port of shipment, or e.g. 'any United Kingdom port'.

[c]Port of destination.

[d]Short description of goods and site work.

[e]This description follows as closely as possible the *Standard Procedure for the Issuing of Documentary Credits by Cable, Telegram or Telex* (App VII *Standard Forms for Issuing Documentary Credits*, ICC Publication No 323).

Conditions of Contract

Where consistent with this tender, we propose that 'Model Form B2/1981', a copy of which is enclosed, shall apply, subject to the following:

The Appendix shall be completed as follows:

Clause

21. (Delay in Completion)
    (a) Percentage to be deducted as damages.
    (b) Maximum percentage which the deductions may not exceed.

25. Percentage on Prime Cost items.

30. The Institution of Mechanical Engineers.*
    The Institution of Electrical Engineers.*

*Delete one line.

Clause 27 (Terms of Payment) – the rate of interest shall apply to all delayed payments from the Purchaser, and shall be 2% (two per cent) per annum over the * Bank base lending rate from time to time in force on the amount of the delayed payment for the period of the delay. Clause 27(i) (a) shall not apply.

*Insert name of bank.

Clause 32 (Variation in Costs) shall not apply.

Contract Price Adjustment

We propose that contract price adjustment in respect of the price for goods shall apply in accordance with the BEAMA *Contract Price Adjustment Clause and Formula for use with Export Contracts – Electrical Machinery: (for which there is no other specific formula)*, edition January 1979, corrected edition,* a copy of which is enclosed.

We further propose that contract price adjustment in respect of the price for site work shall apply in accordance with the same BEAMA Clause and Formula subject to the amendment that adjustment shall take place at the rate of 0.95 per cent of such price per 1.0 per cent difference between the Labour Cost Index published for the month in which the tender date falls and the average of the Index figures published for the last two-thirds of the period during which the Contractor is carrying out site work, this difference being expressed as a percentage of the former Index figure.

The BEAMA Labour Cost Index referred to shall be the Labour Cost Index for Electrical Engineering.*

References to the *Trade and Industry Journal* shall be read as references to *British Business*, which is the new name of the journal.†

*This is only one of several BEAMA CPA Clauses and Formulae. If another is adopted, this form may need consequential amendment. In particular, reference to the BEAMA CPA Clause applicable to Mechanical Plant, and to the Labour Cost Index for Mechanical Engineering, may be appropriate. See p. 16.

†The above form requires that the separate prices to be adjusted shall be identifiable from the contract documents.

Trade Terms

The trade term 'C & F' shall be defined by the current edition of Incoterms, where consistent with this tender. A copy of the definition is enclosed.

Validity

This tender is valid for acceptance until and including *, but shall, after that date, be null and void unless extended by us in writing.

*Date in full, e.g. 'Monday, 19 January 1981'.

Yours faithfully

(Signature)

FORM 7

CIF Tender based on IMechE Model Form B2/1981 Export Contracts, Delivery FOB, CIF or FOR, With Supervision of Erection

Dear Sirs

(heading)

Prices

We offer to supply CIF ,* and supervise the erection on Site at

of the items described in the accompanying Specification for:
£
(pounds sterling)

*Port of destination.

Programme

Our best estimate is that delivery CIF will be achieved within of receiving your order, and that we could complete supervision of erection within a further of receiving your order, making a total of
We enclose * showing the basis of this estimate.

*Appropriate document, e.g. bar chart or critical path network.

Terms of Payment

All prices are payable in pounds sterling in the United Kingdom.

We propose that payment shall be made as follows:

(a) %(per cent) in advance.

(b) The balance through a Confirmed Irrevocable Letter of Credit providing for payment in full upon presentation of documents, to be increased according to any contractual increases in the price, to be extended according to any increase in the period required for execution of our respective obligations under the Contract, and otherwise complying with the following description:

(i) Confirmed by an acceptable bank in London.
(ii) Irrevocably issued, confirmed and advised to us by .[a]
(iii) Date and place of expiry – [a] in London.
(iv) Beneficiary – ourselves.
(v) Amount – the balance of the price.
(vi) Credit available with confirming bank by payment.
(vii) Partial shipments allowed.
(viii) Transhipment allowed.
(ix) Shipment from .[b]
(x) For transportation to .[c]
(xi) Shipment of .[d]
(xii) Documents required:

Commercial Invoice

In respect of goods, full set of clean on board marine Bills of Lading (which shall be acceptable although to the effect that the goods have been shipped on deck or unprotected*) and Insurance Certificate.

In respect of other work, your or your Engineer's Certificate.†

*Delete if inappropriate.

†If this sentence is omitted, payment for such other work should be made out of the letter of credit on presentation only of commercial invoice.

(xiii) Period after issuance of shipping documents for presentation – days.

(xiv) Subject otherwise to the *Uniform Customs and Practice for Documentary Credits*, 1974, ICC Publication No 290.[e]

[a]Date in full, e.g. 'Monday, 19 January 1981'.

[b]Port of shipment, or e.g. 'any United Kingdom port'.

[c]Port of destination.

[d]Short description of goods and site work.

[e]This description follows as closely as possible the *Standard Procedure for the Issuing of Documentary Credits by Cable, Telegram or Telex* (App VII *Standard Forms for Issuing Documentary Credits*, ICC Publication No 323).

Conditions of Contract

Where consistent with this tender, we propose that 'Model Form B2/1981', a copy of which is enclosed, shall apply, subject to the following:

The Appendix shall be completed as follows:

Clause

21. (Delay in Completion)
 (a) Percentage to be deducted as damages.
 (b) Maximum percentage which the deductions may not exceed.

25. Percentage on Prime Cost items.

30. The Institution of Mechanical Engineers.*
 The Institution of Electrical Engineers.*

*Delete one line.

Clause 27 (Terms of Payment) – the rate of interest shall apply to all delayed payments from the Purchaser, and shall be 2% (two per cent) per annum over the * Bank base lending rate from time to time in force on the amount of the delayed payment for the period of the delay. Clause 27(i) (a) shall not apply.

*Insert name of bank.

Clause 32 (Variation in Costs) shall not apply.

Contract Price Adjustment

We propose that contract price adjustment in respect of the price for goods shall apply in accordance with the BEAMA *Contract Price Adjustment Clause and Formula for use with Export Contracts – Electrical Machinery: (for which there is no other specific formula)*, edition January 1979, corrected edition,* a copy of which is enclosed.

We further propose that contract price adjustment in respect of the price for site work shall apply in accordance with the same BEAMA Clause and Formula subject to the amendment that adjustment shall take place at the rate of 0.95 per cent of such price per 1.0 per cent difference between the Labour Cost Index published for the month in which the tender date falls and the average of the Index figures published for the last two-thirds of the period during which the Contractor is carrying out site work, this difference being expressed as a percentage of the former Index figure.

The BEAMA Labour Cost Index referred to shall be the Labour Cost Index for Electrical Engineering.*

References to the *Trade and Industry Journal* shall be read as references to *British Business,* which is the new name of the journal.†

*This is only one of several BEAMA CPA Clauses and Formulae. If another is adopted, this form may need consequential amendment. In particular, reference to the BEAMA CPA Clause applicable to Mechanical Plant, and to the Labour Cost Index for Mechanical Engineering, may be appropriate. See p. 16.

†The above form requires that the separate prices to be adjusted shall be identifiable from the contract documents.

Trade Terms

The trade term 'CIF' shall be defined by the current edition of Incoterms, where consistent with this tender. A copy of the definition is enclosed.

Validity

This tender is valid for acceptance until and including *, but shall, after that date, be null and void unless extended by us in writing.

*Date in full, e.g. 'Monday, 19 January 1981'.

Yours faithfully

(Signature)

FORM 8

Tender based on IMechE Model Form B3/1980 Export Contracts (Including Delivery to and Erection on Site)

Dear Sirs

(heading)

Prices
We offer to supply, deliver to and erect on Site at

the items described in the accompanying Specification for:
£
(pounds sterling)

Programme
Our best estimate is that completion of our work will be achieved within of receiving your order. (We enclose * showing the basis of this estimate.)†
*Appropriate document, e.g. bar chart or critical path network.
†Delete if inappropriate.

Terms of Payment
All prices are payable in pounds sterling in the United Kingdom.
We propose that payment shall be made as follows:
(a) %(per cent) in advance.
(b) The balance through a Confirmed Irrevocable Letter of Credit providing for payment in full upon presentation of documents, to be increased according to any contractual increases in the price, to be extended according to any increase in the period required for execution of our respective obligations under the Contract, and otherwise complying with the following description:
(i) Confirmed by an acceptable bank in London.
(ii) Irrevocably issued, confirmed and advised to us by .[a]
(iii) Date and place of expiry – [a] in London.
(iv) Beneficiary – ourselves.
(v) Amount – the balance of the price.
(vi) Credit available with confirming bank by payment.
(vii) Partial shipments allowed.
(viii) Transhipment allowed.
(ix) Shipment from .[b]
(x) For transportation to .[c]
(xi) Shipment of .[d]
(xii) Documents required:

Commercial Invoice

In respect of goods, full set of clean on board marine Bills of Lading (which shall be acceptable although to the effect that the goods have been shipped on deck or unprotected*) and Insurance Certificate.

In respect of other work, your or your Engineer's Certificate.†

*Delete if inappropriate.

†If this sentence is omitted, payment for such other work should be made out of the letter of credit on presentation only of commercial invoice.

(xiii) Period after issuance of shipping documents for presentation – days.

(xiv) Subject otherwise to the *Uniform Customs and Practice for Documentary Credits*, 1974, ICC Publication No 290.[e]

[a]Date in full, e.g. 'Monday, 19 January 1981'.

[b]Port of shipment, or e.g. 'any United Kingdom port'.

[c]Port of destination.

[d]Short description of goods and site work.

[e]This description follows as closely as possible the *Standard Procedure for the Issuing of Documentary Credits by Cable, Telegram or Telex* (App VII *Standard Forms for Issuing Documentary Credits*, ICC Publication No 323).

Conditions of Contract

Where consistent with this tender, we propose that 'Model Form B3/1980', including Erratum, a copy of which is enclosed, shall apply, subject to the following:

The Appendix shall be completed as follows:

Clause

24. Additional risks to be covered by insurance.
27. (Delay in Completion)
    (a) Percentage of Contract Value to be deducted as damages.
    (b) Maximum percentage of Contract Value which the deductions may not exceed.
33. Percentage on Prime Cost items.
38. The Institution of Mechanical Engineers.*
    The Institution of Electrical Engineers.*

*Delete one line.

Clause 35 (Terms of Payment) – the rate of interest shall apply to all delayed payments from the Purchaser, and shall be 2% (two per cent) per annum over the * Bank base lending rate from time to time in force on the amount of the delayed payment for the period of the delay. Clause 35(i) (a) shall not apply.

*Insert name of bank.

Clause 40 (Variation in Costs) shall not apply.

Contract Price Adjustment

We propose that contract price adjustment in respect of the price for goods shall apply in accordance with the BEAMA *Contract Price Adjustment Clause and Formula for use with Export Contracts – Electrical Machinery: (for which there is no other specific formula)*, edition January 1979, corrected edition,* a copy of which is enclosed.

We further propose that contract price adjustment in respect of the price for site work shall apply in accordance with the same BEAMA Clause and

Formula subject to the amendment that adjustment shall take place at the rate of 0.95 per cent of such price per 1.0 per cent difference between the Labour Cost Index published for the month in which the tender date falls and the average of the Index figures published for the last two-thirds of the period during which the Contractor is carrying out site work, this difference being expressed as a percentage of the former Index figure.

The BEAMA Labour Cost Index referred to shall be the Labour Cost Index for Electrical Engineering.*

References to the *Trade and Industry Journal* shall be read as references to *British Business*, which is the new name of the journal.†

*This is only one of several BEAMA CPA Clauses and Formulae. If another is adopted, this form may need consequential amendment. In particular, reference to the BEAMA CPA Clause applicable to Mechanical Plant, and to the Labour Cost Index for Mechanical Engineering, may be appropriate. See p. 16.

†The above form requires that the separate prices to be adjusted shall be identifiable from the contract documents.

Validity

This tender is valid for acceptance until and including *, but shall, after that date, be null and void unless extended by us in writing.

*Date in full, e.g. 'Monday, 19 January 1981'.

Yours faithfully

(Signature)

FORM 9

Tender based on IMechE Model Form C/1975 for the Supply of Electrical and Mechanical Goods, other than Electric Cables (Home – without erection)

Dear Sirs

(heading)

Prices
We offer to supply * the items described in the accompanying Specification for:
£
(pounds)
*Appropriate trade term, e.g. 'ex works'.

Programme
Our best estimate is that delivery * will be achieved within of receiving your order.
*Appropriate trade term, e.g. 'ex works'.

Terms of Payment
We propose that payment shall be made as follows:
(a) % (per cent) in advance.
(b) The balance in respect of each item of the goods upon delivery * of that item.

*Appropriate trade term, e.g. 'ex works'.

Conditions of Contract
Where consistent with this tender, we propose that 'Model Form C/1975, Home – without erection' with September 1978 amendments, a copy of which is enclosed, shall apply, subject to the following:

Clause 10 (Terms of Payment) – If the payment of any sum payable to us shall be delayed, interest at the rate of 2% (two per cent) per annum over the * Bank base lending rate from time to time in force on the amount of the delayed payment for the period of the delay shall be added to the price.
*Insert name of bank.

Clause 15 (Arbitration) – The Institution referred to shall be The Institution of (Electrical)* (Mechanical)* Engineers.

*Delete alternative.

The Supplementary Clause (Variations in Costs) shall not apply.

Contract Price Adjustment
We propose that contract price adjustment shall apply in accordance with the BEAMA *Contract Price Adjustment Clause and Formula for use with Home Contracts –*

Electrical Machinery: (for which there is no other specific formula), edition January 1979, corrected edition,* a copy of which is enclosed.

The BEAMA Labour Cost Index referred to shall be the Labour Cost Index for Electrical Engineering.*

References to the *Trade and Industry Journal* shall be read as references to *British Business*, which is the new name of the journal.

*This is only one of several BEAMA CPA Clauses and Formulae. If another is adopted, this form may need consequential amendment. In particular, reference to the BEAMA CPA Clause applicable to Mechanical Plant, and to the Labour Cost Index for Mechanical Engineering, may be appropriate. See p. 16.

Trade Terms

The trade term * shall be defined by the current edition of Incoterms, where consistent with this tender. A copy of the definition is enclosed.

*Appropriate trade term, e.g. 'ex works'.

Validity

This tender is valid for acceptance until and including *, but shall, after that date, be null and void unless extended by us in writing.

*Date in full, e.g. 'Monday, 19 January 1981'.

Yours faithfully

(Signature)

FORM 10

Tender based on IEE Model Form E/1973 Home Cable Contracts – with installation

Dear Sirs

(heading)

Prices

We offer to supply, deliver to and install on Site at

the items described in the accompanying Specification for:

£

(pounds)

Programme

Our best estimate is that completion of our work will be achieved within of receiving your order. (We enclose * showing the basis of this estimate).†

*Appropriate document, e.g. bar chart or critical path network.

†Delete if inappropriate.

Terms of Payment

We propose that payment shall be made as follows:

(a) %(per cent) in advance.

(b) The balance in accordance with the Conditions of Contract referred to in this tender.

Conditions of Contract

Where consistent with this tender, we propose that 'Model Form E/1973', with July 1976 and September 1978 amendments, a copy of which is enclosed, shall apply, subject to the following:

The Appendix shall be completed as follows:

Clause

30. (Delay in Completion)
    (a) Percentage of Contract Value to be deducted as damages.
    (b) Maximum percentage of Contract Value which the deductions may not exceed.
36. Percentage on Prime Cost items.

Clause 40 (Terms of Payment) – the rate of interest shall apply to all delayed payments from the Purchaser, and shall be 2% (two per cent) per annum over the * Bank base lending rate from time to time in force on the amount of the delayed payment for the period of the delay.

*Insert name of bank.

Clause 44 (Variation in Costs) shall not apply.

Contract Price Adjustment

We propose that contract price adjustment shall apply in accordance with the BEAMA *Contract Price Adjustment Clause and Formula for use with Home Contracts – Electrical Machinery: (for which there is no other specific formula)*, edition January 1979, corrected edition,* a copy of which is enclosed.

The provisions of the Clause and Formulae applicable to contracts including erection shall apply.

The BEAMA Labour Cost Index referred to shall be the Labour Cost Index for Electrical Engineering.*

References to the *Trade and Industry Journal* shall be read as references to *British Business*, which is the new name of the journal.

*This is only one of several BEAMA CPA Clauses and Formulae. If another is adopted, this form may need consequential amendment. In particular, reference to the BEAMA CPA Clause applicable to Mechanical Plant, and to the Labour Cost Index for Mechanical Engineering, may be appropriate. See p. 16.

Validity

This tender is valid for acceptance until and including   *, but shall, after that date, be null and void unless extended by us in writing.

*Date in full, e.g. 'Monday, 19 January 1981'.

Yours faithfully

(Signature)

FORM 11

Tender based on FIDIC Conditions of Contract (International) for Electrical and Mechanical Works (Including Erection on Site) second edition, 1980

Dear Sirs

(heading)

Prices

We offer to supply, deliver to and erect on Site at

the items described in the accompanying Specification for:
£
(pounds sterling)

Programme

Our best estimate is that completion of our work will be achieved within of receiving your order. (We enclose * showing the basis of this estimate.)†

*Appropriate document, e.g. bar chart or critical path network.
†Delete if inappropriate.

Terms of Payment

All prices are payable in pounds sterling in the United Kingdom.

We propose that payment shall be made as follows:

(a) %(per cent) in advance.

(b) The balance through a Confirmed Irrevocable Letter of Credit providing for payment in full upon presentation of documents, to be increased according to any contractual increases in the price, to be extended according to any increase in the period required for execution of our respective obligations under the Contract, and otherwise complying with the following description:

(i) Confirmed by an acceptable bank in London.
(ii) Irrevocably issued, confirmed and advised to us by .[a]
(iii) Date and place of expiry – [a] in London.
(iv) Beneficiary – ourselves.
(v) Amount – the balance of the price.
(vi) Credit available with confirming bank by payment.
(vii) Partial shipments allowed.
(viii) Transhipment allowed.
(ix) Shipment from .[b]
(x) For transportation to .[c]
(xi) Shipment of .[d]
(xii) Documents required:

Commercial Invoice

In respect of goods, full set of clean on board marine Bills of Lading (which shall be acceptable although to the effect that the goods have been shipped on deck or unprotected*) and Insurance Certificate.

In respect of other work, your or your Engineer's Certificate.†

*Delete if inappropriate.

†If this sentence is omitted, payment for such other work should be made out of the letter of credit on presentation only of commercial invoice.

(xiii) Period after issuance of shipping documents for presentation – days.

(xiv) Subject otherwise to the *Uniform Customs and Practice for Documentary Credits*, 1974, ICC Publication No 290.[e]

[a] Date in full, e.g. 'Monday, 19 January 1981'.

[b] Port of shipment, or e.g. 'any United Kingdom port'.

[c] Port of destination.

[d] Short description of goods and site work.

[e] This description follows as closely as possible the *Standard Procedure for the Issuing of Documentary Credits by Cable, Telegram or Telex* (App VII *Standard Forms for Issuing Documentary Credits*, ICC Publication No 323).

Conditions of Contract

Where consistent with this tender, we propose that FIDIC *Conditions of Contract (International) for Electrical and Mechanical Works (Including Erection on Site)*, second edition 1980, a copy of which is enclosed, shall apply, subject to the following:

Part II – Conditions of Particular Application

Please refer to pages 38, 39 and 40 of the enclosed FIDIC Conditions of Contract. The relevant text of these pages is set out below, and with regard to the Clauses identified on those pages, we propose:

	FIDIC Text	*Tenderer's Proposals*
Clause 1	*Definitions* Employer. The Employer is Engineer. The Engineer is	
Clause 2.1	*Engineer's Duties* The Employer's specific approval is required for the following:	We make no proposals in this respect
Clause 4.1	*Languages* Language. The languages The Ruling Language is	are English and English
Clause 4.2	(b) *Law* The Law to which the Contract is to be subject is	English law
Clause 9	*Performance Bond or Surety* Period of Validity Procedure if forfeit Arrangements for release Currency of Bond	Pounds sterling – A proposed form of Performance Bond or Surety is enclosed

Clause 12.1	*Programme* Time limit for submission of programme	
Clause 14.1	*Contractor's Equipment* The Employer shall be responsible for the provision of the following:	
Clause 14.3	*Electricity, Water and Gas* The Employer shall provide the following electricity, water and gas facilities	
Clause 14.4	*Employer's Lifting Equipment* Details of lifting equipment to be provided by the Employer	
Clause 16.4	*Limitation of Liability* Contractor's liability shall not exceed the sum of	£ (pounds sterling)
	Contractor's liability shall expire	12 months after taking over
Clause 17.1	*Insurance of Works* Amount of insurance during Defects Liability Period	£ (pounds sterling)
Clause 22.9	*Labour – Other Conditions* Permits for imported labour, control, health, hours and conditions, rates of pay, compliance with labour legislation	We make no proposals in this respect
Clause 31.3	*Bonus*	We make no proposals in this respect
Clause 33.1	*Defects Liability Period*	12 months
Clauses 33.3 and 33.4	*Defects Liability Period* Maximum permitted extension	Two years from the date of taking over
Clause 36.3	*Vesting of Plant and Contractor's Equipment* Other conditions concerning Contractor's Equipment	We make no proposals in this respect
Clause 37.6	*Advance and Progress Payments* Provision for progress payments during manufacture at Contractor's Works	As stated in our tender. Clause 37.6 shall not apply
Clause 41.1	*Interest on Delayed Payment* Amount by which interest to be increased if delay exceeds 60 days	2% (two per cent) per annum. Interest shall apply to all delayed payments from the Employer
Clause 41.4	*Payment in Foreign Currencies* Foreign currencies in which payment to be made, proportions, rate of exchange and conditions applicable thereto	As may be stated in our tender
Clause 50.2	*Notices* The Employer's address The Engineer's address	

Clause 52.1	*Changes in Cost and Legislation* In appropriate cases this Clause should cover such matters as: adjustment of Contract Sum by reason of alteration in rates of wages and allowances payable to labour and local staff; change in conditions of employment of labour and local staff; change in cost of materials for permanent or temporary works or in consumable stores, fuel and power; variation in freight and insurance rates; customs or other import duties; the operation of any law, statute, etc. This Clause should also specify when payments thereof should be made	As stated in our tender
Clause 53.1	*Customs and Import Duties* Method of payment whether by Employer or Contractor. Whether included in or excluded from Contract Sum	The Employer shall promptly pay all customs and import duties, which are excluded from the Contract Sum
Clause 54	*Taxation* Payment of or exemption from local income or other taxes both as regards the Contractor, his staff and his expatriate operatives	All prices under the Contract exclude all foreign taxes. The Employer shall indemnify the Contractor and his personnel against all such taxes. The amount or amounts required in respect of such indemnity shall be notified by the Contractor to the Employer from time to time as and when the Contractor, or his personnel, are assessed for, or pay, any foreign tax or taxes. The expression 'foreign taxes' includes all present and future taxes in force in the country where the Site is located,* and all other present and future income, capital or other taxes, levies, contributions, duties, impositions and compulsory payments of all names, descriptions and kinds, payable in any country other than the United Kingdom *It would be as well to list by name all such known taxes

Clause 55	*Miscellaneous* In certain cases it may be desirable to insert clauses to cover (and number accordingly) such matters as: (a) bribery and corruption; (b) photographs of the Works and advertising; (c) undertakings regarding non-disclosure of secret information; (e) any other matters special to the Contract	We make no proposals in this respect

With regard to the other Clauses mentioned below, we propose:

Clause

4.3 *Documents Mutually Explanatory*
The first sentence of this Clause in Part I, General Conditions, shall be deleted, and the following substituted:

'Unless otherwise provided in the Contract, in case of conflict Part II shall prevail over Part I of these Conditions. The Contractor's tender and all other documents agreed between the parties shall prevail both over Part I and Part II of the Conditions.'

5.6 *Operating and Maintenance Instructions*
Add the following to this Clause in Part I, General Conditions:

'The Operating and Maintenance Instructions shall be in the English and languages. All words written on the Drawings furnished to the Employer under this Clause shall be in the English and languages.'

Appendix to Tender

Please refer to page 42 of the enclosed FIDIC Conditions of Contract. The relevant text of that page is set out below, and with regard to the Clauses identified on that page, we propose:

	Clause		
Time for Completion	7.1		days
Amount of Bond or Guarantee (if any)	9.1	£ (	pounds sterling)
Amount of Insurance during Defects Liability Period	17.1	£ (	pounds sterling)
Minimum Amount of Third Party Insurance	17.2	£ (	pounds sterling)
Amount of reduction per week of delay	31.1	%(of Contract Price	per cent)
Maximum reduction	31.1	%(of Contract Price	per cent)
Limit of Liability for delay	31.2	£ (	pounds sterling)
Defects Liability Period	33.1	12 months	
Use of Works assumed		hours per day	
Reduction of period for more intensive use		days for each extra hour of daily use	
Percentage for adjustment of Provisional Sums	39.4	%(	per cent)
Name and address of Contractor's Bank	41.1		

Firm Prices*

The prices quoted are firm, and are not subject to adjustment for changes in our costs.

*Delete this section if a Contract Price Adjustment section is to be included in the tender.

Contract Price Adjustment*

We propose that contract price adjustment in respect of the price for goods shall apply in accordance with the BEAMA *Contract Price Adjustment Clause and Formula for use with Export Contracts – Electrical Machinery: (for which there is no other specific formula)*, edition January 1979, corrected edition,† a copy of which is enclosed.

We further propose that contract price adjustment in respect of the price for site work shall apply in accordance with the same BEAMA Clause and Formula subject to the amendment that adjustment shall take place at the rate of 0.95 per cent of such price per 1.0 per cent difference between the Labour Cost Index published for the month in which the tender date falls and the average of the Index figures published for the last two-thirds of the period during which the Contractor is carrying out site work, this difference being expressed as a percentage of the former Index figure.

The BEAMA Labour Cost Index referred to shall be the Labour Cost Index for Electrical Engineering.†

References to the *Trade and Industry Journal* shall be read as references to *British Business*, which is the new name of the journal.‡

*Delete this section if prices quoted are firm, in which case Firm Prices section should be included.

†This is only one of several BEAMA CPA Clauses and Formula. If another is adopted, this form may need consequential amendment. In particular, reference to the BEAMA CPA Clause applicable to Mechanical Plant, and to the Labour Cost Index for Mechanical Engineering, may be appropriate. See p. 16.

‡The above form requires that the separate prices to be adjusted shall be identifiable from the contract documents.

Validity

This tender is valid for acceptance until and including *, but shall, after that date, be null and void unless extended by us in writing.

*Date in full, e.g. 'Monday, 19 January 1981'.

Yours faithfully

(Signature)

FORM 12

Tender based on BEAMA Conditions of Sale (A) for Machinery and Equipment (Exclusive of Erection) United Kingdom, edition June 1979

Dear Sirs

(heading)

Prices
We offer to supply * the items described in the accompanying Specification for:
£
(pounds)
*Appropriate trade term, e.g. 'ex works'.

Programme
Our best estimate is that delivery * will be achieved within of receiving your order.
*Appropriate trade term, e.g. 'ex works'.

Terms of Payment
We propose that payment shall be made as follows:
(a) %(per cent) in advance.
(b) The balance in respect of each item of the goods upon notification by us that it is ready for despatch.

Conditions of Contract
Where consistent with this tender, we propose that BEAMA *Conditions of Sale (A) for Machinery and Equipment (Exclusive of Erection) United Kingdom,* edition June 1979, a copy of which is enclosed, shall apply, subject to the following:

Value Added Tax (VAT) – Prices exclude VAT. To the extent that VAT is properly chargeable on the supply to you of any goods or services under the Contract, you shall pay such VAT as an addition to payments otherwise due under the Contract.

Clause 9 (Liability for Delay) – The figures to be inserted shall be ½% (one-half per cent) and 5% (five per cent) respectively.

Interest – If the payment of any sum payable to us shall be delayed, interest at the rate of 2% (two per cent) per annum over the * Bank base lending rate from time to time in force on the amount of the delayed payment for the period of the delay shall be added to the price.
*Insert name of bank.

Contract Price Adjustment
We propose that contract price adjustment shall apply in accordance with the

BEAMA *Contract Price Adjustment Clause and Formula for use with Home Contracts – Electrical Machinery: (for which there is no other specific formula)*, edition January 1979, corrected edition,* a copy of which is enclosed.

The BEAMA Labour Cost Index referred to shall be the Labour Cost Index for Electrical Engineering.*

References to the *Trade and Industry Journal* shall be read as references to *British Business*, which is the new name of the journal.

*This is only one of several BEAMA CPA Clauses and Formulae. If another is adopted, this form may need consequential amendment. In particular, reference to the BEAMA CPA Clause applicable to Mechanical Plant, and to the Labour Cost Index for Mechanical Engineering, may be appropriate. See p. 16.

Trade Terms

The trade term * shall be defined by the current edition of Incoterms, where consistent with this tender. A copy of the definition is enclosed.

*Appropriate trade term, e.g. 'ex works'.

Validity

This tender is valid for acceptance until and including *, but shall, after that date, be null and void unless extended by us in writing.

*Date in full, e.g. 'Monday, 19 January 1981'.

Yours faithfully

(Signature)

FORM 13

FOB Tender based on BEAMA Conditions of Sale (AE) for Machinery and Equipment (Exclusive of Erection) Export FOB, FOR and FOT, edition December 1980

Dear Sirs

(heading)

Prices
We offer to supply FOB * the items described in the accompanying Specification for:
£
(pounds sterling)
*Port of shipment.

Programme
Our best estimate is that delivery FOB will be achieved within of receiving your order.

Terms of Payment
All prices are payable in pounds sterling in the United Kingdom.

We propose that payment shall be made as follows:

(a) %(per cent) in advance.

(b) The balance through a Confirmed Irrevocable Letter of Credit providing for payment in full upon presentation of documents, to be increased according to any contractual increases in the price, to be extended according to any increase in the period required for execution of our respective obligations under the Contract, and otherwise complying with the following description:

(i) Confirmed by an acceptable bank in London.
(ii) Irrevocably issued, confirmed and advised to us by .[a]
(iii) Date and place of expiry – [a] in London.
(iv) Beneficiary – ourselves.
(v) Amount – the balance of the price.
(vi) Credit available with confirming bank by payment.
(vii) Partial shipments allowed.
(viii) Transhipment allowed.
(ix) Shipment from .[b]
(x) For transportation to .[c]
(xi) Shipment of .[d]
(xii) Documents required:

Commercial Invoice

Full set of clean on board marine Bills of Lading (which shall be acceptable although to the effect that the goods have been shipped on deck or unprotected*).

*Delete if inappropriate.

(xiii) Period after issuance of shipping documents for presentation – days.

(xiv) Subject otherwise to the *Uniform Customs and Practice for Documentary Credits*, 1974, ICC Publication No 290.[e]

[a]Date in full, e.g. 'Monday, 19 January 1981'.
[b]Port of shipment, or e.g. 'any United Kingdom port'.
[c]Port of destination.
[d]Short description of goods.
[e]This description follows as closely as possible the *Standard Procedure for the Issuing of Documentary Credits by Cable, Telegram or Telex* (App VII *Standard Forms for Issuing Documentary Credits*, ICC Publication No 323).

Conditions of Contract

Where consistent with this tender, we propose that BEAMA *Conditions of Sale (AE) for Machinery and Equipment (Exclusive of Erection) Export FOB, FOR and FOT*, edition December 1980, a copy of which is enclosed, shall apply, subject to the following:

Clause 11 (Liability for Delay) – The figures to be inserted shall be ½% (one-half per cent) and 5% (five per cent) respectively.

Interest – If the payment of any sum payable to us shall be delayed, interest at the rate of 2% (two per cent) per annum over the * Bank base lending rate from time to time in force on the amount of the delayed payment for the period of the delay shall be added to the price.

*Insert name of bank.

Contract Price Adjustment

We propose that contract price adjustment shall apply in accordance with the BEAMA *Contract Price Adjustment Clause and Formula for use with Export Contracts – Electrical Machinery: (for which there is no other specific formula)*, edition January 1979, corrected edition,* a copy of which is enclosed.

The BEAMA Labour Cost Index referred to shall be the Labour Cost Index for Electrical Engineering.*

References to the *Trade and Industry Journal* shall be read as references to *British Business*, which is the new name of the journal.

*This is only one of several BEAMA CPA Clauses and Formulae. If another is adopted, this form may need consequential amendment. In particular, reference to the BEAMA CPA Clause applicable to Mechanical Plant, and to the Labour Cost Index for Mechanical Engineering, may be appropriate. See p. 16.

Trade Terms

The trade term 'FOB' shall be defined by the current edition of Incoterms, where consistent with this tender. A copy of the definition is enclosed.

Validity

This tender is valid for acceptance until and including *, but shall, after that date, be null and void unless extended by us in writing.

*Date in full, e.g. 'Monday, 19 January 1981'.

Yours faithfully

(Signature)

FORM 14

CIF Tender based on BEAMA Conditions of Sale (AEC) for Machinery and Equipment (Exclusive of Erection) Export CIF and C&F, edition December 1980

Dear Sirs

(heading)

Prices
We offer to supply CIF * the items described in the accompanying Specification for:
£
(pounds sterling)
*Port of destination.

Programme
Our best estimate is that delivery CIF will be achieved within of receiving your order.

Terms of Payment
All prices are payable in pounds sterling in the United Kingdom.

We propose that payment shall be made as follows:

(a) %(per cent) in advance.

(b) The balance through a Confirmed Irrevocable Letter of Credit providing for payment in full upon presentation of documents, to be increased according to any contractual increases in the price, to be extended according to any increase in the period required for execution of our respective obligations under the Contract, and otherwise complying with the following description:

(i) Confirmed by an acceptable bank in London.
(ii) Irrevocably issued, confirmed and advised to us by .[a]
(iii) Date and place of expiry – [a] in London.
(iv) Beneficiary – ourselves.
(v) Amount – the balance of the price.
(vi) Credit available with confirming bank by payment.
(vii) Partial shipments allowed.
(viii) Transhipment allowed.
(ix) Shipment from .[b]
(x) For transportation to .[c]
(xi) Shipment of .[d]
(xii) Documents required:

Commercial Invoice

Full set of clean on board marine Bills of Lading (which shall be acceptable although to the effect that the goods have been shipped on deck or unprotected*).

Insurance Certificate

*Delete if inappropriate.

(xiii) Period after issuance of shipping documents for presentation – days.

(xiv) Subject otherwise to the *Uniform Customs and Practice for Documentary Credits*, 1974, ICC Publication No 290.[e]

[a] Date in full, e.g. 'Monday, 19 January 1981'.
[b] Port of shipment, or e.g. 'any United Kingdom port'.
[c] Port of destination.
[d] Short description of goods.
[e] This description follows as closely as possible the *Standard Procedure for the Issuing of Documentary Credits by Cable, Telegram or Telex* (App VII *Standard Forms for Issuing Documentary Credits*, ICC Publication No 323).

Conditions of Contract

Where consistent with this tender, we propose that BEAMA *Conditions of Sale (AEC) for Machinery and Equipment (Exclusive of Erection) Export CIF and C&F*, edition December 1980, a copy of which is enclosed, shall apply, subject to the following:

Clause 11 (Liability for Delay) – The figures to be inserted shall be ½% (one-half per cent) and 5% (five per cent) respectively.

Interest – If the payment of any sum payable to us shall be delayed, interest at the rate of 2% (two per cent) per annum over the * Bank base lending rate from time to time in force on the amount of the delayed payment for the period of the delay shall be added to the price.

*Insert name of bank.

Contract Price Adjustment

We propose that contract price adjustment shall apply in accordance with the BEAMA *Contract Price Adjustment Clause and Formula for use with Export Contracts – Electrical Machinery: (for which there is no other specific formula)*, edition January 1979, corrected edition,* a copy of which is enclosed.

The BEAMA Labour Cost Index referred to shall be the Labour Cost Index for Electrical Engineering.*

References to the *Trade and Industry Journal* shall be read as references to *British Business*, which is the new name of the journal.

*This is only one of several BEAMA CPA Clauses and Formulae. If another is adopted, this form may need consequential amendment. In particular, reference to the BEAMA CPA Clause applicable to Mechanical Plant, and to the Labour Cost Index for Mechanical Engineering, may be appropriate. See p. 16.

Trade Terms

The trade term 'CIF' shall be defined by the current edition of Incoterms, where consistent with this tender. A copy of the definition is enclosed.

Validity

This tender is valid for acceptance until and including *, but shall, after that date, be null and void unless extended by us in writing.

*Date in full, e.g. 'Monday, 19 January 1981'.

Yours faithfully

(Signature)

FORM 15

C & F Tender based on BEAMA Conditions of Sale (AEC) for Machinery and Equipment (Exclusive of Erection) Export CIF and C & F, edition December 1980

Dear Sirs

(heading)

Prices

We offer to supply C & F * the items described in the accompanying Specification for:

£

(pounds sterling)

*Port of destination.

Programme

Our best estimate is that delivery C & F will be achieved within of receiving your order.

Terms of Payment

All prices are payable in pounds sterling in the United Kingdom.

We propose that payment shall be made as follows:

(a) % (per cent) in advance.

(b) The balance through a Confirmed Irrevocable Letter of Credit providing for payment in full upon presentation of documents, to be increased according to any contractual increases in the price, to be extended according to any increase in the period required for execution of our respective obligations under the Contract, and otherwise complying with the following description:

(i) Confirmed by an acceptable bank in London.
(ii) Irrevocably issued, confirmed and advised to us by .[a]
(iii) Date and place of expiry – [a] in London.
(iv) Beneficiary – ourselves.
(v) Amount – the balance of the price.
(vi) Credit available with confirming bank by payment.
(vii) Partial shipments allowed.
(viii) Transhipment allowed.
(ix) Shipment from .[b]
(x) For transportation to .[c]
(xi) Shipment of .[d]
(xii) Documents required:

Commercial Invoice

Full set of clean on board marine Bills of Lading (which shall be acceptable although to the effect that the goods have been shipped on deck or unprotected*).

*Delete if inappropriate.

(xiii) Period after issuance of shipping documents for presentation – days.

(xiv) Subject otherwise to the *Uniform Customs and Practice for Documentary Credits*, 1974, ICC Publication No 290.[e]

[a] Date in full, e.g. 'Monday, 19 January 1981'.
[b] Port of shipment, or e.g. 'any United Kingdom port'.
[c] Port of destination.
[d] Short description of goods.
[e] This description follows as closely as possible the *Standard Procedure for the Issuing of Documentary Credits by Cable, Telegram or Telex* (App VII *Standard Forms for Issuing Documentary Credits*, ICC Publication No 323).

Conditions of Contract

Where consistent with this tender, we propose that BEAMA *Conditions of Sale (AEC) for Machinery and Equipment (Exclusive of Erection) Export CIF and C&F*, edition December 1980, a copy of which is enclosed, shall apply, subject to the following:

Clause 11 (Liability for Delay) – The figures to be inserted shall be ½% (one-half per cent) and 5% (five per cent) respectively.

Interest– If the payment of any sum payable to us shall be delayed, interest at the rate of 2% (two per cent) per annum over the * Bank base lending rate from time to time in force on the amount of the delayed payment for the period of the delay shall be added to the price.

*Insert name of bank.

Contract Price Adjustment

We propose that contract price adjustment shall apply in accordance with the BEAMA *Contract Price Adjustment Clause and Formula for use with Export Contracts – Electrical Machinery: (for which there is no other specific formula)*, edition January 1979, corrected edition,* a copy of which is enclosed.

The BEAMA Labour Cost Index referred to shall be the Labour Cost Index for Electrical Engineering.*

References to the *Trade and Industry Journal* shall be read as references to *British Business*, which is the new name of the journal.

*This is only one of several BEAMA CPA Clauses and Formulae. If another is adopted, this form may need consequential amendment. In particular, reference to the BEAMA CPA Clause applicable to Mechanical Plant, and to the Labour Cost Index for Mechanical Engineering, may be appropriate. See p. 16.

Trade Terms

The trade term 'C & F' shall be defined by the current edition of Incoterms, where consistent with this tender. A copy of the definition is enclosed.

Validity

This tender is valid for acceptance until and including *, but shall, after that date, be null and void unless extended by us in writing.

*Date in full, e.g. 'Monday, 19 January 1981'.

Yours faithfully

(Signature)

FORM 16

Tender based on BEAMA Conditions of Sale (B) for Machinery and Equipment Including Supervision of Erection, United Kingdom, edition June 1979

Dear Sirs

(heading)

Prices
We offer to supply *, and supervise the erection on Site at

of the items described in the accompanying Specification for:
£
(pounds)
*Appropriate trade term, e.g. 'ex works'.

Programme
Our best estimate is that delivery * will be achieved within of receiving your order, and that we could complete supervision of erection within a further of receiving your order, making a total of .

We enclose † showing the basis of this estimate.
*Appropriate trade term, e.g. 'ex works'.
†Appropriate document, e.g. bar chart or critical path network.

Terms of Payment
We propose that payment shall be made as follows:
(a) %(per cent) in advance.
(b) The balance in accordance with the Conditions of Contract referred to in this tender.

Conditions of Contract
Where consistent with this tender, we propose that BEAMA *Conditions of Sale (B) for Machinery and Equipment Including Supervision of Erection, United Kingdom*, edition June 1979, a copy of which is enclosed, shall apply, subject to the following:

Value Added Tax (VAT) – Prices exclude VAT. To the extent that VAT is properly chargeable on the supply to you of any goods or services under the Contract, you shall pay such VAT as an addition to payments otherwise due under the Contract.

Clause 10 (Liability for Delay) – The figures to be inserted shall be ½% (one-half per cent) and 5% (five per cent) respectively.

Interest– If the payment of any sum payable to us shall be delayed, interest at the rate of 2% (two per cent) per annum over the * Bank base lending rate from time to time in force on the amount of the delayed payment for the period of the delay shall be added to the price.

*Insert name of bank.

Contract Price Adjustment

We propose that contract price adjustment shall apply in accordance with the BEAMA *Contract Price Adjustment Clause and Formula for use with Home Contracts – Electrical Machinery: (for which there is no other specific formula)*, edition January 1979, corrected edition,* a copy of which is enclosed.

The provisions of the Clause and Formulae applicable to contracts including erection shall apply.

The BEAMA Labour Cost Index referred to shall be the Labour Cost Index for Electrical Engineering.*

References to the *Trade and Industry Journal* shall be read as references to *British Business*, which is the new name of the journal.

*This is only one of several BEAMA CPA Clauses and Formulae. If another is adopted, this form may need consequential amendment. In particular, reference to the BEAMA CPA Clause applicable to Mechanical Plant, and to the Labour Cost Index for Mechanical Engineering, may be appropriate. See p. 16.

Trade Terms

The trade term   * shall be defined by the current edition of Incoterms, where consistent with this tender. A copy of the definition is enclosed.

*Appropriate trade term, e.g. 'ex works'.

Validity

This tender is valid for acceptance until and including   *, but shall, after that date, be null and void unless extended by us in writing.

*Date in full, e.g. 'Monday, 19 January 1981'.

Yours faithfully

(Signature)

FORM 17

FOB Tender based on BEAMA Conditions of Sale (BE) for Machinery and Equipment, Including Supervision of Erection, Export FOB, edition June 1979

Dear Sirs

(heading)

Prices
We offer to supply FOB * and supervise the erection on Site at

of the items described in the accompanying Specification for:
£
(pounds sterling)
*Port of shipment.

Programme
Our best estimate is that delivery FOB will be achieved within of receiving your order, and that we could complete supervision of erection within a further of receiving your order, making a total of .
We enclose * showing the basis of this estimate.
*Appropriate document, e.g. bar chart or critical path network.

Terms of Payment
All prices are payable in pounds sterling in the United Kingdom.

We propose that payment shall be made as follows:

(a) %(per cent) in advance.

(b) The balance through a Confirmed Irrevocable Letter of Credit providing for payment in full upon presentation of documents, to be increased according to any contractual increases in the price, to be extended according to any increase in the period required for execution of our respective obligations under the Contract, and otherwise complying with the following description:

(i) Confirmed by an acceptable bank in London.
(ii) Irrevocably issued, confirmed and advised to us by .[a]
(iii) Date and place of expiry– [a] in London.
(iv) Beneficiary – ourselves.
(v) Amount – the balance of the price.
(vi) Credit available with confirming bank by payment.
(vii) Partial shipments allowed.
(viii) Transhipment allowed.
(ix) Shipment from .[b]
(x) For transportation to .[c]
(xi) Shipment of .[d]
(xii) Documents required:

Commercial Invoice

In respect of goods, full set of clean on board marine Bills of Lading (which shall be acceptable although to the effect that the goods have been shipped on deck or unprotected*).

In respect of other work, your or your Engineer's Certificate.†

*Delete if inappropriate.
†If this sentence is omitted, payment for such other work should be made out of the letter of credit on presentation only of commercial invoice.

(xiii) Period after issuance of shipping documents for presentation – days.

(xiv) Subject otherwise to the *Uniform Customs and Practice for Documentary Credits*, 1974, ICC Publication No 290.[e]

[a]Date in full, e.g. 'Monday, 19 January 1981'.
[b]Port of shipment, or e.g. 'any United Kingdom port'.
[c]Port of destination.
[d]Short description of goods and site work.
[e]This description follows as closely as possible the *Standard Procedure for the Issuing of Documentary Credits by Cable, Telegram or Telex* (App VII *Standard Forms for Issuing Documentary Credits*, ICC Publication No 323).

Conditions of Contract

Where consistent with this tender, we propose that BEAMA *Conditions of Sale (BE) for Machinery and Equipment Including Supervision of Erection, Export FOB*, edition June 1979, a copy of which is enclosed, shall apply, subject to the following:

Clause 11 (Liability for Delay) – The figures to be inserted shall be ½% (one-half per cent) and 5% (five per cent) respectively.

Interest– If the payment of any sum payable to us shall be delayed, interest at the rate of 2% (two per cent) per annum over the * Bank base lending rate from time to time in force on the amount of the delayed payment for the period of the delay shall be added to the price.
*Insert name of bank.

Contract Price Adjustment

We propose that contract price adjustment in respect of the price for goods shall apply in accordance with the BEAMA *Contract Price Adjustment Clause and Formula for use with Export Contracts – Electrical Machinery: (for which there is no other specific formula)*, edition January 1979, corrected edition,* a copy of which is enclosed.

We further propose that contract price adjustment in respect of the price for site work shall apply in accordance with the same BEAMA Clause and Formula subject to the amendment that adjustment shall take place at the rate of 0.95 per cent of such price per 1.0 per cent difference between the Labour Cost Index published for the month in which the tender date falls and the average of the Index figures published for the last two-thirds of the period during which the Contractor is carrying out site work, this difference being expressed as a percentage of the former Index figure.

The BEAMA Labour Cost Index referred to shall be the Labour Cost Index for Electrical Engineering.*

References to the *Trade and Industry Journal* shall be read as references to *British Business*, which is the new name of the journal.†

*This is only one of several BEAMA CPA Clauses and Formulae. If another is adopted, this form may need consequential amendment. In particular, reference to the BEAMA CPA Clause applicable to Mechanical Plant, and to the Labour Cost Index for Mechanical Engineering, may be appropriate. See p. 16).

†The above form requires that the separate prices to be adjusted shall be identifiable from the contract documents.

Trade Terms

The trade term 'FOB' shall be defined by the current edition of Incoterms, where consistent with this tender. A copy of the definition is enclosed.

Validity

This tender is valid for acceptance until and including *, but shall, after that date, be null and void unless extended by us in writing.

*Date in full, e.g. 'Monday, 19 January 1981'.

Yours faithfully

(Signature)

FORM 18

C & F Tender based on BEAMA Conditions of Sale (BE) for Machinery and Equipment, Including Supervision of Erection, Export FOB, edition June 1979

Dear Sirs

(heading)

Prices

We offer to supply C & F *, and supervise the erection on Site at

of the items described in the accompanying Specification for:

£

(pounds sterling)

*Port of destination.

Programme

Our best estimate is that delivery C & F will be achieved within of receiving your order, and that we could complete supervision of erection within a further of receiving your order, making a total of .

We enclose * showing the basis of this estimate.

*Appropriate document, e.g. bar chart or critical path network.

Terms of Payment

All prices are payable in pounds sterling in the United Kingdom.

We propose that payment shall be made as follows:

(a) % (per cent) in advance.

(b) The balance through a Confirmed Irrevocable Letter of Credit providing for payment in full upon presentation of documents, to be increased according to any contractual increases in the price, to be extended according to any increase in the period required for execution of our respective obligations under the Contract, and otherwise complying with the following description:

(i) Confirmed by an acceptable bank in London.
(ii) Irrevocably issued, confirmed and advised to us by .[a]
(iii) Date and place of expiry – [a] in London.
(iv) Beneficiary – ourselves.
(v) Amount – the balance of the price.
(vi) Credit available with confirming bank by payment.
(vii) Partial shipments allowed.
(viii) Transhipment allowed.
(ix) Shipment from .[b]
(x) For transportation to .[c]
(xi) Shipment of .[d]
(xii) Documents required:

Commercial Invoice

In respect of goods, full set of clean on board marine Bills of Lading (which shall be acceptable although to the effect that the goods have been shipped on deck or unprotected*).

In respect of other work, your or your Engineer's Certificate.†

*Delete if inappropriate.

†If this sentence is omitted, payment for such other work should be made out of the letter of credit on presentation only of commercial invoice.

(xiii) Period after issuance of shipping documents for presentation – days.

(xiv) Subject otherwise to the *Uniform Customs and Practice for Documentary Credits*, 1974, ICC Publication No 290.[e]

[a]Date in full, e.g. 'Monday, 19 January 1981'.

[b]Port of shipment, or e.g. 'any United Kingdom port'.

[c]Port of destination.

[d]Short description of goods and site work.

[e]This description follows as closely as possible the *Standard Procedure for the Issuing of Documentary Credits by Cable, Telegram or Telex* (App VII *Standard Forms for Issuing Documentary Credits,* ICC Publication No 323).

Conditions of Contract

Where consistent with this tender, we propose that BEAMA *Conditions of Sale (BE) for Machinery and Equipment Including Supervision of Erection, Export FOB*, edition June 1979, a copy of which is enclosed, shall apply, subject to the following:

Clause 11 (Liability for Delay) – The figures to be inserted shall be ½% (one-half per cent) and 5% (five per cent) respectively.

Interest – If the payment of any sum payable to us shall be delayed, interest at the rate of 2% (two per cent) per annum over the * Bank base lending rate from time to time in force on the amount of the delayed payment for the period of the delay shall be added to the price.

*Insert name of bank.

Contract Price Adjustment

We propose that contract price adjustment in respect of the price for goods shall apply in accordance with the BEAMA *Contract Price Adjustment Clause and Formula for use with Export Contracts – Electrical Machinery: (for which there is no other specific formula)*, edition January 1979, corrected edition,* a copy of which is enclosed.

We further propose that contract price adjustment in respect of the price for site work shall apply in accordance with the same BEAMA Clause and Formula subject to the amendment that adjustment shall take place at the rate of 0.95 per cent of such price per 1.0 per cent difference between the Labour Cost Index published for the month in which the tender date falls and the average of the Index figures published for the last two-thirds of the period during which the Contractor is carrying out site work, this difference being expressed as a percentage of the former Index figure.

The BEAMA Labour Cost Index referred to shall be the Labour Cost Index for Electrical Engineering.*

References to the *Trade and Industry Journal* shall be read as references to *British Business*, which is the new name of the journal.†

*This is only one of several BEAMA CPA Clauses and Formulae. If another is adopted, this form may need consequential amendment. In particular, reference to the BEAMA CPA Clause applicable to Mechanical Plant, and to the Labour Cost Index for Mechanical Engineering, may be appropriate. See p. 16.

†The above form requires that the separate prices to be adjusted shall be identifiable from the contract documents.

Trade Terms

The trade term 'C & F' shall be defined by the current edition of Incoterms, where consistent with this tender. A copy of the definition is enclosed.

Validity

This tender is valid for acceptance until and including *, but shall, after that date, be null and void unless extended by us in writing.

*Date in full, e.g. 'Monday, 19 January 1981'.

Yours faithfully

(Signature)

FORM 19

CIF Tender based on BEAMA Conditions of Sale (BE) for Machinery and Equipment, Including Supervision of Erection, Export FOB, edition June 1979

Dear Sirs

(heading)

Prices

We offer to supply CIF *, and supervise the erection on Site at

of the items described in the accompanying Specification for:

£

(pounds sterling)

*Port of destination.

Programme

Our best estimate is that delivery CIF will be achieved within of receiving your order, and that we could complete supervision of erection within a further of receiving your order, making a total of .

We enclose * showing the basis of this estimate.

*Appropriate document, e.g. bar chart or critical path network.

Terms of Payment

All prices are payable in pounds sterling in the United Kingdom.

We propose that payment shall be made as follows:

(a) % (per cent) in advance.

(b) The balance through a Confirmed Irrevocable Letter of Credit providing for payment in full upon presentation of documents, to be increased according to any contractual increases in the price, to be extended according to any increase in the period required for execution of our respective obligations under the Contract, and otherwise complying with the following description:

(i) Confirmed by an acceptable bank in London.

(ii) Irrevocably issued, confirmed and advised to us by .[a]

(iii) Date and place of expiry– [a] in London.

(iv) Beneficiary – ourselves.

(v) Amount – the balance of the price.

(vi) Credit available with confirming bank by payment.

(vii) Partial shipments allowed.

(viii) Transhipment allowed.

(ix) Shipment from .[b]

(x) For transportation to .[c]

(xi) Shipment of .[d]

(xii) Documents required:

Commercial Invoice

In respect of goods, full set of clean on board marine Bills of Lading (which shall be acceptable although to the effect that the goods have been shipped on deck or unprotected*) and Insurance Certificate.

In respect of other work, your or your Engineer's Certificate.†

*Delete if inappropriate.

†If this sentence is omitted, payment for such other work should be made out of the letter of credit on presentation only of commercial invoice.

(xiii) Period after issuance of shipping documents for presentation – days.

(xiv) Subject otherwise to the *Uniform Customs and Practice for Documentary Credits*, 1974, ICC Publication No 290.[e]

[a]Date in full, e.g. 'Monday, 19 January 1981'.

[b]Port of shipment, or e.g. 'any United Kingdom port'.

[c]Port of destination.

[d]Short description of goods and site work.

[e]This description follows as closely as possible the *Standard Procedure for the Issuing of Documentary Credits by Cable, Telegram or Telex* (App VII *Standard Forms for Issuing Documentary Credits*, ICC Publication No 323).

Conditions of Contract

Where consistent with this tender, we propose that BEAMA *Conditions of Sale (BE) for Machinery and Equipment Including Supervision of Erection, Export FOB*, edition June 1979, a copy of which is enclosed, shall apply, subject to the following:

Clause 11 (Liability for Delay) – The figures to be inserted shall be ½% (one-half per cent) and 5% (five per cent) respectively.

Interest– If the payment of any sum payable to us shall be delayed, interest at the rate of 2% (two per cent) per annum over the * Bank base lending rate from time to time in force on the amount of the delayed payment for the period of the delay shall be added to the price.

*Insert name of bank.

Contract Price Adjustment

We propose that contract price adjustment in respect of the price for goods shall apply in accordance with the BEAMA *Contract Price Adjustment Clause and Formula for use with Export Contracts – Electrical Machinery: (for which there is no other specific formula)*, edition January 1979, corrected edition,* a copy of which is enclosed.

We further propose that contract price adjustment in respect of the price for site work shall apply in accordance with the same BEAMA Clause and Formula subject to the amendment that adjustment shall take place at the rate of 0.95 per cent of such price per 1.0 per cent difference between the Labour Cost Index published for the month in which the tender date falls and the average of the Index figures published for the last two-thirds of the period during which the Contractor is carrying out site work, this difference being expressed as a percentage of the former Index figure.

The BEAMA Labour Cost Index referred to shall be the Labour Cost Index for Electrical Engineering.*

References to the *Trade and Industry Journal* shall be read as references to *British Business*, which is the new name of the journal.†

*This is only one of several BEAMA CPA Clauses and Formulae. If another is adopted, this form may need consequential amendment. In particular, reference to the BEAMA CPA Clause applicable to Mechanical Plant, and to the Labour Cost Index for Mechanical Engineering, may be appropriate. See p. 16.

†The above form requires that the separate prices to be adjusted shall be identifiable from the contract documents.

Trade Terms

The trade term 'CIF' shall be defined by the current edition of Incoterms, where consistent with this tender. A copy of the definition is enclosed.

Validity

This tender is valid for acceptance until and including *, but shall, after that date, be null and void unless extended by us in writing.

*Date in full, e.g. 'Monday, 19 January 1981'.

Yours faithfully

(Signature)

FORM 20

Tender based on BEAMA Conditions of Sale (C) for Electronic Equipment Including Installation, United Kingdom, edition June 1979

Dear Sirs

(heading)

Prices

We offer to supply, deliver and install on Site at

the items described in the accompanying Specification for:

£

(pounds)

Programme

Our best estimate is that completion of our work will be achieved within of receiving your order. (We enclose * showing the basis of this estimate.)†

*Appropriate document, e.g. bar chart or critical path network.

†Delete if inappropriate.

Terms of Payment

We propose that payment shall be made in accordance with the Conditions of Contract referred to in this tender, with *Clause 15* (Terms of Payment) completed as follows:

(i) per cent of the contract price at the time of placing the order.

(ii) per cent of the contract value of equipment as and when delivered from time to time on site and of work done on site respectively.

(iii) 2½ per cent of the contract value of equipment as and when this has been taken over, or has been deemed to have been taken over by you.

(iv) The balance of the contract price one calendar month after payment of the above 2½ per cent has become due.

Conditions of Contract

Where consistent with this tender, we propose that BEAMA *Conditions of Sale (C) for Electronic Equipment Including Installation, United Kingdom,* edition June 1979, a copy of which is enclosed, shall apply, subject to the following:

Value Added Tax (VAT) – Prices exclude VAT. To the extent that VAT is properly chargeable on the supply to you of any goods or services under the Contract, you shall pay such VAT as an addition to payments otherwise due under the Contract.

Clause 11 (Liability for Delay) – The figures to be inserted shall be ½% (one-half per cent) and 5% (five per cent) respectively.

Interest– If the payment of any sum payable to us shall be delayed, interest at the rate of 2% (two per cent) per annum over the * Bank base lending rate from time to time in force on the amount of the delayed payment for the period of the delay shall be added to the price.

*Insert name of bank.

Contract Price Adjustment

We propose that contract price adjustment shall apply in accordance with the BEAMA *Contract Price Adjustment Clause and Formula for use with Home Contracts – Electrical Machinery: (for which there is no other specific formula)*, edition January 1979, corrected edition,* a copy of which is enclosed.

The provisions of the Clause and Formulae applicable to contracts including erection shall apply.

The BEAMA Labour Cost Index referred to shall be the Labour Cost Index for Electrical Engineering.*

References to the *Trade and Industry Journal* shall be read as references to *British Business,* which is the new name of the journal.

*This is only one of several BEAMA CPA Clauses and Formulae. If another is adopted, this form may need consequential amendment. In particular, reference to the BEAMA CPA Clause applicable to Mechanical Plant, and to the Labour Cost Index for Mechanical Engineering, may be appropriate. See p. 16.

Validity

This tender is valid for acceptance until and including *, but shall, after that date, be null and void unless extended by us in writing.

*Date in full, e.g. 'Monday, 19 January 1981'.

Yours faithfully

(Signature)

FORM 21

Home Tender based on BEAMA Conditions of Contract (E) for Erection of Electrical Plant and Machinery (Home or Export), edition November 1980

Dear Sirs

(heading)

Prices
We offer to erect on Site at

the Plant and Machinery described in the accompanying Specification for (the approximate price of*):
£
(pounds)
(Our services are offered on a time basis. The above approximate price is computed at the rates and for the estimated duration shown in the accompanying Specification. The sum actually payable for our services shall be computed on a time basis, and may be more or less than the above approximate price.)*
*Delete if lump sum.

Programme
Our best estimate is that completion of our work will be achieved within of receiving your order. (We enclose * showing the basis of this estimate.)†

*Appropriate document, e.g. bar chart or critical path network.
†Delete if inappropriate.

Terms of Payment
We propose that payment shall be made as follows:
(a) %(per cent) in advance.
(b) The balance in accordance with the Conditions of Contract referred to in this tender. Payment in respect of site work shall be due monthly at the end of each month.

Conditions of Contract
Where consistent with this tender, we propose that BEAMA *Conditions of Contract (E) for Erection of Electrical Plant and Machinery (home or export)*, edition November 1980, a copy of which is enclosed, shall apply, subject to the following:

Value Added Tax (VAT) – Prices exclude VAT. To the extent that VAT is properly chargeable on the supply to you of any goods or services under the Contract, you shall pay such VAT as an addition to payments otherwise due under the Contract.

Clause 10 (Liability for Delay) – The figures to be inserted shall be ½% (one-half per cent) and 5% (five per cent) respectively.

Interest– If the payment of any sum payable to us shall be delayed, interest at the rate of 2% (two per cent) per annum over the * Bank base lending rate from time to time in force on the amount of the delayed payment for the period of the delay shall be added to the price.
*Insert name of bank.

Clause 19 (Variations in Costs) shall not apply.

Contract Price Adjustment

We propose that contract price adjustment shall apply in accordance with the BEAMA *Contract Price Adjustment Clause and Formula for use with Home Contracts – Electrical Machinery: (for which there is no other specific formula)*, edition January 1979, corrected edition,* a copy of which is enclosed, subject to the amendment that adjustment shall take place at the rate of 0.95 per cent of such price per 1.0 per cent difference between the Labour Cost Index published for the month in which the tender date falls and the average of the Index figures published for the last two-thirds of the period during which the Contractor is carrying out site work, this difference being expressed as a percentage of the former Index figure.

The BEAMA Labour Cost Index referred to shall be the Labour Cost Index for Electrical Engineering.*

References to the *Trade and Industry Journal* shall be read as references to *British Business*, which is the new name of the journal.

*This is only one of several BEAMA CPA Clauses and Formulae. If another is adopted, this form may need consequential amendment. In particular, reference to the BEAMA CPA Clause applicable to Mechanical Plant, and to the Labour Cost Index for Mechanical Engineering, may be appropriate. See p. 16.

Validity

This tender is valid for acceptance until and including *, but shall, after that date, be null and void unless extended by us in writing.
*Date in full, e.g. 'Monday, 19 January 1981'.

Yours faithfully

(Signature)

FORM 22

Export Tender based on BEAMA Conditions of Contract (E) for Erection of Electrical Plant and Machinery (Home or Export), edition November 1980

Dear Sirs

(heading)

Prices
We offer to erect on Site at

the Plant and Machinery described in the accompanying Specification for (the approximate price of*):
£
(pounds sterling)
(Our services are offered on a time basis. The above approximate price is computed at the rates and for the estimated duration shown in the accompanying Specification. The sum actually payable for our services shall be computed on a time basis, and may be more or less than the above approximate price.)*

*Delete if lump sum.

Programme
Our best estimate is that completion of our work will be achieved within of receiving your order. (We enclose * showing the basis of this estimate.)†

* Appropriate document, e.g. bar chart or critical path network.
† Delete if inappropriate.

Terms of Payment
All prices are payable in pounds sterling in the United Kingdom.

We propose that payment shall be made as follows:

(a) %(per cent) in advance.

(b) The balance through a Confirmed Irrevocable Letter of Credit providing for payment in full upon presentation of documents, to be increased according to any contractual increases in the price, to be extended according to any increase in the period required for execution of our respective obligations under the Contract, and otherwise complying with the following description:
   (i) Confirmed by an acceptable bank in London.
   (ii) Irrevocably issued, confirmed and advised to us by .[a]
   (iii) Date and place of expiry – [a] in London.
   (iv) Beneficiary – ourselves.
   (v) Amount – the balance of the price.
   (vi) Credit available with confirming bank by payment.

(vii) Documents to relate to .[b]

(viii) Documents required:

(Certified)* Commercial Invoice

Your or your Engineer's Certificate†

*Payment would, in fact, be made out of the letter of credit even if certified only by the exporter. Where no further definition is given 'banks will accept such documents as tendered'. See *Uniform Customs and Practice for Documentary Credits*, 1974, Article 33, ICC Publication No. 290.

†If this sentence is omitted, payment should be made out of the letter of credit on presentation only of commercial invoice.

(ix) Subject otherwise to the *Uniform Customs and Practice for Documentary Credits*, 1974, ICC Publication No 290.[c]

[a]Date in full, e.g. 'Monday, 19 January 1981'.

[b]Short description of site work.

[c]This description follows as closely as possible the *Standard Procedure for the Issuing of Documentary Credits by Cable, Telegram or Telex* (App VII *Standard Forms for Issuing Documentary Credits*, ICC Publication No 323).

Payment in respect of site work shall be due monthly at the end of each month.

Conditions of Contract

Where consistent with this tender, we propose that BEAMA *Conditions of Contract (E) for Erection of Electrical Plant and Machinery (home or export)*, edition November 1980, a copy of which is enclosed, shall apply, subject to the following:

Clause 10 (Liability for Delay) – The figures to be inserted shall be ½% (one-half per cent) and 5% (five per cent) respectively.

Interest – If the payment of any sum payable to us shall be delayed, interest at the rate of 2% (two per cent) per annum over the * Bank base lending rate from time to time in force on the amount of the delayed payment for the period of the delay shall be added to the price.

*Insert name of bank.

Clause 19 (Variations in Costs) shall not apply.

Contract Price Adjustment

We propose that contract price adjustment shall apply in accordance with the BEAMA *Contract Price Adjustment Clause and Formula for use with Export Contracts – Electrical Machinery: (for which there is no other specific formula)*, edition January 1979, corrected edition,* a copy of which is enclosed, subject to the amendment that adjustment shall take place at the rate of 0.95 per cent of such price per 1.0 per cent difference between the Labour Cost Index published for the month in which the tender date falls and the average of the Index figures published for the last two-thirds of the period during which the Contractor is carrying out site work, this difference being expressed as a percentage of the former Index figure.

The BEAMA Labour Cost Index referred to shall be the Labour Cost Index for Electrical Engineering.*

References to the *Trade and Industry Journal* shall be read as references to *British Business*, which is the new name of the journal.

*This is only one of several BEAMA CPA Clauses and Formulae. If another is adopted, this form may need consequential amendment. In particular, reference to the BEAMA CPA Clause applicable to Mechanical Plant, and to the Labour Cost Index for Mechanical Engineering, may be appropriate. See p. 16.

Validity

This tender is valid for acceptance until and including *, but shall, after that date, be null and void unless extended by us in writing.

*Date in full, e.g. 'Monday, 19 January 1981'.

Yours faithfully

(Signature)

FORM 23

Tender based on BEAMA Conditions (R) for the Repair of Machinery and Equipment, United Kingdom, edition June 1979

Dear Sirs

(heading)

Prices*
We offer to carry out at our works at

the repairs described in the accompanying Specification for the approximate price of:
£
(pounds)
Our offer (includes)† (excludes)† dismantling
on Site at
loading, transport to our works at
,
return to Site, off-loading and re-erection on Site.
*This section and the following section are alternatives.
†Delete alternatives.

Prices*
We offer to carry out on Site at
the repairs described in the accompanying Specification for the approximate price of:
£
(pounds)
Our offer (includes)† (excludes)† dismantling
on Site at
loading, transport to our works at
,
return to Site, off-loading and re-erection on Site.
*This section and the preceding section are alternatives.
†Delete alternatives.

Programme
Our best estimate is that completion of our work will be achieved within
of receiving your order.

Terms of Payment
We propose that payment shall be made as follows:
(a) %(per cent) in advance.
(b) The balance in accordance with the Conditions of Contract referred to in this tender.

Conditions of Contract

Where consistent with this tender, we propose that BEAMA *Conditions (R) for the Repair of Machinery and Equipment, United Kingdom,* edition June 1979, a copy of which is enclosed, shall apply, subject to the following:

Value Added Tax (VAT)–Prices exclude VAT. To the extent that VAT is properly chargeable on the supply to you of any goods or services under the Contract, you shall pay such VAT as an addition to payments otherwise due under the Contract.

Interest – If the payment of any sum payable to us shall be delayed, interest at the rate of 2% (two per cent) per annum over the * Bank base lending rate from time to time in force on the amount of the delayed payment for the period of the delay shall be added to the price.
*Insert name of bank.

Contract Price Adjustment

We propose that contract price adjustment shall apply in accordance with the BEAMA *Contract Price Adjustment Clause and Formula for use with Home Contracts – Electrical Machinery: (for which there is no other specific formula),* edition January 1979, corrected edition,* a copy of which is enclosed.

The provisions of the Clause and Formulae applicable to contracts including erection shall apply.

The BEAMA Labour Cost Index referred to shall be the Labour Cost Index for Electrical Engineering.*

References to the *Trade and Industry Journal* shall be read as references to *British Business,* which is the new name of the journal.
*This is only one of several BEAMA CPA Clauses and Formulae. If another is adopted, this form may need consequential amendment. In particular, reference to the BEAMA CPA Clause applicable to Mechanical Plant, and to the Labour Cost Index for Mechanical Engineering, may be appropriate.

Validity

This tender is valid for acceptance until and including *, but shall, after that date, be null and void unless extended by us in writing.
*Date in full, e.g. 'Monday, 19 January 1981'.

Yours faithfully

(Signature)

FORM 24

FOB Tender based on BEAMA Conditions (RE) for the Repair of Machinery and Equipment, Export FOB, edition June 1979

Dear Sirs

(heading)

Prices
We offer to carry out at our works at

the repairs described in the accompanying Specification, and to supply the repaired items FOB *, for the approximate price of:
£
(pounds sterling)
Our offer (includes)† (excludes)† dismantling
on Site at
loading and transport to our works.

*Port of shipment.
†Delete alternatives.

Programme
Our best estimate is that delivery FOB will be achieved within of receiving your order.

Terms of Payment
All prices are payable in pounds sterling in the United Kingdom.

We propose that payment shall be made as follows:

(a) %(per cent) in advance.
(b) The balance through a Confirmed Irrevocable Letter of Credit providing for payment in full upon presentation of documents, to be increased according to any contractual increases in the price, to be extended according to any increase in the period required for execution of our respective obligations under the Contract, and otherwise complying with the following description:
   (i) Confirmed by an acceptable bank in London.
   (ii) Irrevocably issued, confirmed and advised to us by .[a]
   (iii) Date and place of expiry – [a] in London.
   (iv) Beneficiary – ourselves.
   (v) Amount – the balance of the price.
   (vi) Credit available with confirming bank by payment.
   (vii) Partial shipments allowed.
   (viii) Transhipment allowed.
   (ix) Shipment from .[b]
   (x) For transportation to .[c]

(xi) Shipment of .[d]

(xii) Documents required:

Commercial Invoice

Full set of clean on board marine Bills of Lading (which shall be acceptable although to the effect that the goods have been shipped on deck or unprotected*).

*Delete if inappropriate.

(xiii) Period after issuance of shipping documents for presentation – days.

(xiv) Subject otherwise to the *Uniform Customs and Practice for Documentary Credits*, 1974, ICC Publication No 290.[e]

[a]Date in full, e.g. 'Monday, 19 January 1981'.
[b]Port of shipment, or e.g. 'any United Kingdom port'.
[c]Port of destination.
[d]Short description of goods.
[e]This description follows as closely as possible the *Standard Procedure for the Issuing of Documentary Credits by Cable, Telegram or Telex* (App VII *Standard Forms for Issuing Documentary Credits*, ICC Publication No 323).

Conditions of Contract

Where consistent with this tender, we propose that BEAMA *Conditions (RE) for the Repair of Machinery and Equipment, Export FOB*, edition June 1979, a copy of which is enclosed, shall apply, subject to the following:

Interest – If the payment of any sum payable to us shall be delayed, interest at the rate of 2% (two per cent) per annum over the * Bank base lending rate from time to time in force on the amount of the delayed payment for the period of the delay shall be added to the price.

*Insert name of bank.

Contract Price Adjustment

We propose that contract price adjustment shall apply in accordance with the BEAMA *Contract Price Adjustment Clause and Formula for use with Export Contracts – Electrical Machinery: (for which there is no other specific formula)*, edition January 1979, corrected edition,* a copy of which is enclosed.

The BEAMA Labour Cost Index referred to shall be the Labour Cost Index for Electrical Engineering.*

References to the *Trade and Industry Journal* shall be read as references to *British Business*, which is the new name of the journal.

*This is only one of several BEAMA CPA Clauses and Formulae. If another is adopted, this form may need consequential amendment. In particular, reference to the BEAMA CPA Clause applicable to Mechanical Plant, and to the Labour Cost Index for Mechanical Engineering, may be appropriate. See p. 16.

Trade Terms

The trade term 'FOB' shall be defined by the current edition of Incoterms, where consistent with this tender. A copy of the definition is enclosed.

Validity

This tender is valid for acceptance until and including *, but shall, after that date, be null and void unless extended by us in writing.

*Date in full, e.g. 'Monday, 19 January 1981'.

Yours faithfully

(Signature)

FORM 25

C & F Tender based on BEAMA Conditions (RE) for the Repair of Machinery and Equipment, Export FOB, edition June 1979

Dear Sirs

(heading)

Prices

We offer to carry out at our works at

the repairs described in the accompanying Specification, and to supply the repaired items C & F *, for the approximate price of:

£

(pounds sterling)

Our offer (includes)† (excludes)† dismantling on Site at loading and transport to our works.

*Port of destination.

†Delete alternatives.

Programme

Our best estimate is that delivery C & F will be achieved within of receiving your order.

Terms of Payment

All prices are payable in pounds sterling in the United Kingdom.

We propose that payment shall be made as follows:

(a) %(per cent) in advance.

(b) The balance through a Confirmed Irrevocable Letter of Credit providing for payment in full upon presentation of documents, to be increased according to any contractual increases in the price, to be extended according to any increase in the period required for execution of our respective obligations under the Contract, and otherwise complying with the following description:

(i) Confirmed by an acceptable bank in London.

(ii) Irrevocably issued, confirmed and advised to us by .[a]

(iii) Date and place of expiry – [a] in London.

(iv) Beneficiary – ourselves.

(v) Amount – the balance of the price.

(vi) Credit available with confirming bank by payment.

(vii) Partial shipments allowed.

(viii) Transhipment allowed.

(ix) Shipment from .[b]

(x) For transportation to .[c]

(xi) Shipment of .[d]

(xii) Documents required:

Commercial Invoice

Full set of clean on board marine Bills of Lading (which shall be acceptable although to the effect that the goods have been shipped on deck or unprotected*).

*Delete if inappropriate.

(xiii) Period after issuance of shipping documents for presentation – days.

(xiv) Subject otherwise to the *Uniform Customs and Practice for Documentary Credits,* 1974, ICC Publication No 290.[e]

[a]Date in full, e.g. 'Monday, 19 January 1981'.
[b]Port of shipment, or e.g. 'any United Kingdom port'.
[c]Port of destination.
[d]Short description of goods.
[e]This description follows as closely as possible the *Standard Procedure for the Issuing of Documentary Credits by Cable, Telegram or Telex* (App VII *Standard Forms for Issuing Documentary Credits,* ICC Publication No 323).

Conditions of Contract

Where consistent with this tender, we propose that BEAMA *Conditions (RE) for the Repair of Machinery and Equipment, Export FOB,* edition June 1979, a copy of which is enclosed, shall apply, subject to the following:

Interest – If the payment of any sum payable to us shall be delayed, interest at the rate of 2% (two per cent) per annum over the * Bank base lending rate from time to time in force on the amount of the delayed payment for the period of the delay shall be added to the price.

*Insert name of bank.

Contract Price Adjustment

We propose that contract price adjustment shall apply in accordance with the BEAMA *Contract Price Adjustment Clause and Formula for use with Export Contracts – Electrical Machinery: (for which there is no other specific formula),* edition January 1979, corrected edition,* a copy of which is enclosed.

The BEAMA Labour Cost Index referred to shall be the Labour Cost Index for Electrical Engineering.*

References to the *Trade and Industry Journal* shall be read as references to *British Business,* which is the new name of the journal.

*This is only one of several BEAMA CPA Clauses and Formulae. If another is adopted, this form may need consequential amendment. In particular, reference to the BEAMA CPA Clause applicable to Mechanical Plant, and to the Labour Cost Index for Mechanical Engineering, may be appropriate. See p. 16.

Trade Terms

The trade term 'C & F' shall be defined by the current edition of Incoterms, where consistent with this tender. A copy of the definition is enclosed.

Validity
This tender is valid for acceptance until and including *, but shall, after that date, be null and void unless extended by us in writing.

*Date in full, e.g. 'Monday, 19 January 1981'.

Yours faithfully

(Signature)

FORM 26

CIF Tender based on BEAMA Conditions (RE) for the Repair of Machinery and Equipment, Export FOB, edition June 1979

Dear Sirs

(heading)

Prices

We offer to carry out at our works at

the repairs described in the accompanying Specification, and to supply the repaired items CIF *, for the approximate price of:

£

(pounds sterling)

Our offer (includes)† (excludes)† dismantling on Site at

loading and transport to our works.

*Port of destination.
†Delete alternatives.

Programme

Our best estimate is that delivery CIF will be achieved within of receiving your order.

Terms of Payment

All prices are payable in pounds sterling in the United Kingdom.

We propose that payment shall be made as follows:

(a) % (per cent) in advance.

(b) The balance through a Confirmed Irrevocable Letter of Credit providing for payment in full upon presentation of documents, to be increased according to any contractual increases in the price, to be extended according to any increase in the period required for execution of our respective obligations under the Contract, and otherwise complying with the following description:

(i) Confirmed by an acceptable bank in London.
(ii) Irrevocably issued, confirmed and advised to us by [a]
(iii) Date and place of expiry– [a] in London.
(iv) Beneficiary – ourselves.
(v) Amount – the balance of the price.
(vi) Credit available with confirming bank by payment.
(vii) Partial shipments allowed.
(viii) Transhipment allowed.
(ix) Shipment from . [b]

(x) For transportation to .[c]
(xi) Shipment of .[d]
(xii) Documents required:

Commercial Invoice

Full set of clean on board marine Bills of Lading (which shall be acceptable although to the effect that the goods have been shipped on deck or unprotected*).

Insurance Certificate

*Delete if inappropriate.

(xiii) Period after issuance of shipping documents for presentation – days.
(xiv) Subject otherwise to the *Uniform Customs and Practice for Documentary Credits*, 1974, ICC Publication No 290.[e]

[a] Date in full, e.g. 'Monday, 19 January 1981'.
[b] Port of shipment, or e.g. 'any United Kingdom port'.
[c] Port of destination.
[d] Short description of goods.
[e] This description follows as closely as possible the *Standard Procedure for the Issuing of Documentary Credits by Cable, Telegram or Telex* (App VII *Standard Forms for Issuing Documentary Credits*, ICC Publication No 323).

Conditions of Contract

Where consistent with this tender, we propose that BEAMA *Conditions (RE) for the Repair of Machinery and Equipment, Export FOB*, edition June 1979, a copy of which is enclosed, shall apply, subject to the following:

Interest – If the payment of any sum payable to us shall be delayed, interest at the rate of 2% (two per cent) per annum over the * Bank base lending rate from time to time in force on the amount of the delayed payment for the period of the delay shall be added to the price.

*Insert name of bank.

Contract Price Adjustment

We propose that contract price adjustment shall apply in accordance with the BEAMA *Contract Price Adjustment Clause and Formula for use with Export Contracts – Electrical Machinery: (for which there is no other specific formula)*, edition January 1979, corrected edition,* a copy of which is enclosed.

The BEAMA Labour Cost Index referred to shall be the Labour Cost Index for Electrical Engineering.*

References to the *Trade and Industry Journal* shall be read as references to *British Business*, which is the new name of the journal.

*This is only one of several BEAMA CPA Clauses and Formulae. If another is adopted, this form may need consequential amendment. In particular, reference to the BEAMA CPA Clause applicable to Mechanical Plant, and to the Labour Cost Index for Mechanical Engineering, may be appropriate. See p. 16.

Trade Terms

The trade term 'CIF' shall be defined by the current edition of Incoterms, where consistent with this tender. A copy of the definition is enclosed.

Validity

This tender is valid for acceptance until and including *, but shall, after that date, be null and void unless extended by us in writing.

*Date in full, e.g. 'Monday, 19 January 1981'.

Yours faithfully

(Signature)

FORM 27

Tender based on BEAMA Conditions (RC) for the Reconstruction, Modification or Repair of Plant and Equipment in the United Kingdom, Involving Work on Site, edition 1978, first published Summer 1979, with revised Clause 21.

Dear Sirs

(heading)

Prices

We offer to carry out on Site at

the reconstruction, modification and/or repairs described in the accompanying Specification for:

£

(pounds)

Programme

Our best estimate is that completion of our work will be achieved within of receiving your order. (We enclose * showing the basis of this estimate.)†

*Appropriate document, e.g. bar chart or critical path network.

†Delete if inappropriate.

Terms of Payment

We propose that payment shall be made as follows:

(a) % (per cent) in advance.

(b) The balance in accordance with the Conditions of Contract referred to in this tender.

Conditions of Contract

Where consistent with this tender, we propose that BEAMA *Conditions (RC) for the Reconstruction, Modification or Repair of Plant and Equipment in the United Kingdom, Involving Work on Site,* edition 1978, first published Summer 1979, with revised Clause 21, a copy of which is enclosed, shall apply, subject to the following:

The Appendix shall be completed as follows:

Clause

22. Additional risks to be insured – None.
25. (Delay in Completion)
    (a) Percentage of Contract price to be deducted as damages.
    (b) Maximum percentage of Contract Price which the deductions may not exceed.

Recovery of Additional Costs
Percentage to be added per cent.
Clauses to which this percentage is applicable:
Clause
5. (Mistakes in Information) Sub-Clause (iii).
13. (Delivery of Plant).
14. (Access to and Possession of Site) Sub-Clause (iv).
27. (Taking Over) Sub-Clause (iii).
28. (Recovery of Additional Costs) Sub-Clause (iii).
35. (Arbitration) Sub-Clause (iii).

Clause 36 (Variation in Costs and Tax Fluctuations) shall not apply.

Value Added Tax (VAT) – Prices exclude VAT. To the extent that VAT is properly chargeable on the supply to you of any goods or services under the Contract, you shall pay such VAT as an addition to payments otherwise due under the Contract.

Interest – If the payment of any sum payable to us shall be delayed, interest at the rate of 2% (two per cent) per annum over the *Bank base lending rate from time to time in force on the amount of the delayed payment for the period of the delay shall be added to the price.
*Insert name of bank.

Contract Price Adjustment*
We propose that contract price adjustment shall apply in accordance with the BEAMA *Contract Price Adjustment Clause and Formula for use with Home Contracts – Electrical Machinery: (for which there is no other specific formula)*, edition January 1979, corrected edition†, a copy of which is enclosed.

The provisions of the Clause and Formulae applicable to contracts including erection shall apply.

The BEAMA Labour Cost Index referred to shall be the Labour Cost Index for Electrical Engineering.†

References to the *Trade and Industry Journal* shall be read as references to *British Business*, which is the new name of the journal.

*This section and the following section are alternatives.
†This is only one of several BEAMA CPA Clauses and Formulae. If another is adopted, this form may need consequential amendment. In particular, reference to the BEAMA CPA Clause applicable to Mechanical Plant, and to the Labour Cost Index for Mechanical Engineering, may be appropriate. See p. 16.

Contract Price Adjustment*
We propose that contract price adjustment shall apply in accordance with the BEAMA *Contract Price Adjustment Clause and Formulae for use with Home Contracts – Electrical Machinery: (for which there is no other specific formula)*, edition January 1979, corrected edition,† a copy of which is enclosed, subject to the amendment that adjustment shall take place at the rate of 0.95 per cent of such price per 1.0 per cent difference between the Labour Cost Index published for the month in which the tender date falls and the average of the Index figures published for the last two-thirds of the period during which the Contractor is carrying out site work, this difference being expressed as a percentage of the former Index figure.

The BEAMA Labour Cost Index referred to shall be the Labour Cost Index for Electrical Engineering.†

References to the *Trade and Industry Journal* shall be read as references to *British Business*, which is the new name of the journal.

*This section and the preceding section are alternatives.
†This is only one of several BEAMA CPA Clauses and Formulae. If another is adopted, this form may need consequential amendment. In particular, reference to the BEAMA CPA Clause applicable to Mechanical Plant, and to the Labour Cost Index for Mechanical Engineering, may be appropriate. See p. 16.

Validity

This tender is valid for acceptance until and including *, but shall, after that date, be null and void unless extended by us in writing.

*Date in full, e.g. 'Monday, 19 January 1981'.

Yours faithfully

(Signature)

FORM 28

Tender based on BEAMA Conditions of Sale (SA) for Stock and Catalogue Articles, United Kingdom, edition March 1978

Dear Sirs

(heading)

Prices
We offer to supply *, the items described in the accompanying Specification for:

£
(pounds)
*Appropriate trade term, e.g. 'ex works'.

Programme
Our best estimate is that delivery * will be achieved within of receiving your order.
*Appropriate trade term, e.g. 'ex works'.

Terms of Payment
We propose that payment shall be made as follows:
(a) % (per cent) in advance.
(b) The balance in accordance with the Conditions of Contract referred to in this tender.

Conditions of Contract
Where consistent with this tender, we propose that BEAMA *Conditions of Sale (SA) for Stock and Catalogue Articles, United Kingdom*, edition March 1978, a copy of which is enclosed, shall apply, subject to the following:

Clause 6 (Delivery) – shall be completed as follows:
On orders below £ in value (exclusive of VAT) carriage will be charged except where delivery is made by our own transport.

Value Added Tax (VAT) – Prices exclude VAT. To the extent that VAT is properly chargeable on the supply to you of any goods or services under the Contract, you shall pay such VAT as an addition to payments otherwise due under the Contract.

Interest – If the payment of any sum payable to us shall be delayed, interest at the rate of 2% (two per cent) per annum over the * Bank base lending rate from time to time in force on the amount of the delayed payment for the period of the delay shall be added to the price.
*Insert name of bank.

Contract Price Adjustment

We propose that contract price adjustment shall apply in accordance with the BEAMA *Contract Price Adjustment Clause and Formula for use with Home Contracts – Electrical Machinery: (for which there is no other specific formula)*, edition January 1979, corrected edition,* a copy of which is enclosed.

The BEAMA Labour Cost Index referred to shall be the Labour Cost Index for Electrical Engineering.*

References to the *Trade and Industry Journal* shall be read as references to *British Business*, which is the new name of the journal.

*This is only one of several BEAMA CPA Clauses and Formulae. If another is adopted, this form may need consequential amendment. In particular, reference to the BEAMA CPA Clause applicable to Mechanical Plant, and to the Labour Cost Index for Mechanical Engineering, may be appropriate. See p. 16.

Trade Terms

The trade term * shall be defined by the current edition of Incoterms, where consistent with this tender. A copy of the definition is enclosed.

*Appropriate trade term, e.g. 'ex works'.

Validity

This tender is valid for acceptance until and including *, but shall, after that date, be null and void unless extended by us in writing.

*Date in full, e.g. 'Monday, 19 January 1981'.

Yours faithfully

(Signature)

FORM 29

FOB Tender based on BEAMA Conditions of Sale (SAE) for Stock and Catalogue Articles, Export FOB, edition December 1980

Dear Sirs

(heading)

Prices
We offer to supply FOB * the items described in the accompanying Specification for:
£
(pounds sterling)
*Port of shipment.

Programme
Our best estimate is that delivery FOB will be achieved within of receiving your order.

Terms of Payment
All prices are payable in pounds sterling in the United Kingdom.

We propose that payment shall be made as follows:

(a) %(per cent) in advance.

(b) The balance through a Confirmed Irrevocable Letter of Credit providing for payment in full upon presentation of documents, to be increased according to any contractual increases in the price, to be extended according to any increase in the period required for execution of our respective obligations under the Contract, and otherwise complying with the following description:

(i) Confirmed by an acceptable bank in London.
(ii) Irrevocably issued, confirmed and advised to us by .[a]
(iii) Date and place of expiry – [a] in London.
(iv) Beneficiary – ourselves.
(v) Amount – the balance of the price.
(vi) Credit available with confirming bank by payment.
(vii) Partial shipments allowed.
(viii) Transhipment allowed.
(ix) Shipment from .[b]
(x) For transportation to .[c]
(xi) Shipment of .[d]
(xii) Documents required:

Commercial Invoice

Full set of clean on board marine Bills of Lading (which shall be acceptable although to the effect that the goods have been shipped on deck or unprotected*).

*Delete if inappropriate.

(xiii) Period after issuance of shipping documents for presentation – days.

(xiv) Subject otherwise to the *Uniform Customs and Practice for Documentary Credits*, 1974, ICC Publication No 290.[e]

[a]Date in full, e.g. 'Monday, 19 January 1981'.
[b]Port of shipment, or e.g. 'any United Kingdom port'.
[c]Port of destination.
[d]Short description of goods.
[e]This description follows as closely as possible the *Standard Procedure for the Issuing of Documentary Credits by Cable, Telegram or Telex* (App VII *Standard Forms for Issuing Documentary Credits*, ICC Publication No 323).

Conditions of Contract

Where consistent with this tender, we propose that BEAMA *Conditions of Sale (SAE) for Stock and Catalogue Articles, Export FOB*, edition December 1980, a copy of which is enclosed, shall apply, subject to the following:

Interest – If the payment of any sum payable to us shall be delayed, interest at the rate of 2% (two per cent) per annum over the * Bank base lending rate from time to time in force on the amount of the delayed payment for the period of the delay shall be added to the price.

*Insert name of bank.

Contract Price Adjustment

We propose that contract price adjustment shall apply in accordance with the BEAMA *Contract Price Adjustment Clause and Formula for use with Export Contracts – Electrical Machinery: (for which there is no other specific formula)*, edition January 1979, corrected edition,* a copy of which is enclosed.

The BEAMA Labour Cost Index referred to shall be the Labour Cost Index for Electrical Engineering.*

References to the *Trade and Industry Journal* shall be read as references to *British Business*, which is the new name of the journal.

*This is only one of several BEAMA CPA Clauses and Formulae. If another is adopted, this form may need consequential amendment. In particular, reference to the BEAMA CPA Clause applicable to Mechanical Plant, and to the Labour Cost Index for Mechanical Engineering, may be appropriate. See p. 16.

Trade Terms

The trade term 'FOB' shall be defined by the current edition of Incoterms, where consistent with this tender. A copy of the definition is enclosed.

Validity

This tender is valid for acceptance until and including *, but shall, after that date, be null and void unless extended by us in writing.

*Date in full, e.g. 'Monday, 19 January 1981'.

Yours faithfully

(Signature)

FORM 30

C & F Tender based on BEAMA Conditions of Sale (SAE) for Stock and Catalogue Articles, Export FOB, edition December 1980

Dear Sirs

(heading)

Prices
We offer to supply C & F * the items described in the accompanying Specification for:
£
(pounds sterling)

*Port of destination.

Programme
Our best estimate is that delivery C & F will be achieved within of receiving your order.

Terms of Payment
All prices are payable in pounds sterling in the United Kingdom.

We propose that payment shall be made as follows:

(a) % (per cent) in advance.

(b) The balance through a Confirmed Irrevocable Letter of Credit providing for payment in full upon presentation of documents, to be increased according to any contractual increases in the price, to be extended according to any increase in the period required for execution of our respective obligations under the Contract, and otherwise complying with the following description:

(i) Confirmed by an acceptable bank in London.
(ii) Irrevocably issued, confirmed and advised to us by .[a]
(iii) Date and place of expiry – [a] in London.
(iv) Beneficiary – ourselves.
(v) Amount – the balance of the price.
(vi) Credit available with confirming bank by payment.
(vii) Partial shipments allowed.
(viii) Transhipment allowed.
(ix) Shipment from .[b]
(x) For transportation to .[c]
(xi) Shipment of .[d]
(xii) Documents required:

Commercial Invoice

Full set of clean on board marine Bills of Lading (which shall be acceptable although to the effect that the goods have been shipped on deck or unprotected*).

*Delete if inappropriate.

(xiii) Period after issuance of shipping documents for presentation – days.

(xiv) Subject otherwise to the *Uniform Customs and Practice for Documentary Credits*, 1974, ICC Publication No 290.[e]

[a] Date in full, e.g. 'Monday, 19 January 1981'.
[b] Port of shipment, or e.g. 'any United Kingdom port'.
[c] Port of destination.
[d] Short description of goods.
[e] This description follows as closely as possible the *Standard Procedure for the Issuing of Documentary Credits by Cable, Telegram or Telex* (App VII *Standard Forms for Issuing Documentary Credits*, ICC Publication No 323).

Conditions of Contract

Where consistent with this tender, we propose that BEAMA *Conditions of Sale (SAE) for Stock and Catalogue Articles, Export FOB*, edition December 1980, a copy of which is enclosed, shall apply, subject to the following:

Interest – If the payment of any sum payable to us shall be delayed, interest at the rate of 2% (two per cent) per annum over the * Bank base lending rate from time to time in force on the amount of the delayed payment for the period of the delay shall be added to the price.
*Insert name of bank.

Contract Price Adjustment

We propose that contract price adjustment shall apply in accordance with the BEAMA *Contract Price Adjustment Clause and Formula for use with Export Contracts – Electrical Machinery: (for which there is no other specific formula)*, edition January 1979, corrected edition,* a copy of which is enclosed.

The BEAMA Labour Cost Index referred to shall be the Labour Cost Index for Electrical Engineering.*

References to the *Trade and Industry Journal* shall be read as references to *British Business*, which is the new name of the journal.

*This is only one of several BEAMA CPA Clauses and Formulae. If another is adopted, this form may need consequential amendment. In particular, reference to the BEAMA CPA Clause applicable to Mechanical Plant, and to the Labour Cost Index for Mechanical Engineering, may be appropriate. See p. 16.

Trade Terms

The trade term 'C & F' shall be defined by the current edition of Incoterms, where consistent with this tender. A copy of the definition is enclosed.

Validity

This tender is valid for acceptance until and including *, but shall, after that date, be null and void unless extended by us in writing.
*Date in full, e.g. 'Monday, 19 January 1981'.

Yours faithfully

(Signature)

FORM 31

CIF Tender based on BEAMA Conditions of Sale (SAE) for Stock and Catalogue Articles, Export FOB, edition December 1980

Dear Sirs

(heading)

Prices
We offer to supply CIF * the items described in the accompanying Specification for:
£
(pounds sterling)
*Port of destination.

Programme
Our best estimate is that delivery CIF will be achieved within of receiving your order.

Terms of Payment
All prices are payable in pounds sterling in the United Kingdom.

We propose that payment shall be made as follows:

(a) %(per cent) in advance.
(b) The balance through a Confirmed Irrevocable Letter of Credit providing for payment in full upon presentation of documents, to be increased according to any contractual increases in the price, to be extended according to any increase in the period required for execution of our respective obligations under the Contract, and otherwise complying with the following description:
  (i) Confirmed by an acceptable bank in London.
  (ii) Irrevocably issued, confirmed and advised to us by .[a]
  (iii) Date and place of expiry – [a] in London.
  (iv) Beneficiary – ourselves.
  (v) Amount – the balance of the price.
  (vi) Credit available with confirming bank by payment.
  (vii) Partial shipments allowed.
  (viii) Transhipment allowed.
  (ix) Shipment from .[b]
  (x) For transportation to .[c]
  (xi) Shipment of .[d]
  (xii) Documents required:

  Commercial Invoice

  Full set of clean on board marine Bills of Lading (which shall be acceptable although to the effect that the goods have been shipped on deck or unprotected*).

Insurance Certificate

*Delete if inappropriate.

(xiii) Period after issuance of shipping documents for presentation – days.

(xiv) Subject otherwise to the *Uniform Customs and Practice for Documentary Credits*, 1974, ICC Publication No 290.[e]

[a]Date in full, e.g. 'Monday, 19 January 1981'.

[b]Port of shipment, or e.g. 'any United Kingdom port'.

[c]Port of destination.

[d]Short description of goods.

[e]This description follows as closely as possible the *Standard Procedure for the Issuing of Documentary Credits by Cable, Telegram or Telex* (App VII *Standard Forms for Issuing Documentary Credits*, ICC Publication No 323).

Conditions of Contract

Where consistent with this tender, we propose that BEAMA *Conditions of Sale (SAE) for Stock and Catalogue Articles, Export FOB*, edition December 1980, a copy of which is enclosed, shall apply, subject to the following:

Interest – If the payment of any sum payable to us shall be delayed, interest at the rate of 2% (two per cent) per annum over the * Bank base lending rate from time to time in force on the amount of the delayed payment for the period of the delay shall be added to the price.

*Insert name of bank.

Contract Price Adjustment

We propose that contract price adjustment shall apply in accordance with the BEAMA *'Contract Price Adjustment Clause and Formula for use with Export Contracts – Electrical Machinery: (for which there is no other specific formula)*, edition January 1979, corrected edition,* a copy of which is enclosed.

The BEAMA Labour Cost Index referred to shall be the Labour Cost Index for Electrical Engineering.*

References to the *Trade and Industry Journal* shall be read as references to *British Business*, which is the new name of the journal.

*This is only one of several BEAMA CPA Clauses and Formulae. If another is adopted, this form may need consequential amendment. In particular, reference to the BEAMA CPA Clause applicable to Mechanical Plant, and to the Labour Cost Index for Mechanical Engineering, may be appropriate. See p. 16.

Trade Terms

The trade term 'CIF' shall be defined by the current edition of Incoterms, where consistent with this tender. A copy of the definition is enclosed.

Validity

This tender is valid for acceptance until and including *, but shall, after that date, be null and void unless extended by us in writing.

*Date in full, e.g. 'Monday, 19 January 1981'.

Yours faithfully

(Signature)

FORM 32

Tender based on BEAMA Conditions of Contract for Commissioning Electronic Equipment, United Kingdom, edition June 1979

Dear Sirs

(heading)

Prices

We offer to carry out on Site at

the commissioning work described in the accompanying Specification for:
£
(pounds)

Programme

Our best estimate is that completion of commissioning will be achieved within of receiving your order.

Terms of Payment

We propose that payment shall be made as follows:

(a) % (per cent) in advance.

(b) The balance in accordance with the Conditions of Contract referred to in this tender.

Conditions of Contract

Where consistent with this tender, we propose that BEAMA *Conditions of Contract for Commissioning Electronic Equipment, United Kingdom,* edition June 1979, a copy of which is enclosed, shall apply, subject to the following:

Value Added Tax (VAT) – Prices exclude VAT. To the extent that VAT is properly chargeable on the supply to you of any goods or services under the Contract, you shall pay such VAT as an addition to payments otherwise due under the Contract.

Clause 9 (Liability for Delay) – The figures to be inserted shall be ½% (one-half per cent) and 5% (five per cent) respectively.

Interest – If the payment of any sum payable to us shall be delayed, interest at the rate of 2% (two per cent) per annum over the * Bank base lending rate from time to time in force on the amount of the delayed payment for the period of the delay shall be added to the price.

*Insert name of bank.

Contract Price Adjustment

We propose that contract price adjustment shall apply in accordance with the BEAMA *Contract Price Adjustment Clause and Formula for use with Home Contracts – Electrical Machinery: (for which there is no other specific formula)*, edition January 1979, corrected edition,* a copy of which is enclosed, subject to the amendment that adjustment shall take place at the rate of 0.95 per cent of such price per 1.0 per cent difference between the Labour Cost Index published for the month in which the tender date falls and the average of the Index figures published for the last two-thirds of the period during which the Contractor is carrying out site work, this difference being expressed as a percentage of the former Index figure.

The BEAMA Labour Cost Index referred to shall be the Labour Cost Index for Electrical Engineering.*

References to the *Trade and Industry Journal* shall be read as references to *British Business*, which is the new name of the journal.

*This is only one of several BEAMA CPA Clauses and Formulae. If another is adopted, this form may need consequential amendment. In particular, reference to the BEAMA CPA Clause applicable to Mechanical Plant, and to the Labour Cost Index for Mechanical Engineering, may be appropriate. See p. 16.

Validity

This tender is valid for acceptance until and including *, but shall, after that date, be null and void unless extended by us in writing.

*Date in full, e.g. 'Monday, 19 January 1981'.

Yours faithfully

(Signature)

FORM 33

Tender based on BEAMA Conditions of Contract for Systems Incorporating Electronic Equipment (Including Installation), United Kingdom, edition June 1979

Dear Sirs

(heading)

Prices

We offer to supply, deliver to and install on Site at

the items described in the accompanying Specification for:
£
(pounds)

Programme

Our best estimate is that completion of our work will be achieved within of receiving your order. (We enclose * showing the basis of this estimate.)†

*Appropriate document, e.g. bar chart or critical path network.
†Delete if inappropriate.

Terms of Payment

We propose that payment shall be made in accordance with the Conditions of Contract referred to in this tender, with Clause 16.1 (Terms of Payment) completed as follows:

(a) per cent of the Contract Price at the time of placing the order or at the time when we receive your written instructions to put the work in hand, whichever is the earlier.
(b) A further per cent of the Contract Price as progress payments at the times and percentages specified in our tender. When no such times and percentages are specified the payments shall be made on delivery of the Equipment or portions thereof from time to time on site.
(c) A further per cent of the Contract price (or *pro rata* for a portion) when the Equipment or any portion thereof has passed or is deemed to have passed the Tests on Installation.
(d) A further per cent of the Contract Price when the System has passed or is deemed to have passed the System Tests.
(e) The balance of the Contract Price 30 days after the payment under (d) falls due.
(f) Payments for any work undertaken at an agreed rate shall be made monthly in arrears.

Conditions of Contract

Where consistent with this tender, we propose that BEAMA *Conditions of Contract for Systems Incorporating Electronic Equipment (Including Installation) United*

Kingdom, edition June 1979, a copy of which is enclosed, shall apply, subject to the following:

Value Added Tax (VAT) – Prices exclude VAT. To the extent that VAT is properly chargeable on the supply to you of any goods or services under the Contract, you shall pay such VAT as an addition to payments otherwise due under the Contract.

Interest – If the payment of any sum payable to us shall be delayed, interest at the rate of 2% (two per cent) per annum over the * Bank base lending rate from time to time in force on the amount of the delayed payment for the period of the delay shall be added to the price.

*Insert name of bank.

Clause 18 (Variation in Costs) shall not apply.

Clause 19 (Liability for Delay) – The figures to be inserted shall be ½% (one-half per cent) and 5% (five per cent) respectively.

Appendix A shall be completed as follows:*

*Delete if not required.

Contract Price Adjustment

We propose that contract price adjustment shall apply in accordance with the BEAMA *Contract Price Adjustment Clause and Formula for use with Home Contracts – Electrical Machinery: (for which there is no other specific formula)*, edition January 1979, corrected edition,* a copy of which is enclosed.

The provisions of the Clause and Formulae applicable to contracts including erection shall apply.

The BEAMA Labour Cost Index referred to shall be the Labour Cost Index for Electrical Engineering.*

References to the *Trade and Industry Journal* shall be read as references to *British Business*, which is the new name of the journal.

*This is only one of several BEAMA CPA Clauses and Formulae. If another is adopted, this form may need consequential amendment. In particular, reference to the BEAMA CPA Clause applicable to Mechanical Plant, and to the Labour Cost Index for Mechanical Engineering, may be appropriate. See p. 16.

Validity

This tender is valid for acceptance until and including *, but shall, after that date, be null and void unless extended by us in writing.

*Date in full, e.g. 'Monday, 19 January 1981'.

Yours faithfully

(Signature)

FORM 34

FOB Tender based on General Conditions No 188 for the Supply of Plant and Machinery for Export, prepared under the auspices of the United Nations Economic Commission for Europe, Geneva, March 1953

Dear Sirs

(heading)

Prices
We offer to supply FOB * the items described in the accompanying Specification for:
£
(pounds sterling)
*Port of shipment.

Programme
Our best estimate is that delivery FOB will be achieved within of receiving your order.

Terms of Payment
All prices are payable in pounds sterling in the United Kingdom.
We propose that payment shall be made as follows:
(a) % (per cent) in advance.
(b) The balance through a Confirmed Irrevocable Letter of Credit providing for payment in full upon presentation of documents, to be increased according to any contractual increases in the price, to be extended according to any increase in the period required for execution of our respective obligations under the Contract, and otherwise complying with the following description:
(i) Confirmed by an acceptable bank in London.
(ii) Irrevocably issued, confirmed and advised to us by .[a]
(iii) Date and place of expiry – [a] in London.
(iv) Beneficiary – ourselves.
(v) Amount – the balance of the price.
(vi) Credit available with confirming bank by payment.
(vii) Partial shipments allowed.
(viii) Transhipment allowed.
(ix) Shipment from .[b]
(x) For transportation to .[c]
(xi) Shipment of .[d]
(xii) Documents required:
Commercial Invoice
Full set of clean on board marine Bills of Lading (which shall be

acceptable although to the effect that the goods have been shipped on deck or unprotected*).

*Delete if inappropriate.

(xiii) Period after issuance of shipping documents for presentation – days.

(xiv) Subject otherwise to the *Uniform Customs and Practice for Documentary Credits*, 1974, ICC Publication No 290.[e]

[a]Date in full, e.g. 'Monday, 19 January 1981'.
[b]Port of shipment, or e.g. 'any United Kingdom port'.
[c]Port of destination.
[d]Short description of goods.
[e]This description follows as closely as possible the *Standard Procedure for the Issuing of Documentary Credits by Cable, Telegram or Telex* (App VII *Standard Forms for Issuing Documentary Credits*, ICC Publication No 323).

Conditions of Contract

Where consistent with this tender, we propose that *General Conditions No 188 for the Supply of Plant and Machinery for Export*, prepared under the auspices of the United Nations Economic Commission for Europe, Geneva, March 1953, a copy of which is enclosed, shall apply, subject to the following:

The Appendix shall be completed as follows:

Appendix to UNECE Conditions 188

	Clause	
A. Percentage to be deducted for each week's delay	7.3	%(per cent)
B. Maximum percentage which the deductions above may not exceed	7.3	%(per cent)
C. Maximum amount recoverable for non-delivery	7.5	£ (pounds sterling)
D. Maximum amount recoverable on termination by Vendor for failure to take delivery or make payment	7.7 and 8.7	£ (pounds sterling)
E. Rate of interest on overdue payments	8.7	2% (two per cent) per annum over the * Bank base lending rate from time to time in force
F. Period of delay in payment authorizing termination by Vendor	8.7	months
G. Guarantee Period for original Plant and parts replaced or renewed	9.2 and 9.7	months

H.	Maximum extension of Guarantee Period	9.5	months
I.	(1) Daily use of Plant	9.6	hours/day
	(2) Reduction of Guarantee Period for more intensive use	9.6	

*Insert name of bank.

Contract Price Adjustment

We propose that contract price adjustment shall apply in accordance with the BEAMA *Contract Price Adjustment Clause and Formula for use with Export Contracts – Electrical Machinery: (for which there is no other specific formula)*, edition January 1979, corrected edition,* a copy of which is enclosed.

The BEAMA Labour Cost Index referred to shall be the Labour Cost Index for Electrical Engineering.*

References to the *Trade and Industry Journal* shall be read as references to *British Business*, which is the new name of the journal.

*This is only one of several BEAMA CPA Clauses and Formulae. If another is adopted, this form may need consequential amendment. In particular, reference to the BEAMA CPA Clause applicable to Mechanical Plant, and to the Labour Cost Index for Mechanical Engineering, may be appropriate. See p. 16.

Trade Terms

The trade term 'FOB' shall be defined by the current edition of Incoterms, where consistent with this tender. A copy of the definition is enclosed.

Validity

This tender is valid for acceptance until and including *, but shall, after that date, be null and void unless extended by us in writing.

*Date in full, e.g. 'Monday, 19 January 1981'.

Yours faithfully

(Signature)

FORM 35

C & F Tender based on General Conditions No 188 for the Supply of Plant and Machinery for Export, prepared under the auspices of the United Nations Economic Commission for Europe, Geneva, March 1953

Dear Sirs

(heading)

Prices
We offer to supply C & F * the items described in the accompanying Specification for:
£
(pounds sterling)

*Port of destination.

Programme
Our best estimate is that delivery C & F will be achieved within of receiving your order.

Terms of Payment
All prices are payable in pounds sterling in the United Kingdom.

We propose that payment shall be made as follows:

(a) % (per cent) in advance.

(b) The balance through a Confirmed Irrevocable Letter of Credit providing for payment in full upon presentation of documents, to be increased according to any contractual increases in the price, to be extended according to any increase in the period required for execution of our respective obligations under the Contract, and otherwise complying with the following description:

(i) Confirmed by an acceptable bank in London.
(ii) Irrevocably issued, confirmed and advised to us by .[a]
(iii) Date and place of expiry – [a] in London.
(iv) Beneficiary – ourselves.
(v) Amount – the balance of the price.
(vi) Credit available with confirming bank by payment.
(vii) Partial shipments allowed.
(viii) Transhipment allowed.
(ix) Shipment from .[b]
(x) For transportation to .[c]
(xi) Shipment of .[d]
(xii) Documents required:

Commercial Invoice

Full set of clean on board marine Bills of Lading (which shall be

acceptable although to the effect that the goods have been shipped on deck or unprotected*).

*Delete if inappropriate.

(xiii) Period after issuance of shipping documents for presentation – days.

(xiv) Subject otherwise to the *Uniform Customs and Practice for Documentary Credits*, 1974, ICC Publication No 290.[e]

[a]Date in full, e.g. 'Monday, 19 January 1981'.
[b]Port of shipment, or e.g. 'any United Kingdom port'.
[c]Port of destination.
[d]Short description of goods.
[e]This description follows as closely as possible the *Standard Procedure for the Issuing of Documentary Credits by Cable, Telegram or Telex* (App VII *Standard Forms for Issuing Documentary Credits*, ICC Publication No 323).

Conditions of Contract

Where consistent with this tender, we propose that *General Conditions No 188 for the Supply of Plant and Machinery for Export*, prepared under the auspices of the United Nations Economic Commission for Europe, Geneva, March 1953, a copy of which is enclosed, shall apply, subject to the following:

The Appendix shall be completed as follows:

Appendix to UNECE Conditions 188

	Clause	
A. Percentage to be deducted for each week's delay	7.3	%(per cent)
B. Maximum percentage which the deductions above may not exceed	7.3	%(per cent)
C. Maximum amount recoverable for non-delivery	7.5	£ (pounds sterling)
D. Maximum amount recoverable on termination by Vendor for failure to take delivery or make payment	7.7 and 8.7	£ (pounds sterling)
E. Rate of interest on overdue payments	8.7	2% (two per cent) per annum over the * Bank base lending rate from time to time in force
F. Period of delay in payment authorizing termination by Vendor	8.7	months
G. Guarantee Period for original Plant and parts replaced or renewed	9.2 and 9.7	months

H.	Maximum extension of Guarantee Period	9.5	months
I.	(1) Daily use of Plant	9.6	hours/day
	(2) Reduction of Guarantee Period for more intensive use	9.6	

*Insert name of bank.

Contract Price Adjustment

We propose that contract price adjustment shall apply in accordance with the BEAMA *Contract Price Adjustment Clause and Formula for use with Export Contracts – Electrical Machinery: (for which there is no other specific formula)*, edition January 1979, corrected edition,* a copy of which is enclosed.

The BEAMA Labour Cost Index referred to shall be the Labour Cost Index for Electrical Engineering.*

References to the *Trade and Industry Journal* shall be read as references to *British Business*, which is the new name of the journal.

*This is only one of several BEAMA CPA Clauses and Formulae. If another is adopted, this form may need consequential amendment. In particular, reference to the BEAMA CPA Clause applicable to Mechanical Plant, and to the Labour Cost Index for Mechanical Engineering, may be appropriate. See p. 16.

Trade Terms

The trade term 'C & F' shall be defined by the current edition of Incoterms, where consistent with this tender. A copy of the definition is enclosed.

Validity

This tender is valid for acceptance until and including *, but shall, after that date, be null and void unless extended by us in writing.

*Date in full, e.g. 'Monday, 19 January 1981'.

Yours faithfully

(Signature)

FORM 36

CIF Tender based on General Conditions No 188 for the Supply of Plant and Machinery for Export, prepared under the auspices of the United Nations Economic Commission for Europe, Geneva, March 1953

Dear Sirs

(heading)

Prices

We offer to supply CIF * the items described in the accompanying Specification for:

£

(pounds sterling)

*Port of destination.

Programme

Our best estimate is that delivery CIF will be achieved within of receiving your order.

Terms of Payment

All prices are payable in pounds sterling in the United Kingdom.

We propose that payment shall be made as follows:

(a) % (per cent) in advance.

(b) The balance through a Confirmed Irrevocable Letter of Credit providing for payment in full upon presentation of documents, to be increased according to any contractual increases in the price, to be extended according to any increase in the period required for execution of our respective obligations under the Contract, and otherwise complying with the following description:

(i) Confirmed by an acceptable bank in London.
(ii) Irrevocably issued, confirmed and advised to us by [a].
(iii) Date and place of expiry – [a] in London.
(iv) Beneficiary – ourselves.
(v) Amount – the balance of the price.
(vi) Credit available with confirming bank by payment.
(vii) Partial shipments allowed.
(viii) Transhipment allowed.
(ix) Shipment from .[b]
(x) For transportation to .[c]
(xi) Shipment of .[d]
(xii) Documents required:

Commercial Invoice

Full set of clean on board marine Bills of Lading (which shall be

acceptable although to the effect that the goods have been shipped on deck or unprotected*).

Insurance Certificate

*Delete if inappropriate.

(xiii) Period after issuance of shipping documents for presentation – days.

(xiv) Subject otherwise to the *Uniform Customs and Practice for Documentary Credits*, 1974, ICC Publication No 290.[e]

[a] Date in full, e.g. 'Monday, 19 January 1981'.
[b] Port of shipment, or e.g. 'any United Kingdom port'.
[c] Port of destination.
[d] Short description of goods.
[e] This description follows as closely as possible the *Standard Procedure for the Issuing of Documentary Credits by Cable, Telegram or Telex* (App VII *Standard Forms for Issuing Documentary Credits*, ICC Publication No 323).

Conditions of Contract

Where consistent with this tender, we propose that *General Conditions No 188 for the Supply of Plant and Machinery for Export*, prepared under the auspices of the United Nations Economic Commission for Europe, Geneva, March 1953, a copy of which is enclosed, shall apply, subject to the following:

The Appendix shall be completed as follows:

Appendix to UNECE Conditions 188

	Clause	
A. Percentage to be deducted for each week's delay	7.3	%(per cent)
B. Maximum percentage which the deductions above may not exceed	7.3	%(per cent)
C. Maximum amount recoverable for non-delivery	7.5	£ (pounds sterling)
D. Maximum amount recoverable on termination by Vendor for failure to take delivery or make payment	7.7 and 8.7	£ (pounds sterling)
E. Rate of interest on overdue payments	8.7	2% (two per cent) per annum over the * Bank base lending rate from time to time in force
F. Period of delay in payment authorizing termination by Vendor	8.7	months

G. Guarantee Period for original Plant and parts replaced or renewed	9.2 and 9.7	months
H. Maximum extension of Guarantee Period	9.5	months
I. (1) Daily use of Plant	9.6	hours/day
(2) Reduction of Guarantee Period for more intensive use	9.6	

*Insert name of bank.

Contract Price Adjustment

We propose that contract price adjustment shall apply in accordance with the BEAMA *Contract Price Adjustment Clause and Formula for use with Export Contracts – Electrical Machinery: (for which there is no other specific formula)*, edition January 1979, corrected edition,* a copy of which is enclosed.

The BEAMA Labour Cost Index referred to shall be the Labour Cost Index for Electrical Engineering.*

References to the *Trade and Industry Journal* shall be read as references to *British Business*, which is the new name of the journal.

*This is only one of several BEAMA CPA Clauses and Formulae. If another is adopted, this form may need consequential amendment. In particular, reference to the BEAMA CPA Clause applicable to Mechanical Plant, and to the Labour Cost Index for Mechanical Engineering, may be appropriate. See p. 16.

Trade Terms

The trade term 'CIF' shall be defined by the current edition of Incoterms, where consistent with this tender. A copy of the definition is enclosed.

Validity

This tender is valid for acceptance until and including *, but shall, after that date, be null and void unless extended by us in writing.

*Date in full, e.g. 'Monday, 19 January 1981'.

Yours faithfully

(Signature)

FORM 37

FOB Tender based on General Conditions No 188 for the Supply of Plant and Machinery for Export, prepared under the auspices of the United Nations Economic Commission for Europe, Geneva, March 1953, and Additional Clauses for Supervision of Erection of Plant and Machinery Abroad No 188B, prepared under the auspices of the United Nations Economic Commission for Europe, Geneva, April 1964

Dear Sirs

(heading)

Prices
We offer to supply FOB * and supervise the erection on Site at

of the items described in the accompanying Specification for:
£
(pounds sterling)
*Port of shipment.

(Our services in respect of site work are offered on a time basis. The above price includes an estimated sum of £ for such services, computed at the rates and for the estimated duration shown in the accompanying Specification. The sum actually payable for such services shall be computed on a time basis, and may be either more or less than the estimated sum.)*
*Delete if lump sum.

Programme
Our best estimate is that delivery FOB will be achieved within of receiving your order, and that we could complete supervision of erection within a further of receiving your order, making a total of .
We enclose * showing the basis of this estimate.
*Appropriate document, e.g. bar chart or critical path network.

Terms of Payment
All prices are payable in pounds sterling in the United Kingdom.
We propose that payment shall be made as follows:
(a) %(per cent) in advance.
(b) The balance through a Confirmed Irrevocable Letter of Credit providing for payment in full upon presentation of documents, to be increased according to any contractual increases in the price, to be extended according to any increase in the period required for execution of our respective obligations under the Contract, and otherwise complying with the following description:

(i) Confirmed by an acceptable bank in London.
(ii) Irrevocably issued, confirmed and advised to us by .[a]
(iii) Date and place of expiry – [a] in London.
(iv) Beneficiary – ourselves.
(v) Amount – the balance of the price.
(vi) Credit available with confirming bank by payment.
(vii) Partial shipments allowed.
(viii) Transhipment allowed.
(ix) Shipment from .[b]
(x) For transportation to .[c]
(xi) Shipment of .[d]
(xii) Documents required:

Commercial Invoice

In respect of goods, full set of clean on board marine Bills of Lading (which shall be acceptable although to the effect that the goods have been shipped on deck or unprotected*).

In respect of other work, your or your Engineer's Certificate.†

*Delete if inappropriate.

†If this sentence is omitted, payment for such other work should be made out of the letter of credit on presentation only of commercial invoice.

(xiii) Period after issuance of shipping documents for presentation – days.
(xiv) Subject otherwise to the *Uniform Customs and Practice for Documentary Credits*, 1974, ICC Publication No 290.[e]

[a]Date in full, e.g. 'Monday, 19 January 1981'.

[b]Port of shipment, or e.g. 'any United Kingdom port'.

[c]Port of destination.

[d]Short description of goods and site work.

[e]This description follows as closely as possible the *Standard Procedure for the Issuing of Documentary Credits by Cable, Telegram or Telex* (App VII *Standard Forms for Issuing Documentary Credits*, ICC Publication No 323).

Payment in respect of site work shall be due monthly at the end of each month.

Conditions of Contract

Where consistent with this tender, we propose that *General Conditions No 188 for the Supply of Plant and Machinery for Export*, prepared under the auspices of the United Nations Economic Commission for Europe, Geneva, March 1953, and *Additional Clauses for Supervision of Erection of Plant and Machinery Abroad No 188B*, prepared under the auspices of the United Nations Economic Commission for Europe, Geneva, April 1964, a copy of which is enclosed, shall apply, subject to the following:

The Appendix in each case shall be completed as follows:

Appendix to UNECE Conditions 188

	Clause		
A. Percentage to be deducted for each week's delay	7.3	%(	per cent)

B.	Maximum percentage which the deductions above may not exceed	7.3	%(per cent)
C.	Maximum amount recoverable for non-delivery	7.5	£ (pounds sterling)
D.	Maximum amount recoverable on termination by Vendor for failure to take delivery or make payment	7.7 and 8.7	£ (pounds sterling)
E.	Rate of interest on overdue payments	8.7	2% (two per cent) per annum over the * Bank base lending rate from time to time in force
F.	Period of delay in payment authorizing termination by Vendor	8.7	months
G.	Guarantee Period for original Plant and parts replaced or renewed	9.2 and 9.7	months
H.	Maximum extension of Guarantee Period	9.5	months
I.	(1) Daily use of Plant	9.6	hours/day
	(2) Reduction of Guarantee Period for more intensive use	9.6	

*Insert name of bank.

Appendix to UNECE Conditions 188B

J.	Duration of interruption in erection at the expiry of which the Vendor is authorized to recall his supervising engineers	11.1	months
K.	Maximum indemnities payable by the parties	12.4	£ (pounds sterling)

Clause 4 (Charges Payable by the Purchaser) of UNECE Conditions 188B – Supervision of erection being carried out for a lump sum, the quoted price includes all the items mentioned in Clause 4. Provided that if supervision of erection is prolonged for any cause for which the Purchaser or any of his contractors other than the Vendor is responsible and if, as a result, the work of the Vendor's employees is suspended or added to, a charge will be made for any

idle time, any extra work, any extra living expenses of the Vendor's employees and the cost of any extra journey.*

*Delete if erection is on a time basis. This clause is based on Clause 7.2 of UNECE Conditions 188A, by kind permission of the United Nations.

Contract Price Adjustment

We propose that contract price adjustment in respect of the price for goods shall apply in accordance with the BEAMA *Contract Price Adjustment Clause and Formula for use with Export Contracts – Electrical Machinery: (for which there is no other specific formula)*, edition January 1979, corrected edition,* a copy of which is enclosed.

We further propose that contract price adjustment in respect of the price for site work shall apply in accordance with the same BEAMA Clause and Formula subject to the amendment that adjustment shall take place at the rate of 0.95 per cent of such price per 1.0 per cent difference between the Labour Cost Index published for the month in which the tender date falls and the average of the Index figures published for the last two-thirds of the period during which the Contractor is carrying out site work, this difference being expressed as a percentage of the former Index figure.

The BEAMA Labour Cost Index referred to shall be the Labour Cost Index for Electrical Engineering.*

References to the *Trade and Industry Journal* shall be read as references to *British Business*, which is the new name of the journal.†

*This is only one of several BEAMA CPA Clauses and Formulae. If another is adopted, this form may need consequential amendment. In particular, reference to the BEAMA CPA Clause applicable to Mechanical Plant, and to the Labour Cost Index for Mechanical Engineering, may be appropriate. See p. 16.
†The above form requires that the separate prices to be adjusted shall be identifiable from the contract documents.

Trade Terms

The trade term 'FOB' shall be defined by the current edition of Incoterms, where consistent with this tender. A copy of the definition is enclosed.

Validity

This tender is valid for acceptance until and including *, but shall, after that date, be null and void unless extended by us in writing.
*Date in full, e.g. 'Monday, 19 January 1981'.

Yours faithfully

(Signature)

FORM 38

C & F Tender based on General Conditions No 188 for the Supply of Plant and Machinery for Export, prepared under the auspices of the United Nations Economic Commission for Europe, Geneva, March 1953, and Additional Clauses for Supervision of Erection of Plant and Machinery Abroad No 188B, prepared under the auspices of the United Nations Economic Commission for Europe, Geneva, April 1964

Dear Sirs

(heading)

Prices

We offer to supply C & F *, and supervise the erection on Site at

of the items described in the accompanying Specification for:
£
(pounds sterling)
*Port of destination.

(Our services in respect of site work are offered on a time basis. The above price includes an estimated sum of £ for such services, computed at the rates and for the estimated duration shown in the accompanying Specification. The sum actually payable for such services shall be computed on a time basis, and may be either more or less than the estimated sum.)*
*Delete if lump sum.

Programme

Our best estimate is that delivery C & F will be achieved within of receiving your order, and that we could complete supervision of erection within a further of receiving your order, making a total of .
We enclose * showing the basis of this estimate.
*Appropriate document, e.g. bar chart or critical path network.

Terms of Payment

All prices are payable in pounds sterling in the United Kingdom.

We propose that payment shall be made as follows:

(a) % (per cent) in advance.

(b) The balance through a Confirmed Irrevocable Letter of Credit providing for payment in full upon presentation of documents, to be increased according to any contractual increases in the price, to be extended according to any increase in the period required for execution of our respective obligations under the Contract, and otherwise complying with the following description:

(i) Confirmed by an acceptable bank in London.
(ii) Irrevocably issued, confirmed and advised to us by .[a]
(iii) Date and place of expiry – [a] in London.
(iv) Beneficiary – ourselves.
(v) Amount – the balance of the price.
(vi) Credit available with confirming bank by payment.
(vii) Partial shipments allowed.
(viii) Transhipment allowed.
(ix) Shipment from .[b]
(x) For transportation to .[c]
(xi) Shipment of .[d]
(xii) Documents required:

Commercial Invoice

In respect of goods, full set of clean on board marine Bills of Lading (which shall be acceptable although to the effect that the goods have been shipped on deck or unprotected*).

In respect of other work, your or your Engineer's Certificate.†

*Delete if inappropriate.
†If this sentence is omitted, payment for such other work should be made out of the letter of credit on presentation only of commercial invoice.

(xiii) Period after issuance of shipping documents for presentation – days.
(xiv) Subject otherwise to the *Uniform Customs and Practice for Documentary Credits*, 1974, ICC Publication No 290.[e]

[a]Date in full, e.g. 'Monday, 19 January 1981'.
[b]Port of shipment, or e.g. 'any United Kingdom port'.
[c]Port of destination.
[d]Short description of goods.
[e]This description follows as closely as possible the *Standard Procedure for the Issuing of Documentary Credits by Cable, Telegram or Telex* (App VII *Standard Forms for Issuing Documentary Credits*, ICC Publication No 323).

Payment in respect of site work shall be due monthly at the end of each month.

Conditions of Contract

Where consistent with this tender, we propose that *General Conditions No 188 for the Supply of Plant and Machinery for Export*, prepared under the auspices of the United Nations Economic Commission for Europe, Geneva, March 1953, and *Additional Clauses for Supervision of Erection of Plant and Machinery Abroad No 188B*, prepared under the auspices of the United Nations Economic Commission for Europe, Geneva, April 1964, a copy of which is enclosed, shall apply, subject to the following:

The Appendix in each case shall be completed as follows:

Appendix to UNECE Conditions 188

		Clause	
A.	Percentage to be deducted for each week's delay	7.3	%(per cent)
B.	Maximum percentage which the deductions above may not exceed	7.3	%(per cent)
C.	Maximum amount recoverable for non-delivery	7.5	£ (pounds sterling)
D.	Maximum amount recoverable on termination by Vendor for failure to take delivery or make payment	7.7 and 8.7	£ (pounds sterling)
E.	Rate of interest on overdue payments	8.7	2% (two per cent) per annum over the * Bank base lending rate from time to time in force
F.	Period of delay in payment authorizing termination by Vendor	8.7	months
G.	Guarantee Period for original Plant and parts replaced or renewed	9.2 and 9.7	months
H.	Maximum extension of Guarantee Period	9.5	months
I.	(1) Daily use of Plant	9.6	hours/day
	(2) Reduction of Guarantee Period for more intensive use	9.6	

*Insert name of bank.

Appendix to UNECE Conditions 188B

J.	Duration of interruption in erection at the expiry of which the Vendor is authorized to recall his supervising engineers	11.1	months
K.	Maximum indemnities payable by the parties	12.4	£ (pounds sterling)

Clause 4 (Charges Payable by the Purchaser) of UNECE Conditions 188B – Supervision of erection being carried out for a lump sum, the quoted price

includes all the items mentioned in Clause 4. Provided that if supervision of erection is prolonged for any cause for which the Purchaser or any of his contractors other than the Vendor is responsible and if, as a result, the work of the Vendor's employees is suspended or added to, a charge will be made for any idle time, any extra work, any extra living expenses of the Vendor's employees and the cost of any extra journey.*

*Delete if erection is on a time basis. This clause is based on Clause 7.2 of UNECE Conditions 188A, by kind permission of the United Nations.

Contract Price Adjustment

We propose that contract price adjustment in respect of the price for goods shall apply in accordance with the BEAMA *Contract Price Adjustment Clause and Formula for use with Export Contracts – Electrical Machinery: (for which there is no other specific formula)*, edition January 1979, corrected edition,* a copy of which is enclosed.

We further propose that contract price adjustment in respect of the price for site work shall apply in accordance with the same BEAMA Clause and Formula subject to the amendment that adjustment shall take place at the rate of 0.95 per cent of such price per 1.0 per cent difference between the Labour Cost Index published for the month in which the tender date falls and the average of the Index figures published for the last two-thirds of the period during which the Contractor is carrying out site work, this difference being expressed as a percentage of the former Index figure.

The BEAMA Labour Cost Index referred to shall be the Labour Cost Index for Electrical Engineering.*

References to the *Trade and Industry Journal* shall be read as references to *British Business*, which is the new name of the journal.†

*This is only one of several BEAMA CPA Clauses and Formulae. If another is adopted, this form may need consequential amendment. In particular, reference to the BEAMA CPA Clause applicable to Mechanical Plant, and to the Labour Cost Index for Mechanical Engineering, may be appropriate. See p. 16.
†The above form requires that the separate prices to be adjusted shall be identifiable from the contract documents.

Trade Terms

The trade term 'C & F' shall be defined by the current edition of Incoterms, where consistent with this tender. A copy of the definition is enclosed.

Validity

This tender is valid for acceptance until and including *, but shall, after that date, be null and void unless extended by us in writing.
*Date in full, e.g. 'Monday, 19 January 1981'.

Yours faithfully

(Signature)

FORM 39

CIF Tender based on General Conditions No 188 for the Supply of Plant and Machinery for Export, prepared under the auspices of the United Nations Economic Commission for Europe, Geneva, March 1953, and Additional Clauses for Supervision of Erection of Plant and Machinery Abroad No 188B, prepared under the auspices of the United Nations Economic Commission for Europe, Geneva, April 1964

Dear Sirs

(heading)

Prices
We offer to supply CIF *, and supervise the erection on site at

of the items described in the accompanying Specification for:
£
(pounds sterling)
*Port of destination.

(Our services in respect of site work are offered on a time basis. The above price includes an estimated sum of £ for such services, computed at the rates and for the estimated duration shown in the accompanying Specification. The sum actually payable for such services shall be computed on a time basis, and may be either more or less than the estimated sum.)*
*Delete if lump sum.

Programme
Our best estimate is that delivery CIF will be achieved within of receiving your order, and that we could complete supervision of erection within a further of receiving your order, making a total of .
We enclose * showing the basis of this estimate.
*Appropriate document, e.g. bar chart or critical path network.

Terms of Payment
All prices are payable in pounds sterling in the United Kingdom.
We propose that payment shall be made as follows:
(a) %(per cent) in advance.
(b) The balance through a Confirmed Irrevocable Letter of Credit providing for payment in full upon presentation of documents, to be increased according to any contractual increases in the price, to be extended according to any increase in the period required for execution of our respective obligations under the Contract, and otherwise complying with

the following description:

(i) Confirmed by an acceptable bank in London.
(ii) Irrevocably issued, confirmed and advised to us by [a].
(iii) Date and place of expiry – [a] in London.
(iv) Beneficiary – ourselves.
(v) Amount – the balance of the price.
(vi) Credit available with confirming bank by payment.
(vii) Partial shipments allowed.
(viii) Transhipment allowed.
(ix) Shipment from .[b]
(x) For transportation to .[c]
(xi) Shipment of .[d]
(xii) Documents required:

Commercial Invoice

In respect of goods, full set of clean on board marine Bills of Lading (which shall be acceptable although to the effect that the goods have been shipped on deck or unprotected*) and Insurance Certificate.

In respect of other work, your or your Engineer's Certificate.†

*Delete if inappropriate.

†If this sentence is omitted, payment for such other work should be made out of the letter of credit on presentation only of commercial invoice.

(xiii) Period after issuance of shipping documents for presentation – days.
(xiv) Subject otherwise to the *Uniform Customs and Practice for Documentary Credits*, 1974, ICC Publication No 290.[e]

[a]Date in full, e.g. 'Monday, 19 January 1981'.

[b]Port of shipment, or e.g. 'any United Kingdom port'.

[c]Port of destination.

[d]Short description of goods and site work.

[e]This description follows as closely as possible the *Standard Procedure for the Issuing of Documentary Credits by Cable, Telegram or Telex* (App VII *Standard Forms for Issuing Documentary Credits*, ICC Publication No 323).

Payment in respect of site work shall be due monthly at the end of each month.

Conditions of Contract

Where consistent with this tender, we propose that *General Conditions No 188 for the Supply of Plant and Machinery for Export*, prepared under the auspices of the United Nations Economic Commission for Europe, Geneva, March 1953, and *Additional Clauses for Supervision of Erection of Plant and Machinery Abroad No 188B*, prepared under the auspices of the United Nations Economic Commission for Europe, Geneva, April 1964, a copy of which is enclosed, shall apply, subject to the following:

The Appendix in each case shall be completed as follows:

Appendix to UNECE Conditions 188

	Clause		
A. Percentage to be deducted for each week's delay	7.3	%(	per cent)
B. Maximum percentage which the deductions above may not exceed	7.3	%(	per cent)
C. Maximum amount recoverable for non-delivery	7.5	£ (	 pounds sterling)
D. Maximum amount recoverable on termination by Vendor for failure to take delivery or make payment	7.7 and 8.7	£ (	 pounds sterling)
E. Rate of interest on overdue payments	8.7	2% (two per cent) per annum over the * Bank base lending rate from time to time in force	
F. Period of delay in payment authorizing termination by Vendor	8.7	months	
G. Guarantee Period for original Plant and parts replaced or renewed	9.2 and 9.7	months	
H. Maximum extension of Guarantee Period	9.5	months	
I. (1) Daily use of Plant	9.6	hours/day	
(2) Reduction of Guarantee Period for more intensive use	9.6		

*Insert name of bank.

Appendix to UNECE Conditions 188B

J. Duration of interruption in erection at the expiry of which the Vendor is authorized to recall his supervising engineers	11.1	months	
K. Maximum indemnities payable by the parties	12.4	£ (	 pounds sterling)

Clause 4 (Charges Payable by the Purchaser) of UNECE Conditions 188B – Supervision of erection being carried out for a lump sum, the quoted price

includes all the items mentioned in Clause 4. Provided that if supervision of erection is prolonged for any cause for which the Purchaser or any of his contractors other than the Vendor is responsible and if, as a result, the work of the Vendor's employees is suspended or added to, a charge will be made for any idle time, any extra work, any extra living expenses of the Vendor's employees and the cost of any extra journey.*

*Delete if erection is on a time basis. This clause is based on Clause 7.2 of UNECE Conditions 188A, by kind permission of the United Nations.

Contract Price Adjustment

We propose that contract price adjustment in respect of the price for goods shall apply in accordance with the BEAMA *Contract Price Adjustment Clause and Formula for use with Export Contracts – Electrical Machinery: (for which there is no other specific formula)*, edition January 1979, corrected edition,* a copy of which is enclosed.

We further propose that contract price adjustment in respect of the price for site work shall apply in accordance with the same BEAMA Clause and Formula subject to the amendment that adjustment shall take place at the rate of 0.95 per cent of such price per 1.0 per cent difference between the Labour Cost Index published for the month in which the tender date falls and the average of the Index figures published for the last two-thirds of the period during which the Contractor is carrying out site work, this difference being expressed as a percentage of the former Index figure.

The BEAMA Labour Cost Index referred to shall be the Labour Cost Index for Electrical Engineering.*

References to the *Trade and Industry Journal* shall be read as references to *British Business*, which is the new name of the journal.†

*This is only one of several BEAMA CPA Clauses and Formulae. If another is adopted, this form may need consequential amendment. In particular, reference to the BEAMA CPA Clause applicable to Mechanical Plant, and to the Labour Cost Index for Mechanical Engineering, may be appropriate. See p. 16.

†The above form requires that the separate prices to be adjusted shall be identifiable from the contract documents.

Trade Terms

The trade term 'CIF' shall be defined by the current edition of Incoterms, where consistent with this tender. A copy of the definition is enclosed.

Validity

This tender is valid for acceptance until and including *, but shall, after that date, be null and void unless extended by us in writing.

*Date in full, e.g. 'Monday, 19 January 1981'.

Yours faithfully

(Signature)

FORM 40

Tender based on General Conditions No 188A for the Supply and Erection of Plant and Machinery for Import and Export, prepared under the auspices of the United Nations Economic Commission for Europe, Geneva, March 1957, providing for Delivery to and Erection on Site

Dear Sirs

(heading)

Prices

We offer to supply, deliver to and erect on Site at

the items described in the accompanying Specification for:
£
(pounds sterling)
(Our services in respect of site work are offered on a time basis. The above price includes an estimated sum of £ for such services, computed at the rates and for the estimated duration shown in the accompanying Specification. The sum actually payable for such services shall be computed on a time basis, and may be either more or less than the estimated sum.)*

*Delete if lump sum.

Programme

Our best estimate is that completion of our work will be achieved within of receiving your order. (We enclose * showing the basis of this estimate.)†

*Appropriate document, e.g. bar chart or critical path network.
†Delete if inappropriate.

Terms of Payment

All prices are payable in pounds sterling in the United Kingdom.

We propose that payment shall be made as follows:

(a) %(per cent) in advance.

(b) The balance through a Confirmed Irrevocable Letter of Credit providing for payment in full upon presentation of documents, to be increased according to any contractual increases in the price, to be extended according to any increase in the period required for execution of our respective obligations under the Contract, and otherwise complying with the following description:
  - (i) Confirmed by an acceptable bank in London.
  - (ii) Irrevocably issued, confirmed and advised to us by .[a]
  - (iii) Date and place of expiry – [a] in London.
  - (iv) Beneficiary – ourselves.
  - (v) Amount – the balance of the price.

(vi) Credit available with confirming bank by payment.
(vii) Partial shipments allowed.
(viii) Transhipment allowed.
(ix) Shipment from .[b]
(x) For transportation to .[c]
(xi) Shipment of .[d]
(xii) Documents required:

Commercial Invoice

In respect of goods, full set of clean on board marine Bills of Lading (which shall be acceptable although to the effect that the goods have been shipped on deck or unprotected*) and Insurance Certificate.

In respect of other work, your or your Engineer's Certificate.†

*Delete if inappropriate.
†If this sentence is omitted, payment for such other work should be made out of the letter of credit on presentation only of commercial invoice.

(xiii) Period after issuance of shipping documents for presentation – days.
(xiv) Subject otherwise to the *Uniform Customs and Practice for Documentary Credits,* 1974, ICC Publication No 290.[e]

[a]Date in full, e.g. 'Monday, 19 January 1981'.
[b]Port of shipment, or e.g. 'any United Kingdom port'.
[c]Port of destination.
[d]Short description of goods and site work.
[e]This description follows as closely as possible the *Standard Procedure for the Issuing of Documentary Credits by Cable, Telegram or Telex* (App VII *Standard Forms for Issuing Documentary Credits,* ICC Publication No 323).

Payment in respect of site work shall be due monthly at the end of each month.

Conditions of Contract

Where consistent with this tender, we propose that *General Conditions No 188A for the Supply and Erection of Plant and Machinery for Import and Export,* prepared under the auspices of the United Nations Economic Commission for Europe, Geneva, March 1957, a copy of which is enclosed, shall apply, subject to the following:

The Appendix shall be completed as follows:

Appendix to UNECE Conditions 188A

	Clause	
A. Maximum amount recoverable on termination by Contractor for failure to take delivery or make payment	10.2 and 11.7	£ (pounds sterling)
B. Rate of interest on overdue payments	11.7	2% (two per cent) per annum over the * Bank base lending rate from time to time in force

C.	Period of delay in payment authorizing termination by Contractor	11.7		months
D.	Percentage to be deducted for each week's delay	20.3		% (per cent)
E.	Maximum percentage which the deductions above may not exceed	20.3		% (per cent)
F.	Maximum amount recoverable for non-completion	20.5	£ (	 pounds sterling)
G.	Maximum post-ponement of taking over tests by Contractor	22.3		weeks
H.	Guarantee Period for original Works and parts replaced or renewed	23.2 and 23.5		months
I.	Maximum indemnities for personal injury or damage	24.3	£ (	 pounds sterling)
J.	(1) Daily use of Plant	23.4		hours/day
	(2) Reduction of Guarantee Period for more intensive use	23.4		

*Insert name of bank.

Clause 7 (Erection on a Time Basis and Lump-Sum Erection) – Erection is carried out (on a time basis)* (for a lump sum).*

*Delete alternatives.

Clause 9 (Passing of Risk) – Risk shall pass upon delivery to Site.

Customs and Import Duties – The Purchaser shall in good time procure any necessary import permits and promptly pay all customs and import duties which may become payable upon the importation of the Plant into the country in which the Plant is to be erected.

Contract Price Adjustment

We propose that contract price adjustment in respect of the price for goods shall apply in accordance with the BEAMA *Contract Price Adjustment Clause and Formula for use with Export Contracts – Electrical Machinery: (for which there is no other specific formula)*, edition January 1979, corrected edition,* a copy of which is enclosed.

We further propose that contract price adjustment in respect of the price for site work shall apply in accordance with the same BEAMA Clause and Formula subject to the amendment that adjustment shall take place at the rate of 0.95 per cent of such price per 1.0 per cent difference between the Labour Cost Index published for the month in which the tender date falls and the average of the Index figures published for the last two-thirds of the period during which the Contractor is carrying out site work, this difference being expressed as a percentage of the former Index figure.

The BEAMA Labour Cost Index referred to shall be the Labour Cost Index for Electrical Engineering.*

References to the *Trade and Industry Journal* shall be read as references to *British Business,* which is the new name of the journal.†

*This is only one of several BEAMA CPA Clauses and Formulae. If another is adopted, this form may need consequential amendment. In particular, reference to the BEAMA CPA Clause applicable to Mechanical Plant, and to the Labour Cost Index for Mechanical Engineering, may be appropriate. See p. 16.

†The above form requires that the separate prices to be adjusted shall be identifiable from the contract documents.

Validity

This tender is valid for acceptance until and including *, but shall, after that date, be null and void unless extended by us in writing.

*Date in full, e.g. 'Monday, 19 January 1981'.

Yours faithfully

(Signature)

FORM 41

Tender based on General Conditions No 188A for the Supply and Erection of Plant and Machinery for Import and Export, prepared under the auspices of the United Nations Economic Commission for Europe, Geneva, March 1957, providing for Delivery FOB and Erection on Site

Dear Sirs

(heading)

Prices
We offer to supply FOB * and erect on Site at

the items described in the accompanying Specification for:
£
(pounds sterling)
*Port of shipment.
(Our services in respect of site work are offered on a time basis. The above price includes an estimated sum of £ for such services, computed at the rates and for the estimated duration shown in the accompanying Specification. The sum actually payable for such services shall be computed on a time basis, and may be either more or less than the estimated sum.)*
*Delete if lump sum.

Programme
Our best estimate is that delivery FOB will be achieved within of receiving your order, and that we could complete erection within a further of receiving your order, making a total of .
(We enclose * showing the basis of this estimate.)†
*Appropriate document, e.g. bar chart or critical path network.
†Delete if inappropriate.

Terms of Payment
All prices are payable in pounds sterling in the United Kingdom.
We propose that payment shall be made as follows:
(a) %(per cent) in advance.
(b) The balance through a Confirmed Irrevocable Letter of Credit providing for payment in full upon presentation of documents, to be increased according to any contractual increases in the price, to be extended according to any increase in the period required for execution of our respective obligations under the Contract, and otherwise complying with the following description:
   (i) Confirmed by an acceptable bank in London.
   (ii) Irrevocably issued, confirmed and advised to us by .[a]
   (iii) Date and place of expiry – [a] in London.

(iv) Beneficiary – ourselves.
(v) Amount – the balance of the price.
(vi) Credit available with confirming bank by payment.
(vii) Partial shipments allowed.
(viii) Transhipment allowed.
(ix) Shipment from .[b]
(x) For transportation to .[c]
(xi) Shipment of .[d]
(xii) Documents required:

Commercial Invoice

In respect of goods, full set of clean on board marine Bills of Lading (which shall be acceptable although to the effect that the goods have been shipped on deck or unprotected*).

In respect of other work, your or your Engineer's Certificate.†

*Delete if inappropriate.

†If this sentence is omitted, payment for such other work should be made out of the letter of credit on presentation only of commercial invoice.

(xiii) Period after issuance of shipping documents for presentation – days.
(xiv) Subject otherwise to the *Uniform Customs and Practice for Documentary Credits,* 1974, ICC Publication No 290.[e]

[a]Date in full, e.g. 'Monday, 19 January 1981'.
[b]Port of shipment, or e.g. 'any United Kingdom port'.
[c]Port of destination.
[d]Short description of goods and site work.
[e]This description follows as closely as possible the *Standard Procedure for the Issuing of Documentary Credits by Cable, Telegram or Telex* (App VII *Standard Forms for Issuing Documentary Credits,* ICC Publication No 323).

Payment in respect of site work shall be due monthly at the end of each month.

Conditions of Contract

Where consistent with this tender, we propose that *General Conditions No 188A for the Supply and Erection of Plant and Machinery for Import and Export,* prepared under the auspices of the United Nations Economic Commission for Europe, Geneva, March 1957, a copy of which is enclosed, shall apply, subject to the following:

The Appendix shall be completed as follows:

Appendix to UNECE Conditions 188A

	Clause	
A. Maximum amount recoverable on termination by Contractor for failure to take delivery or make payment	10.2 and 11.7	£ (pounds sterling)

B.	Rate of interest on overdue payments	11.7	2% (two per cent) per annum over the * Bank base lending rate from time to time in force
C.	Period of delay in payment authorizing termination by Contractor	11.7	months
D.	Percentage to be deducted for each week's delay	20.3	%(per cent)
E.	Maximum percentage which the deductions above may not exceed	20.3	%(per cent)
F.	Maximum amount recoverable for non-completion	20.5	£ (pounds sterling)
G.	Maximum post-ponement of taking over tests by Contractor	22.3	months
H.	Guarantee Period for original Works and parts replaced or renewed	23.2 and 23.5	months
I.	Maximum indemnities for personal injury or damage	24.3	£ (pounds sterling)
J.	(1) Daily use of Plant	23.4	hours/day
	(2) Reduction of Guarantee Period for more intensive use	23.4	

*Insert name of bank.

Clause 7 (Erection on a Time Basis and Lump-Sum Erection) – Erection is carried out (on a time basis)* (for a lump sum).*

*Delete alternatives.

Contract Price Adjustment

We propose that contract price adjustment in respect of the price for goods shall apply in accordance with the BEAMA *Contract Price Adjustment Clause and Formula for use with Export Contracts – Electrical Machinery: (for which there is no other specific formula)*, edition January 1979, corrected edition,* a copy of which is enclosed.

We further propose that contract price adjustment in respect of the price for site work shall apply in accordance with the same BEAMA Clause and Formula subject to the amendment that adjustment shall take place at the rate of 0.95 per cent of such price per 1.0 per cent difference between the Labour Cost

Index published for the month in which the tender date falls and the average of the Index figures published for the last two-thirds of the period during which the Contractor is carrying out site work, this difference being expressed as a percentage of the former Index figure.

The BEAMA Labour Cost Index referred to shall be the Labour Cost Index for Electrical Engineering.*

References to the *Trade and Industry Journal* shall be read as references to *British Business*, which is the new name of the journal.†

*This is only one of several BEAMA CPA Clauses and Formulae. If another is adopted, this form may need consequential amendment. In particular, reference to the BEAMA CPA Clause applicable to Mechanical Plant, and to the Labour Cost Index for Mechanical Engineering, may be appropriate. See p. 16.

†The above form requires that the separate prices to be adjusted shall be identifiable from the contract documents.

Trade Terms

The trade term 'FOB' shall be defined by the current edition of Incoterms, where consistent with this tender. A copy of the definition is enclosed.

Validity

This tender is valid for acceptance until and including *, but shall, after that date, be null and void unless extended by us in writing.

*Date in full, e.g. 'Monday, 19 January 1981'.

Yours faithfully

(Signature)

FORM 42

Tender based on General Conditions No 188A for the Supply and Erection of Plant and Machinery for Import and Export, prepared under the auspices of the United Nations Economic Commission for Europe, Geneva, March 1957, providing for Delivery C & F and Erection on Site

Dear Sirs

(heading)

Prices

We offer to supply C & F * and erect on Site at

the items described in the accompanying Specification for:
£
(pounds sterling)

*Port of destination.

(Our services in respect of site work are offered on a time basis. The above price includes an estimated sum of £ for such services, computed at the rates and for the estimated duration shown in the accompanying Specification. The sum actually payable for such services shall be computed on a time basis, and may be either more or less than the estimated sum.)*

*Delete if lump sum.

Programme

Our best estimate is that delivery C & F will be achieved within of receiving your order, and that we could complete erection within a further of receiving your order, making a total of .
(We enclose * showing the basis of this estimate.)†

*Appropriate document, e.g. bar chart or critical path network.

†Delete if inappropriate.

Terms of Payment

All prices are payable in pounds sterling in the United Kingdom.

We propose that payment shall be made as follows:

(a) %(per cent) in advance.

(b) The balance through a Confirmed Irrevocable Letter of Credit providing for payment in full upon presentation of documents, to be increased according to any contractual increases in the price, to be extended according to any increase in the period required for execution of our respective obligations under the Contract, and otherwise complying with the following description:

(i) Confirmed by an acceptable bank in London.

(ii) Irrevocably issued, confirmed and advised to us by .[a]

(iii) Date and place of expiry – [a] in London.
(iv) Beneficiary – ourselves.
(v) Amount – the balance of the price.
(vi) Credit available with confirming bank by payment.
(vii) Partial shipments allowed.
(viii) Transhipment allowed.
(ix) Shipment from .[b]
(x) For transportation to .[c]
(xi) Shipment of .[d]
(xii) Documents required:

Commercial Invoice

In respect of goods, full set of clean on board marine Bills of Lading (which shall be acceptable although to the effect that the goods have been shipped on deck or unprotected*).

In respect of other work, your or your Engineer's Certificate.†

*Delete if inappropriate.

†If this sentence is omitted, payment for such other work should be made out of the letter of credit on presentation only of commercial invoice.

(xiii) Period after issuance of shipping documents for presentation – days.
(xiv) Subject otherwise to the *Uniform Customs and Practice for Documentary Credits,* 1974, ICC Publication No 290.[e]

[a] Date in full, e.g. 'Monday, 19 January 1981'.

[b] Port of shipment, or e.g. 'any United Kingdom port'.

[c] Port of destination.

[d] Short description of goods and site work.

[e] This description follows as closely as possible the *Standard Procedure for the Issuing of Documentary Credits by Cable, Telegram or Telex* (App VII *Standard Forms for Issuing Documentary Credits,* ICC Publication No 323).

Payment in respect of site work shall be due monthly at the end of each month.

Conditions of Contract

Where consistent with this tender, we propose that *General Conditions No 188A for the Supply and Erection of Plant and Machinery for Import and Export,* prepared under the auspices of the United Nations Economic Commission for Europe, Geneva, March 1957, a copy of which is enclosed, shall apply, subject to the following:

The Appendix shall be completed as follows:

Appendix to UNECE Conditions 188A

	Clause	
A. Maximum amount recoverable on termination by Contractor for failure to take delivery or make payment	10.2 and 11.7	£ (pounds sterling)

B.	Rate of interest on overdue payments	11.7	2% (two per cent) per annum over the * Bank base lending rate from time to time in force
C.	Period of delay in payment authorizing termination by Contractor	11.7	months
D.	Percentage to be deducted for each week's delay	20.3	% (per cent)
E.	Maximum percentage which the deductions above may not exceed	20.3	% (per cent)
F.	Maximum amount recoverable for non-completion	20.5	£ (pounds sterling)
G.	Maximum post-ponement of taking over tests by Contractor	22.3	weeks
H.	Guarantee period for original Works and parts replaced or renewed	23.2 and 23.5	months
I.	Maximum indemnities for personal injury or damage	24.3	£ (pounds sterling)
J.	(1) Daily use of Plant	23.4	hours/day
	(2) Reduction of Guarantee Period for more intensive use	23.4	

*Insert name of bank.

Clause 7 (Erection on a Time Basis and Lump-Sum Erection) – Erection is carried out (on a time basis)* (for a lump sum).*

*Delete alternatives.

Contract Price Adjustment

We propose that contract price adjustment in respect of the price for goods shall apply in accordance with the BEAMA *Contract Price Adjustment Clause and Formula for use with Export Contracts – Electrical Machinery: (for which there is no other specific formula)*, edition January 1979, corrected edition,* a copy of which is enclosed.

We further propose that contract price adjustment in respect of the price for site work shall apply in accordance with the same BEAMA Clause and Formula subject to the amendment that adjustment shall take place at the rate of 0.95 per cent of such price per 1.0 per cent difference between the Labour Cost

Index published for the month in which the tender date falls and the average of the Index figures published for the last two-thirds of the period during which the Contractor is carrying out site work, this difference being expressed as a percentage of the former Index figure.

The BEAMA Labour Cost Index referred to shall be the Labour Cost Index for Electrical Engineering.*

References to the *Trade and Industry Journal* shall be read as references to *British Business,* which is the new name of the journal.†

*This is only one of several BEAMA CPA Clauses and Formulae. If another is adopted, this form may need consequential amendment. In particular, reference to the BEAMA CPA Clause applicable to Mechanical Plant, and to the Labour Cost Index for Mechanical Engineering, may be appropriate. See p. 16.
†The above form requires that the separate prices to be adjusted shall be identifiable from the contract documents.

Trade Terms
The trade term 'C & F' shall be defined by the current edition of Incoterms, where consistent with this tender. A copy of the definition is enclosed.

Validity
This tender is valid for acceptance until and including *, but shall, after that date, be null and void unless extended by us in writing.
*Date in full, e.g. 'Monday, 19 January 1981'.

Yours faithfully

(Signature)

FORM 43

Tender based on General Conditions No 188A for the Supply and Erection of Plant and Machinery for Import and Export, prepared under the auspices of the United Nations Economic Commission for Europe, Geneva, March 1957, providing for Delivery CIF and Erection on Site

Dear Sirs

(heading)

Prices
We offer to supply CIF * and erect on Site at

the items described in the accompanying Specification for:
£
(pounds sterling)
*Port of destination.
(Our services in respect of site work are offered on a time basis. The above price includes an estimated sum of £ for such services, computed at the rates and for the estimated duration shown in the accompanying Specification. The sum actually payable for such services shall be computed on a time basis, and may be either more or less than the estimated sum.)*
*Delete if lump sum.

Programme
Our best estimate is that delivery CIF will be achieved within of receiving your order, and that we could complete erection within a further of receiving your order, making a total of .
(We enclose * showing the basis of this estimate.)†
*Appropriate document, e.g. bar chart or critical path network.
†Delete if inappropriate.

Terms of Payment
All prices are payable in pounds sterling in the United Kingdom.
We propose that payment shall be made as follows:
(a) %(per cent) in advance.
(b) The balance through a Confirmed Irrevocable Letter of Credit providing for payment in full upon presentation of documents, to be increased according to any contractual increases in the price, to be extended according to any increase in the period required for execution of our respective obligations under the Contract, and otherwise complying with the following description:
  (i) Confirmed by an acceptable bank in London.
  (ii) Irrevocably issued, confirmed and advised to us by .[a]
  (iii) Date and place of expiry – [a] in London.

(iv) Beneficiary – ourselves.
(v) Amount – the balance of the price.
(vi) Credit available with confirming bank by payment.
(vii) Partial shipments allowed.
(viii) Transhipment allowed.
(ix) Shipment from .[b]
(x) For transportation to .[c]
(xi) Shipment of .[d]
(xii) Documents required:

Commercial Invoice

In respect of goods, full set of clean on board marine Bills of Lading (which shall be acceptable although to the effect that the goods have been shipped on deck or unprotected*) and Insurance Certificate.

In respect of other work, your or your Engineer's Certificate.†

*Delete if inappropriate.

†If this sentence is omitted, payment for such other work should be made out of the letter of credit on presentation only of commercial invoice.

(xiii) Period after issuance of shipping documents for presentation – days.
(xiv) Subject otherwise to the *Uniform Customs and Practice for Documentary Credits*, 1974, ICC Publication No 290.[e]

[a] Date in full, e.g. 'Monday, 19 January 1981'.

[b] Port of shipment, or e.g. 'any United Kingdom port'.

[c] Port of destination.

[d] Short description of goods, and site work.

[e] This description follows as closely as possible the *Standard Procedure for the Issuing of Documentary Credits by Cable, Telegram or Telex* (App VII *Standard Forms for Issuing Documentary Credits*, ICC Publication No 323).

Payment in respect of site work shall be due monthly at the end of each month.

Conditions of Contract

Where consistent with this tender, we propose that *General Conditions No 188A for the Supply and Erection of Plant and Machinery for Import and Export*, prepared under the auspices of the United Nations Economic Commission for Europe, Geneva, March 1957, a copy of which is enclosed, shall apply, subject to the following:

The Appendix shall be completed as follows:

Appendix to UNECE Conditions 188A

	Clause	
A. Maximum amount recoverable on termination by Contractor for failure to take delivery or make payment	10.2 and 11.7	£ (pounds sterling)

		Clause	
B.	Rate of interest on overdue payments	11.7	2% (two per cent) per annum over the * Bank base lending rate from time to time in force
C.	Period of delay in payment authorizing termination by Contractor	11.7	months
D.	Percentage to be deducted for each week's delay	20.3	% (per cent)
E.	Maximum percentage which the deductions above may not exceed	20.3	% (per cent)
F.	Maximum amount recoverable for non-completion	20.5	£ (pounds sterling)
G.	Maximum post-ponement of taking over tests by Contractor	22.3	weeks
H.	Guarantee Period for original Works and parts replaced or renewed	23.2 and 23.5	months
I.	Maximum indemnities for personal injury or damage	24.3	£ (pounds sterling)
J.	(1) Daily use of Plant	23.4	hours/day
	(2) Reduction of Guarantee Period for more intensive use	23.4	

*Insert name of bank.

Clause 7 (Erection on a Time Basis and Lump-Sum Erection) – Erection is carried out (on a time basis)* (for a lump sum).*

*Delete alternatives.

Contract Price Adjustment

We propose that contract price adjustment in respect of the price for goods shall apply in accordance with the BEAMA *Contract Price Adjustment Clause and Formula for use with Export Contracts – Electrical Machinery: (for which there is no other specific formula)*, edition January 1979, corrected edition,* a copy of which is enclosed.

We further propose that contract price adjustment in respect of the price for site work shall apply in accordance with the same BEAMA Clause and Formula subject to the amendment that adjustment shall take place at the rate of 0.95 per cent of such price per 1.0 per cent difference between the Labour Cost

Index published for the month in which the tender date falls and the average of the Index figures published for the last two-thirds of the period during which the Contractor is carrying out site work, this difference being expressed as a percentage of the former Index figure.

The BEAMA Labour Cost Index referred to shall be the Labour Cost Index for Electrical Engineering.*

References to the *Trade and Industry Journal* shall be read as references to *British Business,* which is the new name of the journal.†

*This is only one of several BEAMA CPA Clauses and Formulae. If another is adopted, this form may need consequential amendment. In particular, reference to the BEAMA CPA Clause applicable to Mechanical Plant, and to the Labour Cost Index for Mechanical Engineering, may be appropriate. See p. 16.

†The above form requires that the separate prices to be adjusted shall be identifiable from the contract documents.

Trade Terms

The trade term 'CIF' shall be defined by the current edition of Incoterms, where consistent with this tender. A copy of the definition is enclosed.

Validity

This tender is valid for acceptance until and including *, but shall, after that date, be null and void unless extended by us in writing.

*Date in full, e.g. 'Monday, 19 January 1981'.

Yours faithfully

(Signature)

FORM 44

Tender based on General Conditions No 188A for the Supply and Erection of Plant and Machinery for Import and Export, prepared under the auspices of the United Nations Economic Commission for Europe, Geneva, March 1957, providing for Delivery to and Supervision of Erection on Site

Dear Sirs

(heading)

Prices

We offer to supply, deliver to, and supervise the erection on Site at

of the items described in the accompanying Specification for:
£
(pounds sterling)
(Our services in respect of site work are offered on a time basis. The above price includes an estimated sum of £ for such services, computed at the rates and for the estimated duration shown in the accompanying Specification. The sum actually payable for such services shall be computed on a time basis, and may be either more or less than the estimated sum.)*

*Delete if lump sum.

Programme

Our best estimate is that delivery to Site will be achieved within of receiving your order, and that we could complete supervision of erection within a further of receiving your order, making a total of
(We enclose * showing the basis of this estimate.)†

*Appropriate document, e.g. bar chart or critical path network.
†Delete if inappropriate.

Terms of Payment

All prices are payable in pounds sterling in the United Kingdom.

We propose that payment shall be made as follows:

(a) %(per cent) in advance.

(b) The balance through a Confirmed Irrevocable Letter of Credit providing for payment in full upon presentation of documents, to be increased according to any contractual increases in the price, to be extended according to any increase in the period required for execution of our respective obligations under the Contract, and otherwise complying with the following description:

  (i) Confirmed by an acceptable bank in London.
  (ii) Irrevocably issued, confirmed and advised to us by [a]
  (iii) Date and place of expiry – [a] in London.

(iv) Beneficiary – ourselves.
(v) Amount – the balance of the price.
(vi) Credit available with confirming bank by payment.
(vii) Partial shipments allowed.
(viii) Transhipment allowed.
(ix) Shipment from .[b]
(x) For transportation to .[c]
(xi) Shipment of .[d]
(xii) Documents required:

Commercial Invoice

In respect of goods, full set of clean on board marine Bills of Lading (which shall be acceptable although to the effect that the goods have been shipped on deck or unprotected*) and Insurance Certificate.

In respect of other work, your or your Engineer's Certificate.†

*Delete if inappropriate.
†If this sentence is omitted, payment for such other work should be made out of the letter of credit on presentation only of commercial invoice.

(xiii) Period after issuance of shipping documents for presentation – days.
(xiv) Subject otherwise to the *Uniform Customs and Practice for Documentary Credits,* 1974, ICC Publication No 290.[e]

[a]Date in full, e.g. 'Monday, 19 January 1981'.
[b]Port of shipment, or e.g. 'any United Kingdom port'.
[c]Port of destination.
[d]Short description of goods and site work.
[e]This description follows as closely as possible the *Standard Procedure for the Issuing of Documentary Credits by Cable, Telegram or Telex* (App VII *Standard Forms for Issuing Documentary Credits,* ICC Publication No 323).

Payment in respect of site work shall be due monthly at the end of each month.

Conditions of Contract

Where consistent with this tender, we propose that *General Conditions No 188A for the Supply and Erection of Plant and Machinery for Import and Export,* prepared under the auspices of the United Nations Economic Commission for Europe, Geneva, March 1957, a copy of which is enclosed, shall apply, subject to the following:

The Appendix shall be completed as follows:

Appendix to UNECE Conditions 188A

	Clause	
A. Maximum amount recoverable on termination by Contractor for failure to take delivery or make payment	10.2 and 11.7	£ (pounds sterling)

B.	Rate of interest on overdue payments	11.7	2% (two per cent) per annum over the * Bank base lending rate from time to time in force
C.	Period of delay in payment authorizing termination by Contractor	11.7	months
D.	Percentage to be deducted for each week's delay	20.3	%(per cent)
E.	Maximum percentage which the deductions above may not exceed	20.3	%(per cent)
F.	Maximum amount recoverable for non-completion	20.5	£ (pounds sterling)
G.	Maximum postponement of taking over tests by Contractor	22.3	weeks
H.	Guarantee Period for original Works and parts replaced or renewed	23.2 and 23.5	months
I.	Maximum indemnities for personal injury or damage	24.3	£ (pounds sterling)
J.	(1) Daily use of Plant	23.4	hours/day
	(2) Reduction of Guarantee Period for more intensive use	23.4	

*Insert name of bank.

Clause 7 (Erection on a Time Basis and Lump-Sum Erection) – Supervision of erection is carried out (on a time basis)* (for a lump sum).*

*Delete alternatives.

Erection will be carried out by the Purchaser, who shall, at his own expense, provide the skilled and unskilled labour, all equipment and everything necessary for the erection of the Plant.

References to 'erection' in Clause 7 shall be read as references to 'supervision of erection'. The Contractor's obligation in this respect will be to provide the services of one or more competent engineers

(a) to give to the Purchaser or his representative mentioned in Clause 13.1 the necessary instructions for the erection of the Plant by the Purchaser and, if provided in the accompanying Specification or other documents, for its commissioning by him; and

(b) to supervise the manner in which the Contractor's instructions have been carried out.

The number and qualifications of the Contractor's staff, and the estimated duration of erection, shall be as specified in this tender or in the accompanying Specification or other documents.

The date on which the Contractor's staff should arrive on Site shall be as provided in the accompanying Specification or other documents; if not so provided, the Purchaser shall give the Contractor not less than one month's notice requiring such arrival.*

*This clause is based on Clause 2 of UNECE Conditions 188B, by kind permission of the United Nations.

Clause 20 (Time for Completion) – As the Contract will only include supervision of erection, the Contractor cannot undertake that erection will be completed within a specific time. Therefore, for the purposes of this Clause, completion of delivery to Site shall be deemed to constitute completion of the Works.

Customs and Import Duties – The Purchaser shall in good time procure any necessary import permits and promptly pay all customs and import duties which may become payable upon the importation of the Plant into the country in which the Plant is to be erected.

Contract Price Adjustment

We propose that contract price adjustment in respect of the price for goods shall apply in accordance with the BEAMA *Contract Price Adjustment Clause and Formula for use with Export Contracts – Electrical Machinery: (for which there is no other specific formula)*, edition January 1979, corrected edition,* a copy of which is enclosed.

We further propose that contract price adjustment in respect of the price for site work shall apply in accordance with the same BEAMA Clause and Formula subject to the amendment that adjustment shall take place at the rate of 0.95 per cent of such price per 1.0 per cent difference between the Labour Cost Index published for the month in which the tender date falls and the average of the Index figures published for the last two-thirds of the period during which the Contractor is carrying out site work, this difference being expressed as a percentage of the former Index figure.

The BEAMA Labour Cost Index referred to shall be the Labour Cost Index for Electrical Engineering.*

References to the *Trade and Industry Journal* shall be read as references to *British Business*, which is the new name of the journal.†

*This is only one of several BEAMA CPA Clauses and Formulae. If another is adopted, this form may need consequential amendment. In particular, reference to the BEAMA CPA Clause applicable to Mechanical Plant, and to the Labour Cost Index for Mechanical Engineering, may be appropriate. See p. 16.

†The above form requires that the separate prices to be adjusted shall be identifiable from the contract documents.

Validity

This tender is valid for acceptance until and including *, but shall, after that date, be null and void unless extended by us in writing.

*Date in full, e.g. 'Monday, 19 January 1981'.

Yours faithfully

(Signature)

FORM 45

Tender based on General Conditions No 188A for the Supply and Erection of Plant and Machinery for Import and Export, prepared under the auspices of the United Nations Economic Commission for Europe, Geneva, March 1957, providing for delivery FOB and Supervision of Erection on Site

Dear Sirs

(heading)

Prices

We offer to supply FOB * and supervise the erection on Site at

of the items described in the accompanying Specification for:

£

(pounds sterling)

*Port of shipment

(Our services in respect of site work are offered on a time basis. The above price includes an estimated sum of £ for such services, computed at the rates and for the estimated duration shown in the accompanying Specification. The sum actually payable for such services shall be computed on a time basis, and may be either more or less than the estimated sum.)*

*Delete if lump sum.

Programme

Our best estimate is that delivery FOB will be achieved within of receiving your order, and that we could complete supervision of erection within a further of receiving your order, making a total of .
(We enclose * showing the basis of this estimate.)†

*Appropriate document, e.g. bar chart or critical path network.
†Delete if inappropriate.

Terms of Payment

All prices are payable in pounds sterling in the United Kingdom.

We propose that payment shall be made as follows:

(a) % (per cent) in advance.
(b) The balance through a Confirmed Irrevocable Letter of Credit providing for payment in full upon presentation of documents, to be increased according to any contractual increases in the price, to be extended according to any increase in the period required for execution of our respective obligations under the Contract, and otherwise complying with the following description:
   (i) Confirmed by an acceptable bank in London.
   (ii) Irrevocably issued, confirmed and advised to us by .[a]

(iii) Date and place of expiry – [a] in London.
(iv) Beneficiary – ourselves.
(v) Amount – the balance of the price.
(vi) Credit available with confirming bank by payment.
(vii) Partial shipments allowed.
(viii) Transhipment allowed.
(ix) Shipment from .[b]
(x) For transportation to .[c]
(xi) Shipment of .[d]
(xii) Documents required:

Commercial Invoice

In respect of goods, full set of clean on board marine Bills of Lading (which shall be acceptable although to the effect that the goods have been shipped on deck or unprotected*).

In respect of other work, your or your Engineer's Certificate.†

*Delete if inappropriate.

†If this sentence is omitted, payment for such other work should be made out of the letter of credit on presentation only of commercial invoice.

(xiii) Period after issuance of shipping documents for presentation – days.
(xiv) Subject otherwise to the *Uniform Customs and Practice for Documentary Credits*, 1974, ICC Publication No 290. [e]

[a] Date in full, e.g. 'Monday, 19 January 1981'.

[b] Port of shipment, or e.g. 'any United Kingdom port'.

[c] Port of destination.

[d] Short description of goods and site work.

[e] This description follows as closely as possible the *Standard Procedure for the Issuing of Documentary Credits by Cable, Telegram or Telex* (App VII *Standard Forms for Issuing Documentary Credits*, ICC Publication No 323).

Payment in respect of site work shall be due monthly at the end of each month.

Conditions of Contract

Where consistent with this tender, we propose that *General Conditions No 188A for the Supply and Erection of Plant and Machinery for Import and Export*, prepared under the auspices of the United Nations Economic Commission for Europe, Geneva, March 1957, a copy of which is enclosed, shall apply, subject to the following:

The Appendix shall be completed as follows:

Appendix to UNECE Conditions 188A

	Clause	
A. Maximum amount recoverable on termination by Contractor for failure to take delivery or make payment	10.2 and 11.7	£ (pounds sterling)

B.	Rate of interest on overdue payments	11.7	2% (two per cent) per annum over the * Bank base lending rate from time to time in force
C.	Period of delay in payment authorizing termination by Contractor	11.7	months
D.	Percentage to be deducted for each week's delay	20.3	% (per cent)
E.	Maximum percentage which the deductions above may not exceed	20.3	% (per cent)
F.	Maximum amount recoverable for non-completion	20.5	£ (pounds sterling)
G.	Maximum post-ponement of taking over tests by Contractor	22.3	weeks
H.	Guarantee Period for original Works and parts replaced or renewed	23.2 and 23.5	months
I.	Maximum indemnities for personal injury or damage	24.3	£ (pounds sterling)
J.	(1) Daily use of Plant	23.4	hours/day
	(2) Reduction of Guarantee Period for more intensive use	23.4	

*Insert name of bank.

Clause 7 (Erection on a Time Basis and Lump-Sum Erection) – Supervision of erection is carried out (on a time basis)* (for a lump sum).*

*Delete alternatives.

Erection will be carried out by the Purchaser, who shall, at his own expense, provide the skilled and unskilled labour, all equipment and everything necessary for the erection of the Plant.

References to 'erection' in Clause 7 shall be read as references to 'supervision of erection'. The Contractor's obligation in this respect will be to provide the services of one or more competent engineers

(a) to give to the Purchaser or his representative mentioned in Clause 13.1 the necessary instructions for the erection of the Plant by the Purchaser and, if provided in the accompanying Specification or other documents, for its commissioning by him; and

(b) to supervise the manner in which the Contractor's instructions have been carried out.

The number and qualifications of the Contractor's staff, and the estimated duration of erection, shall be as specified in this tender or in the accompanying Specification or other documents.

The date on which the Contractor's staff should arrive on Site shall be as provided in the accompanying Specification or other documents; if not so provided, the Purchaser shall give the Contractor not less than one month's notice requiring such arrival.*

*This clause is based on Clause 2 of UNECE Conditions 188B, by kind permission of the United Nations.

Clause 20 (Time for Completion) – As the Contract will only include supervision of erection, the Contractor cannot undertake that erection will be completed within a specific time. Therefore, for the purposes of this Clause, completion of delivery FOB shall be deemed to constitute completion of the Works.

Contract Price Adjustment

We propose that contract price adjustment in respect of the price for goods shall apply in accordance with the BEAMA *Contract Price Adjustment Clause and Formula for use with Export Contracts – Electrical Machinery: (for which there is no other specific formula)*, edition January 1979, corrected edition,* a copy of which is enclosed.

We further propose that contract price adjustment in respect of the price for site work shall apply in accordance with the same BEAMA Clause and Formula subject to the amendment that adjustment shall take place at the rate of 0.95 per cent of such price per 1.0 per cent difference between the Labour Cost Index published for the month in which the tender date falls and the average of the Index figures published for the last two-thirds of the period during which the Contractor is carrying out site work, this difference being expressed as a percentage of the former Index figure.

The BEAMA Labour Cost Index referred to shall be the Labour Cost Index for Electrical Engineering.*

References to the *Trade and Industry Journal* shall be read as references to *British Business*, which is the new name of the journal.†

*This is only one of several BEAMA CPA Clauses and Formulae. If another is adopted, this form may need consequential amendment. In particular, reference to the BEAMA CPA Clause applicable to Mechanical Plant, and to the Labour Cost Index for Mechanical Engineering, may be appropriate. See p. 16.

†The above form requires that the separate prices to be adjusted shall be identifiable from the contract documents.

Trade Terms

The trade term 'FOB' shall be defined by the current edition of Incoterms, where consistent with this tender. A copy of the definition is enclosed.

Validity

This tender is valid for acceptance until and including *, but shall, after that date, be null and void unless extended by us in writing.

*Date in full, e.g. 'Monday, 19 January 1981'.

Yours faithfully

(Signature)

FORM 46

Tender based on General Conditions No 188A for the Supply and Erection of Plant and Machinery for Import and Export, prepared under the auspices of the United Nations Economic Commission for Europe, Geneva, March 1957, providing for Delivery C & F and Supervision of Erection on Site

Dear Sirs

(heading)

Prices
We offer to supply C & F *, and supervise the erection on Site at

of the items described in the accompanying Specification for:
£
(pounds sterling)

*Port of destination.

(Our services in respect of site work are offered on a time basis. The above price includes an estimated sum of £ for such services, computed at the rates and for the estimated duration shown in the accompanying Specification. The sum actually payable for such services shall be computed on a time basis, and may be either more or less than the estimated sum.)*

*Delete if lump sum.

Programme
Our best estimate is that delivery C & F will be achieved within of receiving your order, and that we could complete suervision of erection within a further of receiving your order, making a total of .
(We enclose * showing the basis of this estimate.)†

*Appropriate document, e.g. bar chart or critical path network.
†Delete if inappropriate.

Terms of Payment
All prices are payable in pounds sterling in the United Kingdom.

We propose that payment shall be made as follows:

(a) % (per cent) in advance.

(b) The balance through a Confirmed Irrevocable Letter of Credit providing for payment in full upon presentation of documents, to be increased according to any contractual increases in the price, to be extended according to any increase in the period required for execution of our respective obligations under the Contract, and otherwise complying with the following description:
   (i) Confirmed by an acceptable bank in London.
   (ii) Irrevocably issued, confirmed and advised to us by .[a]

(iii) Date and place of expiry – [a] in London.
(iv) Beneficiary – ourselves.
(v) Amount – the balance of the price.
(vi) Credit available with confirming bank by payment.
(vii) Partial shipments allowed.
(viii) Transhipment allowed.
(ix) Shipment from .[b]
(x) For transportation to .[c]
(xi) Shipment of .[d]
(xii) Documents required:

Commercial Invoice

In respect of goods, full set of clean on board marine Bills of Lading (which shall be acceptable although to the effect that the goods have been shipped on deck or unprotected*).

In respect of other work, your or your Engineer's Certificate.†

*Delete if inappropriate.
†If this sentence is omitted, payment for such other work should be made out of the letter of credit on presentation only of commercial invoice.

(xiii) Period after issuance of shipping documents for presentation – days.
(xiv) Subject otherwise to the *Uniform Customs and Practice for Documentary Credits*, 1974, ICC Publication No 290.[e]

[a] Date in full, e.g. 'Monday, 19 January 1981'.
[b] Port of shipment, or e.g. 'any United Kingdom port'.
[c] Port of destination.
[d] Short description of goods and site work.
[e] This description follows as closely as possible the *Standard Procedure for the Issuing of Documentary Credits by Cable, Telegram or Telex* (App VII *Standard Forms for Issuing Documentary Credits*, ICC Publication No 323).

Payment in respect of site work shall be due monthly at the end of each month.

Conditions of Contract

Where consistent with this tender, we propose that *General Conditions No 188A for the Supply and Erection of Plant and Machinery for Import and Export,* prepared under the auspices of the United Nations Economic Commission for Europe, Geneva, March 1957, a copy of which is enclosed, shall apply, subject to the following:

The Appendix shall be completed as follows:

Appendix to UNECE Conditions 188A

	Clause	
A. Maximum amount recoverable on termination by Contractor for failure to take delivery or make payment	10.2 and 11.7	£ (pounds sterling)

B.	Rate of interest on overdue payments	11.7	2% (two per cent) per annum over the * Bank base lending rate from time to time in force
C.	Period of delay in payment authorizing termination by Contractor	11.7	months
D.	Percentage to be deducted for each week's delay	20.3	% (per cent)
E.	Maximum percentage which the deductions above may not exceed	20.3	% (per cent)
F.	Maximum amount recoverable for non-completion	20.5	£ (pounds sterling)
G.	Maximum post-ponement of taking over tests by Contractor	22.3	weeks
H.	Guarantee Period for original Works and parts replaced or renewed	23.2 and 23.5	months
I.	Maximum indemnities for personal injury or damage	24.3	£ (pounds sterling)
J.	(1) Daily use of Plant	23.4	hours/day
	(2) Reduction of Guarantee Period for more intensive use	23.4	

*Insert name of bank.

Clause 7 (Erection on a Time Basis and Lump-Sum Erection) – Supervision of erection is carried out (on a time basis)* (for a lump sum).*

*Delete alternatives.

Erection will be carried out by the Purchaser, who shall, at his own expense, provide the skilled and unskilled labour, all equipment and everything necessary for the erection of the Plant.

References to 'erection' in Clause 7 shall be read as references to 'supervision of erection'. The Contractor's obligation in this respect will be to provide the services of one or more competent engineers

(a) to give to the Purchaser or his representative mentioned in Clause 13.1 the necessary instructions for the erection of the Plant by the Purchaser and, if provided in the accompanying Specification or other documents, for its commissioning by him; and
(b) to supervise the manner in which the Contractor's instructions have been carried out.

The number and qualifications of the Contractor's staff, and the estimated duration of erection, shall be as specified in this tender or in the accompanying Specification or other documents.

The date on which the Contractor's staff should arrive on Site shall be as provided in the accompanying Specification or other documents; if not so provided, the Purchaser shall give the Contractor not less than one month's notice requiring such arrival.*

*This clause is based on Clause 2 of UNECE Conditions 188B, by kind permission of the United Nations.

Clause 20 (Time for Completion) – As the Contract will only include supervision of erection, the Contractor cannot undertake that erection will be completed within a specific time. Therefore, for the purposes of this Clause, completion of delivery C & F shall be deemed to constitute completion of the Works.

Contract Price Adjustment

We propose that contract price adjustment in respect of the price for goods shall apply in accordance with the BEAMA *Contract Price Adjustment Clause and Formula for use with Export Contracts – Electrical Machinery: (for which there is no other specific formula)*, edition January 1979, corrected edition,* a copy of which is enclosed.

We further propose that contract price adjustment in respect of the price for site work shall apply in accordance with the same BEAMA Clause and Formula subject to the amendment that adjustment shall take place at the rate of 0.95 per cent of such price per 1.0 per cent difference between the Labour Cost Index published for the month in which the tender date falls and the average of the Index figures published for the last two-thirds of the period during which the Contractor is carrying out site work, this difference being expressed as a percentage of the former Index figure.

The BEAMA Labour Cost Index referred to shall be the Labour Cost Index for Electrical Engineering.*

References to the *Trade and Industry Journal* shall be read as references to *British Business,* which is the new name of the journal.†

*This is only one of several BEAMA CPA Clauses and Formulae. If another is adopted, this form may need consequential amendment. In particular, reference to the BEAMA CPA Clause applicable to Mechanical Plant, and to the Labour Cost Index for Mechanical Engineering, may be appropriate. See p. 16.

†The above form requires that the separate prices to be adjusted shall be identifiable from the contract documents.

Trade Terms

The trade term 'C & F' shall be defined by the current edition of Incoterms, where consistent with this tender. A copy of the definition is enclosed.

Validity

This tender is valid for acceptance until and including *, but shall, after that date, be null and void unless extended by us in writing.

*Date in full, e.g. 'Monday, 19 January 1981'.

Yours faithfully

(Signature)

FORM 47

Tender based on General Conditions No 188A for the Supply and Erection of Plant and Machinery for Import and Export, prepared under the auspices of the United Nations Economic Commission for Europe, Geneva, March 1957, providing for Delivery CIF and Supervision of Erection on Site

Dear Sirs

(heading)

Prices

We offer to supply CIF * and supervise the erection on Site at

of the items described in the accompanying Specification for:
£
(pounds sterling)
*Port of destination.

(Our services in respect of site work are offered on a time basis. The above price includes an estimated sum of £ for such services, computed at the rates and for the estimated duration shown in the accompanying Specification. The sum actually payable for such services shall be computed on a time basis, and may be either more or less than the estimated sum.)*
*Delete if lump sum.

Programme

Our best estimate is that delivery CIF will be achieved within of receiving your order, and that we could complete supervision of erection within a further of receiving your order, making a total of .
(We enclose * showing the basis of this estimate.)†
*Appropriate document, e.g. bar chart or critical path network.
†Delete if inappropriate.

Terms of Payment

All prices are payable in pounds sterling in the United Kingdom.

We propose that payment shall be made as follows:

(a) %(per cent) in advance.
(b) The balance through a Confirmed Irrevocable Letter of Credit providing for payment in full upon presentation of documents, to be increased according to any contractual increases in the price, to be extended according to any increase in the period required for execution of our respective obligations under the Contract, and otherwise complying with the following description:
    (i) Confirmed by an acceptable bank in London.
    (ii) Irrevocably issued, confirmed and advised to us by .[a]

(iii) Date and place of expiry – [a] in London.
(iv) Beneficiary – ourselves.
(v) Amount – the balance of the price.
(vi) Credit available with confirming bank by payment.
(vii) Partial shipments allowed.
(viii) Transhipment allowed.
(ix) Shipment from .[b]
(x) For transportation to .[c]
(xi) Shipment of .[d]
(xii) Documents required:

Commercial Invoice

In respect of goods, full set of clean on board marine Bills of Lading (which shall be acceptable although to the effect that the goods have been shipped on deck or unprotected*) and Insurance Certificate.

In respect of other work, your or your Engineer's Certificate.†

*Delete if inappropriate.

†If this sentence is omitted, payment for such other work should be made out of the letter of credit on presentation only of commercial invoice.

(xiii) Period after issuance of shipping documents for presentation – days.
(xiv) Subject otherwise to the *Uniform Customs and Practice for Documentary Credits,* 1974, ICC Publication No 290.[e]

[a]Date in full, e.g. 'Monday, 19 January 1981'.
[b]Port of shipment, or e.g. 'any United Kingdom port'.
[c]Port of destination.
[d]Short description of goods and site work.
[e]This description follows as closely as possible the *Standard Procedure for the Issuing of Documentary Credits by Cable, Telegram or Telex* (App VII *Standard Forms for Issuing Documentary Credits,* ICC Publication No 323).

Payment in respect of site work shall be due monthly at the end of each month.

Conditions of Contract

Where consistent with this tender, we propose that *General Conditions No 188A for the Supply and Erection of Plant and Machinery for Import and Export,* prepared under the auspices of the United Nations Economic Commission for Europe, Geneva, March 1957, a copy of which is enclosed, shall apply, subject to the following:

The Appendix shall be completed as follows:

Appendix to UNECE Conditions 188A

	Clause	
A. Maximum amount recoverable on termination by Contractor for failure to take delivery or make payment	10.2 and 11.7	£ (pounds sterling)

B.	Rate of interest on overdue payments	11.7	2% (two per cent) per annum over the * Bank base lending rate from time to time in force
C.	Period of delay in payment authorizing termination by Contractor	11.7	months
D.	Percentage to be deducted for each week's delay	20.3	% (per cent)
E.	Maximum percentage which the deductions above may not exceed	20.3	% (per cent)
F.	Maximum amount recoverable for non-completion	20.5	£ (pounds sterling)
G.	Maximum post-ponement of taking over tests by Contractor	22.3	weeks
H.	Guarantee Period for original Works and parts replaced or renewed	23.2 and 23.5	months
I.	Maximum indemnities for personal injury or damage	24.3	£ (pounds sterling)
J.	(1) Daily use of Plant	23.4	hours/day
	(2) Reduction of Guarantee Period for more intensive use	23.4	

*Insert name of bank.

Clause 7 (Erection on a Time Basis and Lump-Sum Erection) – Supervision of erection is carried out (on a time basis)* (for a lump sum).*

*Delete alternatives.

Erection will be carried out by the Purchaser, who shall, at his own expense, provide the skilled and unskilled labour, all equipment and everything necessary for the erection of the Plant.

References to 'erection' in Clause 7 shall be read as references to 'supervision of erection'. The Contractor's obligation in this respect will be to provide the services of one or more competent engineers

(a) to give to the Purchaser or his representative mentioned in Clause 13.1 the necessary instructions for the erection of the Plant by the Purchaser and, if provided in the accompanying Specification or other documents, for its commissioning by him; and

(b) to supervise the manner in which the Contractor's instructions have been carried out.

The number and qualifications of the Contractor's staff, and the estimated duration of erection, shall be as specified in this tender or in the accompanying Specification or other documents.

The date on which the Contractor's staff should arrive on Site shall be as provided in the accompanying Specification or other documents; if not so provided, the Purchaser shall give the Contractor not less than one month's notice requiring such arrival.*

*This clause is based on Clause 2 of UNECE Conditions 188B, by kind permission of the United Nations.

Clause 20 (Time for Completion) – As the Contract will only include supervision of erection, the Contractor cannot undertake that erection will be completed within a specific time. Therefore, for the purposes of this Clause, completion of delivery CIF shall be deemed to constitute completion of the Works.

Contract Price Adjustment

We propose that contract price adjustment in respect of the price for goods shall apply in accordance with the BEAMA *Contract Price Adjustment Clause and Formula for use with Export Contracts – Electrical Machinery: (for which there is no other specific formula)*, edition January 1979, corrected edition,* a copy of which is enclosed.

We further propose that contract price adjustment in respect of the price for site work shall apply in accordance with the same BEAMA Clause and Formula subject to the amendment that adjustment shall take place at the rate of 0.95 per cent of such price per 1.0 per cent difference between the Labour Cost Index published for the month in which the tender date falls and the average of the Index figures published for the last two-thirds of the period during which the Contractor is carrying out site work, this difference being expressed as a percentage of the former Index figure.

The BEAMA Labour Cost Index referred to shall be the Labour Cost Index for Electrical Engineering.*

References to the *Trade and Industry Journal* shall be read as references to *British Business*, which is the new name of the journal.†

*This is only one of several BEAMA CPA Clauses and Formulae. If another is adopted, this form may need consequential amendment. In particular, reference to the BEAMA CPA Clause applicable to Mechanical Plant, and to the Labour Cost Index for Mechanical Engineering, may be appropriate. See p. 16.

†The above form requires that the separate prices to be adjusted shall be identifiable from the contract documents.

Trade Terms

The trade term 'CIF' shall be defined by the current edition of Incoterms, where consistent with this tender. A copy of the definition is enclosed.

Validity

This tender is valid for acceptance until and including *, but shall, after that date, be null and void unless extended by us in writing.

*Date in full, e.g. 'Monday, 19 January 1981'.

Yours faithfully

(Signature)

FORM 48

Tender based on General Conditions for the Erection of Plant and Machinery Abroad No 188D, prepared by the Secretariat of the United Nations Economic Commission for Europe, Geneva, August 1963

Dear Sirs

(heading)

Prices

We offer to erect on Site at

the Plant and Machinery described in the accompanying Specification for (the approximate price of*):

£

(pounds sterling)

(Our services are offered on a time basis. The above approximate price is computed at the rates and for the estimated duration shown in the accompanying Specification. The sum actually payable for such services shall be computed on a time basis, and may be more or less than the above approximate price.)*

*Delete if lump sum.

Programme

Our best estimate is that completion of our work will be achieved within of receiving your order. (We enclose * showing the basis of this estimate.)†

*Appropriate document, e.g. bar chart or critical path network.

†Delete if inappropriate.

Terms of Payment

All prices are payable in pounds sterling in the United Kingdom.

We propose that payment shall be made as follows:

(a) %(per cent) in advance.

(b) The balance through a Confirmed Irrevocable Letter of Credit providing for payment in full upon presentation of documents, to be increased according to any contractual increases in the price, to be extended according to any increase in the period required for execution of our respective obligations under the Contract, and otherwise complying with the following description:

- (i) Confirmed by an acceptable bank in London.
- (ii) Irrevocably issued, confirmed and advised to us by .[a]
- (iii) Date and place of expiry – [a] in London.
- (iv) Beneficiary – ourselves.
- (v) Amount – the balance of the price.

(vi) Credit available with confirming bank by payment.
(vii) Documents to relate to .[b]
(viii) Documents required:

(Certified)* Commercial Invoice

Your or your Engineer's Certificate.†

*Payment would, in fact, be made out of the letter of credit even if certified only by the exporter. Where no further definition is given 'banks will accept such documents as tendered'. See *Uniform Customs and Practice for Documentary Credits*, 1974, Article 33, ICC Publication No. 290.
†If this sentence is omitted, payment should be made out of the letter of credit on presentation only of commercial invoice.

(ix) Subject otherwise to the *Uniform Customs and Practice for Documentary Credits*, 1974, ICC Publication No 290.[c]

[a]Date in full, e.g. 'Monday, 19 January 1981'.
[b]Short description of site work.
[c]This description follows as closely as possible the *Standard Procedure for the Issuing of Documentary Credits by Cable, Telegram or Telex* (App VII *Standard Forms for Issuing Documentary Credits*, ICC Publication No 323).

Payment in respect of site work shall be due monthly at the end of each month.

Conditions of Contract

Where consistent with this tender, we propose that *General Conditions for the Erection of Plant and Machinery Abroad No 188D*, prepared by the Secretariat of the United Nations Economic Commission for Europe, Geneva, August 1963, a copy of which is enclosed, shall apply, subject to the following:

The Appendix shall be completed as follows:

Appendix to UNECE Conditions 188D

	Clause	
A. Percentage of reduction for each week's delay	14.3	% (per cent)
B. Maximum amount of above reduction	14.3	% (per cent)
C. Maximum amount recoverable for non-completion	14.5	£ (pounds sterling)
D. Rate of interest on overdue payments	15.6	2% (two per cent) per annum over the * Bank base lending rate from time to time in force
E. Period of delay in payment authorizing termination by erector	15.6	months

F. Maximum amount recoverable on termination by erector for failure to make payment	15.6	£ (	 pounds sterling)
G. Maximum indemnities payable by erector for repair or replacement of defective parts	18.1	£ (	 pounds sterling)
H. Guarantee Period for erection	18.2		months
I. Maximum indemnities for personal injury or damage	19.3	£ (	 pounds sterling)

*Insert name of bank.

Clause 6 (Erection on a Time Basis and Lump-Sum Erection) – Erection is carried out (on a time basis)* (for a lump sum).*

*Delete alternatives.

Contract Price Adjustment

We propose that contract price adjustment shall apply in accordance with the BEAMA *Contract Price Adjustment Clause and Formula for use with Export Contracts – Electrical Machinery: (for which there is no other specific formula)*, edition January 1979, corrected edition,* a copy of which is enclosed, subject to the amendment that adjustment shall take place at the rate of 0.95 per cent of the price per 1.0 per cent difference between the Labour Cost Index published for the month in which the tender date falls and the average of the Index figures published for the last two-thirds of the period during which the Contractor is carrying out site work, this difference being expressed as a percentage of the former Index figure.

The BEAMA Labour Cost Index referred to shall be the Labour Cost Index for Electrical Engineering.*

References to the *Trade and Industry Journal* shall be read as references to *British Business*, which is the new name of the journal.

*This is only one of several BEAMA CPA Clauses and Formulae. If another is adopted, this form may need consequential amendment. In particular, reference to the BEAMA CPA Clause applicable to Mechanical Plant, and to the Labour Cost Index for Mechanical Engineering, may be appropriate. See p. 16.

Validity

This tender is valid for acceptance until and including *, but shall, after that date, be null and void unless extended by us in writing.

*Date in full, e.g. 'Monday, 19 January 1981'.

Yours faithfully

(Signature)

FORM 49

FOB Tender based on General Conditions No 574 for the Supply of Plant and Machinery for Export, prepared under the auspices of the United Nations Economic Commission for Europe, Geneva, December 1955

Dear Sirs

(heading)

Prices

We offer to supply FOB * the items described in the accompanying Specification for:

£

(pounds sterling)

*Port of shipment.

Programme

Our best estimate is that delivery FOB will be achieved within of receiving your order.

Terms of Payment

All prices are payable in pounds sterling in the United Kingdom.

We propose that payment shall be made as follows:

(a) %(per cent) in advance.

(b) The balance through a Confirmed Irrevocable Letter of Credit providing for payment in full upon presentation of documents, to be increased according to any contractual increases in the price, to be extended according to any increase in the period required for execution of our respective obligations under the Contract, and otherwise complying with the following description:

(i) Confirmed by an acceptable bank in London.
(ii) Irrevocably issued, confirmed and advised to us by .[a]
(iii) Date and place of expiry – [a] in London.
(iv) Beneficiary – ourselves.
(v) Amount – the balance of the price.
(vi) Credit available with confirming bank by payment.
(vii) Partial shipments allowed.
(viii) Transhipment allowed.
(ix) Shipment from .[b]
(x) For transportation to .[c]
(xi) Shipment of .[d]
(xii) Documents required:

Commercial Invoice

Full set of clean on board marine Bills of Lading (which shall be

acceptable although to the effect that the goods have been shipped on deck or unprotected*).

*Delete if inappropriate.

(xiii) Period after issuance of shipping documents for presentation – days.

(xiv) Subject otherwise to the *Uniform Customs and Practice for Documentary Credits*, 1974, ICC Publication No 290.[e]

[a] Date in full, e.g. 'Monday, 19 January 1981'.
[b] Port of shipment, or e.g. 'any United Kingdom port'.
[c] Port of destination.
[d] Short description of goods.
[e] This description follows as closely as possible the *Standard Procedure for the Issuing of Documentary Credits by Cable, Telegram or Telex* (App VII *Standard Forms for Issuing Documentary Credits*, ICC Publication No 323).

Conditions of Contract

Where consistent with this tender, we propose that *General Conditions No 574 for the Supply of Plant and Machinery for Export*, prepared under the auspices of the United Nations Economic Commission for Europe, Geneva, December 1955, a copy of which is enclosed, shall apply, subject to the following:

The Appendix shall be completed as follows:

Appendix to UNECE Conditions 574

	Clause	
A. Percentage to be deducted for each week's delay	7.3	%(per cent)
B. Maximum percentage which the deductions above may not exceed	7.3	%(per cent)
C. Maximum amount recoverable for non-delivery	7.5	£ (pounds sterling)
D. Maximum amount recoverable on termination by Vendor for failure to take delivery or make payment	7.7 and 8.7	£ (pounds sterling)
E. Rate of interest on overdue payments	8.7	2% (two per cent) per annum over the * Bank base lending rate from time to time in force
F. Period of delay in payment authorizing termination by Vendor	8.7	months
G. Guarantee Period for original Plant and parts replaced or renewed	9.2 and 9.7	months

H. Maximum extension of Guarantee Period 9.5 months

*Insert name of bank.

Contract Price Adjustment

We propose that contract price adjustment shall apply in accordance with the BEAMA *Contract Price Adjustment Clause and Formula for use with Export Contracts – Electrical Machinery: (for which there is no other specific formula)*, edition January 1979, corrected edition,* a copy of which is enclosed.

The BEAMA Labour Cost Index referred to shall be the Labour Cost Index for Electrical Engineering.*

References to the *Trade and Industry Journal* shall be read as references to *British Business*, which is the new name of the journal.

*This is only one of several BEAMA CPA Clauses and Formulae. If another is adopted, this form may need consequential amendment. In particular, reference to the BEAMA CPA Clause applicable to Mechanical Plant, and to the Labour Cost Index for Mechanical Engineering, may be appropriate. See p. 16.

Trade Terms

The trade term 'FOB' shall be defined by the current edition of Incoterms, where consistent with this tender. A copy of the definition is enclosed.

Validity

This tender is valid for acceptance until and including *, but shall, after that date, be null and void unless extended by us in writing.

*Date in full, e.g. 'Monday, 19 January 1981'.

Yours faithfully

(Signature)

FORM 50

C & F Tender based on General Conditions No 574 for the Supply of Plant and Machinery for Export, prepared under the auspices of the United Nations Economic Commission for Europe, Geneva, December 1955

Dear Sirs

(heading)

Prices
We offer to supply C & F * the items described in the accompanying Specification for:
£
(pounds sterling)
*Port of destination.

Programme
Our best estimate is that delivery C & F will be achieved within of receiving your order.

Terms of Payment
All prices are payable in pounds sterling in the United Kingdom.

We propose that payment shall be made as follows:

(a) %(per cent) in advance.

(b) The balance through a Confirmed Irrevocable Letter of Credit providing for payment in full upon presentation of documents, to be increased according to any contractual increases in the price, to be extended according to any increase in the period required for execution of our respective obligations under the Contract, and otherwise complying with the following description:

(i) Confirmed by an acceptable bank in London.
(ii) Irrevocably issued, confirmed and advised to us by .[a]
(iii) Date and place of expiry – [a] in London.
(iv) Beneficiary – ourselves.
(v) Amount – the balance of the price.
(vi) Credit available with confirming bank by payment.
(vii) Partial shipments allowed.
(viii) Transhipment allowed.
(ix) Shipment from .[b]
(x) For transportation to .[c]
(xi) Shipment of .[d]
(xii) Documents required:

Commercial Invoice

Full set of clean on board marine Bills of Lading (which shall be

acceptable although to the effect that the goods have been shipped on deck or unprotected*).

*Delete if inappropriate.

(xiii) Period after issuance of shipping documents for presentation – days.

(xiv) Subject otherwise to the *Uniform Customs and Practice for Documentary Credits*, 1974, ICC Publication No 290.[e]

[a]Date in full, e.g. 'Monday, 19 January 1981'.

[b]Port of shipment, or e.g. 'any United Kingdom port'.

[c]Port of destination.

[d]Short description of goods.

[e]This description follows as closely as possible the *Standard Procedure for the Issuing of Documentary Credits by Cable, Telegram or Telex* (App VII *Standard Forms for Issuing Documentary Credits*, ICC Publication No 323).

Conditions of Contract

Where consistent with this tender, we propose that *General Conditions No 574 for the Supply of Plant and Machinery for Export*, prepared under the auspices of the United Nations Economic Commission for Europe, Geneva, December 1955, a copy of which is enclosed, shall apply, subject to the following:

The Appendix shall be completed as follows:

Appendix to UNECE Conditions 574

	Clause	
A. Percentage to be deducted for each week's delay	7.3	%(per cent)
B. Maximum percentage which the deductions above may not exceed	7.3	%(per cent)
C. Maximum amount recoverable for non-delivery	7.5	£ (pounds sterling)
D. Maximum amount recoverable on termination by Vendor for failure to take delivery or make payment	7.7 and 8.7	£ (pounds sterling)
E. Rate of interest on overdue payments	8.7	2% (two per cent) per annum over the * Bank base lending rate from time to time in force
F. Period of delay in payment authorizing termination by Vendor	8.7	months
G. Guarantee Period for original Plant and parts replaced or renewed	9.2 and 9.7	months

H. Maximum extension
of Guarantee Period 9.5 months

*Insert name of bank.

Contract Price Adjustment

We propose that contract price adjustment shall apply in accordance with the BEAMA *Contract Price Adjustment Clause and Formula for use with Export Contracts – Electrical Machinery: (for which there is no other specific formula)*, edition January 1979, corrected edition,* a copy of which is enclosed.

The BEAMA Labour Cost Index referred to shall be the Labour Cost Index for Electrical Engineering.*

References to the *Trade and Industry Journal* shall be read as references to *British Business*, which is the new name of the journal.

*This is only one of several BEAMA CPA Clauses and Formulae. If another is adopted, this form may need consequential amendment. In particular, reference to the BEAMA CPA Clause applicable to Mechanical Plant, and to the Labour Cost Index for Mechanical Engineering, may be appropriate. See p. 16.

Trade Terms

The trade term 'C & F' shall be defined by the current edition of Incoterms, where consistent with this tender. A copy of the definition is enclosed.

Validity

This tender is valid for acceptance until and including *, but shall, after that date, be null and void unless extended by us in writing.

*Date in full, e.g. 'Monday, 19 January 1981'.

Yours faithfully

(Signature)

FORM 51

CIF Tender based on General Conditions No 574 for the Supply of Plant and Machinery for Export, prepared under the auspices of the United Nations Economic Commission for Europe, Geneva, December 1955

Dear Sirs

(heading)

Prices

We offer to supply CIF * the items described in the accompanying Specification for:

£

(pounds sterling)

*Port of destination.

Programme

Our best estimate is that delivery CIF will be achieved within of receiving your order.

Terms of Payment

All prices are payable in pounds sterling in the United Kingdom.

We propose that payment shall be made as follows:

(a) %(per cent) in advance.

(b) The balance through a Confirmed Irrevocable Letter of Credit providing for payment in full upon presentation of documents, to be increased according to any contractual increases in the price, to be extended according to any increase in the period required for execution of our respective obligations under the Contract, and otherwise complying with the following description:

(i) Confirmed by an acceptable bank in London.
(ii) Irrevocably issued, confirmed and advised to us by .[a]
(iii) Date and place of expiry – [a]in London.
(iv) Beneficiary – ourselves.
(v) Amount – the balance of the price.
(vi) Credit available with confirming bank by payment.
(vii) Partial shipments allowed.
(viii) Transhipment allowed.
(ix) Shipment from .[b]
(x) For transportation to .[c]
(xi) Shipment of .[d]
(xii) Documents required:

Commercial Invoice

Full set of clean on board marine Bills of Lading (which shall be

acceptable although to the effect that the goods have been shipped on deck or unprotected*).

Insurance Certificate

*Delete if inappropriate.

(xiii) Period after issuance of shipping documents for presentation – days.

(xiv) Subject otherwise to the *Uniform Customs and Practice for Documentary Credits*, 1974, ICC Publication No 290.[e]

[a] Date in full, e.g. 'Monday, 19 January 1981'.

[b] Port of shipment, or e.g. 'any United Kingdom port'.

[c] Port of destination.

[d] Short description of goods.

[e] This description follows as closely as possible the *Standard Procedure for the Issuing of Documentary Credits by Cable, Telegram or Telex* (App VII *Standard Forms for Issuing Documentary Credits*, ICC Publication No 323).

Conditions of Contract

Where consistent with this tender, we propose that *General Conditions No 574 for the Supply of Plant and Machinery for Export*, prepared under the auspices of the United Nations Economic Commission for Europe, Geneva, December 1955, a copy of which is enclosed, shall apply, subject to the following:

The Appendix shall be completed as follows:

Appendix to UNECE Conditions 574

	Clause	
A. Percentage to be deducted for each week's delay	7.3	% (per cent)
B. Maximum percentage which the deductions above may not exceed	7.3	% (per cent)
C. Maximum amount recoverable for non-delivery	7.5	£ (pounds sterling)
D. Maximum amount recoverable on termination by Vendor for failure to take delivery or make payment	7.7 and 8.7	£ (pounds sterling)
E. Rate of interest on overdue payments	8.7	2% (two per cent) per annum over the * Bank base lending rate from time to time in force
F. Period of delay in payment authorizing termination by Vendor	8.7	months

G. Guarantee Period for original Plant and parts replaced or renewed	9.2 and 9.7	months
H. Maximum extension of Guarantee Period	9.5	months

*Insert name of bank.

Contract Price Adjustment

We propose that contract price adjustment shall apply in accordance with the BEAMA *Contract Price Adjustment Clause and Formula for use with Export Contracts – Electrical Machinery: (for which there is no other specific formula)*, edition January 1979, corrected edition,* a copy of which is enclosed.

The BEAMA Labour Cost Index referred to shall be the Labour Cost Index for Electrical Engineering.*

References to the *Trade and Industry Journal* shall be read as references to *British Business*, which is the new name of the journal.

*This is only one of several BEAMA CPA Clauses and Formulae. If another is adopted, this form may need consequential amendment. In particular, reference to the BEAMA CPA Clause applicable to Mechanical Plant, and to the Labour Cost Index for Mechanical Engineering, may be appropriate. See p. 16.

Trade Terms

The trade term 'CIF' shall be defined by the current edition of Incoterms, where consistent with this tender. A copy of the definition is enclosed.

Validity

This tender is valid for acceptance until and including   *, but shall, after that date, be null and void unless extended by us in writing.

*Date in full, e.g. 'Monday, 19 January 1981'.

Yours faithfully

(Signature)

FORM 52

FOB Tender based on General Conditions No 574 for the Supply of Plant and Machinery for Export, prepared under the auspices of the United Nations Economic Commission for Europe, Geneva, December 1955, and Additional Clauses for Supervision of Erection of Plant and Machinery Abroad No 574B, prepared under the auspices of the United Nations Economic Commission for Europe, Geneva, April 1964

Dear Sirs

(heading)

Prices

We offer to supply FOB * and supervise the erection on Site at

of the items described in the accompanying Specification for:

£

(pounds sterling)

*Port of shipment.

(Our services in respect of site work are offered on a time basis. The above price includes an estimated sum of £ for such services, computed at the rates and for the estimated duration shown in the accompanying Specification. The sum actually payable for such services shall be computed on a time basis, and may be either more or less than the estimated sum.)*

*Delete if lump sum.

Programme

Our best estimate is that delivery FOB will be achieved within of receiving your order, and that we could complete supervision of erection within a further of receiving your order, making a total of .

(We enclose * showing the basis of this estimate.)†

*Appropriate document, e.g. bar chart or critical path network.

†Delete if inappropriate.

Terms of Payment

All prices are payable in pounds sterling in the United Kingdom.

We propose that payment shall be made as follows:

(a) %(per cent) in advance.

(b) The balance through a Confirmed Irrevocable Letter of Credit providing for payment in full upon presentation of documents, to be increased according to any contractual increases in the price, to be extended according to any increase in the period required for execution of our respective obligations under the Contract, and otherwise complying with the following description:

(i) Confirmed by an acceptable bank in London.
(ii) Irrevocably issued, confirmed and advised to us by .[a]
(iii) Date and place of expiry – [a] in London.
(iv) Beneficiary – ourselves.
(v) Amount – the balance of the price.
(vi) Credit available with confirming bank by payment.
(vii) Partial shipments allowed.
(viii) Transhipment allowed.
(ix) Shipment from .[b]
(x) For transportation to .[c]
(xi) Shipment of .[d]
(xii) Documents required:

Commercial Invoice

In respect of goods, full set of clean on board marine Bills of Lading (which shall be acceptable although to the effect that the goods have been shipped on deck or unprotected*).

In respect of other work, your or your Engineer's Certificate.†

*Delete if inappropriate.

†If this sentence is omitted, payment for such other work should be made out of the letter of credit on presentation only of commercial invoice.

(xiii) Period after issuance of shipping documents for presentation – days.
(xiv) Subject otherwise to the *Uniform Customs and Practice for Documentary Credits*, 1974, ICC Publication No 290.[e]

[a]Date in full, e.g. 'Monday, 19 January 1981'.

[b]Port of shipment, or e.g. 'any United Kingdom port'.

[c]Port of destination.

[d]Short description of goods and site work.

[e]This description follows as closely as possible the *Standard Procedure for the Issuing of Documentary Credits by Cable, Telegram or Telex* (App VII *Standard Forms for Issuing Documentary Credits*, ICC Publication No 323).

Payment in respect of site work shall be due monthly at the end of each month.

Conditions of Contract

Where consistent with this tender, we propose that *General Conditions No 574 for the Supply of Plant and Machinery for Import and Export*, prepared under the auspices of the United Nations Economic Commission for Europe, Geneva, December 1955, and *Additional Clauses for Supervision of Erection of Plant and Machinery Abroad No 574B*, prepared under the auspices of the United Nations Economic Commission for Europe, Geneva, April 1964, a copy of which is enclosed, shall apply, subject to the following:

The Appendix shall be completed as follows:

Appendix to UNECE Conditions 574

	Clause		
A. Percentage to be deducted for each week's delay	7.3	% (	per cent)

B.	Maximum percentage which the deductions above may not exceed	7.3	% (per cent)
C.	Maximum amount recoverable for non-delivery	7.5	£ (pounds sterling)
D.	Maximum amount recoverable on termination by Vendor for failure to take delivery or make payment	7.7 and 8.7	£ (pounds sterling)
E.	Rate of interest on overdue payments	8.7	2% (two per cent) per annum over the * Bank base lending rate from time to time in force
F.	Period of delay in payment authorizing termination by Vendor	8.7	months
G.	Guarantee Period for original Plant and parts replaced or renewed	9.2 and 9.7	months
H.	Maximum extension of Guarantee Period	9.5	months

*Insert name of bank.

Appendix to UNECE Conditions 574B

J.	Duration of interruption in erection at the expiry of which the Vendor is authorized to recall his supervising engineers	11.1	months
K.	Maximum indemnities payable by the parties	12.4	£ (pounds sterling)

Clause 4 (Charges Payable by the Purchaser) of UNECE Conditions 574B – Supervision of erection being carried out for a lump sum, the quoted price includes all the items mentioned in Clause 4. Provided that if supervision of erection is prolonged for any cause for which the Purchaser or any of his contractors other than the Vendor is responsible and if, as a result, the work of the Vendor's employees is suspended or added to, a charge will be made for any idle time, any extra work, any extra living expenses of the Vendor's employees and the cost of any extra journey.*

*Delete if erection is on a time basis. This clause is based on Clause 7.2 of UNECE Conditions 188A, by kind permission of the United Nations.

Contract Price Adjustment

We propose that contract price adjustment shall apply in accordance with the

BEAMA *Contract Price Adjustment Clause and Formula for use with Export Contracts – Electrical Machinery: (for which there is no other specific formula)*, edition January 1979, corrected edition,* a copy of which is enclosed.

We further propose that contract price adjustment in respect of the price for site work shall apply in accordance with the same BEAMA Clause and Formula subject to the amendment that adjustment shall take place at the rate of 0.95 per cent of such price per 1.0 per cent difference between the Labour Cost Index published for the month in which the tender date falls and the average of the Index figures published for the last two-thirds of the period during which the Contractor is carrying out site work, this difference being expressed as a percentage of the former Index figure.

The BEAMA Labour Cost Index referred to shall be the Labour Cost Index for Electrical Engineering.*

References to the *Trade and Industry Journal* shall be read as references to *British Business*, which is the new name of the journal.†

*This is only one of several BEAMA CPA Clauses and Formulae. If another is adopted, this form may need consequential amendment. In particular, reference to the BEAMA CPA Clause applicable to Mechanical Plant, and to the Labour Cost Index for Mechanical Engineering, may be appropriate. See p. 16.

†The above form requires that the separate prices to be adjusted shall be identifiable from the contract documents.

Trade Terms

The Trade Term 'FOB' shall be defined by the current edition of Incoterms, where consistent with this tender. A copy of the definition is enclosed.

Validity

This tender is valid for acceptance until and including *, but shall, after that date, be null and void unless extended by us in writing.

*Date in full, e.g. 'Monday, 19 January 1981'.

Yours faithfully

(Signature)

FORM 53

C & F Tender based on General Conditions No 574 for the Supply of Plant and Machinery for Export, prepared under the auspices of the United Nations Economic Commission for Europe, Geneva, December 1955, and Additional Clauses for Supervision of Erection of Plant and Machinery Abroad No 574B, prepared under the auspices of the United Nations Economic Commission for Europe, Geneva, April 1964

Dear Sirs

(heading)

Prices

We offer to supply C & F *, and supervise the erection on Site at

of the items described in the accompanying Specification for:

£

(pounds sterling)

*Port of destination.

(Our services in respect of site work are offered on a time basis. The above price includes an estimated sum of £ for such services, computed at the rates and for the estimated duration shown in the accompanying Specification. The sum actually payable for such services shall be computed on a time basis, and may be either more or less than the estimated sum.)*

*Delete if lump sum.

Programme

Our best estimate is that delivery C & F will be achieved within of receiving your order, and that we could complete supervision of erection within a further of receiving your order, making a total of .

(We enclose * showing the basis of this estimate.)†

*Appropriate document, e.g. bar chart or critical path network.

†Delete if inappropriate.

Terms of Payment

All prices are payable in pounds sterling in the United Kingdom.

We propose that payment shall be made as follows:

(a) % (per cent) in advance.

(b) The balance through a Confirmed Irrevocable Letter of Credit providing for payment in full upon presentation of documents, to be increased according to any contractual increases in the price, to be extended according to any increase in the period required for execution of our respective obligations under the Contract, and otherwise complying with the following description:

(i) Confirmed by an acceptable bank in London.
(ii) Irrevocably issued, confirmed and advised to us by .[a]
(iii) Date and place of expiry – [a] in London.
(iv) Beneficiary – ourselves.
(v) Amount – the balance of the price.
(vi) Credit available with confirming bank by payment.
(vii) Partial shipments allowed.
(viii) Transhipment allowed.
(ix) Shipment from .[b]
(x) For transportation to .[c]
(xi) Shipment of .[d]
(xii) Documents required:

Commercial Invoice

In respect of goods, full set of clean on board marine Bills of Lading (which shall be acceptable although to the effect that the goods have been shipped on deck or unprotected*).

In respect of other work, your or your Engineer's Certificate.†

*Delete if inappropriate.

†If this sentence is omitted, payment for such other work should be made out of the letter of credit on presentation only of commercial invoice.

(xiii) Period after issuance of shipping documents for presentation – days.
(xiv) Subject otherwise to the *Uniform Customs and Practice for Documentary Credits,* 1974, ICC Publication No 290.[e]

[a]Date in full, e.g. 'Monday, 19 January 1981'.
[b]Port of shipment, or e.g. 'any United Kingdom port'.
[c]Port of destination.
[d]Short description of goods and site work.
[e]This description follows as closely as possible the *Standard Procedure for the Issuing of Documentary Credits by Cable, Telegram or Telex* (App VII *Standard Forms for Issuing Documentary Credits,* ICC Publication No 323).

Payment in respect of site work shall be due monthly at the end of each month.

Conditions of Contract

Where consistent with this tender, we propose that *General Conditions No 574 for the Supply of Plant and Machinery for Import and Export,* prepared under the auspices of the United Nations Economic Commission for Europe, Geneva, December 1955, and *Additional Clauses for Supervision of Erection of Plant and Machinery Abroad No 574B,* prepared under the auspices of the United Nations Economic Commission for Europe, Geneva, April 1964, a copy of which is enclosed, shall apply, subject to the following:

The Appendix shall be completed as follows:

Appendix to UNECE Conditions 574

	Clause		
A. Percentage to be deducted for each week's delay	7.3	% (	per cent)

B. Maximum percentage which the deductions above may not exceed	7.3	% (per cent)
C. Maximum amount recoverable for non-delivery	7.5	£ (pounds sterling)
D. Maximum amount recoverable on termination by Vendor for failure to take delivery or make payment	7.7 and 8.7	£ (pounds sterling)
E. Rate of interest on overdue payments	8.7	2% (two per cent) per annum over the * Bank base lending rate from time to time in force
F. Period of delay in payment authorizing termination by Vendor	8.7	months
G. Guarantee Period for original Plant and parts replaced or renewed	9.2 and 9.7	months
H. Maximum extension of Guarantee Period	9.5	months

*Insert name of bank.

Appendix to UNECE Conditions 574B

J. Duration of interruption in erection at the expiry of which the Vendor is authorized to recall his supervising engineers	11.1	months
K. Maximum indemnities payable by the parties	12.4	£ (pounds sterling)

Clause 4 (Charges Payable by the Purchaser) of UNECE Conditions 574B – Supervision of erection being carried out for a lump sum, the quoted price includes all the items mentioned in Clause 4. Provided that if supervision of erection is prolonged for any cause for which the Purchaser or any of his contractors other than the Vendor is responsible and if, as a result, the work of the Vendor's employees is suspended or added to, a charge will be made for any idle time, any extra work, any extra living expenses of the Vendor's employees and the cost of any extra journey.*

*Delete if erection is on a time basis. This clause is based on Clause 7.2 of UNECE Conditions 188A, by kind permission of the United Nations.

Contract Price Adjustment

We propose that contract price adjustment shall apply in accordance with the BEAMA *Contract Price Adjustment Clause and Formula for use with Export Contracts – Electrical Machinery: (for which there is no other specific formula)*, edition January 1979, corrected edition,* a copy of which is enclosed.

We further propose that contract price adjustment in respect of the price for site work shall apply in accordance with the same BEAMA Clause and Formula subject to the amendment that adjustment shall take place at the rate of 0.95 per cent of such price per 1.0 per cent difference between the Labour Cost Index published for the month in which the tender date falls and the average of the Index figures published for the last two-thirds of the period during which the Contractor is carrying out site work, this difference being expressed as a percentage of the former Index figure.

The BEAMA Labour Cost Index referred to shall be the Labour Cost Index for Electrical Engineering.*

References to the *Trade and Industry Journal* shall be read as references to *British Business*, which is the new name of the journal.†

*This is only one of several BEAMA CPA Clauses and Formulae. If another is adopted, this form may need consequential amendment. In particular, reference to the BEAMA CPA Clause applicable to Mechanical Plant, and to the Labour Cost Index for Mechanical Engineering, may be appropriate. See p. 16.

†The above form requires that the separate prices to be adjusted shall be identifiable from the contract documents.

Trade Terms

The Trade Term 'C & F' shall be defined by the current edition of Incoterms, where consistent with this tender. A copy of the definition is enclosed.

Validity

This tender is valid for acceptance until and including *, but shall, after that date, be null and void unless extended by us in writing.

*Date in full, e.g. 'Monday, 19 January 1981'.

Yours faithfully

(Signature)

FORM 54

CIF Tender based on General Conditions No 574 for the Supply of Plant and Machinery for Export, prepared under the auspices of the United Nations Economic Commission for Europe, Geneva, December 1955, and Additional Clauses for Supervision of Erection of Plant and Machinery Abroad No 574B, prepared under the auspices of the United Nations Economic Commission for Europe, Geneva, April 1964

Dear Sirs

(heading)

Prices

We offer to supply CIF *, and supervise the erection on Site at

of the items described in the accompanying Specification for:

£

(pounds sterling)

*Port of destination.

(Our services in respect of site work are offered on a time basis. The above price includes an estimated sum of £ for such services, computed at the rates and for the estimated duration shown in the accompanying Specification. The sum actually payable for such services shall be computed on a time basis, and may be either more or less than the estimated sum.)*

*Delete if lump sum.

Programme

Our best estimate is that delivery CIF will be achieved within of receiving your order, and that we could complete supervision of erection within a further of receiving your order, making a total of .

(We enclose * showing the basis of this estimate.)†

*Appropriate document, e.g. bar chart or critical path network.

†Delete if inappropriate.

Terms of Payment

All prices are payable in pounds sterling in the United Kingdom.

We propose that payment shall be made as follows:

(a) %(per cent) in advance.

(b) The balance through a Confirmed Irrevocable Letter of Credit providing for payment in full upon presentation of documents, to be increased according to any contractual increases in the price, to be extended according to any increase in the period required for execution of our respective obligations under the Contract, and otherwise complying with the following description:

(i) Confirmed by an acceptable bank in London.
(ii) Irrevocably issued, confirmed and advised to us by .[a]
(iii) Date and place of expiry – [a] in London.
(iv) Beneficiary – ourselves.
(v) Amount – the balance of the price.
(vi) Credit available with confirming bank by payment.
(vii) Partial shipments allowed.
(viii) Transhipment allowed.
(ix) Shipment from .[b]
(x) For transportation to .[c]
(xi) Shipment of .[d]
(xii) Documents required:

Commercial Invoice

In respect of goods, full set of clean on board marine Bills of Lading (which shall be acceptable although to the effect that the goods have been shipped on deck or unprotected*) and Insurance Certificate.

In respect of other work, your or your Engineer's Certificate.†

*Delete if inappropriate.

†If this sentence is omitted, payment for such other work should be made out of the letter of credit on presentation only of commercial invoice.

(xiii) Period after issuance of shipping documents for presentation – days.
(xiv) Subject otherwise to the *Uniform Customs and Practice for Documentary Credits*, 1974, ICC Publication No 290.[e]

[a]Date in full, e.g. 'Monday, 19 January 1981'.

[b]Port of shipment, or e.g. 'any United Kingdom port'.

[c]Port of destination.

[d]Short description of goods.

[e]This description follows as closely as possible the *Standard Procedure for the Issuing of Documentary Credits by Cable, Telegram or Telex* (App VII *Standard Forms for Issuing Documentary Credits*, ICC Publication No 323).

Payment in respect of site work shall be due monthly at the end of each month.

Conditions of Contract

Where consistent with this tender, we propose that *General Conditions No 574 for the Supply of Plant and Machinery for Export*, prepared under the auspices of the United Nations Economic Commission for Europe, Geneva, December 1955, and *Additional Clauses for Supervision of Erection of Plant and Machinery Abroad No 574B*, prepared under the auspices of the United Nations Economic Commission for Europe, Geneva, April 1964, a copy of which is enclosed, shall apply, subject to the following:

The Appendix in each case shall be completed as follows:

Appendix to UNECE Conditions 574

	Clause		
A. Percentage to be deducted for each week's delay	7.3	%(	per cent)

B.	Maximum percentage which the deductions above may not exceed	7.3	%(per cent)
C.	Maximum amount recoverable for non-delivery	7.5	£ (pounds sterling)
D.	Maximum amount recoverable on termination by Vendor for failure to take delivery or make payment	7.7 and 8.7	£ (pounds sterling)
E.	Rate of interest on overdue payments	8.7	2% (two per cent) per annum over the * Bank base lending rate from time to time in force
F.	Period of delay in payment authorizing termination by Vendor	8.7	months
G.	Guarantee Period for original Plant and parts replaced or renewed	9.2 and 9.7	months
H.	Maximum extension of Guarantee Period	9.5	months

*Insert name of bank.

Appendix to UNECE Conditions 574B

J.	Duration of interruption in erection at the expiry of which the Vendor is authorized to recall his supervising engineers	11.1	months
K.	Maximum indemnities payable by the parties	12.4	£ (pounds sterling)

Clause 4 (Charges Payable by the Purchaser) of UNECE Conditions 574B – Supervision of erection being carried out for a lump sum, the quoted price includes all the items mentioned in Clause 4. Provided that if supervision of erection is prolonged for any cause for which the Purchaser or any of his contractors other than the Vendor is responsible and if, as a result, the work of the Vendor's employees is suspended or added to, a charge will be made for any idle time, any extra work, any extra living expenses of the Vendor's employees and the cost of any extra journey.*

*Delete if erection is on a time basis. This clause is based on Clause 7.2 of UNECE Conditions 188A, by kind permission of the United Nations.

Contract Price Adjustment

We propose that contract price adjustment in respect of the price for goods shall apply in accordance with the BEAMA *Contract Price Adjustment Clause and Formula for use with Export Contracts – Electrical Machinery: (for which there is no other specific formula)*, edition January 1979, corrected edition,* a copy of which is enclosed.

We further propose that contract price adjustment in respect of the price for site work shall apply in accordance with the same BEAMA Clause and Formula subject to the amendment that adjustment shall take place at the rate of 0.95 per cent of such price per 1.0 per cent difference between the Labour Cost Index published for the month in which the tender date falls and the average of the Index figures published for the last two-thirds of the period during which the Contractor is carrying out site work, this difference being expressed as a percentage of the former Index figure.

The BEAMA Labour Cost Index referred to shall be the Labour Cost Index for Electrical Engineering.*

References to the *Trade and Industry Journal* shall be read as references to *British Business*, which is the new name of the journal.†

*This is only one of several BEAMA CPA Clauses and Formulae. If another is adopted, this form may need consequential amendment. In particular, reference to the BEAMA CPA Clause applicable to Mechanical Plant, and to the Labour Cost Index for Mechanical Engineering, may be appropriate. See p. 16.

†The above form requires that the separate prices to be adjusted shall be identifiable from the contract documents.

Trade Terms

The trade term 'CIF' shall be defined by the current edition of Incoterms, where consistent with this tender. A copy of the definition is enclosed.

Validity

This tender is valid for acceptance until and including   *, but shall, after that date, be null and void unless extended by us in writing.

*Date in full, e.g. 'Monday, 19 January 1981'.

Yours faithfully

(Signature)

FORM 55

Tender based on General Conditions No 574A for the Supply and Erection of Plant and Machinery for Import and Export, prepared under the auspices of the United Nations Economic Commission for Europe, Geneva, March 1957, providing for Delivery to and Erection on Site

Dear Sirs

(heading)

Prices

We offer to supply, deliver to and erect on Site at

the items described in the accompanying Specification for:
£
(pounds sterling)
(Our services in respect of site work are offered on a time basis. The above price includes an estimated sum of £ for such services, computed at the rates and for the estimated duration shown in the accompanying Specification. The sum actually payable for such services shall be computed on a time basis, and may be either more or less than the estimated sum.)*

*Delete if lump sum.

Programme

Our best estimate is that completion of our work will be achieved within of receiving your order. (We enclose * showing the basis of this estimate.)†

*Appropriate document, e.g. bar chart or critical path network.
†Delete if inappropriate.

Terms of Payment

All prices are payable in pounds sterling in the United Kingdom.

We propose that payment shall be made as follows:

(a) % (per cent) in advance.

(b) The balance through a Confirmed Irrevocable Letter of Credit providing for payment in full upon presentation of documents, to be increased according to any contractual increases in the price, to be extended according to any increase in the period required for execution of our respective obligations under the Contract, and otherwise complying with the following description:

(i) Confirmed by an acceptable bank in London.
(ii) Irrevocably issued, confirmed and advised to us by .[a]
(iii) Date and place of expiry – [a] in London.
(iv) Beneficiary – ourselves.
(v) Amount – the balance of the price.

(vi) Credit available with confirming bank by payment.
(vii) Partial shipments allowed.
(viii) Transhipment allowed.
(ix) Shipment from .[b]
(x) For transportation to .[c]
(xi) Shipment of .[d]
(xii) Documents required:

Commercial Invoice

In respect of goods, full set of clean on board marine Bills of Lading (which shall be acceptable although to the effect that the goods have been shipped on deck or unprotected*) and Insurance Certificate.

In respect of other work, your or your Engineer's Certificate.†

*Delete if inappropriate.

†If this sentence is omitted, payment for such other work should be made out of the letter of credit on presentation only of commercial invoice.

(xiii) Period after issuance of shipping documents for presentation – days.
(xiv) Subject otherwise to the *Uniform Customs and Practice for Documentary Credits*, 1974, ICC Publication No 290.[e]

[a] Date in full, e.g. 'Monday, 19 January 1981'.

[b] Port of shipment, or e.g. 'any United Kingdom port'.

[c] Port of destination.

[d] Short description of goods and site work.

[e] This description follows as closely as possible the *Standard Procedure for the Issuing of Documentary Credits by Cable, Telegram or Telex* (App VII *Standard Forms for Issuing Documentary Credits*, ICC Publication No 323).

Payment in respect of site work shall be due monthly at the end of each month.

Conditions of Contract

Where consistent with this tender, we propose that *General Conditions No 574A for the Supply and Erection of Plant and Machinery for Import and Export*, prepared under the auspices of the United Nations Economic Commission for Europe, Geneva, March 1957, a copy of which is enclosed, shall apply, subject to the following:

The Appendix shall be completed as follows:

Appendix to UNECE Conditions 574A

	Clause	
A. Maximum amount recoverable on termination by Contractor for failure to take delivery or make payment	10.2 and 11.7	£ (pounds sterling)
B. Rate of interest on overdue payments	11.7	2% (two per cent) per annum over the * Bank base lending rate from time to time in force

C.	Period of delay in payment authorizing termination by Contractor	11.7		months
D.	Percentage to be deducted for each week's delay	20.3		% (per cent)
E.	Maximum percentage which the deductions above may not exceed	20.3		% (per cent)
F.	Maximum amount recoverable for non-completion	20.5	£ (	 pounds sterling)
G.	Maximum post-ponement of taking over tests by Contractor	22.3		weeks
H.	Guarantee Period for original Works and parts replaced or renewed	23.2 and 23.5		months
I.	Maximum indemnities for personal injury or damage	24.3	£ (	 pounds sterling)

*Insert name of bank.

Clause 7 (Erection on a Time Basis and Lump-Sum Erection) – Erection is carried out (on a time basis)* (for a lump sum).*

*Delete alternatives.

Clause 9 (Passing of Risk) – Risk shall pass upon delivery to Site.

Customs and Import Duties – The Purchaser shall in good time procure any necessary import permits and promptly pay all customs and import duties which may become payable upon the importation of the Plant into the country in which the Plant is to be erected.

Contract Price Adjustment

We propose that contract price adjustment in respect of the price for goods shall apply in accordance with the BEAMA *Contract Price Adjustment Clause and Formula for use with Export Contracts – Electrical Machinery: (for which there is no other specific formula)*, edition January 1979, corrected edition,* a copy of which is enclosed.

We further propose that contract price adjustment in respect of the price for site work shall apply in accordance with the same BEAMA Clause and Formula subject to the amendment that adjustment shall take place at the rate of 0.95 per cent of such price per 1.0 per cent difference between the Labour Cost Index published for the month in which the tender date falls and the average of the Index figures published for the last two-thirds of the period during which the

Contractor is carrying out site work, this difference being expressed as a percentage of the former Index figure.

The BEAMA Labour Cost Index referred to shall be the Labour Cost Index for Electrical Engineering.*

References to the *Trade and Industry Journal* shall be read as references to *British Business,* which is the new name of the journal.†

*This is only one of several BEAMA CPA Clauses and Formulae. If another is adopted, this form may need consequential amendment. In particular, reference to the BEAMA CPA Clause applicable to Mechanical Plant, and to the Labour Cost Index for Mechanical Engineering, may be appropriate. See p. 16.

†The above form requires that the separate prices to be adjusted shall be identifiable from the contract documents.

Validity

This tender is valid for acceptance until and including *, but shall, after that date, be null and void unless extended by us in writing.

*Date in full, e.g. 'Monday, 19 January 1981'.

Yours faithfully

(Signature)

FORM 56

Tender based on General Conditions No 574A for the Supply and Erection of Plant and Machinery for Import and Export, prepared under the auspices of the United Nations Economic Commission for Europe, Geneva, March 1957, providing for Delivery FOB and Erection on Site

Dear Sirs

(heading)

Prices

We offer to supply FOB * and erect on Site at

the items described in the accompanying Specification for:

£

(pounds sterling)

*Port of shipment.

(Our services in respect of site work are offered on a time basis. The above price includes an estimated sum of £ for such services, computed at the rates and for the estimated duration shown in the accompanying Specification. The sum actually payable for such services shall be computed on a time basis, and may be either more or less than the estimated sum.)*

*Delete if lump sum.

Programme

Our best estimate is that delivery FOB will be achieved within of receiving your order, and that we could complete erection within a further of receiving your order, making a total of .

We enclose * showing the basis of this estimate.

*Appropriate document, e.g. bar chart or critical path network.

Terms of Payment

All prices are payable in pounds sterling in the United Kingdom.

We propose that payment shall be made as follows:

(a) %(per cent) in advance.

(b) The balance through a Confirmed Irrevocable Letter of Credit providing for payment in full upon presentation of documents, to be increased according to any contractual increases in the price, to be extended according to any increase in the period required for execution of our respective obligations under the Contract, and otherwise complying with the following description:

(i) Confirmed by an acceptable bank in London.

(ii) Irrevocably issued, confirmed and advised to us by .[a]

(iii) Date and place of expiry – [a] in London.
(iv) Beneficiary – ourselves.
(v) Amount – the balance of the price.
(vi) Credit available with confirming bank by payment.
(vii) Partial shipments allowed.
(viii) Transhipment allowed.
(ix) Shipment from .[b]
(x) For transportation to .[c]
(xi) Shipment of .[d]
(xii) Documents required:

Commercial Invoice

In respect of goods, full set of clean on board marine Bills of Lading (which shall be acceptable although to the effect that the goods have been shipped on deck or unprotected*).

In respect of other work, your or your Engineer's Certificate.†

*Delete if inappropriate.

†If this sentence is omitted, payment for such other work should be made out of the letter of credit on presentation only of commercial invoice.

(xiii) Period after issuance of shipping documents for presentation – days.
(xiv) Subject otherwise to the *Uniform Customs and Practice for Documentary Credits*, 1974, ICC Publication No 290.[e]

[a]Date in full, e.g. 'Monday, 19 January 1981'.

[b]Port of shipment, or e.g. 'any United Kingdom port'.

[c]Port of destination.

[d]Short description of goods and site work.

[e]This description follows as closely as possible the *Standard Procedure for the Issuing of Documentary Credits by Cable, Telegram or Telex* (App VII *Standard Forms for Issuing Documentary Credits*, ICC Publication No 323).

Payment in respect of site work shall be due monthly at the end of each month.

Conditions of Contract

Where consistent with this tender, we propose that *General Conditions No 574A for the Supply and Erection of Plant and Machinery for Import and Export*, prepared under the auspices of the United Nations Economic Commission for Europe, Geneva, March 1957, a copy of which is enclosed, shall apply, subject to the following:

The Appendix shall be completed as follows:

Appendix to UNECE Conditions 574A

	Clause	
A. Maximum amount recoverable on termination by Contractor for failure to take delivery or make payment	10.2 and 11.7	£ (pounds sterling)

B.	Rate of interest on overdue payments	11.7	2% (two per cent) per annum over the * Bank base lending rate from time to time in force
C.	Period of delay in payment authorizing termination by Contractor	11.7	months
D.	Percentage to be deducted for each week's delay	20.3	% (per cent)
E.	Maximum percentage which the deductions above may not exceed	20.3	% (per cent)
F.	Maximum amount recoverable for non-completion	20.5	£ (pounds sterling)
G.	Maximum post-ponement of taking over tests by Contractor	22.3	weeks
H.	Guarantee Period for original Works and parts replaced or renewed	23.2 and 23.5	months
I.	Maximum indemnities for personal injury or damage	24.3	£ (pounds sterling)

*Insert name of bank.

Clause 7 (Erection on a Time Basis and Lump-Sum Erection) – Erection is carried out (on a time basis)* (for a lump sum).*

*Delete alternatives.

Contract Price Adjustment

We propose that contract price adjustment in respect of the price for goods shall apply in accordance with the BEAMA *Contract Price Adjustment Clause and Formula for use with Export Contracts – Electrical Machinery: (for which there is no other specific formula)*, edition January 1979, corrected edition,* a copy of which is enclosed.

We further propose that contract price adjustment in respect of the price for site work shall apply in accordance with the same BEAMA Clause and Formula subject to the amendment that adjustment shall take place at the rate of 0.95 per cent of such price per 1.0 per cent difference between the Labour Cost Index published for the month in which the tender date falls and the average of the Index figures published for the last two-thirds of the period during which the Contractor is carrying out site work, this difference being expressed as a percentage of the former Index figure.

The BEAMA Labour Cost Index referred to shall be the Labour Cost Index for Electrical Engineering.*

References to the *Trade and Industry Journal* shall be read as references to *British Business,* which is the new name of the journal.†

*This is only one of several BEAMA CPA Clauses and Formulae. If another is adopted, this form may need consequential amendment. In particular, reference to the BEAMA CPA Clause applicable to Mechanical Plant, and to the Labour Cost Index for Mechanical Engineering, may be appropriate. See p. 16.

†The above form requires that the separate prices to be adjusted shall be identifiable from the contract documents.

Trade Terms

The trade term 'FOB' shall be defined by the current edition of Incoterms, where consistent with this tender. A copy of the definition is enclosed.

Validity

This tender is valid for acceptance until and including   *, but shall, after that date, be null and void unless extended by us in writing.

*Date in full, e.g. 'Monday, 19 January 1981'.

Yours faithfully

(Signature)

FORM 57

Tender based on General Conditions No 574A for the Supply and Erection of Plant and Machinery for Import and Export, prepared under the auspices of the United Nations Economic Commission for Europe, Geneva, March 1957, providing for Delivery C & F and Erection on Site

Dear Sirs

(heading)

Prices

We offer to supply C & F * and erect on Site at

the items described in the accompanying Specification for:

£

(pounds sterling)

*Port of destination.

(Our services in respect of site work are offered on a time basis. The above price includes an estimated sum of £ for such services, computed at the rates and for the estimated duration shown in the accompanying Specification. The sum actually payable for such services shall be computed on a time basis, and may be either more or less than the estimated sum.)*

*Delete if lump sum.

Programme

Our best estimate is that delivery C & F will be achieved within of receiving your order, and that we could complete erection within a further of receiving your order, making a total of .

We enclose * showing the basis of this estimate.

*Appropriate document, e.g. bar chart or critical path network.

Terms of Payment

All prices are payable in pounds sterling in the United Kingdom.

We propose that payment shall be made as follows:

(a) %(per cent) in advance.

(b) The balance through a Confirmed Irrevocable Letter of Credit providing for payment in full upon presentation of documents, to be increased according to any contractual increases in the price, to be extended according to any increase in the period required for execution of our respective obligations under the Contract, and otherwise complying with the following description:

(i) Confirmed by an acceptable bank in London.

(ii) Irrevocably issued, confirmed and advised to us by .[a]

(iii) Date and place of expiry – [a] in London.
(iv) Beneficiary – ourselves.
(v) Amount – the balance of the price.
(vi) Credit available with confirming bank by payment.
(vii) Partial shipments allowed.
(viii) Transhipment allowed.
(ix) Shipment from .[b]
(x) For transportation to .[c]
(xi) Shipment of .[d]
(xii) Documents required:

Commercial Invoice

In respect of goods, full set of clean on board marine Bills of Lading (which shall be acceptable although to the effect that the goods have been shipped on deck or unprotected*).

In respect of other work, your or your Engineer's Certificate.†

*Delete if inappropriate.

†If this sentence is omitted, payment for such other work should be made out of the letter of credit on presentation only of commercial invoice.

(xiii) Period after issuance of shipping documents for presentation – days.
(xiv) Subject otherwise to the *Uniform Customs and Practice for Documentary Credits*, 1974, ICC Publication No 290.[e]

[a]Date in full, e.g. 'Monday, 19 January 1981'.

[b]Port of shipment, or e.g. 'any United Kingdom port'.

[c]Port of destination.

[d]Short description of goods and site work.

[e]This description follows as closely as possible the *Standard Procedure for the Issuing of Documentary Credits by Cable, Telegram or Telex* (App VII *Standard Forms for Issuing Documentary Credits*, ICC Publication No 323).

Payment in respect of site work shall be due monthly at the end of each month.

Conditions of Contract

Where consistent with this tender, we propose that *General Conditions No 574A for the Supply and Erection of Plant and Machinery for Import and Export*, prepared under the auspices of the United Nations Economic Commission for Europe, Geneva, March 1957, a copy of which is enclosed, shall apply, subject to the following:

The Appendix shall be completed as follows:

Appendix to UNECE Conditions 574A

	Clause	
A. Maximum amount recoverable on termination by Contractor for failure to take delivery or make payment	10.2 and 11.7	£ (pounds sterling)

B.	Rate of interest on overdue payments	11.7	2% (two per cent) per annum over the * Bank base lending rate from time to time in force
C.	Period of delay in payment authorizing termination by Contractor	11.7	months
D.	Percentage to be deducted for each week's delay	20.3	% (per cent)
E.	Maximum percentage which the deductions above may not exceed	20.3	% (per cent)
F.	Maximum amount recoverable for non-completion	20.5	£ (pounds sterling)
G.	Maximum post-ponement of taking over tests by Contractor	22.3	weeks
H.	Guarantee Period for original Works and parts replaced or renewed	23.2 and 23.5	months
I.	Maximum indemnities for personal injury or damage	24.3	£ (pounds sterling)

*Insert name of bank.

Clause 7 (Erection on a Time Basis and Lump-Sum Erection) – Erection is carried out (on a time basis)* (for a lump sum).*

*Delete alternatives.

Contract Price Adjustment

We propose that contract price adjustment in respect of the price for goods shall apply in accordance with the BEAMA *Contract Price Adjustment Clause and Formula for use with Export Contracts – Electrical Machinery: (for which there is no other specific formula)*, edition January 1979, corrected edition,* a copy of which is enclosed.

We further propose that contract price adjustment in respect of the price for site work shall apply in accordance with the same BEAMA Clause and Formula subject to the amendment that adjustment shall take place at the rate of 0.95 per cent of such price per 1.0 per cent difference between the Labour Cost Index published for the month in which the tender date falls and the average of the Index figures published for the last two-thirds of the period during which the Contractor is carrying out site work, this difference being expressed as a percentage of the former Index figure.

The BEAMA Labour Cost Index referred to shall be the Labour Cost Index for Electrical Engineering.*

References to the *Trade and Industry Journal* shall be read as references to *British Business*, which is the new name of the journal.†

*This is only one of several BEAMA CPA Clauses and Formulae. If another is adopted, this form may need consequential amendment. In particular, reference to the BEAMA CPA Clause applicable to Mechanical Plant, and to the Labour Cost Index for Mechanical Engineering, may be appropriate. See p. 16.

†The above form requires that the separate prices to be adjusted shall be identifiable from the contract documents.

Trade Terms

The trade term 'C & F' shall be defined by the current edition of Incoterms, where consistent with this tender. A copy of the definition is enclosed.

Validity

This tender is valid for acceptance until and including __________*, but shall, after that date, be null and void unless extended by us in writing.

*Date in full, e.g. 'Monday, 19 January 1981'.

Yours faithfully

(Signature)

FORM 58

Tender based on General Conditions No 574A for the Supply and Erection of Plant and Machinery for Import and Export, prepared under the auspices of the United Nations Economic Commission for Europe, Geneva, March 1957, providing for Delivery CIF and Erection on Site

Dear Sirs

(heading)

Prices

We offer to supply CIF * and erect on Site at

the items described in the accompanying Specification for:

£

(pounds sterling)

*Port of destination.

(Our services in respect of site work are offered on a time basis. The above price includes an estimated sum of £ for such services, computed at the rates and for the estimated duration shown in the accompanying Specification. The sum actually payable for such services shall be computed on a time basis, and may be either more or less than the estimated sum.)*

*Delete if lump sum.

Programme

Our best estimate is that delivery CIF will be achieved within of receiving your order, and that we could complete erection within a further of receiving your order, making a total of .

We enclose * showing the basis of this estimate.

*Appropriate document, e.g. bar chart or critical path network.

Terms of Payment

All prices are payable in pounds sterling in the United Kingdom.

We propose that payment shall be made as follows:

(a) %(per cent) in advance.

(b) The balance through a Confirmed Irrevocable Letter of Credit providing for payment in full upon presentation of documents, to be increased according to any contractual increases in the price, to be extended according to any increase in the period required for execution of our respective obligations under the Contract, and otherwise complying with the following description:

(i) Confirmed by an acceptable bank in London.

(ii) Irrevocably issued, confirmed and advised to us by .[a]

(iii) Date and place of expiry – [a] in London.
(iv) Beneficiary – ourselves.
(v) Amount – the balance of the price.
(vi) Credit available with confirming bank by payment.
(vii) Partial shipments allowed.
(viii) Transhipment allowed.
(ix) Shipment from .[b]
(x) For transportation to .[c]
(xi) Shipment of .[d]
(xii) Documents required:

Commercial Invoice

In respect of goods, full set of clean on board marine Bills of Lading (which shall be acceptable although to the effect that the goods have been shipped on deck or unprotected*) and Insurance Certificate.

In respect of other work, your or your Engineer's Certificate.†

*Delete if inappropriate.

†If this sentence is omitted, payment for such other work should be made out of the letter of credit on presentation only of commercial invoice.

(xiii) Period after issuance of shipping documents for presentation – days.
(xiv) Subject otherwise to the *Uniform Customs and Practice for Documentary Credits*, 1974, ICC Publication No 290.[e]

[a]Date in full, e.g. 'Monday, 19 January 1981'.

[b]Port of shipment, or e.g. 'any United Kingdom port'.

[c]Port of destination.

[d]Short description of goods and site work.

[e]This description follows as closely as possible the *Standard Procedure for the Issuing of Documentary Credits by Cable, Telegram or Telex* (App VII *Standard Forms for Issuing Documentary Credits*, ICC Publication No 323).

Payment in respect of site work shall be due monthly at the end of each month.

Conditions of Contract

Where consistent with this tender, we propose that *General Conditions No 574A for the Supply and Erection of Plant and Machinery for Import and Export,* prepared under the auspices of the United Nations Economic Commission for Europe, Geneva, March 1957, a copy of which is enclosed, shall apply, subject to the following:

The Appendix shall be completed as follows:

Appendix to UNECE Conditions 574A

	Clause	
A. Maximum amount recoverable on termination by Contractor for failure to take delivery or make payment	10.2 and 11.7	£ (pounds sterling)

B. Rate of interest on overdue payments	11.7	2% (two per cent) per annum over the * Bank base lending rate from time to time in force
C. Period of delay in payment authorizing termination by Contractor	11.7	months
D. Percentage to be deducted for each week's delay	20.3	% (per cent)
E. Maximum percentage which the deductions above may not exceed	20.3	% (per cent)
F. Maximum amount recoverable for non-completion	20.5	£ (pounds sterling)
G. Maximum post-ponement of taking over tests by Contractor	22.3	weeks
H. Guarantee Period for original Works and parts replaced or renewed	23.2 and 23.5	months
I. Maximum indemnities for personal injury or damage	24.3	£ (pounds sterling)

*Insert name of bank.

Clause 7 (Erection on a Time Basis and Lump-Sum Erection) – Erection is carried out (on a time basis)* (for a lump sum).*

*Delete alternatives.

Contract Price Adjustment

We propose that contract price adjustment in respect of the price for goods shall apply in accordance with the BEAMA *Contract Price Adjustment Clause and Formula for use with Export Contracts – Electrical Machinery: (for which there is no other specific formula)*, edition January 1979, corrected edition,* a copy of which is enclosed.

We further propose that contract price adjustment in respect of the price for site work shall apply in accordance with the same BEAMA Clause and Formula subject to the amendment that adjustment shall take place at the rate of 0.95 per cent of such price per 1.0 per cent difference between the Labour Cost Index published for the month in which the tender date falls and the average of the Index figures published for the last two-thirds of the period during which the Contractor is carrying out site work, this difference being expressed as a percentage of the former Index figure.

The BEAMA Labour Cost Index referred to shall be the Labour Cost Index for Electrical Engineering.*

References to the *Trade and Industry Journal* shall be read as references to *British Business,* which is the new name of the journal.†

*This is only one of several BEAMA CPA Clauses and Formulae. If another is adopted, this form may need consequential amendment. In particular, reference to the BEAMA CPA Clause applicable to Mechanical Plant, and to the Labour Cost Index for Mechanical Engineering, may be appropriate. See p. 16.

†The above form requires that the separate prices to be adjusted shall be identifiable from the contract documents.

Trade Terms

The trade term 'CIF' shall be defined by the current edition of Incoterms, where consistent with this tender. A copy of the definition is enclosed.

Validity

This tender is valid for acceptance until and including *, but shall, after that date, be null and void unless extended by us in writing.

*Date in full, e.g. 'Monday, 19 January 1981'.

Yours faithfully

(Signature)

FORM 59

Tender based on General Conditions No 574A for the Supply and Erection of Plant and Machinery for Import and Export, prepared under the auspices of the United Nations Economic Commission for Europe, Geneva, March 1957, providing for Delivery to and Supervision of Erection on Site

Dear Sirs

(heading)

Prices

We offer to supply, deliver to, and supervise the erection on Site at

of the items described in the accompanying Specification for:

£

(pounds sterling)

(Our services in respect of site work are offered on a time basis. The above price includes an estimated sum of £ for such services, computed at the rates and for the estimated duration shown in the accompanying Specification. The sum actually payable for such services shall be computed on a time basis, and may be either more or less than the estimated sum.)*

*Delete if lump sum.

Programme

Our best estimate is that delivery to Site will be achieved within of receiving your order, and that we could complete supervision of erection within a further of receiving your order, making a total of .

We enclose * showing the basis of this estimate.

*Appropriate document, e.g. bar chart or critical path network.

Terms of Payment

All prices are payable in pounds sterling in the United Kingdom.

We propose that payment shall be made as follows:

(a) % (per cent) in advance.

(b) The balance through a Confirmed Irrevocable Letter of Credit providing for payment in full upon presentation of documents, to be increased according to any contractual increases in the price, to be extended according to any increase in the period required for execution of our respective obligations under the Contract, and otherwise complying with the following description:

(i) Confirmed by an acceptable bank in London.

(ii) Irrevocably issued, confirmed and advised to us by .[a]

(iii) Date and place of expiry – [a] in London.
(iv) Beneficiary – ourselves.
(v) Amount – the balance of the price.
(vi) Credit available with confirming bank by payment.
(vii) Partial shipments allowed.
(viii) Transhipment allowed.
(ix) Shipment from .[b]
(x) For transportation to .[c]
(xi) Shipment of .[d]
(xii) Documents required:

Commercial Invoice

In respect of goods, full set of clean on board marine Bills of Lading (which shall be acceptable although to the effect that the goods have been shipped on deck or unprotected*) and Insurance Certificate.

In respect of other work, your or your Engineer's Certificate.†

*Delete if inappropriate.

†If this sentence is omitted, payment for such other work should be made out of the letter of credit on presentation only of commercial invoice.

(xiii) Period after issuance of shipping documents for presentation – days.
(xiv) Subject otherwise to the *Uniform Customs and Practice for Documentary Credits,* 1974, ICC Publication No 290.[e]

[a]Date in full, e.g. 'Monday, 19 January 1981'.

[b]Port of shipment, or e.g. 'any United Kingdom port'.

[c]Port of destination.

[d]Short description of goods and site work.

[e]This description follows as closely as possible the *Standard Procedure for the Issuing of Documentary Credits by Cable, Telegram or Telex* (App VII *Standard Forms for Issuing Documentary Credits,* ICC Publication No 323).

Payment in respect of site work shall be due monthly at the end of each month.

Conditions of Contract

Where consistent with this tender, we propose that *General Conditions No 574A for the Supply and Erection of Plant and Machinery for Import and Export,* prepared under the auspices of the United Nations Economic Commission for Europe, Geneva, March 1957, a copy of which is enclosed, shall apply, subject to the following:

The Appendix shall be completed as follows:

Appendix to UNECE Conditions 574A

	Clause	
A. Maximum amount recoverable on termination by Contractor for failure to take delivery or make payment	10.2 and 11.7	£ (pounds sterling)

B.	Rate of interest on overdue payments	11.7	2% (two per cent) per annum over the * Bank base lending rate from time to time in force
C.	Period of delay in payment authorizing termination by Contractor	11.7	months
D.	Percentage to be deducted for each week's delay	20.3	% (per cent)
E.	Maximum percentage which the deductions above may not exceed	20.3	% (per cent)
F.	Maximum amount recoverable for non-completion	20.5	£ (pounds sterling)
G.	Maximum post-ponement of taking over tests by Contractor	22.3	weeks
H.	Guarantee Period for original Works and parts replaced or renewed	23.2 and 23.5	months
I.	Maximum indemnities for personal injury or damage	24.3	£ (pounds sterling)

*Insert name of bank.

Clause 7 (Erection on a Time Basis and Lump-Sum Erection) – Supervision of erection is carried out (on a time basis)* (for a lump sum).*

*Delete alternatives.

Erection will be carried out by the Purchaser, who shall, at his own expense, provide the skilled and unskilled labour, all equipment and everything necessary for the erection of the Plant.

References to 'erection' in Clause 7 shall be read as references to 'supervision of erection'. The Contractor's obligation in this respect will be to provide the services of one or more competent engineers

(a) to give to the Purchaser or his representative mentioned in Clause 13.1 the necessary instructions for the erection of the Plant by the Purchaser and, if provided in the accompanying Specification or other documents, for its commissioning by him; and
(b) to supervise the manner in which the Contractor's instructions have been carried out.

The number and qualifications of the Contractor's staff, and the estimated duration of erection, shall be as specified in this tender or in the accompanying Specification or other documents.

The date on which the Contractor's staff should arrive on Site shall be as provided in the accompanying Specification or other documents; if not so provided, the Purchaser shall give the Contractor not less than one month's notice requiring such arrival.*

*This clause is based on Clause 2 of UNECE Conditions 188B, by kind permission of the United Nations.

Clause 20 (Time for Completion) – As the Contract will only include supervision of erection, the Contractor cannot undertake that erection will be completed within a specific time. Therefore, for the purposes of this Clause, completion of delivery to Site shall be deemed to constitute completion of the Works.

Customs and Import Duties – The Purchaser shall in good time procure any necessary import permits and promptly pay all customs and import duties which may become payable upon the importation of the Plant into the country in which the Plant is to be erected.

Contract Price Adjustment

We propose that contract price adjustment in respect of the price for goods shall apply in accordance with the BEAMA *Contract Price Adjustment Clause and Formula for use with Export Contracts – Electrical Machinery: (for which there is no other specific formula)*, edition January 1979, corrected edition,* a copy of which is enclosed.

We further propose that contract price adjustment in respect of the price for site work shall apply in accordance with the same BEAMA Clause and Formula subject to the amendment that adjustment shall take place at the rate of 0.95 per cent of such price per 1.0 per cent difference between the Labour Cost Index published for the month in which the tender date falls and the average of the Index figures published for the last two-thirds of the period during which the Contractor is carrying out site work, this difference being expressed as a percentage of the former Index figure.

The BEAMA Labour Cost Index referred to shall be the Labour Cost Index for Electrical Engineering.*

References to the *Trade and Industry Journal* shall be read as references to *British Business*, which is the new name of the journal.†

*This is only one of several BEAMA CPA Clauses and Formulae. If another is adopted, this form may need consequential amendment. In particular, reference to the BEAMA CPA Clause applicable to Mechanical Plant, and to the Labour Cost Index for Mechanical Engineering, may be appropriate. See p. 16.

†The above form requires that the separate prices to be adjusted shall be identifiable from the contract documents.

Validity

This tender is valid for acceptance until and including *, but shall, after that date, be null and void unless extended by us in writing.

*Date in full, e.g. 'Monday, 19 January 1981'.

Yours faithfully

(Signature)

FORM 60

Tender based on General Conditions No 574A for the Supply and Erection of Plant and Machinery for Import and Export, prepared under the auspices of the United Nations Economic Commission for Europe, Geneva, March 1957, providing for Delivery FOB and Supervision of Erection on Site

Dear Sirs

(heading)

Prices
We offer to supply FOB * and supervise the erection on Site at

of the items described in the accompanying Specification for:
£
(pounds sterling)
*Port of shipment.
(Our services in respect of site work are offered on a time basis. The above price includes an estimated sum of £ for such services, computed at the rates and for the estimated duration shown in the accompanying Specification. The sum actually payable for such services shall be computed on a time basis, and may be either more or less than the estimated sum.)*
*Delete if lump sum.

Programme
Our best estimate is that delivery FOB will be achieved within of receiving your order, and that we could complete supervision of erection within a further of receiving your order, making a total of .
We enclose * showing the basis of this estimate.
*Appropriate document, e.g. bar chart or critical path network.

Terms of Payment
All prices are payable in pounds sterling in the United Kingdom.
We propose that payment shall be made as follows:

(a) %(per cent) in advance.

(b) The balance through a Confirmed Irrevocable Letter of Credit providing for payment in full upon presentation of documents, to be increased according to any contractual increases in the price, to be extended according to any increase in the period required for execution of our respective obligations under the Contract, and otherwise complying with the following description:
    (i) Confirmed by an acceptable bank in London.
    (ii) Irrevocably issued, confirmed and advised to us by .[a]

(iii) Date and place of expiry – [a] in London.
(iv) Beneficiary – ourselves.
(v) Amount – the balance of the price.
(vi) Credit available with confirming bank by payment.
(vii) Partial shipments allowed.
(viii) Transhipment allowed.
(ix) Shipment from .[b]
(x) For transportation to .[c]
(xi) Shipment of .[d]
(xii) Documents required:

Commercial Invoice

In respect of goods, full set of clean on board marine Bills of Lading (which shall be acceptable although to the effect that the goods have been shipped on deck or unprotected*).

In respect of other work, your or your Engineer's Certificate.†

*Delete if inappropriate.

†If this sentence is omitted, payment for such other work should be made out of the letter of credit on presentation only of commercial invoice.

(xiii) Period after issuance of shipping documents for presentation – days.
(xiv) Subject otherwise to the *Uniform Customs and Practice for Documentary Credits*, 1974, ICC Publication No 290.[e]

[a]Date in full, e.g. 'Monday, 19 January 1981'.

[b]Port of shipment, or e.g. 'any United Kingdom port'.

[c]Port of destination.

[d]Short description of goods and site work.

[e]This description follows as closely as possible the *Standard Procedure for the Issuing of Documentary Credits by Cable, Telegram or Telex* (App VII *Standard Forms for Issuing Documentary Credits*, ICC Publication No 323).

Payment in respect of site work shall be due monthly at the end of each month.

Conditions of Contract

Where consistent with this tender, we propose that *General Conditions No 574A for the Supply and Erection of Plant and Machinery for Import and Export*, prepared under the auspices of the United Nations Economic Commission for Europe, Geneva, March 1957, a copy of which is enclosed, shall apply, subject to the following:

The Appendix shall be completed as follows:

Appendix to UNECE Conditions 574A

	Clause	
A. Maximum amount recoverable on termination by Contractor for failure to take delivery or make payment	10.2 and 11.7	£ (pounds sterling)

B. Rate of interest on overdue payments	11.7	2% (two per cent) per annum over the * Bank base lending rate from time to time in force
C. Period of delay in payment authorizing termination by Contractor	11.7	months
D. Percentage to be deducted for each week's delay	20.3	% (per cent)
E. Maximum percentage which the deductions above may not exceed	20.3	% (per cent)
F. Maximum amount recoverable for non-completion	20.5	£ (pounds sterling)
G. Maximum post-ponement of taking over tests by Contractor	22.3	weeks
H. Guarantee Period for original Works and parts replaced or renewed	23.2 and 23.5	months
I. Maximum indemnities for personal injury or damage	24.3	£ (pounds sterling)

*Insert name of bank.

Clause 7 (Erection on a Time Basis and Lump-Sum Erection) – Supervision of erection is carried out (on a time basis)* (for a lump sum).*

*Delete alternatives.

Erection will be carried out by the Purchaser, who shall, at his own expense, provide the skilled and unskilled labour, all equipment and everything necessary for the erection of the Plant.

References to 'erection' in Clause 7 shall be read as references to 'supervision of erection'. The Contractor's obligation in this respect will be to provide the services of one or more competent engineers

(a) to give to the Purchaser or his representative mentioned in Clause 13.1 the necessary instructions for the erection of the Plant by the Purchaser and, if provided in the accompanying Specification or other documents, for its commissioning by him; and

(b) to supervise the manner in which the Contractor's instructions have been carried out.

The number and qualifications of the Contractor's staff, and the estimated duration of erection, shall be as specified in this tender or in the accompanying Specification or other documents.

The date on which the Contractor's staff should arrive on Site shall be as provided in the accompanying Specification or other documents; if not so provided, the Purchaser shall give the Contractor not less than one month's notice requiring such arrival.*

*This clause is based on Clause 2 of UNECE Conditions 188B, by kind permission of the United Nations.

Clause 20 (Time for Completion) – As the Contract will only include supervision of erection, the Contractor cannot undertake that erection will be completed within a specific time. Therefore, for the purposes of this Clause, completion of delivery FOB shall be deemed to constitute completion of the Works.

Contract Price Adjustment

We propose that contract price adjustment in respect of the price for goods shall apply in accordance with the BEAMA *Contract Price Adjustment Clause and Formula for use with Export Contracts – Electrical Machinery: (for which there is no other specific formula)*, edition January 1979, corrected edition,* a copy of which is enclosed.

We further propose that contract price adjustment in respect of the price for site work shall apply in accordance with the same BEAMA Clause and Formula subject to the amendment that adjustment shall take place at the rate of 0.95 per cent of such price per 1.0 per cent difference between the Labour Cost Index published for the month in which the tender date falls and the average of the Index figures published for the last two-thirds of the period during which the Contractor is carrying out site work, this difference being expressed as a percentage of the former Index figure.

The BEAMA Labour Cost Index referred to shall be the Labour Cost Index for Electrical Engineering.*

References to the *Trade and Industry Journal* shall be read as references to *British Business*, which is the new name of the journal.†

*This is only one of several BEAMA CPA Clauses and Formulae. If another is adopted, this form may need consequential amendment. In particular, reference to the BEAMA CPA Clause applicable to Mechanical Plant, and to the Labour Cost Index for Mechanical Engineering, may be appropriate. See p. 16.

†The above form requires that the separate prices to be adjusted shall be identifiable from the contract documents.

Trade Terms

The trade term 'FOB' shall be defined by the current edition of Incoterms, where consistent with this tender. A copy of the definition is enclosed.

Validity

This tender is valid for acceptance until and including *, but shall, after that date, be null and void unless extended by us in writing.

*Date in full, e.g. 'Monday, 19 January 1981'.

Yours faithfully

(Signature)

FORM 61

Tender based on General Conditions No 574A for the Supply and Erection of Plant and Machinery for Import and Export, prepared under the auspices of the United Nations Economic Commission for Europe, Geneva, March 1957, providing for Delivery C & F and Supervision of Erection on Site

Dear Sirs

(heading)

Prices

We offer to supply C & F * and supervise the erection on Site at

of the items described in the accompanying Specification for:
£
(pounds sterling)
*Port of destination.
(Our services in respect of site work are offered on a time basis. The above price includes an estimated sum of £ for such services, computed at the rates and for the estimated duration shown in the accompanying Specification. The sum actually payable for such services shall be computed on a time basis, and may be either more or less than the estimated sum.)*
*Delete if lump sum.

Programme

Our best estimate is that delivery C & F will be achieved within of receiving your order, and that we could complete supervision of erection within a further of receiving your order, making a total of .
We enclose * showing the basis of this estimate.
*Appropriate document, e.g. bar chart or critical path network.

Terms of Payment

All prices are payable in pounds sterling in the United Kingdom.

We propose that payment shall be made as follows:

(a) %(per cent) in advance.

(b) The balance through a Confirmed Irrevocable Letter of Credit providing for payment in full upon presentation of documents, to be increased according to any contractual increases in the price, to be extended according to any increase in the period required for execution of our respective obligations under the Contract, and otherwise complying with the following description:

(i) Confirmed by an acceptable bank in London.

(ii) Irrevocably issued, confirmed and advised to us by .[a]

(iii) Date and place of expiry – [a] in London.
(iv) Beneficiary – ourselves.
(v) Amount – the balance of the price.
(vi) Credit available with confirming bank by payment.
(vii) Partial shipments allowed.
(viii) Transhipment allowed.
(ix) Shipment from .[b]
(x) For transportation to .[c]
(xi) Shipment of .[d]
(xii) Documents required:

Commercial Invoice

In respect of goods, full set of clean on board marine Bills of Lading (which shall be acceptable although to the effect that the goods have been shipped on deck or unprotected*).

In respect of other work, your or your Engineer's Certificate.†

*Delete if inappropriate.

†If this sentence is omitted, payment for such other work should be made out of the letter of credit on presentation only of commercial invoice.

(xiii) Period after issuance of shipping documents for presentation – days.
(xiv) Subject otherwise to the *Uniform Customs and Practice for Documentary Credits*, 1974, ICC Publication No 290.[e]

[a]Date in full, e.g. 'Monday, 19 January 1981'.

[b]Port of shipment, or e.g. 'any United Kingdom port'.

[c]Port of destination.

[d]Short description of goods and site work.

[e]This description follows as closely as possible the *Standard Procedure for the Issuing of Documentary Credits by Cable, Telegram or Telex* (App VII *Standard Forms for Issuing Documentary Credits*, ICC Publication No 323).

Payment in respect of site work shall be due monthly at the end of each month.

Conditions of Contract

Where consistent with this tender, we propose that *General Conditions No 574A for the Supply and Erection of Plant and Machinery for Import and Export*, prepared under the auspices of the United Nations Economic Commission for Europe, Geneva, March 1957, a copy of which is enclosed, shall apply, subject to the following:

The Appendix shall be completed as follows:

Appendix to UNECE Conditions 574A

	Clause	
A. Maximum amount recoverable on termination by Contractor for failure to take delivery or make payment	10.2 and 11.7	£ (pounds sterling)

B.	Rate of interest on overdue payments	11.7	2% (two per cent) per annum over the * Bank base lending rate from time to time in force
C.	Period of delay in payment authorizing termination by Contractor	11.7	months
D.	Percentage to be deducted for each week's delay	20.3	% (per cent)
E.	Maximum percentage which the deductions above may not exceed	20.3	% (per cent)
F.	Maximum amount recoverable for non-completion	20.5	£ (pounds sterling)
G.	Maximum post-ponement of taking over tests by Contractor	22.3	weeks
H.	Guarantee Period for original Works and parts replaced or renewed	23.2 and 23.5	months
I.	Maximum indemnities for personal injury or damage	24.3	£ (pounds sterling)

*Insert name of bank.

Clause 7 (Erection on a Time Basis and Lump-Sum Erection) – Supervision of erection is carried out (on a time basis)* (for a lump sum).*

*Delete alternatives.

Erection will be carried out by the Purchaser, who shall, at his own expense, provide the skilled and unskilled labour, all equipment and everything necessary for the erection of the Plant.

References to 'erection' in Clause 7 shall be read as references to 'supervision of erection'. The Contractor's obligation in this respect will be to provide the services of one or more competent engineers

(a) to give to the Purchaser or his representative mentioned in Clause 13.1 the necessary instructions for the erection of the Plant by the Purchaser and, if provided in the accompanying Specification or other documents, for its commissioning by him; and

(b) to supervise the manner in which the Contractor's instructions have been carried out.

The number and qualifications of the Contractor's staff, and the estimated duration of erection, shall be as specified in this tender or in the accompanying Specification or other documents.

The date on which the Contractor's staff should arrive on Site shall be as provided in the accompanying Specification or other documents; if not so provided, the Purchaser shall give the Contractor not less than one month's notice requiring such arrival.*

*This clause is based on Clause 2 of UNECE Conditions 188B, by kind permission of the United Nations.

Clause 20 (Time for Completion) – As the Contract will only include supervision of erection, the Contractor cannot undertake that erection will be completed within a specific time. Therefore, for the purposes of this Clause, completion of delivery C & F shall be deemed to constitute completion of the Works.

Contract Price Adjustment

We propose that contract price adjustment in respect of the price for goods shall apply in accordance with the BEAMA *Contract Price Adjustment Clause and Formula for use with Export Contracts – Electrical Machinery: (for which there is no other specific formula)*, edition January 1979, corrected edition,* a copy of which is enclosed.

We further propose that contract price adjustment in respect of the price for site work shall apply in accordance with the same BEAMA Clause and Formula subject to the amendment that adjustment shall take place at the rate of 0.95 per cent of such price per 1.0 per cent difference between the Labour Cost Index published for the month in which the tender date falls and the average of the Index figures published for the last two-thirds of the period during which the Contractor is carrying out site work, this difference being expressed as a percentage of the former Index figure.

The BEAMA Labour Cost Index referred to shall be the Labour Cost Index for Electrical Engineering.*

References to the *Trade and Industry Journal* shall be read as references to *British Business*, which is the new name of the journal.†

*This is only one of several BEAMA CPA Clauses and Formulae. If another is adopted, this form may need consequential amendment. In particular, reference to the BEAMA CPA Clause applicable to Mechanical Plant, and to the Labour Cost Index for Mechanical Engineering, may be appropriate. See p. 16.

†The above form requires that the separate prices to be adjusted shall be identifiable from the contract documents.

Trade Terms

The trade term 'C & F' shall be defined by the current edition of Incoterms, where consistent with this tender. A copy of the definition is enclosed.

Validity

This tender is valid for acceptance until and including *, but shall, after that date, be null and void unless extended by us in writing.

*Date in full, e.g. 'Monday, 19 January 1981'.

Yours faithfully

(Signature)

FORM 62

Tender based on General Conditions No 574A for the Supply and Erection of Plant and Machinery for Import and Export, prepared under the auspices of the United Nations Economic Commission for Europe, Geneva, March 1957, providing for Delivery CIF and Supervision of Erection on Site

Dear Sirs

(heading)

Prices

We offer to supply CIF ,* and supervise the erection on Site at

of the items described in the accompanying Specification for:

£

(pounds sterling)

*Port of destination.

(Our services in respect of site work are offered on a time basis. The above price includes an estimated sum of £ for such services, computed at the rates and for the estimated duration shown in the accompanying Specification. The sum actually payable for such services shall be computed on a time basis, and may be either more or less than the estimated sum.)*

*Delete if lump sum.

Programme

Our best estimate is that delivery CIF will be achieved within of receiving your order, and that we could complete supervision of erection within a further of receiving your order, making a total of .

We enclose * showing the basis of this estimate.

*Appropriate document, e.g. bar chart or critical path network.

Terms of Payment

All prices are payable in pounds sterling in the United Kingdom.

We propose that payment shall be made as follows:

(a) %(per cent) in advance.

(b) The balance through a Confirmed Irrevocable Letter of Credit providing for payment in full upon presentation of documents, to be increased according to any contractual increases in the price, to be extended according to any increase in the period required for execution of our respective obligations under the Contract, and otherwise complying with the following description:

  (i) Confirmed by an acceptable bank in London.

  (ii) Irrevocably issued, confirmed and advised to us by .[a]

(iii) Date and place of expiry – [a] in London.
(iv) Beneficiary – ourselves.
(v) Amount – the balance of the price.
(vi) Credit available with confirming bank by payment.
(vii) Partial shipments allowed.
(viii) Transhipment allowed.
(ix) Shipment from .[b]
(x) For transportation to .[c]
(xi) Shipment of .[d]
(xii) Documents required:

Commercial Invoice

In respect of goods, full set of clean on board marine Bills of Lading (which shall be acceptable although to the effect that the goods have been shipped on deck or unprotected*) and Insurance Certificate.

In respect of other work, your or your Engineer's Certificate.†

*Delete if inappropriate.

†If this sentence is omitted, payment for such other work should be made out of the letter of credit on presentation only of commercial invoice.

(xiii) Period after issuance of shipping documents for presentation – days.
(xiv) Subject otherwise to the *Uniform Customs and Practice for Documentary Credits,* 1974, ICC Publication No 290.[e]

[a]Date in full, e.g. 'Monday, 19 January 1981'.

[b]Port of shipment, or e.g. 'any United Kingdom port'.

[c]Port of destination.

[d]Short description of goods and site work.

[e]This description follows as closely as possible the *Standard Procedure for the Issuing of Documentary Credits by Cable, Telegram or Telex* (App VII *Standard Forms for Issuing Documentary Credits,* ICC Publication No 323).

Payment in respect of site work shall be due monthly at the end of each month.

Conditions of Contract

Where consistent with this tender, we propose that *General Conditions No 574A for the Supply and Erection of Plant and Machinery for Import and Export,* prepared under the auspices of the United Nations Economic Commission for Europe, Geneva, March 1957, a copy of which is enclosed, shall apply, subject to the following:

The Appendix shall be completed as follows:

Appendix to UNECE Conditions 574A

	Clause	
A. Maximum amount recoverable on termination by Contractor for failure to take delivery or make payment	10.2 and 11.7	£ (pounds sterling)

B. Rate of interest on overdue payments	11.7	2% (two per cent) per annum over the * Bank base lending rate from time to time in force
C. Period of delay in payment authorizing termination by Contractor	11.7	months
D. Percentage to be deducted for each week's delay	20.3	% (per cent)
E. Maximum percentage which the deductions above may not exceed	20.3	% (per cent)
F. Maximum amount recoverable for non-completion	20.5	£ (pounds sterling)
G. Maximum post-ponement of taking over tests by Contractor	22.3	weeks
H. Guarantee Period for original Works and parts replaced or renewed	23.2 and 23.5	months
I. Maximum indemnities for personal injury or damage	24.3	£ (pounds sterling)

*Insert name of bank.

Clause 7 (Erection on a Time Basis and Lump-Sum Erection) – Supervision of erection is carried out (on a time basis)* (for a lump sum).*

*Delete alternatives.

Erection will be carried out by the Purchaser, who shall, at his own expense, provide the skilled and unskilled labour, all equipment and everything necessary for the erection of the Plant.

References to 'erection' in Clause 7 shall be read as references to 'supervision of erection'. The Contractor's obligation in this respect will be to provide the services of one or more competent engineers

(a) to give to the Purchaser or his representative mentioned in Clause 13.1 the necessary instructions for the erection of the Plant by the Purchaser and, if provided in the accompanying Specification or other documents, for its commissioning by him; and
(b) to supervise the manner in which the Contractor's instructions have been carried out.

The number and qualifications of the Contractor's staff, and the estimated duration of erection, shall be as specified in this tender or in the accompanying Specification or other documents.

The date on which the Contractor's staff should arrive on Site shall be as provided in the accompanying Specification or other documents; if not so provided, the Purchaser shall give the Contractor not less than one month's notice requiring such arrival.*

*This clause is based on Clause 2 of UNECE Conditions 188B, by kind permission of the United Nations.

Clause 20 (Time for Completion) – As the Contract will only include supervision of erection, the Contractor cannot undertake that erection will be completed within a specific time. Therefore, for the purposes of this Clause, completion of delivery CIF shall be deemed to constitute completion of the Works.

Contract Price Adjustment

We propose that contract price adjustment in respect of the price for goods shall apply in accordance with the BEAMA *Contract Price Adjustment Clause and Formula for use with Export Contracts – Electrical Machinery: (for which there is no other specific formula)*, edition January 1979, corrected edition,* a copy of which is enclosed.

We further propose that contract price adjustment in respect of the price for site work shall apply in accordance with the same BEAMA Clause and Formula subject to the amendment that adjustment shall take place at the rate of 0.95 per cent of such price per 1.0 per cent difference between the Labour Cost Index published for the month in which the tender date falls and the average of the Index figures published for the last two-thirds of the period during which the Contractor is carrying out site work, this difference being expressed as a percentage of the former Index figure.

The BEAMA Labour Cost Index referred to shall be the Labour Cost Index for Electrical Engineering.*

References to the *Trade and Industry Journal* shall be read as references to *British Business*, which is the new name of the journal.†

*This is only one of several BEAMA CPA Clauses and Formulae. If another is adopted, this form may need consequential amendment. In particular, reference to the BEAMA CPA Clause applicable to Mechanical Plant, and to the Labour Cost Index for Mechanical Engineering, may be appropriate. See p. 16.

†The above form requires that the separate prices to be adjusted shall be identifiable from the contract documents.

Trade Terms

The trade term 'CIF' shall be defined by the current edition of Incoterms, where consistent with this tender. A copy of the definition is enclosed.

Validity

This tender is valid for acceptance until and including *, but shall, after that date, be null and void unless extended by us in writing.

*Date in full, e.g. 'Monday, 19 January 1981'.

Yours faithfully

(Signature)

FORM 63

Tender based on General Conditions for the Erection of Plant and Machinery Abroad No 574D, prepared by the Secretariat of the United Nations Economic Commission for Europe, Geneva, August 1963

Dear Sirs

(heading)

Prices
We offer to erect on Site at

the Plant and Machinery described in the accompanying Specification for (the approximate price of*):
£
(pounds sterling)
(Our services are offered on a time basis. The above approximate price is computed at the rates and for the estimated duration shown in the accompanying Specification. The sum actually payable for such services shall be computed on a time basis, and may be more or less than the above approximate price.)*

*Delete if lump sum.

Programme
Our best estimate is that completion of our work will be achieved within of receiving your order. (We enclose * showing the basis of this estimate.)†

*Appropriate document, e.g. bar chart or critical path network.
†Delete if inappropriate.

Terms of Payment
All prices are payable in pounds sterling in the United Kingdom.

We propose that payment shall be made as follows:

(a) % (per cent) in advance.
(b) The balance through a Confirmed Irrevocable Letter of Credit providing for payment in full upon presentation of documents, to be increased according to any contractual increases in the price, to be extended according to any increase in the period required for execution of our respective obligations under the Contract, and otherwise complying with the following description:
 (i) Confirmed by an acceptable bank in London.
 (ii) Irrevocably issued, confirmed and advised to us by [a]
 (iii) Date and place of expiry – [a] in London.
 (iv) Beneficiary – ourselves.
 (v) Amount – the balance of the price.

(vi) Credit available with confirming bank by payment.
(vii) Documents to relate to .[b]
(viii) Documents required.

(Certified)* Commercial Invoice

Your or your Engineer's Certificate.†

*Payment would, in fact, be made out of the letter of credit even if certified only by the exporter. Where no further definition is given 'banks will accept such documents as tendered'. See *Uniform Customs and Practice for Documentary Credits*, 1974, Article 33, ICC Publication No. 290.

†If this sentence is omitted, payment should be made out of the letter of credit on presentation only of commercial invoice.

(ix) Subject otherwise to the *Uniform Customs and Practice for Documentary Credits*, 1974, ICC Publication No 290.[c]

[a]Date in full, e.g. 'Monday, 19 January 1981'.
[b] Short description of site work.
[c] This description follows as closely as possible the *Standard Procedure for the Issuing of Documentary Credits by Cable, Telegram or Telex* (App VII *Standard Forms for Issuing Documentary Credits*, ICC Publication No 323).

Payment in respect of site work shall be due monthly at the end of each month.

Conditions of Contract

Where consistent with this tender, we propose that *General Conditions for the Erection of Plant and Machinery Abroad No 574D*, prepared by the Secretariat of the United Nations Economic Commission for Europe, Geneva, August 1963, a copy of which is enclosed, shall apply, subject to the following:

The Appendix shall be completed as follows:

Appendix to UNECE Conditions 574D

	Clause		
A. Percentage of reduction for each week's delay	14.3	% (	per cent)
B. Maximum amount of above reduction	14.3	% (	per cent)
C. Maximum amount recoverable for non-completion	14.5	£ (	pounds sterling)
D. Rate of interest on overdue payments	15.6	2% (two per cent) per annum over the * Bank base lending rate from time to time in force	
E. Period of delay in payment authorizing termination by erector	15.6		months

F. Maximum amount recoverable on termination by erector for failure to make payment	15.6	£ (	pounds sterling)
G. Maximum indemnities payable by erector for for repair or replacement of defective parts	18.1	£ (	pounds sterling)
H. Guarantee period for erection	18.2		months
I. Maximum indemnities for personal injury or damage	19.3	£ (	pounds sterling)

*Insert name of bank.

Clause 6 (Erection on a Time Basis and Lump-Sum Erection) – Erection is carried out (on a time basis)* (for a lump sum)*

*Delete alternatives.

Contract Price Adjustment

We propose that contract price adjustment shall apply in accordance with the BEAMA *Contract Price Adjustment Clause and Formula for use with Export Contracts – Electrical Machinery: (for which there is no other specific formula)*, edition January 1979, corrected edition,* a copy of which is enclosed, subject to the amendment that adjustment shall take place at the rate of 0.95 per cent of the price per 1.0 per cent difference between the Labour Cost Index published for the month in which the tender date falls and the average of the Index figures published for the last two-thirds of the period during which the Contractor is carrying out site work, this difference being expressed as a percentage of the former Index figure.

The BEAMA Labour Cost Index referred to shall be the Labour Cost Index for Electrical Engineering.*

References to the *Trade and Industry Journal* shall be read as references to *British Business*, which is the new name of the journal.

*This is only one of several BEAMA CPA Clauses and Formulae. If another is adopted, this form may need consequential amendment. In particular, reference to the BEAMA CPA Clause applicable to Mechanical Plant, and to the Labour Cost Index for Mechanical Engineering, may be appropriate. See p. 16.

Validity

This tender is valid for acceptance until and including *, but shall, after that date, be null and void unless extended by us in writing.

*Date in full, e.g. 'Monday, 19 January 1981'.

Yours faithfully

(Signature)

FORM 64

FOB Tender based on General Conditions of Sale No 730 for the Import and Export of Durable Consumer Goods and of other Engineering Stock Articles, prepared under the auspices of the United Nations Economic Commission for Europe, Geneva, March 1961

Dear Sirs

(heading)

Prices
We offer to supply FOB * the items described in the accompanying Specification for:
£
(pounds sterling)
*Port of shipment.

Programme
Our best estimate is that delivery FOB will be achieved within of receiving your order.

Terms of Payment
All prices are payable in pounds sterling in the United Kingdom.
We propose that payment shall be made as follows:
(a) %(per cent) in advance.
(b) The balance through a Confirmed Irrevocable Letter of Credit providing for payment in full upon presentation of documents, to be increased according to any contractual increases in the price, to be extended according to any increase in the period required for execution of our respective obligations under the Contract, and otherwise complying with the following description:
(i) Confirmed by an acceptable bank in London.
(ii) Irrevocably issued, confirmed and advised to us by .[a]
(iii) Date and place of expiry – [a] in London.
(iv) Beneficiary – ourselves.
(v) Amount – the balance of the price.
(vi) Credit available with confirming bank by payment.
(vii) Partial shipments allowed.
(viii) Transhipment allowed.
(ix) Shipment from .[b]
(x) For transportation to .[c]
(xi) Shipment of .[d]
(xii) Documents required:
Commercial Invoice

Full set of clean on board marine Bills of Lading (which shall be acceptable although to the effect that the goods have been shipped on deck or unprotected*).

*Delete if inappropriate.

(xiii) Period after issuance of shipping documents for presentation – days.

(xiv) Subject otherwise to the *Uniform Customs and Practice for Documentary Credits*, 1974, ICC Publication No 290.[e]

[a] Date in full, e.g. 'Monday, 19 January 1981'.

[b] Port of shipment, or e.g. 'any United Kingdom port'.

[c] Port of destination.

[d] Short description of goods.

[e] This description follows as closely as possible the *Standard Procedure for the Issuing of Documentary Credits by Cable, Telegram or Telex* (App VII *Standard Forms for Issuing Documentary Credits*, ICC Publication No 323).

Conditions of Contract

Where consistent with this tender, we propose that *General Conditions of Sale No 730 for the Import and Export of Durable Consumer Goods and of Other Engineering Stock Articles*, prepared under the auspices of the United Nations Economic Commission for Europe, Geneva, March 1961, a copy of which is enclosed, shall apply, subject to the following:

The Appendix shall be completed as follows:

Appendix to UNECE Conditions 730

	Paragraphs of General Conditions
A. Period after which the parties are entitled to consider the contract as never having been formed if the necessary licence or authorization cannot be obtained	2.3
B. Length of the period of grace for delivery	6.2
C. Period of delay in payment authorizing termination by the Vendor	7.3
D. Period for exercise of the Purchaser's right of rejection	8.1
E. Guarantee period starting on passing of the risk	9.2
F. Guarantee period from sale of goods to first end user	9.2

G. Designation of arbitral body specified by the parties for the purpose of settling disputes arising out of or in connection with the contract — 11.1

Clause 7.2 (b) – Interest – if the payment of any sum payable to us shall be delayed, interest at the rate of 2% (two per cent) per annum over the * Bank base lending rate from time to time in force on the amount of the delayed payment for the period of the delay shall be added to the price.

*Insert name of bank.

Contract Price Adjustment

We propose that contract price adjustment shall apply in accordance with the BEAMA *Contract Price Adjustment Clause and Formula for use with Export Contracts – Electrical Machinery: (for which there is no other specific formula)*, edition January 1979, corrected edition,* a copy of which is enclosed.

The BEAMA Labour Cost Index referred to shall be the Labour Cost Index for Electrical Engineering.*

References to the *Trade and Industry Journal* shall be read as references to *British Business*, which is the new name of the journal.

*This is only one of several BEAMA CPA Clauses and Formulae. If another is adopted, this form may need consequential amendment. In particular, reference to the BEAMA CPA Clause applicable to Mechanical Plant, and to the Labour Cost Index for Mechanical Engineering, may be appropriate. See p. 16.

Trade Terms

The trade term 'FOB' shall be defined by the current edition of Incoterms, where consistent with this tender. A copy of the definition is enclosed.

Validity

This tender is valid for acceptance until and including *, but shall, after that date, be null and void unless extended by us in writing.

*Date in full, e.g. 'Monday, 19 January 1981'.

Yours faithfully

(Signature)

FORM 65

C & F Tender based on General Conditions of Sale No 730 for the Import and Export of Durable Consumer Goods and of other Engineering Stock Articles, prepared under the auspices of the United Nations Economic Commission for Europe, Geneva, March 1961

Dear Sirs

(heading)

Prices
We offer to supply C & F * the items described in the accompanying Specification for:
£
(pounds sterling)
*Port of destination.

Programme
Our best estimate is that delivery C & F will be achieved within of receiving your order.

Terms of Payment
All prices are payable in pounds sterling in the United Kingdom.
We propose that payment shall be made as follows:

(a) % (per cent) in advance.

(b) The balance through a Confirmed Irrevocable Letter of Credit providing for payment in full upon presentation of documents, to be increased according to any contractual increases in the price, to be extended according to any increase in the period required for execution of our respective obligations under the Contract, and otherwise complying with the following description:

(i) Confirmed by an acceptable bank in London.
(ii) Irrevocably issued, confirmed and advised to us by .[a]
(iii) Date and place of expiry – [a] in London.
(iv) Beneficiary – ourselves.
(v) Amount – the balance of the price.
(vi) Credit available with confirming bank by payment.
(vii) Partial shipments allowed.
(viii) Transhipment allowed.
(ix) Shipment from .[b]
(x) For transportation to .[c]
(xi) Shipment of .[d]
(xii) Documents required:

Commercial Invoice

Full set of clean on board marine Bills of Lading (which shall be acceptable although to the effect that the goods have been shipped on deck or unprotected*).

*Delete if inappropriate.

(xiii) Period after issuance of shipping documents for presentation – days.

(xiv) Subject otherwise to the *Uniform Customs and Practice for Documentary Credits*, 1974, ICC Publication No 290.[e]

[a]Date in full, e.g. 'Monday, 19 January 1981'.

[b]Port of shipment, or e.g. 'any United Kingdom port'.

[c]Port of destination.

[d]Short description of goods.

[e]This description follows as closely as possible the *Standard Procedure for the Issuing of Documentary Credits by Cable, Telegram or Telex* (App VII *Standard Forms for Issuing Documentary Credits*, ICC Publication No 323).

Conditions of Contract

Where consistent with this tender, we propose that *General Conditions of Sale No 730 for the Import and Export of Durable Consumer Goods and of Other Engineering Stock Articles*, prepared under the auspices of the United Nations Economic Commission for Europe, Geneva, March 1961, a copy of which is enclosed, shall apply, subject to the following:

The Appendix shall be completed as follows:

Appendix to UNECE Conditions 730

	Paragraphs of General Conditions
A. Period after which the parties are entitled to consider the contract as never having been formed if the necessary licence or authorization cannot be obtained	2.3
B. Length of the period of grace for delivery	6.2
C. Period of delay in payment authorizing termination by the Vendor	7.3
D. Period for exercise of the Purchaser's right of rejection	8.1
E. Guarantee period starting on passing of the risk	9.2
F. Guarantee period from sale of goods to first end user	9.2

G. Designation of arbitral body specified by the parties for the purpose of settling disputes arising out of or in connection with the contract — 11.1

Clause 7.2 (b) – Interest – If the payment of any sum payable to us shall be delayed, interest at the rate of 2% (two per cent) per annum over the * Bank base lending rate from time to time in force on the amount of the delayed payment for the period of the delay shall be added to the price.

*Insert name of bank.

Contract Price Adjustment

We propose that contract price adjustment shall apply in accordance with the BEAMA *Contract Price Adjustment Clause and Formula for use with Export Contracts – Electrical Machinery: (for which there is no other specific formula)*, edition January 1979, corrected edition,* a copy of which is enclosed.

The BEAMA Labour Cost Index referred to shall be the Labour Cost Index for Electrical Engineering.*

References to the *Trade and Industry Journal* shall be read as references to *British Business*, which is the new name of the journal.

*This is only one of several BEAMA CPA Clauses and Formulae. If another is adopted, this form may need consequential amendment. In particular, reference to the BEAMA CPA Clause applicable to Mechanical Plant, and to the Labour Cost Index for Mechanical Engineering, may be appropriate. See p. 16.

Trade Terms

The trade term 'C & F' shall be defined by the current edition of Incoterms, where consistent with this tender. A copy of the definition is enclosed.

Validity

This tender is valid for acceptance until and including *, but shall, after that date, be null and void unless extended by us in writing.

*Date in full, e.g. 'Monday, 19 January 1981'.

Yours faithfully

(Signature)

FORM 66

CIF Tender based on General Conditions of Sale No 730 for the Import and Export of Durable Consumer Goods and of other Engineering Stock Articles, prepared under the auspices of the United Nations Economic Commission for Europe, Geneva, March 1961

Dear Sirs

(heading)

Prices
We offer to supply CIF * the items described in the accompanying Specification for:
£
(pounds sterling)
*Port of destination.

Programme
Our best estimate is that delivery CIF will be achieved within of receiving your order.

Terms of Payment
All prices are payable in pounds sterling in the United Kingdom.
We propose that payment shall be made as follows:
(a) % (per cent) in advance.
(b) The balance through a Confirmed Irrevocable Letter of Credit providing for payment in full upon presentation of documents, to be increased according to any contractual increases in the price, to be extended according to any increase in the period required for execution of our respective obligations under the Contract, and otherwise complying with the following description:
 (i) Confirmed by an acceptable bank in London.
 (ii) Irrevocably issued, confirmed and advised to us by .[a]
 (iii) Date and place of expiry – [a] in London.
 (iv) Beneficiary – ourselves.
 (v) Amount – the balance of the price.
 (vi) Credit available with confirming bank by payment.
 (vii) Partial shipments allowed.
 (viii) Transhipment allowed.
 (ix) Shipment from .[b]
 (x) For transportation to .[c]
 (xi) Shipment of .[d]
 (xii) Documents required:
 Commercial Invoice

Full set of clean on board marine Bills of Lading (which shall be acceptable although to the effect that the goods have been shipped on deck or unprotected*).

Insurance Certificate

*Delete if inappropriate.

(xiii) Period after issuance of shipping documents for presentation – days.

(xiv) Subject otherwise to the *Uniform Customs and Practice for Documentary Credits*, 1974, ICC Publication No 290.[e]

[a] Date in full, e.g. 'Monday, 19 January 1981'.

[b] Port of shipment, or e.g. 'any United Kingdom port'.

[c] Port of destination.

[d] Short description of goods.

[e] This description follows as closely as possible the *Standard Procedure for the Issuing of Documentary Credits by Cable, Telegram or Telex* (App VII *Standard Forms for Issuing Documentary Credits*, ICC Publication No 323).

Conditions of Contract

Where consistent with this tender, we propose that *General Conditions of Sale No 730 for the Import and Export of Durable Consumer Goods and of Other Engineering Stock Articles*, prepared under the auspices of the United Nations Economic Commission for Europe, Geneva, March 1961, a copy of which is enclosed, shall apply, subject to the following:

The Appendix shall be completed as follows:

Appendix to UNECE Conditions 730

	Paragraphs of General Conditions
A. Period after which the parties are entitled to consider the contract as never having been formed if the necessary licence or authorization cannot be obtained	2.3
B. Length of the period of grace for delivery	6.2
C. Period of delay in payment authorizing termination by the Vendor	7.3
D. Period for exercise of the Purchaser's right of rejection	8.1
E. Guarantee period starting on passing of the risk	9.2
F. Guarantee period from sale of goods to first end user	9.2

G. Designation of arbitral body specified by the parties for the purpose of settling disputes arising out of or in connection with the contract — 11.1

Clause 7.2 (b) – Interest – If the payment of any sum payable to us shall be delayed, interest at the rate of 2% (two per cent) per annum over the * Bank base lending rate from time to time in force on the amount of the delayed payment for the period of the delay shall be added to the price.

*Insert name of bank.

Contract Price Adjustment

We propose that contract price adjustment shall apply in accordance with the BEAMA *Contract Price Adjustment Clause and Formula for use with Export Contracts – Electrical Machinery: (for which there is no other specific formula)*, edition January 1979, corrected edition,* a copy of which is enclosed.

The BEAMA Labour Cost Index referred to shall be the Labour Cost Index for Electrical Engineering.*

References to the *Trade and Industry Journal* shall be read as references to *British Business*, which is the new name of the journal.

*This is only one of several BEAMA CPA Clauses and Formulae. If another is adopted, this form may need consequential amendment. In particular, reference to the BEAMA CPA Clause applicable to Mechanical Plant, and to the Labour Cost Index for Mechanical Engineering, may be appropriate. See p. 16.

Trade Terms

The trade term 'CIF' shall be defined by the current edition of Incoterms, where consistent with this tender. A copy of the definition is enclosed.

Validity

This tender is valid for acceptance until and including *, but shall, after that date, be null and void unless extended by us in writing.

*Date in full, e.g. 'Monday, 19 January 1981'.

Yours faithfully

(Signature)

FORM 67

Additional section for incorporation, in Forms 37–48 and 52–63, of Part 1 of ORGALIME Conditions for the Provision of Technical Personnel Abroad, January 1970 (reprint February 1977)

Add at the end of the Conditions of Contract section:

Part 1 of ORGALIME Conditions for the Provision of Technical Personnel Abroad, January 1970 (reprint February 1977) shall apply. A copy of the Conditions is enclosed. In case of any inconsistency, such Conditions shall prevail over the other Conditions of Contract referred to in this tender.

FORM 68

Tender for the Provision of Technical Personnel Abroad based on ORGALIME Conditions for the Provision of Technical Personnel Abroad, Parts 1 and 2, January 1970 (reprint February 1977)

Dear Sirs

(heading)

Prices

We offer to carry out on Site at

the work described in the accompanying Specification for the approximate price of:

£

(pounds sterling)

Our services in respect of site work are offered on a time basis. The above approximate price is computed at the rates and for the estimated duration shown in the accompanying Specification. The sum actually payable for our services shall be computed on a time basis, and may be more or less than the above approximate price.

Programme

Our best estimate is that completion of our work will be achieved within of receiving your order.

Terms of Payment

All prices are payable in pounds sterling in the United Kingdom.

We propose that payment shall be made as follows:

(a) %(per cent) in advance.

(b) The balance through a Confirmed Irrevocable Letter of Credit providing for payment in full upon presentation of documents, to be increased according to any contractual increases in the price, to be extended according to any increase in the period required for execution of our respective obligations under the Contract, and otherwise complying with the following description:

(i) Confirmed by an acceptable bank in London.

(ii) Irrevocably issued, confirmed and advised to us by .[a]

(iii) Date and place of expiry – [a] in London.

(iv) Beneficiary – ourselves.

(v) Amount – the balance of the price.

(vi) Credit available with confirming bank by payment.

(vii) Documents to relate to .[b]

(viii) Documents required:

(Certified)* Commercial Invoice

Your or your Engineer's Certificate.†

*Payment would, in fact, be made out of the letter of credit even if certified only by the exporter. Where no further definition is given 'banks will accept such documents as tendered'. See *Uniform Customs and Practice for Documentary Credits*, 1974, Article 33, ICC Publication No. 290.

†If this sentence is omitted, payment should be made out of the letter of credit on presentation only of commercial invoice.

(ix) Subject otherwise to the *Uniform Customs and Practice for Documentary Credits*, 1974, ICC Publication No. 290.[c]

[a]Date in full, e.g. 'Monday, 19 January 1981'.

[b] Short description of site work.

[c]This description follows as closely as possible the *Standard Procedure for the Issuing of Documentary Credits by Cable, Telegram or Telex* (App VII *Standard Forms for Issuing Documentary Credits*, ICC Publication No 323).

Conditions of Contract

Where consistent with this tender, we propose that ORGALIME *Conditions for the Provision of Technical Personnel Abroad*, Parts 1 and 2, January 1970 (reprint February 1977), a copy of which is enclosed duly completed, shall apply.

Contract Price Adjustment

We propose that contract price adjustment shall apply in accordance with the BEAMA *Contract Price Adjustment Clause and Formula for use with Export Contracts – Electrical Machinery: (for which there is no other specific formula)*, edition January 1979, corrected edition,* a copy of which is enclosed, subject to the amendment that adjustment shall take place at the rate of 0.95 per cent of such price per 1.0 per cent difference between the Labour Cost Index published for the month in which the tender date falls and the average of the Index figures published for the last two-thirds of the period during which the Contractor is carrying out site work, this difference being expressed as a percentage of the former Index figure.

The BEAMA Labour Cost Index referred to shall be the Labour Cost Index for Electrical Engineering.*

References to the *Trade and Industry Journal* shall be read as references to *British Business*, which is the new name of the journal.

*This is only one of several BEAMA CPA Clauses and Formulae. If another is adopted, this form may need consequential amendment. In particular, reference to the BEAMA CPA Clause applicable to Mechanical Plant, and to the Labour Cost Index for Mechanical Engineering, may be appropriate. See p. 16.

Validity

This tender is valid for acceptance until and including *, but shall, after that date, be null and void unless extended by us in writing.

*Date in full, e.g. 'Monday, 19 January 1981'.

Yours faithfully

(Signature)

FORM 69

FOB Tender based on ORGALIME General Conditions for the Import and Export of Semi-Processed Goods and Components for Incorporation in Other Goods, February 1964

Dear Sirs

(heading)

Prices

We offer to supply FOB * the items described in the accompanying Specification for:

£

(pounds sterling)

*Port of shipment.

Programme

Our best estimate is that delivery FOB will be achieved within of receiving your order.

Terms of Payment

All prices are payable in pounds sterling in the United Kingdom.

We propose that payment shall be made as follows:

(a) % (per cent) in advance.

(b) The balance through a Confirmed Irrevocable Letter of Credit providing for payment in full upon presentation of documents, to be increased according to any contractual increases in the price, to be extended according to any increase in the period required for execution of our respective obligations under the Contract, and otherwise complying with the following description:

(i) Confirmed by an acceptable bank in London.
(ii) Irrevocably issued, confirmed and advised to us by .[a]
(iii) Date and place of expiry – [a] in London.
(iv) Beneficiary – ourselves.
(v) Amount – the balance of the price.
(vi) Credit available with confirming bank by payment.
(vii) Partial shipments allowed.
(viii) Transhipment allowed.
(ix) Shipment from .[b]
(x) For transportation to .[c]
(xi) Shipment of .[d]
(xii) Documents required:

Commercial Invoice

Full set of clean on board marine Bills of Lading (which shall be

acceptable although to the effect that the goods have been shipped on deck or unprotected*).

*Delete if inappropriate.

(xiii) Period after issuance of shipping documents for presentation – days.

(xiv) Subject otherwise to the *Uniform Customs and Practice for Documentary Credits*, 1974, ICC Publication No 290.[e]

[a] Date in full, e.g. 'Monday, 19 January 1981'.
[b] Port of shipment, or e.g. 'any United Kingdom port'.
[c] Port of destination.
[d] Short description of goods.
[e] This description follows as closely as possible the *Standard Procedure for the Issuing of Documentary Credits by Cable, Telegram or Telex* (App VII *Standard Forms for Issuing Documentary Credits*, ICC Publication No 323).

Conditions of Contract

Where consistent with this tender, we propose that ORGALIME *General Conditions for the Import and Export of Semi-Processed Goods and Components for Incorporation in other Goods*, February 1964, a copy of which is enclosed, shall apply, subject to the following:

Interest – If the payment of any sum payable to us shall be delayed, interest at the rate of 2% (two per cent) per annum over the * Bank base lending rate from time to time in force on the amount of the delayed payment for the period of the delay shall be added to the price.

*Insert name of bank.

Contract Price Adjustment

We propose that contract price adjustment shall apply in accordance with the BEAMA *Contract Price Adjustment Clause and Formula for use with Export Contracts – Electrical Machinery: (for which there is no other specific formula)*, edition January 1979, corrected edition,* a copy of which is enclosed.

The BEAMA Labour Cost Index referred to shall be the Labour Cost Index for Electrical Engineering.*

References to the *Trade and Industry Journal* shall be read as references to *British Business*, which is the new name of the journal.

*This is only one of several BEAMA CPA Clauses and Formulae. If another is adopted, this form may need consequential amendment. In particular, reference to the BEAMA CPA Clause applicable to Mechanical Plant, and to the Labour Cost Index for Mechanical Engineering, may be appropriate. See p. 16.

Trade Terms

The trade term 'FOB' shall be defined by the current edition of Incoterms, where consistent with this tender. A copy of the definition is enclosed.

Validity

This tender is valid for acceptance until and including *, but shall, after that date, be null and void unless extended by us in writing.

*Date in full, e.g. 'Monday, 19 January 1981'.

Yours faithfully

(Signature)

FORM 70

C & F Tender based on ORGALIME General Conditions for the Import and Export of Semi-Processed Goods and Components for Incorporation in Other Goods, February 1964

Dear Sirs

(heading)

Prices
We offer to supply C & F * the items described in the accompanying Specification for:
£
(pounds sterling)
*Port of destination.

Programme
Our best estimate is that delivery C & F will be achieved within of receiving your order.

Terms of Payment
All prices are payable in pounds sterling in the United Kingdom.

We propose that payment shall be made as follows:

(a) % (per cent) in advance.

(b) The balance through a Confirmed Irrevocable Letter of Credit providing for payment in full upon presentation of documents, to be increased according to any contractual increases in the price, to be extended according to any increase in the period required for execution of our respective obligations under the Contract, and otherwise complying with the following description:

(i) Confirmed by an acceptable bank in London.
(ii) Irrevocably issued, confirmed and advised to us by .[a]
(iii) Date and place of expiry – [a] in London.
(iv) Beneficiary – ourselves.
(v) Amount – the balance of the price.
(vi) Credit available with confirming bank by payment.
(vii) Partial shipments allowed.
(viii) Transhipment allowed.
(ix) Shipment from .[b]
(x) For transportation to .[c]
(xi) Shipment of .[d]
(xii) Documents required:

Commercial Invoice

Full set of clean on board marine Bills of Lading (which shall be acceptable although to the effect that the goods have been shipped on deck or unprotected*).

*Delete if inappropriate.

(xiii) Period after issuance of shipping documents for presentation – days.

(xiv) Subject otherwise to the *Uniform Customs and Practice for Documentary Credits*, 1974, ICC Publication No 290.[e]

[a]Date in full, e.g. 'Monday, 19 January 1981'.
[b]Port of shipment, or e.g. 'any United Kingdom port'.
[c]Port of destination.
[d]Short description of goods.
[e]This description follows as closely as possible the *Standard Procedure for the Issuing of Documentary Credits by Cable, Telegram or Telex* (App VII *Standard Forms for Issuing Documentary Credits*, ICC Publication No 323).

Conditions of Contract

Where consistent with this tender, we propose that ORGALIME *General Conditions for the Import and Export of Semi-Processed Goods and Components for Incorporation in other Goods*, February 1964, a copy of which is enclosed, shall apply, subject to the following:

Interest – If the payment of any sum payable to us shall be delayed, interest at the rate of 2% (two per cent) per annum over the * Bank base lending rate from time to time in force on the amount of the delayed payment for the period of the delay shall be added to the price.

*Insert name of bank.

Contract Price Adjustment

We propose that contract price adjustment shall apply in accordance with the BEAMA *Contract Price Adjustment Clause and Formula for use with Export Contracts – Electrical Machinery: (for which there is no other specific formula)*, edition January 1979, corrected edition,* a copy of which is enclosed.

The BEAMA Labour Cost Index referred to shall be the Labour Cost Index for Electrical Engineering.*

References to the *Trade and Industry Journal* shall be read as references to *British Business*, which is the new name of the journal.

*This is only one of several BEAMA CPA Clauses and Formulae. If another is adopted, this form may need consequential amendment. In particular, reference to the BEAMA CPA Clause applicable to Mechanical Plant, and to the Labour Cost Index for Mechanical Engineering, may be appropriate. See p. 16.

Trade Terms

The trade term 'C & F' shall be defined by the current edition of Incoterms, where consistent with this tender. A copy of the definition is enclosed.

Validity

This tender is valid for acceptance until and including *, but shall, after that date, be null and void unless extended by us in writing.

*Date in full, e.g. 'Monday, 19 January 1981'.

Yours faithfully

(Signature)

FORM 71

CIF Tender based on ORGALIME General Conditions for the Import and Export of Semi-Processed Goods and Components for Incorporation in Other Goods, February 1964

Dear Sirs

(heading)

Prices
We offer to supply CIF * the items described in the accompanying Specification for:
£
(pounds sterling)
*Port of destination.

Programme
Our best estimate is that delivery CIF will be achieved within of receiving your order.

Terms of Payment
All prices are payable in pounds sterling in the United Kingdom.

We propose that payment shall be made as follows:

(a) %(per cent) in advance.

(b) The balance through a Confirmed Irrevocable Letter of Credit providing for payment in full upon presentation of documents, to be increased according to any contractual increases in the price, to be extended according to any increase in the period required for execution of our respective obligations under the Contract, and otherwise complying with the following description:

(i) Confirmed by an acceptable bank in London.
(ii) Irrevocably issued, confirmed and advised to us by .[a]
(iii) Date and place of expiry – [a] in London.
(iv) Beneficiary – ourselves.
(v) Amount – the balance of the price.
(vi) Credit available with confirming bank by payment.
(vii) Partial shipments allowed.
(viii) Transhipment allowed.
(ix) Shipment from .[b]
(x) For transportation to .[c]
(xi) Shipment of .[d]
(xii) Documents required:

Commercial Invoice

Full set of clean on board marine Bills of Lading (which shall be acceptable although to the effect that the goods have been shipped on deck or unprotected*).

Insurance Certificate

*Delete if inappropriate.

(xiii) Period after issuance of shipping documents for presentation – days.

(xiv) Subject otherwise to the *Uniform Customs and Practice for Documentary Credits*, 1974, ICC Publication No 290.[e]

[a]Date in full, e.g. 'Monday, 19 January 1981'.

[b]Port of shipment, or e.g. 'any United Kingdom port'.

[c]Port of destination.

[d]Short description of goods.

[e]This description follows as closely as possible the *Standard Procedure for the Issuing of Documentary Credits by Cable, Telegram or Telex* (App VII *Standard Forms for Issuing Documentary Credits*, ICC Publication No 323).

Conditions of Contract

Where consistent with this tender, we propose that ORGALIME *General Conditions for the Import and Export of Semi-Processed Goods and Components for Incorporation in other Goods*, February 1964, a copy of which is enclosed, shall apply, subject to the following:

Interest – If the payment of any sum payable to us shall be delayed, interest at the rate of 2% (two per cent) per annum over the * Bank base lending rate from time to time in force on the amount of the delayed payment for the period of the delay shall be added to the price.

*Insert name of bank.

Contract Price Adjustment

We propose that contract price adjustment shall apply in accordance with the BEAMA *Contract Price Adjustment Clause and Formula for use with Export Contracts – Electrical Machinery: (for which there is no other specific formula)*, edition January 1979, corrected edition,* a copy of which is enclosed.

The BEAMA Labour Cost Index referred to shall be the Labour Cost Index for Electrical Engineering.*

References to the *Trade and Industry Journal* shall be read as references to *British Business*, which is the new name of the journal.

*This is only one of several BEAMA CPA Clauses and Formulae. If another is adopted, this form may need consequential amendment. In particular, reference to the BEAMA CPA Clause applicable to Mechanical Plant, and to the Labour Cost Index for Mechanical Engineering, may be appropriate. See p. 16.

Trade Terms

The trade term 'CIF' shall be defined by the current edition of Incoterms, where consistent with this tender. A copy of the definition is enclosed.

Validity

This tender is valid for acceptance until and including *, but shall, after that date, be null and void unless extended by us in writing.

*Date in full, e.g. 'Monday, 19 January 1981'.

Yours faithfully

(Signature)

FORM 72

Tender based on ORGALIME Model Form of Maintenance Contract, January 1980

Dear Sirs

(heading)

Prices

We offer to carry out maintenance on your installation
in
for an initial fee per (month)* (quarter)* (year)* in respect of normal maintenance of:
£
(pounds sterling)

The accompanying Specification describes in detail our proposals in respect of normal and additional maintenance services, and in respect of price adjustment. These aspects are covered in outline by Appendices 1, 2 and 3 of the Conditions of Contract referred to in this tender.

*Delete alternatives.

Terms of Payment

All prices are payable in pounds sterling in the United Kingdom.

We propose that payment of the fee for normal maintenance shall be made in advance for each (month)* (quarter)* (year)*. We reserve the right to payment in advance in respect of all other payments under the Contract.

*Delete alternatives.

Conditions of Contract

Where consistent with this tender, we propose that the ORGALIME *Model Form of Maintenance Contract*, January 1980, a copy of which is enclosed, duly completed, shall apply.

Contract Price Adjustment

We propose that contract price adjustment shall apply in accordance with Appendix 3 of the Conditions of Contract referred to in this tender, which we have duly completed.

Validity

This tender is valid for acceptance until and including *, but shall, after that date, be null and void unless extended by us in writing.

*Date in full, e.g. 'Monday, 19 January 1981'.

Yours faithfully

(Signature)

FORM 72A

Tender based on ORGALIME Model Form of Processing Contract, August 1981

Dear Sirs

(heading)

Prices

We offer to carry out at our works the processing described in the enclosed documents.

*The price for processing each item is:

£

(pounds sterling)

*The lump sum price for the Work is:

£

(pounds sterling)

*The quantity of items to be processed is approximately , and the price payable per item shall be calculated as follows:

(a) for the first , the sum of:

£ each

(pounds sterling each)

(b) for the remaining items processed, the sum of:

£ each

(pounds sterling each)

*These three methods of pricing are alternatives. If the second is chosen, the other contract documents must, of course, specify the number of items to be processed.

Terms of Payment

All prices are payable in pounds sterling in the United Kingdom.

We propose that payment shall be made as follows:

(a) %(per cent) in advance.

(b) The balance in respect of each item on notification by us that it is ready for despatch.

Conditions of Contract

Where consistent with this tender, we propose that the ORGALIME *Model Form of Processing Contract*, August 1981, a copy of which is enclosed, duly completed, shall apply.*

*Article 14 (Exclusivity and Non-Competition) might well be deleted.

Contract Price Adjustment

(*Note:* No contract price adjustment clause is suggested here, because the possible circumstances of processing contracts are so infinitely various.)

Validity

This tender is valid for acceptance until and including *, but shall, after that date, be null and void unless extended by us in writing.

*Date in full, e.g. 'Monday, 19 January 1981'.

Yours faithfully

(Signature)

FORM 73

Tender based on the Form of Sub-Contract (as amended 30 March 1973) Designed for Use in Conjunction with The Institution of Civil Engineers' General Conditions of Contract, published by The Federation of Civil Engineering Contractors (commonly called 'the ICE Sub-Contract')

Dear Sirs

(heading)

Prices
We offer to supply, deliver to and erect on Site at

the items described in the accompanying Specification for:
£
(pounds)

Programme
Our best estimate is that completion of our work will be achieved within of receiving your order. (We enclose * showing the basis of this estimate.)†

*Appropriate document, e.g. bar chart or critical path network.
†Delete if inappropriate.

Terms of Payment
We propose that payment shall be made as follows:

(a) % (per cent) in advance.

(b) The balance in accordance with the Conditions of Contract referred to in this tender.

Conditions of Contract
Where consistent with this tender, we propose that the *Form of Sub-Contract (as amended 30 March 1973) Designed for Use in Conjunction with the ICE General Conditions of Contract,* published by the Federation of Civil Engineering Contractors, a copy of which is enclosed, shall apply, subject to the following:

The Schedules shall be completed as shown on the enclosed copy.

Clause 4(2) (Contractor's Facilities) – The words in line 5 'but the Contractor shall have no liability to the Sub-Contractor' and lines 6, 7 and 8, shall be deleted.

Clause 6(1) (Commencement and Completion) –

(i) The words in line 2 'shall enter upon the Site and' shall be deleted.

(ii) The requisite written instructions to commence the execution of the Sub-Contract Works shall be issued to and received by the Sub-Contractor within days of receiving your order.

Interest – If the payment of any sum payable to us shall be delayed, interest at the rate of 2% (two per cent) per annum over the * Bank base lending rate from time to time in force on the amount of the delayed payment for the period of the delay shall be added to the price.

*Insert name of bank.

Contract Price Adjustment

We propose that contract price adjustment shall apply in accordance with the BEAMA *Contract Price Adjustment Clause and Formula for use with Home Contracts – Electrical Machinery: (for which there is no other specific formula)*, edition January 1979, corrected edition,* a copy of which is enclosed.

The provisions of the Clause and Formulae applicable to contracts including erection shall apply.

The BEAMA Labour Cost Index referred to shall be the Labour Cost Index for Electrical Engineering.*

References to the *Trade and Industry Journal* shall be read as references to *British Business,* which is the new name of the journal.

*This is only one of several BEAMA CPA Clauses and Formulae. If another is adopted, this form may need consequential amendment. In particular, reference to the BEAMA CPA Clause applicable to Mechanical Plant, and to the Labour Cost Index for Mechanical Engineering, may be appropriate. See p.16. Adaptation may well be needed in order to accord with the Main Contract, which may be fixed price, or subject to quite different CPA provisions.

Validity

This tender is valid for acceptance until and including *, but shall, after that date, be null and void unless extended by us in writing.

*Date in full, e.g. 'Monday, 19 January 1981'

Yours faithfully

(Signature)

FORM 74

Tender based on The Institution of Chemical Engineers' Model Form of Conditions of Contract for Process Plants Suitable for Lump Sum Contracts in the United Kingdom, October 1968, revised April 1981

Dear Sirs

(heading)

Prices

We offer to supply, deliver to and erect on Site at

the items described in the accompanying Specification for:
£
(pounds)

Programme

Our best estimate is that completion of our work will be achieved within of receiving your order. (We enclose * showing the basis of this estimate.)†

*Appropriate document, e.g. bar chart or critical path network.
†Delete if inappropriate.

Terms of Payment

We propose that payment shall be made as follows:

(a) 15% (fifteen per cent) in advance.
(b) The balance in accordance with the Conditions of Contract referred to in this tender.

Conditions of Contract

Where consistent with this tender, we propose that *The Institution of Chemical Engineers' Model Form of Conditions of Contract for Process Plants Suitable for Lump Sum Contracts in the United Kingdom,* October 1968, revised April 1981, a copy of which is enclosed, shall apply, subject to the following:

The Schedules 1–8 referred to in the Conditions of Contract are enclosed, duly completed.*

*Examples of Schedule 7 and Schedule 8, Part 1, are given. The other Schedules must be drafted specially for each project.

Clause 28.2 (Site Working Conditions) shall not apply.

Clause 39.5 (Payment) – * Bank base lending rate shall be substituted for Bank of England Minimum Lending Rate.

*Insert name of bank.

Clause 40.3 (Provisional and Prime Cost Sums) – To the amounts, if any, paid or payable to nominated Sub-Contractors by the Contractor shall be added % (per cent), which shall also be added to the Contract Price.

Contract Price Adjustment

We propose that contract price adjustment shall apply in accordance with the BEAMA *Contract Price Adjustment Clause and Formula for use with Home Contracts – Electrical Machinery: (for which there is no other specific formula)*, edition January 1979, corrected edition,* a copy of which is enclosed.

The provisions of the Clause and Formulae applicable to contracts including erection shall apply.

The BEAMA Labour Cost Index referred to shall be the Labour Cost Index for Electrical Engineering.*

References to the *Trade and Industry Journal* shall be read as references to *British Business*, which is the new name of the journal.

*This is only one of several BEAMA CPA Clauses and Formulae. If another is adopted, this form may need consequential amendment. In particular, reference to the BEAMA CPA Clause applicable to Mechanical Plant, and to the Labour Cost Index for Mechanical Engineering, may be appropriate. See p. 16.

Validity

This tender is valid for acceptance until and including *, but shall, after that date, be null and void unless extended by us in writing.

*Date in full, e.g. 'Monday, 19 January 1981'.

Yours faithfully

(Signature)

Schedule 7

(Payment of the Contract Price)

The Contractor shall submit invoices to the Engineer from time to time for Plant delivered to and work executed on the Site. Each invoice shall state the amount claimed, and particulars of work executed on the Site and of Plant delivered to the Site up to a date named in the invoice, and since the period covered by the last preceding Engineer's certificate for payment, if any. The Engineer shall certify for payment in accordance with Clause 39. Such certification shall be in respect of a gross amount, regardless of any advance payment made.

The Purchaser shall pay the amount of every sum so certified within 14 days of receiving the certificate therefore, but shall be entitled to deduct:

(a) 15% (fifteen per cent) in respect of the advance payment made under the Contract provided that such deductions shall cease when they equal such advance payment; and

(b) 10% (ten per cent) retention.

One-half of the retention shall be paid within 14 days of the issue of a Taking-Over Certificate. The balance of the retention shall be paid within 14 days of the issue of a Final Certificate. If a Taking-Over Certificate or Final Certificate relates to a section and not the whole of the Works, the payments of retention shall be made in respect of the relevant section, and the amount of such payments of retention shall, in the absence of agreement, be determined by the Engineer.

This scheme of payment is arithmetically presented as follows:

	% of Contract Price
Advance payment	15
Within 14 days of certification	75
Within 14 days of Taking-Over Certificate	5
Within 14 days of Final Certificate	5
	100%

Schedule 8

Liquidated Damages

Part I: For Delay

Liquidated damages shall be paid at the rate of ½% (one-half per cent) of the Contract Price of that portion of the Plant which cannot because of the Contractor's failure be put to its proper use in respect of each week, with a maximum of 5% (five per cent) of the Contract Price of that portion, and such damages shall be in full and final satisfaction of the Contractor's liability for the delay.

FORM 75

Tender based on The Institution of Chemical Engineers' Model Form of Conditions of Contract for Process Plants Suitable for Reimbursable Contracts in the United Kingdom, July 1976

Dear Sirs

(heading)

Prices
We offer to supply, deliver to and erect on Site at

the items described in the accompanying Specification for:
£
(pounds)

Programme
Our best estimate is that completion of our work will be achieved within of receiving your order. (We enclose * showing the basis of this estimate.)†
*Appropriate document, e.g. bar chart or critical path network.
†Delete if inappropriate.

Terms of Payment
We propose that payment shall be made as follows:
(a) % (per cent) in advance.
(b) The balance in accordance with the Conditions of Contract referred to in this tender.

Conditions of Contract
Where consistent with this tender, we propose that *The Institution of Chemical Engineers' Model Form of Conditions of Contract for Process Plants Suitable for Reimbursable Contracts in the United Kingdom*, July 1976, a copy of which is enclosed, shall apply, subject to the following:

The Schedules listed below and referred to in the Conditions of Contract are enclosed, duly completed:*
*List the Schedules. Some are obligatory, some optional. If a Schedule of Liquidated Damages for Delay is included, it may follow generally Schedule 8, Part 1, of the preceding form.

Clause 27.3 (Site Working Conditions) shall not apply.

Clause 39.4 (Terms of Payment) – * Bank base lending rate shall be substituted for Bank of England Minimum Lending Rate.
*Insert name of bank.

Clause 44 (Limitation on Contractor's Liability) – The Contractor's liability pursuant to Sub-Clause 44.2 shall not exceed £ (pounds).

Value Added Tax (VAT) – Prices exclude VAT. To the extent that VAT is properly chargeable on the supply to you of any goods or services under the Contract, you shall pay such VAT as an addition to payments otherwise due under the Contract.

Contract Price Adjustment

We propose that contract price adjustment shall apply in accordance with the BEAMA *Contract Price Adjustment Clause and Formula for use with Home Contracts – Electrical Machinery: (for which there is no other specific formula)*, edition January 1979, corrected edition,* a copy of which is enclosed.

The provisions of the Clause and Formulae applicable to contracts including erection shall apply.

The BEAMA Labour Cost Index referred to shall be the Labour Cost Index for Electrical Engineering.*

References to the *Trade and Industry Journal* shall be read as references to *British Business*, which is the new name of the journal.

*This is only one of several BEAMA CPA Clauses and Formulae. If another is adopted, this form may need consequential amendment. In particular, reference to the BEAMA CPA Clause applicable to Mechanical Plant, and to the Labour Cost Index for Mechanical Engineering, may be appropriate. See p. 16.

Validity

This tender is valid for acceptance until and including *, but shall, after that date, be null and void unless extended by us in writing.

*Date in full, e.g. 'Monday, 19 January 1981'.

Yours faithfully

(Signature)

PART THREE

Additional Forms

FORM 76

Firm Extra Price for Different Trade Terms

Prices

We offer to supply, FOB * the items described in the accompanying Specification for:

£

(pounds sterling)

If you require delivery CIF † instead of delivery FOB as stated above, our extra price‡ for so doing is:

£

(pounds sterling)

*Port of shipment.

†Port of destination.

‡If the alternative is accepted, this may change the Conditions of Contract which are suitable. Therefore, the Conditions of Contract should, if necessary, be quoted in the alternative, depending on which trade term is accepted by the customer.

FORM 77

Approximate Extra Price for Different Trade Terms

Prices
We offer to supply, FOB * the items described in the accompanying Specification for:
£
(pounds sterling)

If you require delivery CIF † instead of delivery FOB as stated above, we cannot quote a firm extra price for so doing, because of the variable nature of the insurance and freight charges involved.

However, we are prepared to undertake CIF delivery, on the basis of our charging the cost price of the insurance and freight charges plus % (per cent). On that basis, we estimate that our extra price‡ for CIF delivery would be:
£
(pounds sterling)

*Port of shipment.
†Port of destination.
‡See Note ‡ to preceding Form.

FORM 78

Firm Extra Price for Site Services

Prices
We offer to supply, FOB *the items described in the accompanying Specification for:
£
(pounds sterling)
If you require † of such items on Site at our extra price‡ for so doing is:
£
(pounds sterling)

*Port of shipment.
†Identify the site services e.g. 'supervision of erection', 'erection', 'commissioning'.
‡See Note ‡ to Form 76. Further Terms of Payment should be quoted which allow presentation of documents in respect of site work, in case the alternative is accepted.

FORM 79

Daily Extra Price for Site Services

Prices

We offer to supply, FOB *the items described in the accompanying Specification for:

£

(pounds sterling)

If you require † of such items on Site at our extra price per day per man for so doing is:

£

(pounds sterling)

We estimate that such site services will require men for days, and therefore we estimate that the total price‡ for such services will be:

£

(pounds sterling)

The sum actually payable for such services shall be computed on a time basis, and may be either more or less than the estimated sum.

*Port of shipment.

†Identify the site services, e.g. 'supervision of erection', 'erection', 'commissioning'.

‡See Note ‡ to preceding Form.

FORM 80

Firm Commitment by Tenderer to a Programme

Programme

We undertake that delivery * will be achieved within of receiving your order, and that we will complete supervision of erection within a further of receiving your order, making a total of .
We enclose † showing the basis of this programme.‡

*Appropriate trade term, e.g. 'FOB'.
†Appropriate document, e.g. bar chart or critical path network.
‡The use of words such as 'Our best estimate is that . . .' leaves the tenderer's legal commitment to a contractual completion date doubtful, under most published Conditions of Contract. The above form leaves no doubt that the tenderer will be liable for liquidated or other damages for delay after the prescribed time, unless otherwise excused by the terms of the Contract. See p.5.

FORM 81

Prices and Terms of Payment Requiring Payment in Foreign Currency

Prices
We offer to supply, deliver to and erect on Site at

the items described in the accompanying Specification for:
*
(†)

Terms of Payment
All prices are payable in †
in ‡

We propose that payment should be made as follows:

* Identification of foreign currency in figures.
† Identification of foreign currency in words.
‡ Identification of place of payment, e.g. 'the United Kingdom' or 'the Kingdom of Saudi Arabia'. Quotation of payment in foreign currency imposes the risk of fluctuation of exchange rates on the exporter. Quotation of payment in the foreign country imposes the risk and responsibility of remitting the funds from the foreign country to the United Kingdom on the exporter. The customer will have discharged his responsibility by paying in the foreign country, and will not be responsible if, for example, the foreign government prevents remittance to the United Kingdom by means of exchange control. These risks may be mitigated by ECGD or other insurance, and by forward currency operations by the exporter. See p.6.

FORM 82

Example of Prices and Terms of Payment Requiring Payment Partly in Foreign Currency and Partly in Sterling

Prices

We offer to supply, deliver to and erect on Site at Jeddah, Kingdom of Saudi Arabia, the items described in the accompanying Specification for the following prices:

Items	*Currency of Payment*	*Price*
Pumps	Pounds sterling	£ 1 250 000
Electric motors	United States dollars	$ 2 112 000
Erection	Saudi Riyals	SR 1 582 000

Terms of Payment

All prices are payable, in the appropriate currency indicated above, in .*

We propose that payment shall be made as follows:

*See Note‡ to preceding Form.

FORM 83

Tender Comparison with Foreign Currency

Prices
We offer to suply CIF * and supervise the erection on Site at

of the items described in the accompanying Specification for:
£
(pounds sterling)
Converted into † at the rate of exchange with pounds sterling on
‡ namely one pound sterling to §, the above price is equal to:
§
(§)
This conversion is stated for tender comparison purposes only. All prices are payable as stated in this tender, regardless of any rate of exchange fluctuation.

Terms of Payment
All prices are payable in pounds sterling in the United Kingdom.
We propose that payment shall be made as follows:

*Port of destination.
†Name of foreign currency.
‡Date in full, e.g. 'Monday, 19 January 1981'.
§Number and name of foreign currency, in figures and words.

FORM 84

Terms of Payment Requiring Confirmed Irrevocable Letter of Credit by Instalments

Terms of Payment

All prices are payable in pounds sterling in the United Kingdom.

We propose that payment shall be made as follows:

(a) % (per cent) in advance.

(b) The balance through a Confirmed Irrevocable Letter of Credit providing for payment in full upon presentation of documents, to be increased according to any contractual increases in the price, to be extended according to any increase in the period required for execution of our respective obligations under the Contract, and otherwise complying with the following description:

(i) Confirmed by an acceptable bank in London.

(ii) Irrevocably issued, confirmed and advised to us by .[a]

(iii) Date and place of expiry – to be always valid for at least six months into the future at any time, and to expire in London.

(iv) Beneficiary – ourselves.

(v) Amount – initially for a further % (per cent) of the price, to be increased by amendment by a further % (per cent) of the price every months thereafter, until payment in full is achieved.

(vi) Credit available with confirming bank by payment.

(vii) Partial shipments allowed.

(viii) Transhipment allowed.

(ix) Shipment from .[b]

(x) For transportation to .[c]

(xi) Shipment of .[d]

(xii) Documents required:

(xiii) Period after issuance of shipping documents for presentation – days.

(xiv) Subject otherwise to the *Uniform Customs and Practice for Documentary Credits,* 1974, ICC Publication No 290.[e]

[a] Date in full, e.g. 'Monday, 19 January 1981'.

[b] Port of shipment, or e.g. 'any United Kingdom port'.

[c] Port of destination.

[d] Short description of goods and site work, if any.

[e] This description follows as closely as possible the *Standard Procedure for the Issuing of Documentary Credits by Cable, Telegram or Telex* (App VII *Standard Forms for Issuing Documentary Credits,* ICC Publication No 323).

FORM 85

Terms of Payment Requiring Revolving Confirmed Irrevocable Letter of Credit

Terms of Payment

All prices are payable in pounds sterling in the United Kingdom.

We propose that payment shall be made as follows:

(a) % (per cent) in advance.

(b) The balance through a Cumulative Revolving Confirmed Irrevocable Letter of Credit providing for payment in full upon presentation of documents, to be increased according to any contractual increases in the price, to be extended according to any increase in the period required for execution of our respective obligations under the Contract, and otherwise complying with the following description:

(i) Confirmed by an acceptable bank in London.
(ii) Irrevocably issued, confirmed and advised to us by .[a]
(iii) Date and place of expiry – [a] in London.
(iv) Beneficiary – ourselves.
(v) Amount – the balance of the price, to be available cumulatively for £ (pounds sterling) per month for months.
(vi) Credit available with confirming bank by payment.
(vii) Partial shipments allowed.
(viii) Transhipment allowed.
(ix) Shipment from .[b]
(x) For transportation to .[c]
(xi) Shipment of .[d]
(xii) Documents required:

(xiii) Period after issuance of shipping documents for presentation – days.
(xiv) Subject otherwise to the *Uniform Customs and Practice for Documentary Credits*, 1974, ICC Publication No 290.[e]

[a]Date in full, e.g. 'Monday, 19 January 1981'.

[b]Port of shipment, or e.g. 'any United Kingdom port'.

[c]Port of destination.

[d]Short description of goods and site work, if any.

[e]This description follows as closely as possible the *Standard Procedure for the Issuing of Documentary Credits by Cable, Telegram or Telex* (App VII *Standard Forms for Issuing Documentary Credits*, ICC Publication No 323).

FORM 86

Terms of Payment Requiring Unconfirmed Irrevocable Letter of Credit

Terms of Payment

All prices are payable in pounds sterling in the United Kingdom.

We propose that payment shall be made as follows:

(a) % (per cent) in advance.

(b) The balance through an unconfirmed Irrevocable Letter of Credit providing for payment in full upon presentation of documents, to be increased according to any contractual increases in the price, to be extended according to any increase in the period required for execution of our respective obligations under the contract, and otherwise complying with the following description:

- (i) Irrevocably issued by an acceptable bank in your country, and advised to us by a bank in London by .[a]
- (ii) Date and place of expiry – [a] in London.
- (iii) Beneficiary – ourselves.
- (iv) Amount – the balance of the price.
- (v) Credit available with advising bank by payment.
- (vi) Partial shipments allowed.
- (vii) Transhipment allowed.
- (viii) Shipment from .[b]
- (ix) For transportation to .[c]
- (x) Shipment of .[d]
- (xi) Documents required:
- (xii) Period after issuance of shipping documents for presentation – days.
- (xiii) Subject otherwise to the *Uniform Customs and Practice for Documentary Credits*, 1974, ICC Publication No 290.[e]

[a] Date in full, e.g. 'Monday, 19 January 1981'.

[b] Port of shipment, or e.g. 'any United Kingdom port'.

[c] Port of destination.

[d] Short description of goods and site work, if any.

[e] This description follows as closely as possible the *Standard Procedure for the Issuing of Documentary Credits by Cable, Telegram or Telex* (App VII *Standard Forms for Issuing Documentary Credits*, ICC Publication No 323).

FORM 87

Example Allowing Presentation of Mate's Receipt or other Document in place of Bills of Lading under Letter of Credit

Clean document in proof of delivery of the goods on board ship (which shall be acceptable although to the effect that the goods have been shipped on deck or unprotected*).†

*Delete if inappropriate.
†All references to Bills of Lading in 'Documents required' should be deleted, and the above substituted. Under normal shipping procedure, it is necessary to produce the mate's receipt to the shipping company, in order to obtain a Bill of Lading.

FORM 88

Example Allowing Presentation of Warehouse Receipts in place of Bills of Lading under Letter of Credit

Full set of clean on board marine Bills of Lading (which shall be acceptable although to the effect that the goods have been shipped on deck or unprotected*) or, alternatively, warehouse receipts accompanied by our certificate that shipment has been delayed as a result of causes beyond our reasonable control.†

*Delete if inappropriate.

†This protects the exporter against such events as dock strikes. Naturally, customers may not be amenable to this suggestion.

FORM 89

Example Allowing Presentation of Combined Transport Documents under Letter of Credit

Terms of Payment

All prices are payable in pounds sterling in the United Kingdom.

We propose that payment shall be made as follows:

(a) % (per cent) in advance.

(b) The balance through a Confirmed Irrevocable Letter of Credit providing for payment in full upon presentation of documents, to be increased according to any contractual increases in the price, to be extended according to any increase in the period required for execution of our respective obligations under the Contract, and otherwise complying with the following description:

(i) Confirmed by an acceptable bank in London.

(ii) Irrevocably issued, confirmed and advised to us by .[a]

(iii) Date and place of expiry – [a] in London.

(iv) Beneficiary – ourselves.

(v) Amount – the balance of the price.

(vi) Credit available with confirming bank by payment.

(vii) Partial shipments allowed.

(viii) Transhipment allowed.

(ix) [b]Taking in charge from .[c]

(x) [b]For transportation to .[d]

(xi) Transportation of .[e]

(xii) Documents required:

Commercial Invoice

Full set of combined transport documents (which shall be acceptable although to the effect that the goods have been shipped on deck or unprotected[f]).

(xiii) Period after issuance of shipping documents for presentation – days.

(xiv) Subject otherwise to the *Uniform Customs and Practice for Documentary Credits*, 1974, ICC Publication No 290.[g]

[a]Date in full, e.g. 'Monday, 19 January 1981'.

[b]Stipulations as to specific ports of loading or discharge should be avoided if the credit stipulates, or allows, the presentation of a combined transport document (unless the combined transport is to commence and/or finish at a port of loading or discharge). See General Recommendations 21 and 22, *Standard Forms for Issuing Documentary Credits* ICC Publication No 323.

[c]Place of first delivery to combined transport operator, or e.g. 'any United Kingdom port or place'.

[d] Place of destination.

[e] Short description of goods and site work, if any.

[f] Delete if inappropriate.

[g]This description follows as closely as possible the *Standard Procedure for the Issuing of Documentary Credits by Cable, Telegram or Telex* (App VII *Standard Forms for issuing Documentary Credits*, ICC Publication No. 323).

FORM 90

Terms of Payment Requiring Payment by Bills of Exchange Payable at Sight – 'Documents Against Payment' or 'D/P'

Terms of Payment

All prices are payable in pounds sterling.

We propose that payment shall be made as follows:

(a) % (per cent) in advance in the United Kingdom.

(b) The balance through a sterling Bill of Exchange payable at sight, which will be presented to you for payment by a bank in your country. The relevant shipping documents will be released to you by the bank upon payment of the Bill, all bank charges being for your account.

FORM 91

Terms of Payment Requiring Payment by Bills of Exchange Payable after an Interval ('Usance') after Sight – 'Documents Against Acceptance' or 'D/A'

Terms of Payment

All prices are payable in pounds sterling in the United Kingdom.

We propose that payment shall be made as follows:

(a) % (per cent) in advance.

(b) The balance through a sterling Bill of Exchange payable days after sight, which will be presented to you for acceptance by a bank in your country. The relevant shipping documents will be released to you by the bank upon acceptance of the Bill, all bank charges being for your account.

FORM 92

Example of Terms of Payment Requiring Payment by Bills of Exchange over a Credit Period

Terms of Payment

All prices are payable in pounds sterling in the United Kingdom.

We propose that payment shall be made as follows:

(a) 15% (fifteen per cent) in advance.

(b) The balance in respect of each shipment to be payable in four half-yearly instalments, the first such half-yearly instalment to be due six months after the relevant date of shipment. Interest on the outstanding balance shall accrue at 10% (ten per cent) per annum from the date of each shipment. Payment of all such instalments and interest shall be made through four sterling Bills of Exchange in respect of each shipment, which will be presented to you for acceptance by a bank in your country. The relevant shipping documents will be released to you by the bank upon acceptance of the Bills, all bank charges being for your account.

For example, if a shipment with a gross invoice value of £250 000 is made on 3 February 1981, this arrangement will lead to the following result:

	£
Gross invoice value	250 000
Less 15% advance payment	37 500
Balance outstanding at date of shipment	212 500

Due Date of Bill	*Principal* £	*Interest at 10% p.a.* £ p	*Total of Bill* £ p
3 August 1981	53 125	2 656.25	55 781.25
3 February 1982	53 125	5 312.50	58 437.50
3 August 1982	53 125	7 968.75	61 093.75
3 February 1983	53 125	10 625.00	63 750.00
	£212 500	£26 562.50	£239 062.50

If the payment of any sum payable under the Bills shall be delayed, interest at the rate of 2% (two per cent) per annum over the * Bank base lending rate from time to time in force on the amount of the delayed payment for the period of the delay shall be added, and the Bills shall so state, together with a provision that, if any one of the Bills is not paid on the due date, all of the Bills remaining unpaid shall, at the holder's option, become immediately due and payable.†

*Insert name of bank.

†Payment by Bills of Exchange in the manner stated above would often be linked to an ECGD supplier credit arrangement.

FORM 93

Introduction to Qualifications to Customer's Conditions of Contract

Conditions of Contract

We have examined the Conditions of Contract incorporated in your tender documents, and agree that they should apply, where consistent with this tender and subject to the following qualifications:*

*If the Conditions of Contract proposed by the customer are covered in this book, certain qualifications possibly required by the tenderer are indicated by the Standard Tenders relating to those Conditions of Contract.

FORM 94

Offer of Bonds and Financial Guarantees

Bonds and Financial Guarantees

As required by your Instructions to Tenderers,* we offer the following bonds and financial guarantees:

(a) A tender bond, enclosed, for % (per cent) of the tender price.

(b) An advance payment bond, draft enclosed, for the amount of the advance payment to be made under the Contract, to be reduced *pro rata* according to payments becoming due under the Contract.†

(c) A performance bond, draft enclosed, for % (per cent) of the Contract Price to be released upon‡

*A tenderer will not generally volunteer bonds unless the customer requires them in his tender documents.

†If the customer has prescribed the form of the bonds and their amounts and terms, it will be sufficient to state: '(b) an advance payment bond and performance bond in accordance with your Instructions to Tenderers'. If they are not so prescribed, the tenderer may offer bonds which are not 'on demand', possibly specifying that they should be 'subject to the *Uniform Rules for Tender, Performance and Repayment Guarantees ('Contract Guarantees') of the International Chamber of Commerce* (ICC Publication No 325).

‡Specify the contractual point at which the release should take place, e.g. 'final acceptance' or such other phrase as is used in the conditions of contract.

FORM 95

Example of FOB Airport Tender to American Customer – Payment in US Dollars – Unconfirmed Irrevocable Letter of Credit – Fixed Price

Dear Sirs

(heading)

Prices
We offer to supply FOB London Heathrow Airport, or other UK airport selected by us, the items described in the accompanying Specification for:

$152 000 (One hundred and fifty-two thousand United States dollars).

Programme
Our best estimate is that delivery FOB Airport will be achieved within three months of receiving your order.

Terms of payment
All prices are payable in United States dollars in the United Kingdom.

We propose that payment shall be made as follows:

(a) 20% (twenty per cent) namely $30 400 (thirty thousand four hundred United States dollars) in advance to our Account No at *(name)* Bank, *(name)* Branch, *(address)*.

(b) The balance, namely $121 600 (one hundred and twenty-one thousand six hundred United States dollars) through an unconfirmed Irrevocable Letter of Credit providing for payment in full upon presentation of documents, to be increased according to any contractual increases in the price, to be extended according to any increase in the period required for execution of our respective obligations under the Contract, and otherwise complying with the following description:

(i) Irrevocably issued by an acceptable bank in the United States, and advised to us by a bank in London by one month after our receipt of your order.

(ii) Date and place of expiry – six months after our receipt of your order, in London.

(iii) Beneficiary – ourselves.

(iv) Amount – $121 600 (one hundred and twenty-one thousand six hundred United States dollars).

(v) Credit available with advising bank by payment.

(vi) Partial shipments allowed.

(vii) Transhipment allowed.

(viii) Despatch from any UK airport.

(ix) For transportation to New York, John F. Kennedy Airport.

(x) Shipment of .*

(xi) Documents required:
Commercial Invoice
Air Waybill

(xii) Period after issuance of shipping documents for presentation – 21 days.

(xiii) Subject otherwise to the *Uniform Customs and Practice for Documentary Credits,* 1974, ICC Publication No 290.†

*Short description of goods.

†This description follows as closely as possible the *Standard Procedure for the Issuing of Documentary Credits by Cable, Telegram or Telex* (App VII *Standard Forms for Issuing Documentary Credits,* ICC Publication No 323).

Conditions of Contract

Where consistent with this tender, we propose that BEAMA *Conditions of Sale (AE) for Machinery and Equipment (Exclusive of Erection) Export FOB, FOR and FOT,* edition December 1980, a copy of which is enclosed, shall apply, subject to the following:

Clause 11 (Liability for Delay) – The figures to be inserted shall be ½% (one-half per cent) and 5% (five per cent) respectively.

Interest – If the payment of any sum payable to us shall be delayed, interest at the rate of 2% (two per cent) per annum over the * Bank base lending rate from time to time in force on the amount of the delayed payment for the period of the delay shall be added to the price.

*Insert name of bank.

Trade Terms

The trade term 'FOB Airport' shall be defined by the current edition of Incoterms, where consistent with this tender. A copy of the definition is enclosed.

Validity

This tender is valid for acceptance until and including * but shall, after that date, be null and void unless extended by us in writing.

*Date in full, e.g. 'Monday, 19 January 1981'.

Yours faithfully

(Signature)

FORM 96

Example of Unconfirmed Irrevocable Letter of Credit and Notification for Beneficiary Required by Form 95*

*These forms are reproduced from International Chamber of Commerce publication No 323, *Standard Forms for Issuing Documentary Credits*, and copyright © in the publication is held by the International Chamber of Commerce, Paris. ICC Publications are available from the International Chamber of Commerce, 38 Cours Albert 1er, 75008 Paris, The British National Committee of the ICC, Centre Point, 103 New Oxford Street, London WC1A 1QB and from National Committees throughout the world.

Annexe - Appendix I (a)

NOM DE LA BANQUE ÉMETTRICE - NAME OF ISSUING BANK

New York X Bank (full address)
New York City

CRÉDIT DOCUMENTAIRE IRRÉVOCABLE - IRREVOCABLE DOCUMENTARY CREDIT

Numéro - Number: 12345

Lieu et date d'émission - Place and date of issue: New York City - 19 January 1981

Date et lieu de validité - Date and place of expiry: 19 June 1981 in London

Donneur d'ordre - Applicant: Importer, Inc., (full address) New York City

Bénéficiaire - Beneficiary: Exporter Limited (full address) London

Banque Notificatrice - Advising Bank: London Y Bank (full address) London

N° Réf. - Ref. No: 678910

Montant - Amount: $121,600 (One hundred and twenty-one thousand six hundred United States dollars)

Crédit utilisable auprès de - Credit available with: advising bank

par - by: [X] *PAIEMENT* PAYMENT [ ] *ACCEPTATION* ACCEPTANCE [ ] *NÉGOCIATION* NEGOTIATION

contre présentation des documents précisés ci-après against presentation of the documents detailed herein

[ ] ~~et de votre/vos traite(s) à~~ ~~and of your draft(s) at~~

~~tirée (s) sur~~ ~~drawn on~~

Expéditions partielles Partial shipments: [X] *autorisées* allowed [ ] *non autorisées* not allowed

Transbordement Transhipment: [X] *autorisé* allowed [ ] *non autorisé* not allowed

Embarquement/expédition/prise en charge de/à ~~Shipment~~/dispatch/~~taking in charge from~~/at: any UK Airport

à destination de for transportation to: New York Kennedy Airport

Commercial invoice

Air waybill

Covering electric motors for Z Project

Documents à présenter dans les 21 *jours après la date d'émission du/des document(s) d'expédition mais dans la période de validité du crédit.*

Documents to be presented within 21 days after the date of issuance of the shipping document(s) but within the validity of the credit.

Nous émettons par la présente ce crédit documentaire en votre faveur. Il est soumis aux Règles et Usances Uniformes relatives aux Crédits Documentaires (révision 1974, Publication N° 290 de la Chambre de Commerce Internationale, Paris, France), et nous engage selon leurs termes et notamment l'article 3. Le numéro et la date du crédit ainsi que le nom de notre banque devront être mentionnés sur toute traite requise. Si le crédit est utilisable par négociation, chaque présentation devra être inscrite au verso de cet avis par la banque où le crédit est utilisable.

We hereby issue this Documentary Credit in your favour. It is subject to the Uniform Customs and Practice for Documentary Credits (1974 Revision, International Chamber of Commerce, Paris, France, Publication No. 290) and engages us in accordance with the terms thereof, and especially in accordance with the terms of Article 3 thereof. The number and the date of the credit and the name of our bank must be quoted on all drafts required. If the credit is available by negotiation, each presentation must be noted on the reverse of this advice by the bank where the credit is available.

New York X Bank
(signature)

Ce document consiste en 1 *page(s) signée(s)*

This document consists of 1 signed page(s)

Avis destiné au bénéficiaire / Advice for the beneficiary

NAME OF ADVISING BANK
London Y Bank
(full address) London

Ref. No. of Advising Bank 678910
Place and date of notification London - 21 January 1981

NOTIFICATION OF IRREVOCABLE DOCUMENTARY CREDIT

Number 12345

Issuing Bank	Beneficiary
New York X Bank (full address) New York City	Exporter Limited (full address) London
Ref. No. of Issuing Bank	Amount $121,600 (One hundred and twenty-one thousand six hundred United States dollars)

We have been informed by our aforementioned correspondent that the abovementioned documentary credit has been issued in your favour. Please find enclosed the advice intended for you.

Please check the credit terms carefully. In the event that you do not agree with the terms and conditions or if you feel unable to comply with any of the terms and conditions, please arrange an amendment of the credit through your contracting party (the applicant for the credit).

[X] This notification and the enclosed advice are sent to you without engagement on our part.

[ ] ~~As requested by our correspondent, we hereby confirm the abovementioned Credit.~~

London Y Bank
(signature)

FORM 97

Examples of Specific Exclusions of Items of Work or Responsibility

Exclusions*

The following are excluded from our offer:

Erection
Commissioning
Painting
Storage on Site
Civil engineering or building work
Off-loading on Site
Reception and unpacking of deliveries to Site
Scaffolding
Spare parts
Consumables required for Plant operation.

*The tenderer should specifically exclude items of work or responsibility in any doubtful cases. Naturally, depending on the type of work undertaken by the tenderer, the possible list of exclusions will vary enormously.

FORM 98

Clause in Tender by Sub-Contractor to UK Main Contractor providing for ECGD or other Credit Insurance to protect both Main and Sub-Contractor

Credit Insurance

We are tendering to you as a sub-contractor on the basis that, should our tender be successful, ECGD or other credit insurance cover will be taken out in agreed terms, in order to protect our interests as well as yours. We will contribute an appropriate and reasonable proportion of the relevant premiums.*

*This clause may be used when the terms of payment between the Main Contractor and the Sub-Contractor are partly or wholly 'as and when' the Main Contractor is paid by the Employer. If the Main Contractor is foreign, the tenderer should see to his own ECGD or other credit insurance cover. See p.33.

FORM 99

Example of Letter of Intent

Dear Sirs

(heading)

With reference to your tender dated , it is our intention to award the Contract to you, subject to agreement being reached in further negotiations.

However, as the matter is urgent, we authorize you to begin design work. We will pay you for such design work at the rate of £ per hour for each of your design staff, payable monthly, subject to a maximum limit of £ .

We reserve the right to require the termination of such design work at any time.

Yours faithfully

(Customer's signature)

FORM 100

Example of Notarial Certificate

I of Notary Public by Royal Authority duly admitted and sworn CERTIFY:

(A) THAT LIMITED (referred to in this Certificate as 'the Company') is a company duly incorporated and existing under English law;

(B) THAT I was present on the date of this Certificate at and saw the Common Seal of the Company affixed to the annexed

(description of document, e.g. 'Power of Attorney in favour of John Smith')

in the presence of one of the Directors of the Company and the Secretary of the Company both personally known to me who then and there in my presence signed the annexed document in their respective capacities; and

(C) THAT the annexed document so sealed and signed is duly executed by the Company in accordance with the form prescribed by its Articles of Association and is binding on the Company in accordance with the provisions of English law relating to companies

IN FAITH AND TESTIMONY of which I have signed and affixed my Seal of Office to this Certificate at this day of . One thousand nine hundred and .

(Signature and Seal of Notary Public)

FORM 101

Foreign Tax Indemnity Clause

Taxation—all prices under the Contract exlude all foreign taxes. This tender is based on your indemnifying us and our personnel against all such taxes. The amount or amounts required in respect of such indemnity will be notified to you from time to time as and when we, or our personnel, are assessed for, or pay, any foreign tax or taxes. The expression 'foreign taxes' includes all present and future taxes in force in the country where the Site is located*, and all other present and future income, capital or other taxes, levies, contributions, duties, impositions, and compulsory payments of all names, descriptions and kinds, payable in any country other than the United Kingdom.

*It would be as well to list by name all such known taxes.

PART FOUR
Appendices

Appendix I: Table of Suitability of Conditions of Contract

Notes

1. Where certain Conditions of Contract are rather specialized, e.g. cable contracts, this is indicated in parentheses.
2. Section 5 of the Preliminary Text on Conditions of Contract should be referred to for further information.

Purpose	*Conditions of Contract*	*Form No*
UK Sales and Projects		
Stock and Catalogue Articles – UK	BEAMA (SA)	28
UK Projects – without erection	IMechE C/1975	9
	BEAMA (A)	12
UK Projects – with supervision of erection	BEAMA (B)	16
UK Projects – with erection	IMechE A/1976/1978	1
	IEE E/1973 (cables)	10
	BEAMA (C) (electronic equipment)	20
	BEAMA Conditions of Contract for Systems Incorporating Electronic Equipment	33
	IChemE (chemical engineering lump sum contracts)	74
	IChemE (chemical engineering reimbursable contracts)	75
Export Sales and Projects		
Export FOB	IMechE B1/1981	2
	BEAMA (AE)	13
	BEAMA (SAE) (stock and catalogue articles)	29
	UNECE 188	34
	UNECE 574	49
	UNECE 730 (stock and catalogue articles)	64
	ORGALIME	69

Export C&F	IMechE B1/1981	3
	BEAMA(AEC)	15
	BEAMA (SAE) (stock and catalogue articles)	30
	UNECE 188	35
	UNECE 574	50
	UNECE 730 (stock and catalogue articles)	65
	ORGALIME	70
Export CIF	IMechE B1/1981	4
	BEAMA (AEC)	14
	BEAMA (SAE) (stock and catalogue articles)	31
	UNECE 188	36
	UNECE 574	51
	UNECE 730 (stock and catalogue articles)	66
	ORGALIME	71
Export FOB Airport	BEAMA (AE)	95
	or other Conditions of Contract listed as suitable for Export FOB, suitably adapted	
Export FOB – with supervision of erection	IMechE B2/1981	5
	BEAMA (BE)	17
	UNECE 188 and 188B	37
	UNECE 188A	45
	UNECE 574 and 574B	52
	UNECE 574A	60
Export C&F – with supervision of erection	IMechE B2/1981	6
	BEAMA (BE)	18
	UNECE 188 and 188B	38
	UNECE 188A	46
	UNECE 574 and 574B	53
	UNECE 574A	61
Export CIF – with supervision of erection	IMechE B2/1981	7
	BEAMA (BE)	19
	UNECE 188 and 188B	39
	UNECE 188A	47
	UNECE 574 and 574B	54
	UNECE 574A	62
Export FOB – with erection	UNECE 188A	41
	UNECE 574A	56
Export C&F – with erection	UNECE 188A	42
	UNECE 574A	57
Export CIF – with erection	UNECE 188A	43
	UNECE 574A	58

Export – with delivery to and supervision of erection on site	UNECE 188A	44
	UNECE 574A	59
Export – with delivery to and erection on site	IMechE B3/1980	8
	FIDIC M&E	11
	UNECE 188A	40
	UNECE 574A	55
Sub-Contracts	ICE*	73

*Please refer to Section 12 of the Preliminary Text on Sub-contracts.

Repairs, Reconstruction, Modification

Repair – UK	BEAMA (R)	23
Reconstruction, Modification or Repair – UK Site	BEAMA (RC)	27
Repair and Export FOB	BEAMA (RE)	24
Repair and Export C&F	BEAMA (RE)	25
Repair and Export CIF	BEAMA (RE)	26

Contracts for Services

Commissioning Electronic Equipment – UK	BEAMA Conditions of Contract for Commissioning Electronic Equipment	32
Erection of Plant and Machinery – UK	BEAMA (E) (electrical)	21
Erection of Plant and Machinery Abroad	BEAMA (E) (electrical)	22
	UNECE 188D	48
	UNECE 574D	63
Provision of Technical Personnel Abroad	ORGALIME Conditions for the Provision of Technical Personnel Abroad – Part I (for use as a supplement)	67
	Parts 1 and 2 (for use as an independent contract)	68
Processing	ORGALIME Model Form of Processing Contract	72A
Maintenance	ORGALIME Model Form of Maintenance Contract	72

Appendices II–XXV

MODEL FORM OF

GENERAL CONDITIONS OF CONTRACT

INCLUDING FORMS OF AGREEMENT AND GUARANTEE

RECOMMENDED BY

THE INSTITUTION OF MECHANICAL ENGINEERS
THE INSTITUTION OF ELECTRICAL ENGINEERS
AND
THE ASSOCIATION OF CONSULTING ENGINEERS

FOR USE IN CONNECTION WITH

HOME CONTRACTS—WITH ERECTION

1976 EDITION
including September 1978 amendments

Published for the Joint Committee on Model Forms of General Conditions of Contract by
THE INSTITUTION OF ELECTRICAL ENGINEERS
and obtainable from
Publications Sales Department, Station House, 70 Nightingale Road, Hitchin, Herts, SG5 1RJ
or to callers, at the Reception Desk, Savoy Place
or from
THE INSTITUTION OF MECHANICAL ENGINEERS
Publications Sales Department, P.O. Box 24, Northgate Avenue, Bury St. Edmunds, Suffolk, IP32 6BW
or to callers, at the Reception Desk, Birdcage Walk, London
or at
THE ASSOCIATION OF CONSULTING ENGINEERS, Hancock House, 87 Vincent Square, London, SW1P 2PH

HISTORICAL NOTE

A "Form of Model Conditions recommended for use in connection with contracts for plant, mains and apparatus for electricity works" was originally drawn up by a Committee convened by the Council of the Institution of Electrical Engineers in 1903. Subsequent revisions were as follows:

1914

Form of Model General Conditions recommended for use in connection with contracts for electrical works.

1921

Model Form of General Conditions "A," Home–with erection.

1926

Model Form of General Conditions "A," Home–with erection.

1929

Model Form of General Conditions "A," Home–with erection.

1938

Model Form of General Conditions "A," Home–with erection.

Following consultations and by agreement between the Council of The Institution of Mechanical Engineers and the Council of The Institution of Electrical Engineers, the scope of the Model Form was enlarged to make it suitable for both the electrical and mechanical engineering industries, and on this basis the Model Form was first issued jointly by the two Council as:

1948

Model Form of General Conditions of Contract "A," Home Contracts–with erection.

As a result of an approach by the Association of Consulting Engineers, it was agreed in 1951 that the Association should adopt and join in recommending this Model Form, simultaneously discontinuing the issue of the "Conditions of Contract" which the Association had hitherto prepared and recommended for use by its members. Accordingly, the name of the Association now appears on the title page of the Model Form.

A Model Form of General Conditions "A," Home - with erection, which incorporated substantial revisions to take account of developments in practice was issued in 1966. This Form was amended in January 1971 to take account of S.E.T. and Metrication, in April 1972 to allow for V.A.T. and in the Autumn of 1973 to take account of V.A.T. (post legislation) and the abolition of S.E.T.

This 1976 Edition incorporates advice received from leading Counsel on certain aspects of Model Form 'B.3' (1971 Edition), an up-dated rate of interest payable by the Purchaser on delayed payments and other revisions to bring the Form up to date in the light of comments received. This Form was amended in September 1978 to take account of the Unfair Contract Terms Act 1977.

This form may be cited as "Model Form A/1976/1978, Home-with erection"

MODEL FORM OF GENERAL CONDITIONS RECOMMENDED BY THE INSTITUTION OF MECHANICAL ENGINEERS, THE INSTITUTION OF ELECTRICAL ENGINEERS AND THE ASSOCIATION OF CONSULTING ENGINEERS FOR USE IN CONNECTION WITH HOME CONTRACTS (WITH ERECTION).

GENERAL CONDITIONS

Definition of Terms

1.–In construing these General Conditions and the Specification, the following words shall have the meanings herein assigned to them unless there is something in the subject matter or context inconsistent with such construction:

The "Contract" shall mean the agreement between the Purchaser and the Contractor for the execution of the Works howsoever made, including therein all documents to which reference may properly be made in order to ascertain the rights and obligations of the parties under the said agreement.

The "Contractor" shall mean the tenderer whose tender has been accepted by the Purchaser, and shall include the Contractor's legal personal representatives, successors, and assigns.

"Contractor's Equipment" shall mean tools, tackle, stores and other things brought upon the Site by the Contractor and required thereon for the purposes of the Works but not for incorporation therein.

The "Contract Price" shall mean the sum named in the Contract as the Contract price.

The "Contract Value" shall mean that part of the Contract Price, adjusted to give effect to such additions or deductions as are provided for in Clause 10 *(Variations and Omissions),* which is properly apportionable to the Plant or work in question, having regard to the state, condition and location of the Plant, the amount of work done and all other relevant circumstances but disregarding any changes pursuant to Clause 39 in the cost of executing the Works.

The "Engineer" shall mean

or the person for the time being or from time to time notified in writing by the Purchaser to the Contractor as the Engineer for the Contract, or in default of any notification the Purchaser.

"Month" shall mean calendar month.

"Plant" shall mean machinery, apparatus, materials, articles, and things of all kinds other than Contractor's Equipment.

The "Purchaser" shall mean

and shall include the Purchaser's legal personal representatives, successors, and assigns.

The "Site" shall mean the actual place to which plant is to be delivered or where work is to be done by the Contractor, together with so much of the area surrounding the said place as the Contractor shall with the consent of the Engineer actually use in connection with the Works otherwise than merely for the purpose of access to the said place.

The "Specification" shall mean the specification annexed to or issued with these General Conditions.

"Sub-contractor" shall mean any person (other than the Contractor) named in the Contract for any part of the Works or any person to whom any part of the Contract has been sub-let with the consent in writing of the Engineer, and the legal representatives, successors, and assigns of such person.

"Tests on Completion" shall mean such tests to be made by the Contractor on completion of erection as are provided for in the Contract or otherwise agreed between the Purchaser and the Contractor.

The "Works" shall mean all Plant to be provided and work to be done by the Contractor under the Contract.

"Writing" shall include any manuscript, type-written, or printed statement, under seal or hand as the case may be.

Words importing persons shall include firms and corporations.

Words importing the singular only shall also include the plural and *vice versa.*

Contractor to inform himself fully.

2.–The Contractor shall be deemed to have examined the Site, if access thereto has been available to him, and the General Conditions and Specification, with such schedules, drawings, and plans as are annexed thereto or referred to therein.

Security for due performance.

3.–(i) If required by the Purchaser, the Contractor shall provide a surety or sureties, or a grantor of an insurance or guarantee policy, who in either case shall be subject to the approval of the Purchaser, which approval shall not be unreasonably withheld, and who shall execute (if two or more jointly and severally) a guarantee, or grant an insurance or guarantee policy to an extent not exceeding 15 per cent of the Contract Price, by way of guarantee for the due and faithful performance of the Contract, such guarantee to be binding notwithstanding such variations, alterations, or extensions of time as may be made, given, conceded, or agreed under these General Conditions. The instrument of guarantee shall be in the form annexed to these General Conditions with such modifications as may be necessary, and the expenses of procuring, preparing, completing and stamping such instrument shall be paid by the Purchaser.

(ii) If the guarantee or other security for the due performance of the Contract required to be furnished pursuant to this clause shall not be duly furnished by the Contractor to the Purchaser within one month after the Contract has been entered into, the Purchaser may, at his option, without prejudice to any rights or claims he may have against the Contractor by reason of the Contractor's non-compliance with any of the provisions of this clause, and within seven days after the expiry of the said period, by notice in writing to the Contractor terminate the Contract forthwith, and the Purchaser shall thereupon not be liable for any claim or demand from the Contractor in respect of any thing then already done or furnished, or in respect of any other matter or thing whatsoever, in connection with the Contract, but the Purchaser shall be entitled to be repaid by the Contractor all out-of-pocket expenses properly incurred by the Purchaser incidental to the obtaining of new tenders.

Expenses of Agreement.

(iii) The expenses of preparing, completing, and stamping the agreement, if any, shall be paid by the Purchaser, and an executed counterpart thereof properly stamped together with copies of all other documents comprising the Contract shall be furnished to the Contractor free of charge.

Drawings.

4.–(i) The Contractor shall submit to the Engineer for approval:

(a) within the time specified in the Specification or if no time is specified then a reasonable time such drawings, samples, patterns, and models as may be called for therein, and

(b) during the progress of the Works, and within such reasonable times as the Engineer may require, such drawings of the general arrangement and of details of the Works or any part thereof as the Engineer may reasonably require, provided that the Contractor shall not be under any obligation to supply copies of shop drawings.

Within a reasonable period after receiving such drawings, samples, patterns, and models the Engineer shall signify his approval or otherwise. If the Engineer shall not approve any drawing, sample, pattern or model thus submitted the same shall be forthwith modified to meet the reasonable requirements of the Engineer and shall be re-submitted. Copies of all drawings which require to be approved by the Engineer shall be provided in triplicate by the Contractor. Drawings, samples, patterns, and models when approved shall if required by either party be signed or identified by both parties and if required by the Contractor one copy shall be returned to him. All dimensions marked on drawings shall be considered correct although measurements by scale may differ therefrom. Detailed drawings shall take precedence where they differ from general arrangement drawings.

(ii) Drawings, samples, patterns, and models approved as above described shall not be departed from except as provided in Clause 10 *(Variations and Omissions).*

(iii) The Engineer shall have the right at all reasonable times to inspect at the premises of the Contractor all drawings of any portion of the Works.

(iv) The Contractor shall, if so desired by the Purchaser, furnish to the Purchaser in writing at the commencement of the maintenance period, or at such earlier times as may be named in the Specification, such information, accompanied by drawings, as may be necessary to enable the Purchaser to operate, maintain, dismantle, reassemble, and adjust all parts of the Works.

(v) Drawings submitted in pursuance of paragraphs *(a)* and *(b)* of Sub-Clause (i) of this clause shall not, without the consent of the Contractor, be used by the Purchaser except for purposes of the Contract, nor shall they without such consent be communicated to third parties save insofar as may be necessary for the proper execution of the Works.

Mistakes in information.

5.–(i) The Contractor shall be responsible for any discrepancies, errors, or omissions in the drawings and information supplied by him, whether they have been approved by the Engineer or not, provided that such discrepancies, errors, or omissions be not due to defective drawings or information furnished to the Contractor by the Purchaser or the Engineer.

(ii) The Contractor shall at his own expense carry out any alterations or remedial work necessitated by reason of such discrepancies, errors, or omissions and modify the drawings and information accordingly, or if the same be done by or on behalf of the Purchaser shall bear all costs reasonably incurred therein. The performance of his obligations under this sub-clause shall be in full satisfaction of the Contractor's liability under Sub-Clause (i) of this clause, but shall not relieve him of his liability under Clause 26 *(Delay in Completion)* insofar as that liability arises as a result of such discrepancies errors or omissions.

(iii) The Purchaser shall be responsible for drawings and information supplied by the Purchaser or the Engineer and for the details of special work specified by either of them. The Purchaser shall pay to the Contractor for alterations of the work necessitated by reason of defective drawings or information so supplied to the Contractor a sum ascertained and determined in like manner to the valuation of variations under Clause 10 *(Variations and Omissions).*

Assignment.

6.–(i) The Contractor shall not, without the consent in writing of the Purchaser, which shall not be unreasonably withheld, assign or transfer the Contract or the benefits or obligations thereof or any part thereof to any other person, provided that this shall not affect any right of the Contractor to assign, either absolutely or by way of charge, any moneys due or to become due to him, or which may become payable to him under the Contract.

Sub-letting.

(ii) The Contractor shall not, without the consent in writing of the Engineer, which shall not be unreasonably withheld, sub-let the Contract or any part thereof, or make any sub-contract with any person or persons for the execution of any portion of the Works but the restriction contained in this clause shall not apply to sub-contracts for materials, for minor details, or for any part of the Works of which the makers are named in the Contract. Any such consent shall not relieve the Contractor from his obligations under the Contract.

Patent rights etc.

7.–(i) The Contractor shall fully indemnify the Purchaser against all actions, claims, demands, costs, charges and expenses arising from or incurred by reason of any infringement or alleged infringement of letters patent, registered design, copyright, trade mark or trade name protected in the United Kingdom by the use of any Plant supplied by the Contractor, but such indemnity shall not cover any use of the Works otherwise than for the purpose indicated by or reasonably to be inferred from the Specification or any infringement which is due to the use of any Plant in association or combination with any other plant not supplied by the Contractor.

(ii) In the event of any claim being made or action brought against the Purchaser arising out of the matters referred to in this clause, the Contractor shall be promptly notified thereof and may at his own expense conduct all negotiations for the settlement of the same, and any litigation that may arise therefrom. The Purchaser shall not, unless and until the Contractor shall have failed to take over the conduct of the negotiations or litigation, make any admission which might be prejudicial thereto. The conduct by the Contractor of such negotiations or litigation shall be conditional upon the Contractor having first given to the Purchaser such reasonable security as shall from time to time be required by the Purchaser to cover the amount ascertained or agreed or estimated, as the case may be, of any compensation, damages, expenses and costs for which the Purchaser may become liable. The Purchaser shall, at the request of the Contractor, afford all available assistance for the purpose of contesting any such claim or action, and shall be repaid all reasonable expenses incurred in so doing.

(iii) The Purchaser on his part warrants that any design or instructions furnished or given by him shall not be such as will cause the Contractor to infringe any letters patent, registered design, trade mark, or copyright in the performance of the Contract.

Manner of Execution.

8.–All Plant to be supplied and all work to be done under the Contract shall be manufactured and executed in the manner set out in the Specification or, where not so set out, to the reasonable satisfaction of the Engineer.

Contractor's Equipment, Labour, etc.

9.–(i) Unless specific arrangements be made to the contrary the Contractor shall, at his own expense, provide all Contractor's Equipment and all materials, labour, haulage, and power necessary to execute and complete the Works.

(ii) The Contractor shall be responsible for the proper fencing, guarding, lighting, and watching of all the Works on the Site until taken over under Clause 28 *(Taking Over)* and for the proper provision during a like period of temporary roadways, footways, guards, and fences as far as the same may be rendered necessary by reason of the Works for the accommodation and protection of the owners and occupiers of adjacent property, the public, and others. No naked light shall be used by the Contractor on the Site otherwise than in the open air without special permission in writing from the Engineer.

Electricity, Water and Gas.

(iii) The Contractor shall be entitled to use for the purposes of the Works such supplies of electricity, water, and gas as may be available therefor on the Site and shall pay to the Purchaser for such use such sum as may be reasonable in the circumstances, and shall at his own expense provide any apparatus necessary for such use.

Lifting Equipment.

(iv) The Purchaser shall at the request of the Contractor and for the execution of the Works operate any suitable lifting equipment belonging to the Purchaser that may be available on the Site and the Contractor shall pay a reasonable sum therefor. The Purchaser shall during such operation retain control of and be responsible for the safe working of the lifting equipment but shall not be responsible for any negligence of the Contractor.

Variations and Omissions.

10.–(i) The Contractor shall not alter any of the Works, except as directed in writing by the Engineer; but the Engineer shall have full power, subject to the proviso hereinafter contained, from time to time during the execution of the Contract by notice in writing to direct the Contractor to alter, amend, omit, add to, or otherwise vary any of the Works, and the Contractor shall carry out such variations, and be bound by the same conditions, so far as applicable, as though the said variations were stated in the Specification. Provided that no such variation shall, except with the consent in writing of the Contractor, be such as will, with any variations already directed to be made, involve a net addition to or deduction from the Contract Price of more than 15 per cent thereof. In any case in which the Contractor has received any such direction from the Engineer which either then or later will, in the opinion of the Contractor, involve an addition to or deduction from the Contract Price, the Contractor shall, as soon as reasonably possible, advise the Engineer in writing to that effect. The amount to be added to or deducted from the Contract Price shall be ascertained and determined in accordance with the rates specified in the schedules of prices, so far as the same may be applicable, and where rates are not contained in the said schedules, or are not applicable, such amount shall be such sum as is reasonable in the circumstances. Due account shall be taken of any partial execution of the Works which is rendered useless by any such variation.

(ii) If the Engineer shall make any such variation in any part of the Works, such reasonable notice in writing shall be given to the Contractor as will enable him to make his arrangements accordingly. If in the opinion of the Contractor any such variation is likely to prevent or prejudice the Contractor from or in fulfilling any of his obligations under the Contract, he shall notify the Engineer thereof in writing, and the Engineer shall decide forthwith whether or not the same shall be carried out. If the Engineer confirms his instructions in writing, the said obligations shall be modified to such an extent as may be justified. Until the Engineer so confirms his instructions they shall be deemed not to have been given.

Underground Works.

11.–In the case of work underground or involving excavation where the actual conditions of the ground are not stated in the Contract and could not reasonably have been inferred from an inspection of the Site by the Contractor before he prepared his tender, if rock, rocky soil, solid chalk, water, running sand, slag, pipes, concrete, or other obstructions are found, or if it should be necessary to leave in timber or provide support for existing works (such necessity not having been indicated in the Contract), the Contractor shall inform the Engineer as soon as reasonably practicable of the steps he proposes to take to deal with the hazard. If, as a consequence thereof, extra cost is incurred by the Contractor, a sum ascertained and determined in like manner to the valuation of variations under Clause 10 *(Variations and Omissions)* shall be added to the Contract Price.

Contractor's Default.

12.–If the Contractor shall neglect to execute the Works with due diligence and expedition, or shall refuse or neglect to comply with any reasonable orders given him in writing by the Engineer in connection with the Works, or shall contravene the provisions of the Contract, the Purchaser may give seven days' notice in writing to the Contractor to make good the failure, neglect, or contravention complained of. Should the Contractor fail to comply with the notice within seven days from the date of service thereof in the case of a failure, neglect, or contravention capable of being made good within that time, or otherwise within such time as may be reasonably necessary for making it good, then and in such case the Purchaser shall be at liberty to employ other workmen, and forthwith execute such part of the Works as the Contractor may have neglected to do, or, if the Purchaser shall think fit, it shall be lawful for him, without prejudice to any other rights he may have under the Contract, to take the Works wholly or in part out of the Contractor's hands and either by his own workmen or by re-contracting with any other person or persons to complete the Works or any part thereof, and in either event the Purchaser shall have the free use of all Contractor's Equipment that may be on the Site in connection with the Works, without being responsible to the Contractor for fair wear and tear thereof, and to the exclusion of any right of the Contractor over the same, and the Purchaser shall be entitled to retain and apply any balance which may be otherwise due on the Contract by him to the Contractor, or such part thereof as may be necessary to the payment of the cost of executing the said part of the Works or of completing the Works as the case may be. If the cost of completing the Works or executing a part thereof as aforesaid shall exceed the amount that would otherwise become due to the Contractor in accordance with the Contract, the Contractor shall pay such excess. If the Purchaser pursuant to this clause takes the Works or part thereof out of the Contractor's hands, the Contractor's liability under Clause 26 *(Delay in Completion)* shall immediately cease in respect of the Works or part thereof, without prejudice to any such liability that shall have already accrued.

Bankruptcy

13.– If the Contractor shall become bankrupt or insolvent, or have a receiving order made against him, or compound with this creditors, or being a corporation commence to be wound up, not being a members' voluntary winding up for the purpose of reconstruction or amalgamation, or carry on its business under a receiver for the benefit of its creditors or any of them, the Purchaser shall be at liberty either–

(a) to terminate the Contract forthwith by notice in writing to the Contractor or to the receiver or liquidator or to any person in whom the Contract may become vested, and to act in the manner provided in Clause 12 *(Contractor's Default)* as though the last mentioned notice had been the notice referred to in such clause and the Works had been taken out of the Contractor's hands, or

(b) to give such receiver, liquidator, or other person the option of carrying out the Contract subject to his providing a guarantee for the due and faithful performance of the Contract up to an amount to be agreed.

Inspection, Testing and Rejection of Plant.

14.–(i) The Engineer shall be entitled at all reasonable times during manufacture to inspect, examine, and test on the Contractor's premises the materials and workmanship and performances of all Plant to be supplied under the Contract, and if part of the said Plant is being manufactured on other premises the Contractor shall obtain for the Engineer permission to inspect, examine, and test as if the said Plant were being manufactured on the Contractor's premises. Such inspection, examination, or testing, shall not release the Contractor from any obligation under the Contract.

(ii) The Contractor shall, after consulting the Engineer, give the Engineer reasonable notice in writing of the date on and the place at which any Plant will be ready for testing as provided in the Contract and unless the Engineer shall attend at the place so named on the date which the Contractor has stated in his notice the Contractor may proceed with the tests, which shall be deemed to have been made in the Engineer's presence, and shall forthwith forward to the Engineer duly certified copies of the test readings. The Engineer shall give the Contractor 24 hours' notice in writing of his intention to attend the tests.

(iii) Where the Contract provides for tests on the premises of the Contractor or of any Sub-contractor the Contractor, except where otherwise specified, shall provide free of charge such assistance, labour, materials, electricity, fuel, stores, apparatus, and instruments as may be requisite and as may be reasonably demanded to carry out such tests efficiently.

(iv) Where the Contract provides for tests on the Site, the Purchaser, except where otherwise specified, shall provide free of charge, such labour, materials, electricity, fuel, water, stores, and apparatus, as may be requisite and as may be reasonably demanded to carry out such tests efficiently.

(v) As and when the Engineer is satisfied that any Plant shall have passed the tests referred to in this clause he shall forthwith notify the Contractor in writing to that effect.

(vi) If after inspecting, examining, or testing any Plant the Engineer shall decide that such Plant or any part thereof is defective or not in accordance with the Contract, he may reject the said Plant or part thereof by giving to the Contractor within a reasonable time notice in writing of such rejection, stating therein the grounds upon which the said decision is based.

(vii) The provisions of Clause 27 (v) *(Tests on Completion)* shall relate also to inspections, examinations, and tests carried out under this clause.

Delivery.

15.–(i) No Plant or Contractor's Equipment shall be delivered to the Site until an authorisation in writing has been applied for and obtained by the Contractor from the Engineer that delivery may be made. The Contractor shall be responsible for the reception on the Site of all Plant and Contractor's Equipment delivered for the purposes of the Contract.

(ii) For the purposes of this clause only–

"delayed Plant" means *(a)* Plant which by delay or failure on the part of the Engineer to give such authorisation as is mentioned in Sub-Clause (i) of this clause or from any cause for which the Purchaser or some other contractor employed by him is responsible the Contractor is prevented from delivering to the Site at the time specified for the delivery thereof or, if no time is specified, at the time when it is reasonable for it to be delivered having regard to the date by which the Works ought to be completed: and *(b)* Plant which has been delivered to the Site but which by delay or failure on the part of the Engineer or from any cause for which the Purchaser or some other contractor employed by him is responsible the Contractor is for the time being prevented from erecting.

"the normal delivery date" means the time when but for such delay, failure or other cause as aforesaid delayed Plant would have been delivered to the Site.

"notice to proceed" means notice in writing from the Engineer to the Contractor that delayed Plant may forthwith be delivered to the Site or (as the case may be) erected.

(iii) If delayed Plant is ready for delivery and has been suitably and sufficiently marked as appropriated to the Contract and the Contractor has given to the Engineer an opportunity of inspecting it or if delayed Plant has been delivered to the Site the Contractor may give notice in writing to the Purchaser requiring that the provisions of Sub-Clause (iv) of this clause shall have effect with respect to such delayed Plant.

(iv) Where notice has been given in accordance with Sub-Clause (iii) of this clause—

(a) There shall be added to the Contract Price a sum, ascertained and determined in like manner to the valuation of variations under Clause 10 *(Variations and Omissions)*, for storing and taking reasonable measures to protect and preserve the delayed Plant from and insuring it against loss, deterioration and damage however caused from the date of the said notice or the normal delivery date if this shall be later until the Contractor shall no longer be prevented from delivering the delayed Plant or (as the case may be) erecting it or shall be relieved of responsibility therefor under Sub-Clause (v) of this clause whichever shall first happen.

(b) The Contractor shall after one month from the normal delivery date or from the date of the said notice (whichever shall be the later) be entitled to have the Contract Value of the delayed Plant included in an interim certificate.

(c) If at the expiration of six months from the normal delivery date or from the date of the said notice (whichever shall be the later) the Contractor shall still be prevented from delivering the delayed Plant to the Site or (as the case may be) from erecting it the Engineer shall, on the application of the Contractor, certify accordingly and within one month from the presentation of such certificate the Contractor shall be entitled to be paid 97½ per cent of the Contract Value of the delayed Plant less any sum previously paid to him in respect thereof.

(d) If at any time after the expiration of 12 months from the date of the said notice or at any time after the delayed Plant has been delivered to the Site the Purchaser shall not have assumed responsibility for the storage of the delayed Plant the Contractor may by a further notice in writing expiring 30 days after receipt thereof by the Purchaser require the Purchaser to assume responsibility for storing, protecting and preserving the delayed Plant and upon the expiration of the last mentioned notice the Purchaser shall assume responsibility for storing the delayed Plant and the Contractor shall in any case forthwith be entitled to be paid 100% of the Contract Value of the delayed Plant on the expiration of 12 months from the date of the notice in respect of such delayed Plant given pursuant to Sub-Clause (iii) of this clause, less any sum previously paid to him in respect thereof: provided always that if notice to proceed shall be given within 30 days after receipt by the Purchaser of the aforesaid further notice given by the Contractor this paragraph of this sub-clause shall not operate.

(e) Without prejudice to the provisions of Sub-Clause (vii) of Clause 30 *(Defects after Taking Over)*, the obligations of the Contractor under that clause with respect to delayed Plant shall not apply to any defect that may develop therein after the expiration of three years from the date of the said notice referred to in Sub-Clause (iii) of this clause or the normal delivery date if this shall be later.

(v) If at any time the Purchaser assumes responsibility for storing delayed Plant whether pursuant to paragraph *(d)* of the last preceding sub-clause or otherwise the Contractor shall thereupon be relieved of any responsibility for the delayed Plant until either the expiration of 30 days after the receipt of a notice to proceed or the Contractor having received the notice to proceed resumes possession of the said Plant whichever shall first occur.

(vi) After the receipt of notice to proceed the Contractor, if he has been relieved of responsibility under Sub-Clause (v) of this clause, shall (and in any other case may) after due notice in writing to the Engineer and if required by the Engineer in his presence examine the delayed Plant and any Plant on the Site that has been erected but not taken over under Clause 28 *(Taking Over)* by reason of delay in the delivery or erection of the delayed Plant, and make good any deterioration or defect therein that may have developed or loss thereof that may have occurred after the normal delivery date or (if later) the date when the Contractor was by such delay, failure or other cause as before-mentioned first prevented from erecting the delayed Plant.

(vii) There shall be added to the Contract Price a reasonable sum for making the examination referred to in Sub-Clause (vi) of this clause or in making good any deterioration, defect or loss as therein mentioned except insofar as the same was caused by faulty workmanship or materials or by the Contractor's failure to take the measures referred to in paragraph *(a)* of Sub-Clause (iv) of this clause or in Sub-Clause (i) of Clause 21 *(Liability for Accidents and Damage)*. If the Contractor incurs additional expense in delivering the delayed Plant to the Site or in erecting the same or any other Plant or in carrying out the tests on completion or in performing his obligations under Clause 30 *(Defects after Taking Over)* which would not have been incurred had the delivery or erection of the delayed Plant not been prevented as aforesaid the Purchaser shall pay a reasonable sum in respect thereof which shall be added to the Contract Price.

Access to and possession of the Site.

16.—(i) Subject to Sub-Clause (iv) of this clause, access to and possession of the Site shall be afforded to the Contractor by the Purchaser in reasonable time and, except insofar as the Specification may provide to the contrary, the Purchaser shall provide a road or railway suitable for the transport of all Plant and Contractor's Equipment necessary for the execution of the Works from a convenient point on a public thoroughfare suitable for such transport or on a railway available to the Contractor to the point on the Site where it is to be delivered or used.

(ii) If a building, structure, foundation, or approach is by the Contract to be provided by the Purchaser, such building, structure, foundation, or approach shall be in a condition suitable for the efficient transport, reception, installation, and maintenance of the Works.

(iii) In the execution of the Works, the Contractor shall not authorize or purport to authorize any person other than his employees and Sub-contractors and their employees to come upon the Site, except by the written permission of the Engineer, but facilities to inspect the Works at all times shall be afforded to the Engineer and his representatives and authorized representatives of the Purchaser.

(iv) The access to and possession of the Site referred to in Sub-Clause (i) hereof shall not be exclusive to the Contractor but only such as shall enable him to execute the Works. The Contractor shall afford to the Purchaser and to other contractors whose names shall have been previously communicated in writing to the Contractor by the Engineer every reasonable facility for the execution of work concurrently with his own.

(v) The Purchaser shall give the Contractor facilities for carrying out the Works on the Site continuously during the normal working hours generally recognized in the district. The Engineer may, after consulting with the Contractor, direct that work shall be done at other times if it shall be practicable in the circumstances for work to be so done, and a sum for work so done shall be added to the Contract Price unless such work has, by the default of the Contractor, become necessary for the completion of the Works within the time fixed by the Contract, or, if no time be fixed within a reasonable time. Such sum shall be ascertained and determined in like manner to the valuation of variations under Clause 10 *(Variations and Omissions)*.

Vesting of Plant and Contractor's Equipment.

17.—(i) Plant supplied pursuant to the Contract shall become the property of the Purchaser at whichever is the earlier of the following times, namely—

(a) when the Plant is delivered pursuant to the Contract.

(b) when by virtue of Clause 15 *(Delivery)* or Clause 29 *(Suspension of Works)* the Contractor becomes entitled to require that the Contract Value of the Plant be included in an interim certificate.

(ii) All Contractor's Equipment owned by the Contractor or by any firm or corporation in which the Contractor has a controlling interest shall when brought upon the Site vest in and become the property of the Purchaser, and shall be used solely for the purpose of the Works and shall not be removed by the Contractor without the permission in writing of the Engineer, which permission shall not be unreasonably withheld in the case of Contractor's Equipment not currently required for the purpose of the Works. The Contractor shall be liable for the loss or destruction of such Contractor's Equipment or for damage thereto which may happen otherwise than through the fault of the Purchaser. If there shall be due, owing, or accruing to the Purchaser from the Contractor any moneys under or in respect of the Contract, of which the Purchaser shall be unable to obtain payment, the Purchaser shall be at liberty at the cost of the Contractor to sell and dispose as he shall think fit of such Contractor's Equipment and to apply the proceeds in or towards the satisfaction of such moneys as aforesaid. Subject to the foregoing and to the right of the Purchaser under Clause 12 *(Contractor's Default)* and Clause 13 *(Bankruptcy)* the property in such Contractor's Equipment and in any Plant which is no longer required for completion of the Works shall revert to the Contractor on their proper removal from the Site or on the completion of the Works or on the termination of the Contract, whichever may be the earliest.

(iii) If the Purchaser shall become bankrupt or insolvent, or have a receiving order made against him, or compound with his creditors, or being a corporation commence to be wound up, not being a members' voluntary winding up for the purpose of reconstruction or amalgamation, or carry on its business under a receiver for the benefit of its creditors or any of them, then notwithstanding the provisions of Sub-Clause (ii) of this clause the property in Contractor's Equipment shall forthwith revert to the Contractor.

Engineer's Supervision.

18.—(i) After the tender has been accepted by the Purchaser, all instructions and orders to the Contractor shall, except as herein otherwise provided, be given by the Engineer.

(ii) The Contractor shall be responsible for ensuring that the positions, levels, and dimensions of the Works are correct according to the drawings, notwithstanding that he may have been assisted by the Engineer in setting out the said positions, levels, and dimensions.

(iii) All the Works shall be carried out under the direction and to the reasonable satisfaction of the Engineer.

Engineer's Representative.

(iv) The Engineer may from time to time delegate any of the powers, discretions, functions, and authorities vested in him and may at any time revoke any such delegation. Any such delegation or revocation shall be in writing signed by the Engineer and, in the case of a delegation, shall specify the power, discretions, functions, and authorities thereby delegated and the person or persons to whom the same are delegated. No such delegation or revocation shall have effect until a copy thereof has been delivered to the Contractor.

Clerk of Works.

(v) If a Clerk of the Works be appointed to watch the carrying out of the Contract, the Contractor shall afford him every reasonable facility for so doing, but the Clerk of the Works shall not be authorized to relieve the Contractor in any way of his duties or obligations under the Contract. Any written notice from the Clerk of the Works condemning any Plant or workmanship shall have the effect of a similar notice given by the Engineer under Clause 24 *(Defects Prior to Taking Over)* except that the Contractor may appeal to the Engineer for his decision in the matter.

Engineer's Decisions.

19.—The Contractor shall proceed with the Works in accordance with decisions, instructions, and orders given by the Engineer in accordance with these Conditions, provided always that—

(a) if the Contractor shall, without undue delay after being given any decision, instruction, or order otherwise than in writing, require it to be confirmed in writing, such decision, instruction, or order shall not be effective until written confirmation thereof has been received by the Contractor, and

(b) if the Contractor shall, by written notice to the Engineer within 14 days after receiving any decision, instruction, or order of the Engineer in writing or written confirmation thereof, intimate that he disputes or questions the decision, instruction, or order, giving his reasons for so doing, either party to the Contract shall be at liberty to refer the matter to arbitration pursuant to Clause 37 *(Arbitration)*, but such an intimation shall not relieve the Contractor of his obligation to proceed with the Works in accordance with the decision, instructions, or order in respect of which the intimation has been given. The Contractor shall be at liberty in any such arbitration to rely on reasons additional to the reasons stated in the said intimation.

Contractor's Representatives and Workmen.

20.—(i) The Contractor shall employ one or more competent representatives, whose name or names shall have previously been communicated in writing to the Engineer by the Contractor, to superintend the carrying out of the Works on the Site. The said representative, or if more than one shall be employed, then one of such representatives, shall be present on the Site during working hours, and any orders or instructions which the Engineer may give to the said representative of the Contractor shall be deemed to have been given to the Contractor.

(ii) The Engineer shall be at liberty by notice in writing to the Contractor to object to any representative or person employed by the Contractor in the execution of or otherwise about the Works who shall, in the opinion of the Engineer, misconduct himself or be incompetent or negligent, and the Contractor shall remove such person from the Works.

Liability for Accidents and Damage.

21.—(i) The Contractor shall properly cover up and protect until taken over under Clause 28 *(Taking Over)* any section or portion of the Works liable to injury by exposure to the weather, and shall take every reasonable precaution to protect any section or portion of the Works not taken over against loss or damage from any cause.

(ii) In the case of loss of or damage to the Works on the Site arising from or occasioned by causes for which the Contractor is not responsible under the Contract the same shall, if required by the Purchaser, be made good by the Contractor but at the cost of the Purchaser at a price to be agreed between the Contractor and the Purchaser or in default of agreement to be settled by arbitration and such cost shall be added to the Contract Price.

(iii) Subject to Sub-Clauses (vi) and (vii) of this clause and Clause 22 *(Limitation on Contractor's Liability)*, all losses of and damage to any section or portion of the Works that shall not have been taken over under Clause 28 *(Taking Over)*, which shall arise from or be occasioned by any act of the Contractor or any Sub-contractor or by a failure of the Contractor to comply with any obligation imposed on him by Sub-Clause (i) of this clause, shall be made good by and at the sole cost of the Contractor and to the reasonable satisfaction of the Engineer.

(iv) The Contractor shall, subject to Sub-Clauses (vi) and (vii) of this clause and Clause 22 *(Limitation on Contractor's Liability)*, indemnify the Purchaser in respect of all damage or injury occurring before all the Works shall have been taken over under Clause 28 *(Taking Over)* to any person or to any property (other than property forming part of the Works not yet taken over) and against all actions, suits, claims, demands, costs, charges, and expenses arising in connection therewith which shall be occasioned by the negligence of or breach of statutory duty by the Contractor or any Sub-contractor, or by defective design (other than a design made, furnished, or specified by the Purchaser and for which the Contractor has disclaimed responsibility in writing within a reasonable time after the receipt of the Purchaser's instructions), materials, or workmanship, but not otherwise. Provided that the Contractor shall not be liable by virtue of this sub-clause in respect of damage or injury attributable to defects in any section or portion of the Works taken over under Clause 28.

(v) If there shall occur any loss of or damage to any property or injury to any person while the Contractor is on the Site for the purpose of making good a defect in any section or portion of the Works pursuant to Clause 30 *(Defects after Taking Over)*, or for the purpose of carrying out tests on completion during the period referred to in Clause 30 as provided in Clause 28 *(Interference with Tests)*, the Contractor shall be liable, subject to the provisions of Sub-Clauses (vi) and (vii) of this clause and Clause 22 *(Limitation on Contractor's Liability)* as follows:

(a) In respect of loss of or damage to the said section or portion the Contractor's liability shall be as defined in Clause 30 *(Defects after Taking Over)*

(b) In respect of damage to any other property or injury to any person and of any actions, claims, demands, costs, charges and expenses arising in connection therewith the Contractor shall be liable, subject to the provisions of Sub-Clauses (i) and (iii) of this clause, to the extent that such damage or injury was caused by the negligence or breach of statutory duty of the Contractor or a Sub-contractor while on the Site as aforesaid or by defective materials or workmanship used in making good the said defect but not otherwise.

The said section or portion of the Works shall be defined by reference to the taking over certificate issued in respect thereof pursuant to Clause 28 *(Taking Over).*

(vi) The Contractor shall not be liable to the Purchaser for–

(a) any damage or injury to the extent that it is caused by or arises from the acts or omissions of the Purchaser or of others (not being the Contractor's servants or Sub-contractors).

(b) any loss or damage in circumstances over which the Contractor has no control.

(vii) Except in respect of personal injury or damage to property conferring on a person other than the Purchaser a good cause of action against the Contractor, the liability of the Contractor to the Purchaser under this clause for any one act or default shall not exceed the Contract Price or £100,000 whichever is the greater.

(viii) In the event of any claim being made against the Purchaser arising out of the matters referred to in and in respect of which the Contractor may be liable under this clause, the Contractor shall be promptly notified thereof, and may at his own expense conduct all negotiations for the settlement of the same and any litigation that may arise therefrom. The Purchaser shall not, unless and until the Contractor shall have failed to take over the conduct of the negotiations or litigation make any admission which might be prejudicial thereto. The conduct by the Contractor of such negotiations or litigation shall be conditional upon the Contractor having first given to the Purchaser such reasonable security as shall from time to time be required by the Purchaser to cover the amount ascertained or agreed or estimated, as the case may be, of any compensation, damages, expenses, and costs for which the Purchaser may become liable. The Purchaser shall, at the request of the Contractor, afford all available assistance for any such purpose, and shall be repaid all reasonable expenses incurred in so doing.

Limitations on Contractor's Liability.

22.–Subject as provided in Clause 26 *(Delay in Completion)* for the deduction of liquidated damages for delay, the Contractor shall not be liable to the Purchaser by way of indemnity or by reason of any breach of the Contract for loss of use (whether complete or partial) of the Works or of profit or of any contract that may be suffered by the Purchaser.

Insurance of Works.

23.–Unless the Purchaser shall have approved in writing other arrangements for the insurance hereinafter mentioned the Contractor shall in the joint names of the Contractor and the Purchaser insure, and keep insured until a taking-over certificate has been issued under Clause 28 *(Taking Over)*, such Works as may for the time being be upon the Site against loss, damage or destruction by fire, explosion, lightning, earthquake, malicious damage, theft, flood, storm, tempest, and aircraft and other aerial devices or articles dropped therefrom for the full replacement value thereof, and shall from time to time, when so required by the Engineer, produce the policy and receipts for the premiums. All moneys received under any such policies shall be applied in or towards the replacement and repair of the Works lost, damaged or destroyed but this provision shall not affect the Contractor's liabilities under the Contract.

Defects prior to Taking over.

24.–If, in respect of any section or portion of the Works not yet taken over, the Engineer shall at any time–

(a) decide that any work done or Plant supplied or materials used by the Contractor or any Sub-contractor is or are defective or not in accordance with the Contract or that such section or portion of the Works is defective or does not fulfil the requirements of the Contract (all such matters being hereinafter in this clause called "defects"), and

(b) as soon as reasonably practicable give to the Contractor notice in writing of the said decision specifying particulars of the defects alleged and of where the same are alleged to exist or to have occurred, and

(c) so far as may be necessary place the Plant at the Contractor's disposal,

then the Contractor shall with all speed and, except as provided in Sub-Clause (vii) of Clause 15 *(Delivery)*, at his own expense make good the defects so specified. In case the Contractor shall fail so to do the Purchaser may, provided he does so without undue delay, take, at the cost of the Contractor, such steps as may in all the circumstances be reasonable to make good such defects. All Plant provided by the Purchaser to replace defective Plant shall comply with the Contract and shall be obtained at reasonable prices and where reasonably practicable under competitive conditions. The Contractor shall be entitled to remove and retain all Plant that the Purchaser may have replaced at the Contractor's cost.

Nothing contained in this clause shall affect any claim by the Purchaser under Clause 26 *(Delay in Completion)*.

Extension of time for Completion.

25.–If, by reason of any industrial dispute or any cause beyond the reasonable control of the Contractor arising after the acceptance of the tender, the Contractor shall have been delayed or impeded in the completion of the Works, whether such delay or impediment occur before or after the time (if any) or extended time fixed for completion, provided that the Contractor shall without delay have given to the Purchaser or the Engineer notice in writing of his claim for an extension of time, the Engineer shall on receipt of such notice grant the Contractor from time to time in writing either prospectively or retrospectively such extension of the time fixed by the Contract for the completion of the Works as may be reasonable. Any delay on the part of a Sub-contractor which prevents the Contractor from completing the Works within the time for completion shall entitle the Contractor to an extension thereof provided such delay was due to any cause for which the Contractor himself would have been entitled to an extension of time under this clause.

Delay in Completion.

26.–If the Contractor fail to complete the Works in accordance with the Contract (except the maintenance thereof as provided in Clause 30 *(Defects after Taking Over)* and such tests as are to be made in accordance with Clause 27 *(Tests on Completion)*) within the time fixed by the Contract for the completion of the Works or any extension of such time, or if no time be fixed, within a reasonable time, and the Purchaser shall have suffered any loss from such failure, there shall be deducted from the Contract Price the percentage named in the Appendix of the Contract Value of such portion or portions only of the Works as cannot in consequence of the said failure be put to the use intended for each week between the time for completion of the Works as aforesaid and the actual date of completion, but the amount so deducted shall not in any case exceed the maximum percentage named in the Appendix of the Contract Value of such portion or portion of the Works, and such deduction shall be in full satisfaction of the Contractor's liability for the said failure.

Tests on Completion.

27.–(i) The Contractor shall give to the Engineer in writing 21 days' notice of the date after which he will be ready to make the Tests on Completion. Unless otherwise agreed, the tests shall take place within 10 days after the said date, on such day or days as the Engineer shall in writing notify the Contractor.

(ii) If the Engineer fail to appoint a time after having been asked to do so, or to attend at any time or place duly appointed for making the said tests the Contractor shall be entitled to proceed in his absence, and the said tests shall be deemed to have been made in the presence of the Engineer.

(iii) If, in the opinion of the Engineer, the tests are being unduly delayed, he may by notice in writing call upon the Contractor to make such tests within 10 days from the receipt of the said notice and the Contractor shall make the said tests on such day within the said 10 days as the Contractor may fix and of which he shall give notice to Engineer. If the Contractor fail to make such tests within the time aforesaid the Engineer may himself proceed to make the tests. All tests so made by the Engineer shall be at the risk and expense of the Contractor unless the Contractor shall establish that the tests were not being unduly delayed in which case tests so made shall be at the risk and expense of the Purchaser.

(iv) The Purchaser, except where otherwise specified, shall provide free of charge, subject to the provisions of Sub-Clause (v) of this clause, such labour, materials, electricity, fuel, water, stores, and apparatus, as may be requisite and as may be reasonably demanded to carry out such tests efficiently.

(v) If any portion of the Works fail to pass the tests, tests of the said portion shall, if required by the Engineer or by the Contractor, be repeated within a reasonable time upon the same terms and conditions, save that all reasonable expenses to which the Purchaser may be put by the repetition of the tests shall be deducted from the Contract Price.

Taking Over.

28.–(i) As soon as the Works have been completed in accordance with the Contract (except in minor respects that do not affect their use for the purpose for which they are intended and save for the obligations of the Contractor under Clause 30 *(Defects after Taking Over)*) and have passed the Tests on Completion the Engineer shall issue a certificate (herein called a "taking-over certificate") in which he shall certify the date on which Works have been so completed and have passed the said tests and the Purchaser shall be deemed to have taken over the Works on the date so certified, but the issue of a taking-over certificate shall not operate as an admission that the Works have been completed in every respect. Save as provided in Sub-Clause (iii) of this clause the Purchaser shall not use the Works or any section or portion thereof until a taking-over certificate has been issued in respect thereof. If nevertheless the Purchaser does so use the Works or any section or portion thereof the Works or section or portion shall be deemed to have been taken over.

(ii) If the Works are divided into two or more sections, Sub-Clause (i) hereof shall apply to each section as it applies to the Works. If by agreement between the Purchaser and the Contractor any portion of the Works (other than a section or sections) shall be taken over before the remainder of the Works, the Engineer shall issue a taking-over certificate in respect of that portion.

(iii) If by reason of any default on the part of the Contractor a taking-over certificate has not been issued in respect of every portion of the Works within one month after the date fixed by the Contract for the completion of the Works, or, if no date be fixed, within a reasonable time, the Purchaser shall be at liberty to use the Works or any portion thereof in respect of which a taking-over certificate has not been issued if and so long as the Works or the portion so used as aforesaid shall be reasonably capable of being used provided that the Contractor shall be afforded reasonable opportunity of taking such steps as may be necessary to permit the issue of the taking-over certificate. The provisions of Clause 21 (i) *(Liability for Accidents and Damage)* shall not apply to any portion of the Works while being so used by the Purchaser

Interference with Tests.

(iv) If by reason of any act or omission of the Purchaser, or the Engineer, or some other contractor employed by the Purchaser, the Contractor shall be prevented from carrying out the Tests on Completion as provided in Clause 27(i) *(Tests on Completion)* then, unless in the meantime the Works shall have been proved not to be substantially in accordance with the Contract, the Purchaser shall be deemed to have taken over the Works, and the Engineer shall issue a taking-over certificate accordingly; nevertheless the Contractor shall make the said tests during the period referred to in Sub-Clause (i) of clause 30 *(Defects after Taking Over)* as and when required by the Engineer by 14 days' notice in writing, and Sub-Clauses (ii), (iii), (iv) and (v) of Clause 27 and Sub-Clause (vi) of Clause 14 *(Inspection, Testing and Rejection of Plant)* shall apply. Such allowances shall be made from the performances required to be attained in the said tests as may be reasonable having regard to any use of the Works by the Purchaser prior to the tests and if the Contractor incurs extra expense in making the said tests pursuant to this sub-clause the Purchaser shall pay to the Contractor a sum in respect thereof which shall be ascertained and determined in like manner to the valuation of variations under Clause 10 *(Variations and Omissions)*.

Suspension of Works.

29.–(i) If by reason of the suspension of the Works by the Purchaser or the Engineer (otherwise than in consequence of some default on the part of the Contractor) or by reason of the Contractor being prevented from or delayed in proceeding with the Works by the Purchaser, the Engineer, or some other contractor employed by the Purchaser, the Contractor shall incur additional expense, there shall be added to the Contract Price a sum in respect thereof, such sum to be ascertained and determined in like manner to the valuation of variations under Clause 10 *(Variations and Omissions)*. Provided that no claim shall be made under this clause unless the Contractor has, within a reasonable time after the event giving rise to the claim given notice in writing to the Engineer of his intention to make such claim.

(ii) If work on the Plant or any portion thereof is suspended as aforesaid by the Purchaser or the Engineer before the Plant or such portion thereof is delivered to the Site and the suspension exceeds three months and the Contractor has suitably and sufficiently marked the Plant or such portion thereof as the Purchaser's property and insured it as provided by Clause 23 *(Insurance of Works)* (the provisions of which clause shall thereafter until actual delivery to the Site apply as if the Plant or such portion thereof were for the time being upon the Site) then without prejudice to the provisions of Sub-Clause 15 (iv) *(b) (Delivery)* the Contractor shall be entitled to have the Contract Value thereof as at the commencement of the suspension included in an interim certificate on the expiration of the said three months or (if later) at the time when, but for such suspension, the Plant or such portion thereof would have been delivered: provided that the Contract Value of any Plant that according to the decision of the Engineer, is defective or not in accordance with the Contract shall not be included in any such certificate.

Defects after Taking Over.

30.–(i) The Contractor shall be responsible for making good with all possible speed any defect in or damage to any portion of the Works which may appear or occur during a period of 12 months after that portion shall have been taken over and which arises either–

(a) from defective materials, workmanship or design (other than a design made, furnished or specified by the Purchaser and for which the Contractor has disclaimed responsibility in writing within a reasonable time after receipt of the Purchaser's instructions), or

(b) from any act or omission of the Contractor done or omitted during the said period.

(ii) If any such defect shall appear or damage occur the Engineer shall inform the Contractor thereof stating in writing the nature of the defect or damage. If the Contractor replaces or renews any portion of the Works, the provisions of this clause shall apply to the portion of the Works so replaced or renewed until the expiration of 12 months from the date of such replacement or renewal.

(iii) The period of 12 months mentioned in Sub-Clauses (i) and (ii) of this clause shall be extended by a period equal to the period during which the Works or that portion thereof in which a defect or damage to which this clause applies has appeared or occurred and cannot be used by reason of that defect or damage.

(iv) If any such defect or damage be not remedied within a reasonable time, the Purchaser may proceed to do the work at the Contractor's risk and expense.

(v) If the replacements or renewals are of such a character as may affect the efficiency of the Works or any portion thereof, the Purchaser may within one month of such replacement or renewal give to the Contractor notice in writing requiring that Tests on Completion be made, in which case such tests shall be carried out as provided in Clause 27 *(Tests on Completion)*.

(vi) These General Conditions shall apply to all inspections, adjustments, replacement, and renewals and to all tests occasioned thereby, carried out by the Contractor pursuant to this clause.

(vii) The Contractor's liability under this clause shall be in lieu of any condition or warranty implied by law as to the quality or fitness for any particular purpose of any portion of the Works taken over under Clause 28 *(Taking Over)* and save as in this clause expressed neither the Contractor nor his Sub-contractors, servants or agents shall be liable, whether in contract, tort or otherwise, in respect of defects in or damage to such portion, or for any injury, damage or loss of whatsoever kind attributable to such defects or damage. For the purposes of this sub-clause the Contractor contracts on his own behalf and on behalf of and as trustee for his Sub-contractors, servants and agents. Nothing in this clause shall affect either the liability of the Contractor under Sub-Clauses (i) and (iii) of Clause 21 *(Liability for Accidents and Damage)* in respect of any portion of the Works not yet taken over or his liability for death or personal injury caused by negligence on his part as defined in Section 1 of the Unfair Contract Terms Act 1977.

(viii) Until the final certificate shall have been issued, the Contractor shall have the right of access, at all reasonable working hours, at his own risk and expense, by himself or his duly authorized representatives, whose names shall have previously been communicated in writing to the Engineer, to all parts of the Works for the purpose of inspecting the working thereof and to records of the working and performance thereof for the purpose of inspecting the same and taking notes therefrom. Subject to the Engineer's approval, which shall not be unreasonably withheld, the Contractor may at his own risk and expense make any tests which he considers desirable.

Interim and Final Certificates.

31.–(i) The Contractor may at the times and in the manner following apply for interim and final certificates, as referred to in Clause 34 *(Terms of Payment)*, for Plant delivered to, and work executed on the Site.

Interim Certificates.

(ii) Applications for interim certificates may be made to the Engineer from time to time during the progress of the Works. Each such application shall state the amount claimed and shall set forth in detail in the order of the schedule of prices particulars of the work executed on the Site and of the Plant delivered to the Site pursuant to the Contract to a date named in the application and since the period covered by the last preceding certificate, if any.

(iii) The Engineer shall issue to the Contractor an interim certificate within 14 days after receiving an application therefor which the Contractor was entitled to make. If the Engineer shall fail to issue an interim certificate as provided in this clause or if the Purchaser shall interfere with or obstruct the issue of any such certificate the Contractor may, without prejudice to any other remedy, either–

(a) after giving the Purchaser or the Engineer 14 days' notice of his intention so to do, stop the Works or any part thereof until the said certificate be issued; in which case the expenses of the Contractor occasioned by such stoppage and the subsequent resumption of work shall be added to the Contract Price, or

(b) after giving to the Purchaser or the Engineer one month's notice of his intention so to do, terminate the Contract, whether or not the Contractor shall have stopped the Works in accordance with paragraph *(a)* hereof or have given notice of his intention so to do.

(iv) Every interim certificate shall certify the total value of the work duly executed on the Site and of the Plant delivered to the Site for use in the Works pursuant to the Contract up to the date named in the application for the certificate, less the said total value so certified in the last previous certificate (if any), provided that the value of any Plant that, according to the decision of the Engineer, does not comply with the Contract or has been brought and is at the date of the certificate prematurely upon the Site, shall not be included in any such certificate.

(v) No interim certificate shall be relied on as conclusive evidence of any matter stated therein or prejudice any right of the Purchaser or the Contractor against the other.

Final Certificate.

(vi) Application for the final certificate may be made to the Engineer at any time after the Contractor has ceased to be under any obligation under Clause 30 *(Defects after Taking Over)*, provided that, if a taking-over certificate has been issued in respect of any portion of the Works, the Contractor may apply for a separate final certificate in respect of each such portion at any time after the said obligation has ceased in relation to such portion, and provided also that, if by reason of the fact that it has been necessary for the Contractor to replace or renew any portion of the Works the obligations of the Contractor under Clause 30 shall continue after the period of 12 months first therein mentioned, the right of the Contractor to apply for a final certificate in respect of the Works or portion thereof other than the portions so replaced or renewed shall not be affected by that fact, and after the Contractor has ceased to be under any obligation under Clause 30 in respect of the portions replaced or renewed he may apply for a final certificate in respect thereof.

(vii) The Engineer shall issue to the Contractor a final certificate within 14 days after receiving an application therefor which the Contractor was entitled to make.

(viii) A final certificate shall certify the total of all amounts comprised in interim certificates previously issued in respect of the Works or the portion thereof to which the final certificate relates, subject to such additions thereto or deductions therefrom as may be authorised in Sub-Clause (x) of this clause.

(ix) A final certificate shall, save in the case of fraud or dishonesty relating to or affecting any matter dealt with in the certificate, be conclusive evidence as to the sufficiency of the Works and of the value thereof unless any proceedings arising out of the Contract whether under Clause 37 *(Arbitration)* or otherwise shall have been commenced by either party before the final certificate has been issued or within one month thereafter.

Adjustments to Certificates.

(x) If any sum shall become payable to the Contractor under the Contract otherwise than for work executed or Plant delivered the amount thereof shall be included in the next certificate (interim or final) issued by the Engineer, and if any sum shall become payable under the Contract by the Contractor to the Purchaser, prior to the issue of the final certificate, whether by deduction from the Contract Price or otherwise, the amount thereof shall be deducted in the next certificate.

(xi) The Engineer may in any certificate give effect to any correction or modification that should properly be made in respect of any previous certificate.

Provisional Sums.

32.–(i) A provisional sum included in the Contract Price shall be expended or used as the Engineer may in writing direct and not otherwise. In so far as a provisional sum is not expended or used it shall be deducted from the Contract Price.

P.C. Items.

(ii) All sums included in the Contract Price in respect of P.C. (Prime Cost) items shall be expended or used as the Engineer may in writing direct and not otherwise. To the net amount paid by the Contractor in respect of each P.C. item there shall be added the percentage named in the Appendix of the said amount. The sum by which the net amount so paid in respect of any P.C. item plus the said percentage thereon exceeds or is less than the sum included in the Contract Price in respect of that item shall be added to or deducted from the Contract Price as the case may be.

(iii) The Contractor shall have no responsiblity for work done or Plant supplied by any other person in pursuance of directions given by the Engineer under this clause unless the Contractor shall have approved the person by whom such work is to be done or such Plant is to be supplied and the Plant, if any, to be supplied.

Payments due from the Contractor.

33.–Without prejudice to any other remedy which the Purchaser may have he shall be entitled to deduct from any moneys due, or becoming due to the Contractor under the Contract, all costs, damages or expenses for which under the Contract the Contractor is liable to the Purchaser.

Terms of Payment.

34.–(i) The Purchaser shall pay to the Contractor in the following manner the Contract Price adjusted to give effect to such additions thereto and such deductions therefrom as are provided for in these Conditions:

(a) Within 14 days from the presentation of each interim certificate a sum equal to 95 per cent of the sum certified therein,

(b) 97½ per cent of the Contract Price adjusted as aforesaid within one month from the date certified in the taking-over certificate,

(c) The balance of the Contract Price adjusted as aforesaid within one month after the presentation of the final certificate provided that if the Contractor shall have furnished to the Purchaser a guarantee acceptable to the Purchaser for the repayment on demand of such balance he shall be entitled to payment thereof with or at any time after the payment provided for by paragraph *(b)* hereof.

If any section or portion of the Works shall be taken over separately under Clause 28 *(Taking Over)* the payments herein provided for on or after taking over shall be made in respect of the section or portion taken over and reference to the Contract Price shall mean such part of the Contract Price as shall, in the absence of agreement, be apportioned thereto by the Engineer.

In determining the amount of any payment under this clause in respect of any portion of the Works due account shall be taken of all payments previously made in respect of the same portion whether under this clause or under Clause 15 *(Delivery)*.

(ii) If at any time at which any payment would fall to be made under paragraph *(b)* of Sub-Clause (i) of this clause there shall be any defect in any portion of the Works in respect of which such payment is proposed, the Purchaser may retain the whole of such payment provided that in the event of the said defect being of a minor character and not such as to affect the use of the Works, or the said portion thereof for the purpose intended without serious risk the Purchaser shall not retain a greater sum than represents the cost of making good the said minor defect. Any sum retained by the Purchaser pursuant to the provisions of this sub-clause shall be paid to the Contractor upon the said defect being made good.

(iii) If the payment of any sum payable under Sub-Clause (i) of this clause shall be improperly delayed by the Purchaser or the Engineer interest at the rate of two per cent per annum over the Bank of England minimum lending rate from time to time in force on the amount of the delayed payment for the period of the delay shall be added to the Contract Price.

(iv) If the Purchaser shall fail to make any payment as provided in this clause the Contractor shall have the like remedies, without prejudice to any other, as are provided in Sub-Clause (iii) of Clause 31 *(interim and Final Certificates)*.

Statutory and Other Regulations.

35.–(i) If the cost to the Contractor of performing his obligations under the Contract shall be increased or reduced by reason of the making or amendment after the date of tender of any law or of any order, regulation, or by-law having the force of law in the United Kingdom that shall affect the Contractor in the performance of his obligations under the Contract, the amount of such increase or reduction (to the extent that it arises directly in respect of the Works) shall be added to or deducted from the Contract Price as the case may be.

Value Added Tax.

(ii) Unless otherwise stated in the tender the Contract Price is deemed to exclude Value Added Tax. To the extent that the Tax is properly chargeable on the supply to the Purchaser of any goods or services provided by the Contractor under the Contract, the Purchaser shall pay such Tax as an addition to payments otherwise due to the Contractor under the Contract.

Metrication.

36.–(i) If any plant described in the Contract or the subject of a variation (hereinafter called 'a variation order') under Clause 10 *(Variation and Omissions)* is described by dimensions in the metric or imperial measure and the Contractor cannot procure such Plant in the measure specified in sufficient time to avoid delay in the performance of his other obligations under the Contract, but can obtain such Plant in the other measure to dimensions approximating to those described in the Contract or in the variation order, then the Contractor shall forthwith give notice to the Engineer of the facts stating the dimensions to which such Plant is procurable in the other measure.

(ii) The Engineer shall within 14 days after the receipt of a notice under the preceding sub-clause give notice to the Contractor in writing pursuant to Clause 10 *(Variations and Omissions)*. Such notice shall either–

(a) direct the Contractor to supply such Plant to the dimensions stated in the said notice instead of to the dimensions described in the Contract or variation order as the case may be, or,

(b) direct the Contractor to make some other variation whereby the need to supply such Plant to the dimensions described in the Contract or variation order will be avoided.

The provisions of the Contract shall apply to directions given under this clause as though such directions had been included in the Specification.

Arbitration.

37.–(i) If at any time any question, dispute, or difference shall arise between the Purchaser and the Contractor, either party shall, as soon as reasonably practicable, give to the other notice in writing of the existence of such question, dispute or difference specifying its nature and the point at issue, and the same shall be referred to the arbitration of a person to be agreed upon, or failing such agreement within six weeks, to some person appointed on the application of either of the parties hereto by the President for the time being of the Institution named in the Appendix, provided that a question, dispute or difference relating to a decision, instruction, or order of the Engineer shall not be referred to arbitration unless notice has been given by the Contractor in accordance with Clause 19 *(b) (Engineer's Decisions).*

(ii) Performance of the Contract shall continue during arbitration proceedings unless the Engineer shall order the suspension thereof or of any part thereof, and if any such suspension shall be ordered the reasonable expenses of the Contractor occasioned by such suspension shall be added to the Contract Price. No payments due or payable by the Purchaser shall be withheld on account of pending reference to arbitration.

Construction of Contract.

38.—The Contract shall in all respects be construed and operate as an English Contract and in conformity with English law, and all payments thereunder shall be made in sterling money. The marginal notes hereto shall not affect the construction hereof.

Variation in Costs.

39.*—If the cost to the Contractor of performing his obligations under the Contract shall be increased or reduced by reason of any rise or fall in the rates of wages payable to labour or in the cost of material or transport above or below such rates and costs ruling at the date of tender, the amount of such increase or reduction shall be added to or deducted from the Contract Price as the case may be, provided that no account shall be taken of any amount by which any cost incurred by the Contractor has been increased by the default or negligence of the Contractor. For the purpose of this clause 'the cost of material' shall be construed as including any duty or tax by whomsoever payable which is payable under or by virtue of any Act of Parliament on the import, purchase, sale, appropriation, processing or use of such material.

***NOTE:—Unless this Clause is excluded any quoted price will be subject to adjustment in accordance with its terms.**

APPENDIX

Clause.		
26		Delay in Completion:
	(a)	*(a)* percentage of Contract Value to be deducted as damages.
	(b)	*(b)* maximum percentage of Contract Value which the deductions may not exceed.
32	....	Percentage on Prime Cost items.

37 The Institution of Mechanical Engineers.*
The Institution of Electrical Engineers.*

*Delete one line.

AGREEMENT

This Agreement is made the day of 19

BETWEEN

(hereinafter referred to as the "Contractor") of the one part and

(hereinafter called the "Purchaser") of the other part **Whereas** the Purchaser desires to have provided and executed certain Works mentioned, enumerated, or referred to in certain General Conditions, Specification, Schedules, Drawings, Plans, Schedule of Prices, the Contractor's Tender and covering letter (if any), the acceptance of the Tender, and any other relevant correspondence* which for the purpose of identification have been signed by on behalf of the Contractor and by of (the Engineer of the Purchaser) on behalf of the Purchaser all of which are deemed to form part of this Contract as though separately set out herein and are included in the expression "Contract" whenever herein used. **And Whereas** the Purchaser has accepted the Tender of the Contractor for the provision and execution of the said Works for the sum of

(hereinafter called "the Contract Price) upon the terms and subject to the Conditions hereinafter mentioned: **Now this Deed Witnesseth** and it is hereby agreed and declared as follows, that is to say, in consideration of the payments to be made to the Contractor by the Purchaser as hereinafter mentioned the Contractor hereby covenants with the Purchaser, his legal personal representatives, successors, and assigns that the Contractor shall and will duly provide, execute, complete, and maintain the said Works and shall do and perform all other acts and things in the Contract mentioned or described or which are to be implied therefrom or may be reasonably necessary for the completion of the said Works within and at the times and in the manner and subject to the terms conditions and stipulations mentioned in the Contract.

And in consideration of the due provision, execution and completion of the Works and the maintenance thereof as aforesaid the Purchaser does hereby for himself his legal personal representatives, successors, and assigns covenant with the Contractor that he, the Purchaser, his legal personal representatives, successors, or assigns will pay to the Contractor the Contract Price or such other sum as may become payable to the Contractor under the provisions of the Contract, such payments to be made at such time and in such manner as is provided by the Contract.

In Witness whereof, etc.

*Delete as required; see Memorandum on the use of the General Conditions, paragraph B.

FORM OF GUARANTEE

(When required, and in cases where an Insurance Policy is not used.)

Whereas by an Agreement dated ....and made between .... (hereinafter called "the Purchaser") and .... (hereinafter called "the Contractor") the parties thereto entered into a contract as therein stated: NOW we hereby jointly and severally guarantee to the Purchaser punctual true and faithful performance and observance by the Contractor of the covenant on his part contained in the said Agreement and undertake to be responsible to the Purchaser his legal personal representatives successors or assigns as Sureties for the Contractor for the payment by him of all sums of money losses damages costs charges and expenses that may become due or payable to the Purchaser his legal personal representatives successors or assigns by or from the Contractor by reason or in consequence of the default of the Contractor in the performance or observance of his said covenant but so nevertheless that the total amount to be demanded or recovered by the Purchaser his legal personal representatives successors or assigns of or from us as Sureties shall not exceed 15 per cent of the Contract Price.

This Guarantee shall not be revocable by notice or by reason of the death of us or either of us and our liability as Sureties hereunder shall not be impaired or discharged by any extensions of time or variations or alterations made given conceded or agreed (with or without our knowledge or consent) under the General Conditions referred to in the said Agreement or (where the Purchaser or the Contractor is a firm) by any change in the constitution of the Purchaser's or the Contractor's respective firms.

MEMORANDUM ON THE USE OF THE GENERAL CONDITIONS OF CONTRACT

A. Applicability

The General Conditions are intended to be suitable for the mechanical engineering and electrical engineering industries and to cover the main conditions for Home Contracts for the supply of plant and materials and the erection thereof on Site.

It is hoped that, in order to secure the advantage of uniformity and to avoid trouble and expense, they may be found suitable in their entirety for the majority of contracts. They may, however, be added to or amended as may be required to suit each particular case. It is suggested, in order to facilitate checking, that when reprinted elsewhere any additions or alterations be made in different type or underlined so as to attract attention.

B. Contract documents

Contracts are commonly made in one of three ways:

(a) By a formal agreement, such as the model form supplied with these Conditions, incorporating in the contract the General Conditions, Specification, Schedules, Drawings, etc. In this case all the terms of the contract are contained in the Agreement, the Conditions, Specification, and other incorporated documents on which Tenders were invited, and agreed variations in or additions to the provisions of these documents are represented by suitable amendments.

(b) By a formal agreement, as in the method *(a)*, but incorporating also the Contractor's Tender and covering letter, if any, the Purchaser's acceptance of the Tender, and all correspondence containing terms accepted by the Purchaser and the Contractor. When this method is adopted any modifications of the General Conditions, Specifications, etc., have to be sought in the covering letter and other correspondence.

(c) Without a formal agreement, the Tender and its acceptance, together with other relevant correspondence, being relied on as constituting the contract between the parties.

When either method *(a)* or *(b)* is adopted, a copy of the form of agreement to be used should be included with the documents supplied to or open to the inspection of intending Tenderers.

When method *(a)* is used, the references in the model form of Agreement to the Tender, covering letter, acceptance of the Tender, and other correspondence should be deleted.

C. Progress Payments during Manufacture

Under the General Conditions the Contractor does not become entitled to any payments (except as provided under Clauses 15 and 29) for plant he has manufactured until it has been delivered to the Site. In the case of contracts involving considerable expenditure on manufacture before any delivery can be made, the General Conditions may be modified to provide for progress payments to the Contractor in respect of plant in course of manufacture, in which case the insurance arrangements may also need to be modified.

MEMORANDUM AS TO TENDERS

Tenderer may propose modifications.

The Tenderer is at liberty to add any details and conditions that he may deem desirable, and in the event of his doing so must print or type the same and annex the added matter to the Specification or General Conditions returned by him, but such additional details and conditions will not be binding on the Purchaser unless they are approved by him and incorporated in the Contract.

Where in special circumstances the Tenderer deems it advisable to provide for the nondisclosure of drawings and information of a confidential nature furnished with his Tender he should include in the Tender or covering letter a stipulation to this effect. If the Tender embodying such a stipulation is accepted the stipulation should be incorporated in any subsequent formal agreement.

If the Tenderer has any doubt as to the meaning of any portion of the General Conditions or of the Specification, he should, when submitting his Tender, set out in his covering letter the interpretation upon which he relies.

Tenders.

One copy of the Specification supplied, together with the General Conditions and Drawings, enclosed in a plain sealed cover bearing the words "Tender for *(here follows the short title of the Works)* "

and not bearing any name or mark indicating the sender shall, subject to modifications mentioned above, be returned intact with the Tender Form and Schedules, if any, filled up and signed, addressed to

and must be received by him on or before

The Purchaser does not bind himself to accept the lowest or any Tender, nor will he be responsible for, or pay for, expenses or losses which may be incurred by any Tenderer in the preparation of his Tender.

Copies of Specification etc.

A duplicate copy of the Specification, General Conditions and Drawings will be supplied to intending Tenderers without charge, and extra copies of the Drawings at the cost of reproduction.

The sum deposited by the Tenderer on application for the Specification will be refunded to him within one month of the date of the adjudication upon the Tenders, if all copies of the Specification, Drawings, etc., have been returned by the Tenderer, unless the Tender was not made in good faith, in which case the deposit may be forfeited.

Agreement.

The Tenderer whose Tender is accepted may be required to enter into a formal agreement with the Purchaser.

NOTE

The advertisement inviting Tenders should state whether and if so where the Agreement, General Conditions, Specification, and Drawings may be inspected, and should give such particulars of the class of plant and apparatus required under each Section as will enable contracting firms to decide, without obtaining the Specification, whether they desire to tender, and should also state the amount of the deposit to be paid for the General Conditions, Specification, and Drawings.

INDEX

B1

MODEL FORM OF

GENERAL CONDITIONS OF CONTRACT

INCLUDING FORM OF AGREEMENT

RECOMMENDED BY

THE INSTITUTION OF MECHANICAL ENGINEERS
THE INSTITUTION OF ELECTRICAL ENGINEERS
AND
THE ASSOCIATION OF CONSULTING ENGINEERS

FOR USE IN CONNECTION WITH

EXPORT CONTRACTS FOR SUPPLY OF PLANT AND MACHINERY

FIFTH EDITION 1981

Published for the Joint Committee on Model Forms of General Conditions of Contract by
THE INSTITUTION OF ELECTRICAL ENGINEERS
and obtainable from
Publications Sales Department, Station House, 70 Nightingale Road, Hitchin, Herts, SG5 1RJ
or to callers, at the Reception Desk, Savoy Place
or from
THE INSTITUTION OF MECHANICAL ENGINEERS
Publications Sales Department, P.O. Box 24, Northgate Avenue, Bury St. Edmunds, Suffolk, IP32 6BW
or to callers, at the Reception Desk, Birdcage Walk, London
or at
THE ASSOCIATION OF CONSULTING ENGINEERS, Alliance House, 12, Caxton Street,
London, SW1H 0QL

HISTORICAL NOTE

A Model Form of General Conditions, 'B1' (Export Contracts, delivery f.o.b.) was first published by The Institution of Electrical Engineers in 1925. A revision was issued in 1928.

Following consultations, and by agreement between the Council of The Institution of Mechanical Engineers and the Council of The Institution of Electrical Engineers, the scope of the Model Form was enlarged to make it suitable for both the electrical and mechanical engineering industries, and on this basis the Model Form was first issued jointly by the two Councils as–

1956 Edition

Model Form of General Conditions of Contract 'B1'– Export Contracts with delivery f.o.b. or c.i.f.

Following an approach in 1952 by The Association of Consulting Engineers, it was agreed that the Association should adopt and recommend to its members the use of this Model Form, and the name of the Association accordingly appears on the title page.

The 1973 Edition incorporated substantial revisions to take account of developments in practice.

This 1981 Edition incorporates further revisions to take account of developments in practice, provision for an advance payment and for methods of delivery for which there are now internationally accepted interpretations of the rights and obligations of the parties, and has been re-titled accordingly.

This form may be cited as–Model Form 'B1/1981'

I.MECH.E./I.E.E. MODEL FORM OF GENERAL CONDITIONS OF CONTRACT 'B1/1981'

GENERAL CONDITIONS

Definitions of Terms.

1. – In construing these General Conditions and the Specification, the following words shall have the meanings herein assigned to them unless there is something in the subject matter or context inconsistent with such construction:

The "Contract" shall mean the agreement howsoever made between the Purchaser and the Contractor for the supply of the Plant, including therein all documents to which reference may properly be made in order to ascertain the rights and obligations of the parties under the said agreement.

The "Contractor" shall mean the tenderer whose tender has been accepted by the Purchaser and shall include the Contractor's legal personal representatives, successors and assigns.

The "Contract Price" shall mean the sum named in the Contract adjusted to give effect to such additions and deductions as may be made in accordance with Clause 11 (*Variations*).

The "Engineer" shall mean
or the person for the time being or from time to time notified in writing by the Purchaser to the Contractor as the Engineer for the Contract, or in default of any notification the Purchaser.

"Month" shall mean calendar month.

The "Plant" shall mean all or any part of the machinery, apparatus, materials, articles and things of all kinds to be provided by the Contractor.

The "Purchaser" shall mean
and shall include the Purchaser's legal personal representatives, successors and assigns.

The "Specification" shall mean the specification annexed to or issued with these General Conditions.

"Sub-Contractor" shall mean any person (other than the Contractor) named in the Contract for the supply of any part of the Plant or any person to whom any part of the Contract has been sub-let with the consent in writing of the Engineer, and the legal representatives, successors and assigns of such person.

The phrase "in writing" shall include any manuscript, type-written or printed statement, under seal or hand as the case may be.

Words importing persons shall include firms and corporations.

Words importing the singular only shall also include the plural and *vice-versa*.

Contractor to inform himself fully.

2. – The Contractor shall be deemed to have examined the General Conditions and Specification, with such schedules, drawings and plans as are annexed thereto or referred to therein.

Expenses of Agreement.

3. – The expenses of preparing, completing and stamping the agreement, if any, shall be paid by the Purchaser, and an executed counterpart thereof properly stamped together with copies of all other documents comprising the Contract shall be furnished to the Contractor free of charge.

Drawings.

4. – (i) The Contractor shall submit to the Engineer for approval:

(a) within the time specified in the Specification or, if no time is specified, then a reasonable

time, such drawings, samples, patterns and models as may be called for therein, and

(b) during the progress of the Contract and within such reasonable times as the Engineer may require, such drawings of the general arrangement and of details of the Plant or any part thereof as the Engineer may reasonably require, provided that the Contractor shall not be under any obligation to supply copies of shop drawings.

Within a reasonable period after receiving such drawings, samples, patterns and models the Engineer shall signify his approval or otherwise. If the Engineer shall not approve any drawing, sample, pattern or model thus submitted, the same shall be forthwith modified to meet the reasonable requirements of the Engineer and shall be re-submitted. Copies of all drawings which require to be approved by the Engineer shall be provided in triplicate by the Contractor. Drawings, samples, patterns and models when approved shall, if required by either party, be signed or identified by both parties and, if required by the Contractor, one copy shall be returned to him. All dimensions marked on drawings shall be considered correct although measurements by scale may differ therefrom. Detailed drawings shall take precedence where they differ from general arrangement drawings.

(ii) Drawings, samples, patterns and models approved as above described shall not be departed from except as provided in Clause 11 (*Variations*).

(iii) The Engineer shall have the right at all reasonable times to inspect at the premises of the Contractor all drawings of any portion of the Plant.

(iv) The Contractor shall, if so desired by the Purchaser, furnish to the Purchaser in writing at the commencement of the period referred to in Clause 18 (*Defects after Delivery*), or at such earlier times as may be named in the Specification, such information, accompanied by drawings, as may be necessary to enable the Purchaser to operate, maintain, dismantle, reassemble and adjust all parts of the Plant.

(v) Drawings submitted in pursuance of paragraphs (*a*) and (*b*) of Sub-Clause (i) of this clause shall not, without the consent of the Contractor, be used by the Purchaser except for the purposes of the Contract, nor shall they without such consent be communicated to third parties save insofar as may be necessary for the proper execution of the Contract.

Mistakes in information.

5. – (i) The Contractor shall be responsible for any discrepancies, errors or omissions in the drawings and information supplied by him, whether they have been approved by the Engineer or not, provided that such discrepancies, errors or omissions be not due to inaccurate drawings or information furnished to the Contractor by the Purchaser or the Engineer.

(ii) The Contractor shall at his own expense carry out any alterations or remedial work necessitated by reason of such discrepancies, errors or omissions and modify the drawings and information accordingly, or if the same be done by or on behalf of the Purchaser, shall bear all costs reasonably incurred therein. The performance of his obligations under this sub-clause shall be in full satisfaction of the Contractor's liability under Sub-Clause (i) of this clause, but shall not relieve him of his liability under Clause 17 *(Delay in Completion)* in so far as that liability is the result of a failure of the Contractor to perform his obligations under that sub-clause.

(iii) The Purchaser shall be responsible for drawings, information and details supplied in writing by the Purchaser or the Engineer. The Purchaser shall pay the extra cost reasonably incurred by the Contractor due to alterations of the work necessitated by reason of inaccurate drawings, information or details so supplied to the Contractor.

Assignment.

6. – (i) The Contractor shall not, without the consent in writing of the Purchaser, which shall not be unreasonably withheld, assign or transfer the Contract or the benefits or obligations thereof or any part thereof to any other person, provided that this shall not affect any right of the Contractor to assign, either absolutely or by way of charge, any moneys due or to become due to him, or which may become payable to him under the Contract.

Sub-letting.

(ii) The Contractor shall not, without the consent in writing of the Engineer, which shall not be unreasonably withheld, sub-let the Contract or any part thereof, or make any sub-contract with any person or persons for the execution of any part of the Contract, but the restriction contained in this clause shall not apply to sub-contracts for materials, for minor details, or for any part of the Plant of which the makers are named in the Contract. Any such consent shall not relieve the Contractor from his obligations under the Contract.

Patent Rights, etc.

7. – (i) The Contractor shall indemnify the Purchaser against all actions, claims, demands, costs, charges and expenses arising from or incurred by reason of any infringement or any alleged infringement of letters patent, registered design, copyright, trade mark or trade name protected in the United Kingdom or, where the country in which the Plant is to be used is specified in the Contract, in that country, by the use of any Plant supplied by the Contractor, but such indemnity shall not cover any use of the Plant otherwise than for the purpose indicated by or reasonably to be inferred from the Specification or to any infringement which is due to the use of any Plant in association or combination with any other plant not supplied by the Contractor.

(ii) In the event of any claim being made or action brought against the Purchaser arising out of the matters referred to in this clause, the Contractor shall be promptly notified thereof and may at his own expense conduct all negotiations for the settlement of the same, and any litigation that may arise therefrom. The Purchaser shall not, unless and until the Contractor shall have failed to take over the conduct of the negotiations or litigation, make any admission which might be prejudicial thereto. The conduct by the Contractor of such negotiations or litigation shall be conditional upon the Contractor having first given to the Purchaser such reasonable security as shall from time to time be required by the Purchaser to cover the amount ascertained or agreed or estimated, as the case may be, of any compensation, damages, expenses and costs for which the Purchaser may become liable. The Purchaser shall, at the request of the Contractor, afford all available assistance for the purpose of contesting any such claim or action, and shall be repaid all reasonable expenses incurred in so doing.

(iii) The Purchaser on his part warrants that any design or instructions furnished or given by him shall not be such as will cause the Contractor in the performance of the Contract to infringe any letters patent, registered design, copyright, trade mark or trade name in the country specified in the Contract in which the Plant is to be used.

Manner of Manufacture.

8. – All Plant shall be manufactured, prepared for shipment and delivered in accordance with the Specification or, if not therein specified, to the reasonable satisfaction of the Engineer.

Engineer's Supervision.

9. – (i) All instructions and orders to the Contractor shall, except as herein otherwise provided, be given by the Engineer.

Engineer's Representative.

(ii) The Engineer may from time to time delegate any of the powers, discretions, functions and authorities vested in him and may at any time revoke any such delegation. Any such delegation or revocation shall be in writing signed by the Engineer and, in the case of a delegation, shall specify the powers, discretions, functions and authorities thereby delegated and the person or persons to whom the same are delegated. No such delegation or revocation shall have effect until a copy thereof has been delivered to the Contractor.

Engineer's Decisions.

10. – The Contractor shall proceed with the Contract in accordance with decisions, instructions and orders given by the Engineer in accordance with these Conditions, provided always that–

(a) if the Contractor shall, without undue delay after being given any decision, instruction or order otherwise than in writing, require it to be confirmed in writing, such decision, instruction or order shall not be effective until written confirmation thereof has been received by the Contractor, and

(b) if the Contractor shall, by written notice to the Engineer within 14 days after receiving any decision, instruction or order of the Engineer in writing or written confirmation thereof, intimate that he disputes or questions the decision, instruction or order, giving his reasons for so doing, either party to the Contract shall be at liberty to refer the matter to arbitration pursuant to Clause 24 (*Arbitration*), but such an intimation shall not relieve the Contractor of his obligation to proceed with the Contract in accordance with the deci-

sion, instruction or order in respect of which the intimation has been given. The Contractor shall be at liberty in any such arbitration to rely on reasons additional to the reasons stated in the said intimation.

Variations.

11. – (i) The Contractor shall not alter any of the Plant except as directed in writing by the Engineer, but the Engineer shall have full power, subject to the proviso hereinafter contained, from time to time during the execution of the Contract by notice in writing to direct the Contractor to alter, amend omit, add to or otherwise vary any of the Plant and the Contractor shall carry out such variations, and be bound by the same conditions, so far as applicable, as though the said variations were stated in the Specification: provided that no such variation shall, except with the consent in writing of the Contractor, be such as will, with any variations already directed to be made, involve a net addition to or deduction from the Contract Price of more than 15 per cent thereof, disregarding for this purpose any addition or deduction previously made pursuant to this clause. In any case in which the Contractor has received any such direction from the Engineer which either then or later will, in the opinion of the Contractor, involve an addition to or deduction from the Contract Price, the Contractor shall, as soon as reasonably possible, advise the Engineer in writing to that effect. The amount to be added to or deducted from the Contract Price shall be ascertained and determined in accordance with the rates specified in the schedules of prices, so far as the same may be applicable, and where rates are not contained in the said schedules, or are not applicable, such amount shall be such sum as is reasonable in the circumstances. Due account shall be taken of any partial execution of the Contract which is rendered useless by any such variation.

(ii) If the Engineer shall make any such variation in any part of the Plant, such reasonable notice in writing shall be given to the Contractor as will enable him to make his arrangements accordingly. If in the opinion of the Contractor any such variation is likely to prevent or prejudice the Contractor from or in fulfilling any of his obligations under the Contract, he shall notify the Engineer thereof in writing, and the Engineer shall decide forthwith whether or not the same shall be carried out. If the Engineer confirms his instructions in writing, the said obligations shall be modified to such an extent as may be justified. Until the Engineer so confirms his instructions they shall be deemed not to have been given.

Contractor's Default.

12. – If the Contractor shall neglect to perform the Contract with due diligence and expedition, or shall refuse or neglect to comply with any reasonable orders given him in writing by the Engineer in connection with the performance of the Contract, or shall contravene the provisions thereof, the Purchaser may give notice in writing to the Contractor to make good the neglect, refusal or contravention complained of. Should the Contractor fail to comply with the notice within 14 days from the date of service thereof, it shall be lawful for the Purchaser forthwith to terminate the Contract by notice in writing to the Contractor without prejudice to any rights which may have accrued thereunder to either party prior to such termination. The Purchaser shall be entitled to retain and apply any balance which may be otherwise due on the Contract by him to the Contractor, or such part thereof as may be necessary, to the payment of the cost of completing the Plant or of purchasing equivalent plant from another source. If such cost shall exceed the amount that would otherwise become due to the Contractor in accordance with the Contract, the Contractor shall pay such excess. If the Purchaser pursuant to this clause completes the Plant or obtains equivalent plant from another source, the Contractor's liability under Clause 17 (*Delay in Completion*) shall cease in respect of the Plant concerned with effect from the termination of the Contract under this clause, without prejudice to any such liability as shall have already accrued.

Bankruptcy.

13. – If the Contractor shall become bankrupt or insolvent, or have a receiving order made against him, or compound with his creditors, or being a corporation commence to be wound up, not being a members' voluntary winding up for the purpose of amalgamation or reconstruction, or carry on its business under a receiver for the benefit of its creditors or any of them, the Purchaser shall be at liberty either –

(a) to terminate the Contract forthwith by notice in writing to the Contractor or to the receiver or liquidator or to any person in whom the Contract may become vested, in which case the relevant provisions of Clause 12 *(Contractor's Default)* shall apply as though the Contract had been terminated pursuant thereto, or

(b) to give such receiver, liquidator or other person the option of carrying out the Contract subject to his providing a guarantee for the due and faithful performance of the Contract up

to an amount to be agreed.

Inspection. Testing and Rejection.

14. – (i) The Engineer shall be entitled at all reasonable times during manufacture to inspect, examine and test on the Contractor's premises the materials, workmanship and performances of all Plant to be supplied under the Contract, and if part of the said Plant is being manufactured on other premises, the Contractor shall obtain for the Engineer permission to inspect, examine and test as if the said Plant were being manufactured on the Contractor's premises. Such inspection, examination or testing shall not release the Contractor from any obligation under the Contract.

(ii) Where the Contract provides for tests of the Plant or any part thereof when completely manufactured such tests shall, in the absence of any arrangements to the contrary, take place on the premises of the Contractor.

(iii) The Contractor shall, after consulting the Engineer, give the Engineer reasonable notice in writing of the date on and the place at which any Plant will be ready for testing as provided in the Contract and unless the Engineer shall attend at the place so named on the date which the Contractor has stated in his notice the Contractor may proceed with the tests, which shall be deemed to have been made in the Engineer's presence, and shall forthwith forward to the Engineer duly certified copies of the test readings. The Engineer shall give the Contractor 24 hours' notice in writing of his intention to attend the tests.

(iv) Where the Contract provides for tests on the premises of the Contractor or of any Sub-Contractor the Contractor, except where otherwise specified, shall provide free of charge such assistance, labour, materials, electricity, fuel, stores, apparatus and instruments as may be requisite and as may be reasonably demanded to carry out such tests efficiently.

(v) As and when the Engineer is satisfied that any Plant shall have passed the tests referred to in this clause he shall notify the Contractor in writing to that effect.

(vi) If after inspecting, examining or testing any Plant the Engineer shall decide that such Plant or any part thereof is defective or not in accordance with the Contract, he may reject the said Plant or part thereof by giving to the Contractor within a reasonable time notice in writing of such rejection, stating therein the grounds upon which the said decision is based.

Delivery.

15. –(i) Save as varied by these General Conditions the obligations of the parties in relation to the delivery terms specified in the Contract and the rules determining the passing of property in and the risk of loss of or damage to the Plant shall be fixed in accordance with the International Rules for the Interpretation of Trade Terms (Incoterms) of the International Chamber of Commerce in force at the date of the Contract.

(ii) If the Contract provides for delivery f.o.b. the Contractor shall not be required to give the Purchaser the notice relating to insurance mentioned in Section 32(3) of the Sale of Goods Act, 1979.

Delayed Plant.

(iii) If by delay or failure on the part of the Purchaser to give any necessary instructions or from any cause for which the Purchaser or some other contractor employed by him is responsible, the Contractor shall be prevented, or at the request of the Purchaser refrains, from delivering any Plant at the time specified for delivery thereof or, if no time is specified, within a reasonable time, and shall have given notice in writing to the Purchaser that such Plant (hereinafter referred to as "the delayed Plant") is ready for delivery, and shall have suitably and sufficiently marked the delayed Plant as appropriated to the Contract, and shall have given to the Engineer an opportunity of inspecting the delayed Plant, then and in any such case the following provisions shall have effect–

(a) There shall be added to the Contract Price a reasonable sum for storing and taking reasonable measures to protect and preserve the delayed Plant from, and insuring it against loss, deterioration and damage however caused from the time when but for the said delay, failure or other cause the delayed Plant would have been delivered (hereinafter referred to as "the normal delivery date") until the Contractor shall no longer be prevented from delivering it or shall be relieved of responsibility therefor under paragraph *(b)* of this Sub-Clause, whichever shall first happen.

(b) If at the expiration of two months from the normal delivery date the Contractor shall still be prevented as aforesaid from delivering the delayed Plant, he may by notice in writing expiring 30 days after receipt thereof by the Purchaser require the Purchaser to assume responsibility for storing, protecting, preserving and insuring the delayed Plant. Upon the expiration of the last-mentioned notice the Contractor shall be relieved of any responsibility for the delayed Plant either until the expiration of 30 days after the receipt of notice in writing from the Engineer that the delayed Plant may be delivered (hereinafter referred to as "the notice to deliver") or until the Contractor, having received the notice to deliver, has proceeded to fulfil the obligation imposed on him by paragraph *(c)* of this Sub-Clause, whichever shall first occur, provided always that if the notice to deliver shall be given within 30 days after the receipt of the last-mentioned notice given by the Contract that notice shall not have effect.

(c) After the receipt of the notice to deliver, the Contractor, if he has been relieved of responsibility under the last preceding paragraph of this Sub-Clause, shall (and in any other case may) examine the delayed Plant and make good any deterioration or defect therein that may have developed or loss thereof that may have occurred after the normal delivery date.

(d) There shall be added to the Contract Price a reasonable sum for making the examination referred to in paragraph *(c)* of this Sub-Clause and for making good any deterioration, defect or loss as therein mentioned, except insofar as the same was caused by faulty workmanship or materials or by the Contractor's failure to take the measures referred to in paragraph *(a)* of this Sub-Clause. Any expense to which the Contractor may be put in delivering the delayed Plant or in performing his obligations under Clause 18 *(Defects after Delivery)* which would not have been incurred had the delivery of the delayed Plant not been prevented as aforesaid shall also be added to the Contract Price.

(e) Without prejudice to the provisions of Sub-Clause (vii) of Clause 18 *(Defects after Delivery)*, the obligations of the Contractor under that clause with respect to delayed Plant shall not apply to any defect that may develop therein after the expiration of three years from the normal delivery date.

Extension of Time for Completion.

16. – If, by reason of any industrial dispute or any cause beyond the reasonable control of the Contractor arising after the acceptance of the tender, the Contractor shall have been delayed or impeded in the completion of the Contract, except for his obligations under Clause 18 *(Defects after Delivery)*, whether the delay or impediment occur before or after the time, if any, or extended time fixed therefor, then provided that the Contractor shall without delay have given to the Purchaser or the Engineer notice in writing of his claim for an extension of time, the Engineer shall on receipt of such notice grant the Contractor from time to time in writing either prospectively or retrospectively such extension of time as may be reasonable. Any delay on the part of a Sub-Contractor which prevents the Contractor from completing the Contract within the time for completion shall entitle the Contractor to an extension thereof provided such delay was due to any cause for which the Contractor himself would have been entitled to an extension of time under this clause.

Delay in Completion.

17. – (i) If the Contractor fails to complete the Contract (except for his obligations under Clause 18 *(Defects after Delivery)*) within the time fixed therefor by the Contract, or if no time be fixed, within a reasonable time or within any extension of time granted by the Engineer, there shall be deducted from the Contract Price the percentage named in the Appendix of such fraction of the Contract Price as would, on the due completion of the Contract, be properly attributable to such portion only of the Plant as cannot in consequence of the said failure be put to the use intended for each week between the time fixed for completion of the Contract (except as aforesaid) and the actual date of completion, provided always that the amount so recoverable shall not exceed the maximum percentage specified in the Appendix.

(ii) When the sum deducted under the provisions of Sub-Clause (i) of this clause has amounted to the maximum therein provided, the Purchaser shall be entitled by notice in writing to the Contractor to require the Contractor to complete the Contract (except as mentioned in Sub-Clause (i) of this clause) within such time (not being less than 28 days) as the Purchaser may specify in the notice. If the Contractor shall fail so to complete within the time so specified the Purchaser shall be entitled, after having given the Contractor notice in writing of his intention so to do, to purchase plant in place of those portions of the Plant which the Contractor has failed to complete and there shall be deducted from the Contract Price that part

7 I.MECH.E./I.E.E. MODEL FORM OF GENERAL CONDITIONS OF CONTRACT 'B1/1981'

thereof which is properly apportionable to the uncompleted Plant. The Contractor shall pay to the Purchaser any sum by which the expenditure reasonably incurred by the Purchaser in obtaining plant in place of the uncompleted Plant exceeds the sum deducted. All plant obtained by the Purchaser in place of uncompleted Plant shall comply with the Contract and shall be obtained at reasonable prices and when practicable under competitive conditions.

(iii) The Purchaser's remedies under Sub-Clauses (i) and (ii) of this clause shall be in lieu of any other remedy in respect of the Contractor's failure to complete the Contract within the time fixed thereby, or if no time be fixed, within a reasonable time or within any extension of time granted by the Engineer.

(iv) Unless otherwise specified in the Contract, for the purposes of this clause only the Contract shall be deemed to have been completed at the time when, by reference to the delivery term chosen by the parties, the risk of loss of or damage to the Plant passes to the Purchaser.

Defects after Delivery.

18. – (i) The Contractor shall be responsible for making good with all possible speed any defect in or damage to any portion of the Plant which may appear or occur under proper use during a period of 18 months after that portion has been delivered and which arises either–

(a) from defective materials, workmanship or design (other than a design made, furnished or specified by the Purchaser and for which the Contractor has disclaimed responsibility in writing within a reasonable time after receipt of the Purchaser's instructions), or

(b) from any act or omission of the Contractor done or omitted during the said period.

(ii) If any such defect shall appear or damage occur, the Engineer shall inform the Contractor thereof stating in writing the nature of the defect or damage. If the Contractor replaces or renews any part of the Plant, the provisions of this clause shall apply to the part of the Plant so replaced or renewed, except that the period during which the Contractor's responsibility pursuant to Sub-Clause (i) of this clause shall subsist shall be either 12 months from the date of replacement or renewal or the unexpired portion of the period mentioned in the said Sub-Clause (i) whichever is the later to expire.

(iii) The periods mentioned in Sub-Clauses (i) and (ii) of this clause shall in relation to any portion of the Plant which cannot be used by reason of a defect to which this clause applies be extended by the period during which such portion cannot be used.

(iv) The supply to the Purchaser carriage paid of a defective or damaged part of the Plant properly repaired or of a part in replacement thereof shall constitute fulfilment by the Contractor of his obligation under Sub-Clause (i) of this clause in respect of that defective or damaged part. If it is reasonably practicable for a defective or damaged part to be returned to the Contractor and the Contractor shall call for its return the Purchaser shall cause it to be returned to the Contractor at the Contractor's expense.

(v) Where pursuant to this clause the Contractor supplies a part in replacement of a defective or damaged part the defective or damaged part shall become the property of the Contractor.

(vi) If any such defect or damage be not remedied within a reasonable time, the Purchaser may proceed to do the work at the Contractor's risk and expense.

(vii) The Contractor's liability under this clause shall be in lieu of any condition or warranty implied by law as to the quality or fitness for any particular purpose of any portion of the Plant delivered and save as in this clause expressed neither the Contractor nor his Sub-Contractors, servants or agents shall be liable, whether in contract, tort or otherwise in respect of defects in or damage to such portion, or for any injury, damage or loss of whatsoever kind attributable to such defects or damage. For the purposes of this sub-clause the Contractor contracts on his own behalf and on behalf of and as trustee for his Sub-Contractors, servants and agents.

Interim and Final Certificates.

19. – (i) The Contractor may at the times and in the manner following apply for interim and final certificates, as referred to in Clause 22 (*Terms of Payment*).

Interim Certificates.

(ii) Applications for interim certificates may be made to the Engineer in respect of Plant ready and due for despatch or despatched. Each such application shall identify the Plant in respect of which it is made, and state the amount claimed and be accompanied by such documents as may be required under the Contract.

(iii) The Engineer shall issue to the Contractor an interim certificate within 14 days after receiving an application therefor which the Contractor was entitled to make. If the Engineer shall fail to issue an interim certificate as provided in this clause or if the Purchaser shall interfere with or obstruct the issue of any such certificate, the Contractor may, without prejudice to any other remedy, either–

(a) after giving to the Purchaser or the Engineer 14 days' notice in writing of his intention so to do, cease performance of the Contract until the said certificate be issued; in which case the expenses of the Contractor occasioned by such cessation and the subsequent resumption shall be added to the Contract Price, or

(b) after giving to the Purchaser or the Engineer one month's notice in writing of his intention so to do, terminate the Contract, if the said certificate be not issued by the expiry of that notice, whether or not the Contractor shall have ceased performance of the Contract in accordance with paragraph *(a)* hereof or have given notice of his intention so to do.

(iv) Every interim certificate shall certify the total value of the Plant ready and due for despatch or despatched up to the date named in the application for the certificate, less the total of any sums previously certified in interim certificates, provided that no sum shall be included in any interim certificate in respect of any Plant that, according to the decision of the Engineer, does not comply with the Contract.

(v) No interim certificate shall be relied on as conclusive evidence of any matter stated therein, or prejudice any right of the Purchaser or the Contractor against the other.

Final Certificate.

(vi) Application for a final certificate in respect of the Plant or any section thereof may be made to the Engineer at any time after the Contractor has ceased to be under any obligation in respect thereof under Clause 18 *(Defects after Delivery)*. If in respect of any portion of the Plant replaced or renewed the obligations of the Contractor under Clause 18 continue after the period of eighteen months mentioned in Sub-Clause (i) of that Clause, the right of the Contractor to apply for a final certificate in respect of the Plant or a section thereof other than the portion so replaced or renewed shall not be affected. After the Contractor has ceased to be under any obligation under Clause 18 in respect of any portion replaced or renewed he may apply for a final certificate in respect of such portion.

(vii) The Engineer shall issue to the Contractor a final certificate within 14 days after receiving an application therefor which the Contractor was entitled to make.

(viii) A final certificate shall certify the total of all amounts comprised in interim certificates previously issued in respect of the Plant or the portion thereof to which the final certificate relates, subject to such additions thereto or deductions therefrom as may be authorised in Sub-Clause (x) of this clause.

(ix) A final certificate shall, save in the case of fraud or dishonesty relating to or affecting any matter dealt with in the certificate, be conclusive evidence as to the sufficiency of the Plant and of the value thereof unless any proceedings arising out of the Contract, whether under Clause 24 *(Arbitration)* or otherwise, shall have been commenced by either party before the final certificate has been issued or within one month thereafter.

(x) If any sum shall become payable to the Contractor under the Contract otherwise than for Plant delivered, the amount thereof shall be included in the next certificate (interim or final) issued by the Engineer, and if any sum shall become payable under the Contract by the Contractor to the Purchaser prior to the issue of the final certificate, whether by deduction from the Contract Price or otherwise, the amount thereof shall be deducted in the next certificate.

(xi) The Engineer may in any certificate give effect to any correction or modification that should properly be made in respect of any previous certificate.

Provisional Sums.

20. – (i) A provisional sum included in the Contract Price shall be expended or used as the Engineer may in writing direct and not otherwise. In so far as a provisional sum is not expended or used it shall be deducted from the Contract Price.

Prime Cost Items.

(ii) All sums in respect of Prime Cost (P.C.) items included in the Contract Price shall be expended or used as the Engineer may in writing direct and not otherwise. To the net amount paid by the Contractor in respect of each P.C. item there shall be added the percentage named in the Appendix of the said amount. The sum by which the net amount so paid in respect of any P.C. item plus the said percentage thereof exceeds or is less than the sum included in the Contract Price in respect of that item shall be added to or deducted from the Contract Price as the case may be.

(iii) The Contractor shall have no responsibility for work done or Plant supplied by any other person in pursuance of directions given by the Engineer under this clause unless the Contractor shall have approved the person by whom such work is to be done or such Plant is to be supplied and the Plant, if any, to be supplied.

Payments due from the Contractor.

21. – Without prejudice to any other remedy which the Purchaser may have he shall be entitled to deduct from any moneys due, or becoming due to the Contractor under the Contract, all costs, damages or expenses for which under the Contract the Contractor is liable to the Purchaser.

Terms of Payment.

22. – (i) The Purchaser shall make payment to the Contractor as follows:

(a) 10% of the Contract Price as an advance payment within 14 days after the Contractor has furnished to the Purchaser an irrevocable letter of guarantee from a guarantor or surety acceptable to the Purchaser with a value and currency equivalent to the advance payment. The letter of guarantee shall provide for its value to reduce by an amount equal to 10% of the sum certified in each interim certificate in respect of Plant ready and due for despatch or despatched and be in a form acceptable to the Purchaser.

(b) Within 14 days of the presentation of each interim certificate a sum equal to 85% of the sum certified therein.

(c) Within 14 days from the presentation of the final certificate the balance of the Contract Price. Provided that if the Contractor shall have furnished to the Purchaser a guarantee acceptable to the Purchaser for the payment on demand of such balance, the Contractor shall be entitled to payment thereof with or at any time after the payment provided for by paragraph (b) hereof.

(ii) If a final certificate shall be issued in respect of any portion of the Plant the expression 'the balance of the Contract Price' as used in this clause shall be construed as meaning the balance of such part of the Contract Price as shall be apportioned to such portion of the Plant by agreement or in default of agreement by the Engineer.

(iii) If the payment of any sum payable under sub-clause (i) of this clause shall be improperly delayed by the Purchaser or by reason of the failure of the Engineer to issue the relevant certificate within the time prescribed therefor, interest at the rate of two per cent per annum over the Bank of England minimum lending rate from time to time in force on the amount of the delayed payment for the period of the delay shall be added to the Contract Price.

(iv) If the Purchaser shall fail to make any payment as provided in this clause the Contractor shall have the like remedies, without prejudice to any other, as are provided in Sub-Clause (iii) of Clause 19 *(Interim and Final Certificates).*

Statutory and other Regulations.

23. – If the cost to the Contractor of the performance of the Contract shall be increased or reduced by reason of the making after the date of his tender in the United Kingdom or elsewhere of any law or of any order, regulation or by-law having the force of law, the amount of such increase or reduction shall be added to or deducted from the Contract Price as the case may be.

Arbitration.

24. – (i) If at any time any question, dispute or difference shall arise between the Purchaser and the Contractor, either party shall, as soon as reasonably practicable, give to the other notice in writing of the existence of such question, dispute or difference specifying its nature and the point at issue, and the same shall be referred to the arbitration of a person to be agreed upon, or failing such agreement within six weeks, to some person appointed on the application of either of the parties hereto by the President for the time being of the Institution named in the Appendix, provided that a question, dispute or difference relating to a decision, instruction or order of the Engineer shall not be referred to arbitration unless notice has been given by the Contractor in accordance with Clause 10 (b) (*Engineer's Decisions*).

(ii) Performance of the Contract shall continue during arbitration proceedings unless the Engineer or the Purchaser shall order the suspension thereof or of any part thereof, and if any such suspension shall be ordered the reasonable expenses of the Contractor occasioned by such suspension shall be added to the Contract Price. No payments due or payable by the Purchaser shall be withheld on account of pending reference to arbitration.

Construction of Contract.

25. – Unless otherwise agreed the Contract shall in all respects be construed and operate as an English contract and in conformity with English Law and the English courts shall have exclusive jurisdiction over any matter arising out of or in connection with the provisions for arbitration under Clause 24 *(Arbitration)*. The marginal notes hereto shall not affect the construction hereof.

Variation in Costs.

26. –* If the cost to the Contractor of performing his obligations under the Contract shall be increased or reduced by reason of any rise or fall in the rates of wages payable to labour or in the cost of material or transport above or below such rates and costs ruling at the date of tender, the amount of such increase or reduction shall be added to or deducted from the Contract Price as the case may be, provided that no account shall be taken of any amount by which any cost incurred by the Contractor has been increased by the default or negligence of the Contractor. For the purpose of this clause 'the cost of material' shall be construed as including any duty or tax, by whomsoever payable, which is payable under or by virtue of any Act of Parliament on the import, purchase, sale, appropriation, processing or use of such material.

Metrication.

27. – (i) If any Plant described in the Contract or the subject of a variation (hereinafter called "a variation order") under Clause 11 *(Variations)* is described by dimensions in the metric or imperial measure and the Contractor cannot procure such Plant in the measure specified in sufficient time to avoid delay in the performance of his other obligations under the Contract, but can obtain such Plant in the other measure to dimensions approximating to those described in the Contract or in the variation order, then the Contractor shall forthwith give notice to the Engineer of the facts stating the dimensions to which such Plant is procurable in the other measure.

(ii) The Engineer shall within 14 days after the receipt of a notice under the preceding sub-clause, give notice to the Contractor in writing pursuant to Clause 11 *(Variations)*. Such notice shall either–

(a) direct the Contractor to supply such Plant to the dimensions stated in the said notice instead of to the dimensions described in the Contract or variation order as the case may be, or

(b) direct the Contractor to make some other variation whereby the need to supply such plant to the dimensions described in the Contract or variation order will be avoided.

The provisions of the Contract shall apply to directions given under this clause as though such directions had been included in the Specification.

***NOTE:– Unless this Clause is excluded any quoted price will be subject to adjustment in accordance with its terms.**

11 I.MECH.E./I.E.E. MODEL FORM OF GENERAL CONDITIONS OF CONTRACT 'B1/1981'

APPENDIX

Clause

17. (Delay in Completion)

(a) . . . . .Percentage to be deducted as damages.

(b) . . . . .Maximum percentage which the deductions may not exceed.

20. . . . . .Percentage on Prime Cost items.

24. . . . . .The Institution of Mechanical Engineers*
The Institution of Electrical Engineers*

*Delete one line

I.MECH.E./I.E.E. MODEL FORM OF GENERAL CONDITIONS OF CONTRACT 'B1/1981' 12

FORM OF AGREEMENT

This Agreement is made the day of 19

BETWEEN

(hereinafter referred to as "the Contractor") of the one part and

(hereinafter called "the Purchaser") of the other part **Whereas** the Purchaser desires to have certain Plant provided as mentioned, enumerated or referred to in certain General Conditions, Specification, Schedules, Drawings, Plans, Schedule of Prices, the Contractor's Tender and covering letter (if any), the acceptance of the Tender and any other relevant correspondence* which for the purpose of identification have been signed by

on behalf of the Contractor

and by

on behalf of the Purchaser all of which are deemed to form part of this Contract as though separately set out herein and are included in the expression "Contract" whenever herein used. **And Whereas** the Purchaser has accepted the tender of the Contractor for the provision of the said Plant for the sum of

upon the terms and subject to the Conditions hereinafter mentioned: **Now this Deed Witnesseth** and it is hereby agreed and declared as follows, that is to say, in consideration of the payments to be made to the Contractor by the Purchaser as hereinafter mentioned the Contractor hereby covenants with the Purchaser, his legal personal representatives, successors and assigns that the Contractor shall and will duly provide the said Plant and shall do and perform all other acts and things in the Contract mentioned or described or which are to be implied therefrom or may be reasonably necessary for the provision of the said Plant within and at the times and in the manner and subject to the terms conditions and stipulations mentioned in the Contract.

And in consideration of the due provision of the Plant as aforesaid the Purchaser does hereby for himself, his legal personal representatives, successors and assigns covenant with the Contractor that he, the Purchaser, his legal personal representatives, successors or assigns will pay the Contractor the said sum and such other sums as may become payable to the Contractor under the provisions of the Contract, such payments to be made at such time and in such manner as is provided by the Contract.

In Witness whereof, etc.

*Delete as required

INDEX

Printed by A. McLay & Co. Ltd., Cardiff. 81/5

MEMORANDUM ON THE USE OF THE GENERAL CONDITIONS OF CONTRACT (B 1/1981)

A. Applicability

The General Conditions are intended to be suitable for the mechanical engineering and the electrical engineering industries and cover the main conditions for export contracts for the supply of plant and machinery, the delivery thereof being in accordance with any of the methods detailed in the International Rules for the Interpretation of Trade Terms (INCOTERMS) published by the International Chamber of Commerce. INCOTERMS may be obtained in the United Kingdom from the British National Committee of the International Chamber of Commerce, London.

It is hoped that, in order to secure the advantage of uniformity and to avoid trouble and expense, the General Conditions may be found suitable in their entirety for most contracts. It is, however, inevitable that additions and/or amendments will have to be incorporated to meet the requirements of particular contracts or of the governmental, financial or other authorities of the countries from which plant is to be exported or of the countries into which it is to be imported. It is suggested, in order to facilitate checking, that any additions or alterations be made in such a way as to be clearly identifiable.

B. Contract documents

Contracts are commonly made in one of three ways–

(a) By a formal agreement, such as the model form supplied with these Conditions, incorporating in the contract the General Conditions, Specification, Schedules, Drawings, etc. In this case all the terms of the Contract are contained in the Agreement, the Conditions, Specification, and other incorporated documents on which tenders were invited, and agreed variations in or additions to the provisions of these documents are represented by suitable amendments.

(b) By a formal agreement, as in the method *(a)*, but incorporating also the Contractor's Tender and covering letter, if any, the Purchaser's acceptance of the Tender, and all correspondence containing terms accepted by the Purchaser and the Contractor. When this method is adopted any modifications of the General Conditions, Specifications, etc., have to be sought in the covering letter and other correspondence.

(c) Without a formal agreement, the Tender and its acceptance, together with other relevant correspondence, being relied on as constituting the Contract between the parties.

When either method *(a)* or *(b)* is adopted, a copy of the form of agreement to be used should be included with the documents supplied to or open to the inspection of intending Tenderers.

When method *(a)* is used, the references in the model form of agreement to the Tender, covering letter, acceptance of the Tender, and other correspondence should be deleted.

C. Delivery

It should be made clear in the contract documents whether delivery is to be f.o.b., c.i.f., or by any other method defined in INCOTERMS eg:

(a) f.o.b.–It should be specified in the contract documents at what port or ports the Plant is to be shipped.

(b) c.i.f.–The Purchaser should specify the port to which the Plant is to be shipped and the place where the documents (i.e. invoice, bill of lading and policy of insurance) are to be tendered, if other than the Purchaser's place of business. Any special stipulations with regard to these matters should also be specified. Delivery is not complete until the Plant has been shipped and the documents mentioned above have been tendered to the Purchaser. The Contractor should therefore not delay the tendering of the policy of insurance, bill of lading and invoice

ii I.MECH.E./I.E.E. MODEL FORM OF GENERAL CONDITIONS OF CONTRACT 'B1/1981'

until the interim certificate is available.
The port of destination named in the Contract is often loosely referred to as the place of delivery. The parties should therefore, when referring to delivery in the contract documents, be clear whether they intend to refer to the port of shipment or the port of destination.

(c) f.o.r. and f.o.t. – Completion of delivery requires loading of the Plant on a railway or road vehicle at the place of delivery specified in the Contract. In case of deliveries by rail the place specified is generally on a railway conveniently near to the Contractor's works. Some Purchasers, however, often stipulate delivery f.o.r. either at the port of destination or at the Purchaser's own railhead. Neither of these methods is consistent with the INCOTERMS' definition. If either of these methods is chosen the parties' respective obligations will need to be specified in the Contract.

D. **Advance Payments and Progress Payments During Manufacture**

The General Conditions provide for an advance payment against a guarantee, but the Contractor does not become entitled to any further payments for Plant he has manufactured until he has shipped it. In the case of contracts involving considerable expenditure on manufacture before any shipment can be made, the General Conditions may be modified to provide for progress payments to the Contractor in respect of Plant in the course of manufacture, in which case the insurance arrangements may also need to be modified.

E. **Customs and Imports Duties**

The General Conditions do not deal with the responsibility for payment of customs and import duties and licenses nor with the responsibility for the obtaining of any export licenses or payment of export levies. Care should be taken to ensure that the form of delivery chosen by reference to INCOTERMS reflects the agreed intentions of the parties and if not to deal with the matter specifically in the Contract documents.

F. **Contract Price Adjustment Formula**

If the parties so wish they may operate Clause 26 by means of an agreed formula.

MEMORANDUM AS TO TENDERS

Tenderer may propose modifications.

The Tenderer is at liberty to add any details and conditions that he may deem desirable, and in the event of his doing so must print or type the same and annex the added matter to the Specification or General Conditions returned by him, but such additional details and conditions will not be binding on the Purchaser unless they are approved by him and incorporated in the Contract.

Where in special circumstances the Tenderer deems it advisable to provide for the nondisclosure of drawings and information of a confidential nature furnished with his Tender, he should include in the Tender or covering letter a stipulation to this effect. If the Tender embodying such a stipulation is accepted the stipulation should be incorporated in any subsequent formal agreement.

If the Tenderer has any doubt as to the meaning of any portion of the General Conditions or of the Specification, he should, when submitting his Tender, set out in his covering letter the interpretation upon which he relies.

Tenders.

One copy of the Specification supplied, together with the General Conditions and Drawings, enclosed in a plain sealed cover bearing the words "Tender for *(here follows the short title of the Plant)*"

and not bearing any name or mark indicating the sender shall, subject to modifications mentioned above, be returned intact with the Tender Form and Schedules, if any, filled up and signed, addressed to

and must be received by him on or before

The Purchaser does not bind himself to accept the lowest or any Tender, nor will he be responsible for, or pay for, expenses or losses which may be incurred by any Tenderer in the preparation of his Tender.

Copies of Specification etc.

A duplicate copy of the Specification, General Conditions and Drawings will be supplied to intending Tenderers without charge, and extra copies of the Drawings at the cost of reproduction.

The sum deposited by the Tenderer on application for the Specification will be refunded to him within one month of the date of the adjudication upon the Tenders, if all copies of the Specification, Drawings, etc., have been returned by the Tenderer, unless the Tender was not made in good faith, in which case the deposit may be forfeited.

Agreement.

The Tenderer whose Tender is accepted may be required to enter into a formal agreement with the Purchaser.

NOTE

The advertisement inviting Tenders should state whether and if so where the Agreement, General Conditions, Specification and Drawings may be inspected, and should give such particulars of the class of plant and apparatus required under each Section as will enable contracting firms to decide, without obtaining the Specification, whether they desire to tender, and should also state the amount of the deposit to be paid for the General Conditions, Specification and Drawings.

B2

MODEL FORM OF

GENERAL CONDITIONS OF CONTRACT

INCLUDING FORMS OF AGREEMENT AND GUARANTEE
RECOMMENDED BY

THE INSTITUTION OF MECHANICAL ENGINEERS

THE INSTITUTION OF ELECTRICAL ENGINEERS

AND

THE ASSOCIATION OF CONSULTING ENGINEERS

FOR USE IN CONNECTION WITH

EXPORT CONTRACTS, DELIVERY F.O.B., C.I.F. OR F.O.R., WITH SUPERVISION OF ERECTION

FIFTH EDITION 1981

Published for the Joint Committee on Model Forms of General Conditions of Contract by
THE INSTITUTION OF ELECTRICAL ENGINEERS
and obtainable from
Publications Sales Department, Station House, 70 Nightingale Road, Hitchin, Herts SG5 1RJ
or to callers, at the Reception Desk, Savoy Place
or from
THE INSTITUTION OF MECHANICAL ENGINEERS
Publications Sales Department, P.O. Box 24, Northgate Avenue, Bury St. Edmunds,Suffolk, IP32 6BW
or to callers, at the Reception Desk, Birdcage Walk, London
or at
THE ASSOCIATION OF CONSULTING ENGINEERS, Alliance House, 12, Caxton Street,
London, S.W.1H 0QL.

HISTORICAL NOTE

A Model Form of General Conditions 'B.2' (Export, including Complete Erection or Supervision of Erection) was first published by The Institution of Electrical Engineers in 1925. A revision was issued in 1928.

Following consultations, and by agreement between the Council of The Institution of Mechanical Engineers and the Council of The Institution of Electrical Engineers, a Model Form 'B.2' was prepared in 1960 for Export Contracts including delivery f.o.b., c.i.f. or f.o.r., with supervision of erection, of mechanical or electrical plant.

Following an approach in 1952 by The Association of Consulting Engineers, it was agreed that the Association should adopt and recommend to its members the use of this Model Form, and the name of the Association accordingly appears on the title page.

The 1972 Edition incorporated substantial revisions to take account of developments in practice. In particular, the term "Works", which is used in Model Forms 'A' and 'B.3' to describe the Plant in its entirety, including its installation and erection on site, but which was used in the previous Edition of Model Form 'B.2' to describe only the combination of the delivered Plant and the supervision services, was dispensed with, so as to avoid the possibility of confusion arising between different definitions of the same terms in the various Model Forms.

This 1981 Edition takes account of changes in legislation (the Unfair Contract Terms Act, 1977 and the Arbitration Act, 1979), makes provision for an advance payment and methods of delivery for which there are now internationally accepted interpretations of the rights and obligations of the parties and has been retitled accordingly.

This form may be cited as 'Model Form B2/1981'

GENERAL CONDITIONS

Definition of Terms.

1.—In construing these General Conditions and the Specification, the following words shall have the meanings herein assigned to them unless there is something in the subject matter or context inconsistent with such construction:

The "Contract" shall mean the agreement howsoever made between the Purchaser and the Contractor for the supply of the Plant and the supervison of erection thereof, including therein all documents to which reference may properly be made in order to ascertain the rights and obligations of the parties under the said agreement.

The "Contractor" shall mean the tenderer whose tender has been accepted by the Purchaser and shall include the Contractor's legal representatives, successors and assigns.

The "Contract Price" shall mean the sum named in the Contract as the Contract Price but shall not include the fee payable pursuant to the Contract for the supervision of erection.

The "Contract Value" shall mean that part of the Contract Price, adjusted to give effect to such additions or deductions as are provided for in Clause 11 (*Variations*), which is properly apportionable to the Plant, having regard to the state, condition and location of the Plant, the amount of work done and all other relevant circumstances but disregarding any changes pursuant to Clause 32 in the cost of supplying the Plant.

The "Engineer" shall mean

or the person for the time being or from time to time notified in writing by the Purchaser to the Contractor as the Engineer for the Contract, or in default of any notification the Purchaser.

"Erection Equipment" shall mean such tools, tackle, and stores as are required upon the Site for the satisfactory erection of the Plant otherwise than for incorporation therein.

"Month" shall mean calendar month.

"Plant" shall mean all or any part of the machinery, apparatus, materials, articles, and things of all kinds to be provided by the Contractor, other than Erection Equipment.

The "Purchaser" shall mean

and shall include the Purchaser's legal personal representatives, successors and assigns.

The "Site" shall mean the actual place or places at which the Plant is to be erected.

The "Specification" shall mean the specitifcation annexed to or issued with these General Conditions.

"Sub-Contractor" shall mean any person (other than the Contractor) named in the Contract for the supply of any part of the Plant or any person to whom any part of the Contract has been sub-let with the consent in writing of the Engineer, and the legal representatives, successors and assigns of such person.

"Tests on Completion"' shall mean such tests to be made before the Plant is taken over by the Purchaser as are provided for in the Contract or otherwise agreed between the Purchaser and Contractor.

"Writing" shall include any manuscript, type-written, or printed statement, under seal or hand as the case may be.

Words importing persons shall include firms and corporations.

Words importing the singular only shall also include the plural and *vice-versa*.

Contractor to inform himself fully.

2.—The Contractor when making his tender shall be deemed to have examined the General Conditions and Specification, with such schedules, drawings and plans as are annexed thereto or referred to therein.

Expenses of Agreement.

3.—The expenses of preparing, completing and stamping the Agreement, if any, shall be paid by the Purchaser, and an executed counterpart thereof properly stamped together with copies of all other documents comprising the Contract shall be furnished to the Contractor free of charge.

Drawings.

4.–(i) The Contractor shall submit to the Engineer for approval:

(*a*) within the time specified in the Specification or if no time is specified, then a reasonable time, such drawings, samples, patterns and models as may be called for therein, and

(*b*) during the progress of the Contract and within such reasonable times as the Engineer may require, such drawings of the general arrangement and of details of the Plant or any part thereof as the Engineer may reasonably require, provided that the Contractor shall not be under any obligation to supply copies of shop drawings.

Within a reasonable period after receiving such drawings, samples, patterns and models the Engineer shall signify his approval or otherwise. If the Engineer shall not approve any drawing, sample, pattern or model thus submitted, the same shall be forthwith modified to meet the reasonable requirements of the Engineer and shall be re-submitted. Copies of all drawings which require to be approved by the Engineer shall be provided in triplicate by the Contractor. Drawings, samples, patterns and models when approved shall, if required by either party, be signed or identified by both parties and if required by the Contractor one copy shall be returned to him. All dimensions marked shall be considered correct although measurements by scale may differ therefrom. Detailed drawings shall take precedence where they differ from general arrangement drawings.

(ii) Drawings, samples, patterns and models approved as above described shall not be departed from except as provided in Clause 11 (*Variations*).

(iii) The Engineer shall have the right at all reasonable times to inspect at the premises of the Contractor all drawings of any portion of the Plant.

(iv) The Contractor shall, if so desired by the Purchaser, furnish to the Purchaser in writing at the commencement of the period referred to in Clause 23 (*Defects after Delivery*), or at such earlier times as may be named in the Specification, such information, accompanied by drawings, as may be necessary to enable the Purchaser to operate, maintain, dismantle, reassemble and adjust all parts of the Plant.

(v) Drawings submitted in pursuance of paragraphs (*a*) and (*b*) of Sub-Clause (i) of this clause shall not, without the consent of the Contractor, be used by the Purchaser except for purposes of the Contract, nor shall they without such consent be communicated to third parties save insofar as may be necessary for the proper execution of the Contract.

Mistakes in information.

5.–(i) The Contractor shall be responsible for any discrepancies, errors or omissions in the drawings and information supplied by him, whether they have been approved by the Engineer or not, provided that such discrepancies, errors or omissions be not due to inaccurate drawings or information furnished to the Contractor by the Purchaser or the Engineer.

(ii) The Contractor shall at his own expense carry out any alterations or remedial work necessitated by reason of such discrepancies, errors or omissions and modify the drawings and information accordingly, or if the same be done by or on behalf of the Purchaser, shall bear all costs reasonably incurred therein. The performance of his obligations under this sub-clause shall be in full satisfaction of the Contractor's liability under Sub-Clause (i) of this clause, but shall not relieve him of his liability under Clause 22 (*Delay in Completion*) insofar as that liability is the result of a failure of the Contractor to perform his obligations under that sub-clause.

(iii) The Purchaser shall be responsible for drawings, information and details supplied in writing by the Purchaser or the Engineer. The Purchaser shall pay the extra cost reasonably incurred by the Contractor due to alterations of the work necessitated by reason of inaccurate drawings, information or details so supplied to the Contractor.

Assignment.

6.–(i) The Contractor shall not, without the consent in writing of the Purchaser, which shall not be unreasonably withheld, assign or transfer the Contract or the benefits or obligations thereof or any part thereof to any other person, provided that this shall not affect any right of the Contractor to assign, either absolutely or by way of charge, any moneys due or to become due to him, or which may become payable to him under the Contract.

Sub-letting.

(ii) The Contractor shall not, without the consent in writing of the Engineer, which shall not be unreasonably withheld, sub-let the Contract or any part thereof, or make any sub-contract with any person or persons for the execution of any part of the Contract but the restriction contained in this clause shall not apply to sub-contracts for materials, for minor details or for any part of the Plant of which the makers are named in the Contract. Any such consent shall not relieve the Contractor from his obligations under the Contract.

Patent Rights, etc.

7.–(i) The Contractor shall indemnify the Purchaser against all actions, claims, demands, costs, charges and expenses arising from or incurred by reason of any infringement or alleged infringement of letters patent, registered design, copyright, trade mark or trade name protected in the United Kingdom or, where the country in which the Plant is to be used is specified in the Contract, in that country, by the use of any Plant supplied by the Contractor, but such indemnity shall not cover any use of the Plant otherwise than for the purpose indicated by or reasonably to be inferred from the Specification or to any infringement which is due to the use of any Plant in association or combination with any other plant not supplied by the Contractor.

(ii) In the event of any claim being made or action brought against the Purchaser arising out of the matters referred to in this clause, the Contractor shall be promptly notified thereof and may at his own expense conduct all negotiations for the settlement of the same, and any litigation that may arise therefrom. The Purchaser shall not, unless and until the Contractor shall have failed to take over the conduct of the negotiations or litigation, make any admission which might be prejudicial thereto. The conduct by the Contractor of such negotiations or litigation shall be conditional upon the Contractor having first given to the Purchaser such reasonable security as shall from time to time be required by the Purchaser to cover the amount ascertained or agreed or estimated, as the case may be, of any compensation, damages, expenses and costs for which the Purchaser may become liable. The Purchaser shall, at the request of the Contractor, afford all available assistance for the purpose of contesting any such claim or action, and shall be repaid all reasonable expenses incurred in so doing.

(iii) The Purchaser on his part warrants that any design or instructions furnished or given by him shall not be such as will cause the Contractor in the performance of the Contract to infringe any letters patent, registered design, copyright, trade mark or trade name in the country specified in the Contract in which the Plant is to be used.

Manner of Execution.

8.–All Plant to be supplied and all work to be done under the Contract shall be manufactured and executed in the manner set out in the Specification or, where not so set out, to the reasonable satisfaction of the Engineer, and all work on Site shall be carried out in accordance with such reasonable directions as the Engineer may give.

Engineer's Supervision

9.–(i) All instructions and orders to the Contractor shall, except as herein otherwise provided, be given by the Engineer.

Engineer's Representative.

(ii) The Engineer may from time to time delegate any of the powers, discretions, functions and authorities vested in him and may at any time revoke any such delegation. Any such delegation or revocation shall be in writing signed by the Engineer and, in the case of a delegation, shall specify the powers, discretions, functions and authorities thereby delegated and the person or persons to whom the same are delegated. No such delegation or revocation shall have effect until a copy thereof has been delivered to the Contractor.

Engineer's Decisions.

10.–The Contractor shall proceed with the Contract in accordance with decisions, instructions and orders given by the Engineer in accordance with these Conditions, provided always that–

(*a*) if the Contractor shall, without undue delay after being given any decision, instruction or order otherwise than in writing, require it to be confirmed in writing, such decision, instruction or order shall not be effective until written confirmation thereof has been received by the Contractor, and

(*b*) if the Contractor shall, by written notice to the Engineer within 14 days after receiving any decision, instruction or order of the Engineer in writing or written confirmation thereof, intimate that he disputes or questions the decision, instruction or order, giving his reasons for so doing, either party to the Contract shall be at liberty to refer the matter to arbitration pursuant to Clause 30 (*Arbitration*), but such an initiation shall not relieve the Contractor of his obligation to proceed with the Contract in accordance with the decision, instruction or order in respect of which the intimation has been given. The Contractor shall be at liberty in any such arbitration to rely on reasons additional to the reasons stated in the said intimation.

Variations

11.–(i) The Contractor shall not alter any of the Plant except as directed in writing by the Engineer; but the Engineer shall have full power, subject to the proviso hereinafter contained, from time to time during the execution of the Contract by notice in writing to direct the Contractor to alter, amend, add to or otherwise vary any of the Plant and the Contractor shall carry out such variations, and be bound by the same conditions, so far as applicable, as though the said variations were stated in the Specification: provided that no such variation shall, except with the consent in writing of the Contractor, be such as will, with any variations already directed to be made, involve a net addition to or deduction from the Contract Price of more than 15 per cent thereof. In any case in which the Contractor has received any such direction from the Engineer which either then or later will, in the opinion of the Contractor, involve an addition to or deduction from the Contract Price, the Contractor shall, as soon as reasonably possible, advise the Engineer in writing to that effect. The amount to be added to or deducted from the Contract Price shall be ascertained and determined in accordance with the rates specified in the schedules of prices, so far as the same may be applicable, and where rates are not contained in the said schedules, or are not applicable, such amount shall be such sum as is reasonable in the circumstances. Due account shall be taken of any partial execution of the Contract which is rendered useless by any such variation.

(ii) If the Engineer shall make any such variation in any part of the Plant such reasonable notice in writing shall be given to the Contractor as will enable him to make his arrangements accordingly. If in the opinion of the Contractor any such variation is likely to prevent or prejudice the Contractor from or in fulfilling any of his obligations under the Contract, he shall notify the Engineer thereof in writing, and the Engineer shall decide forthwith whether or not the same shall be carried out. If the Engineer confirms his instructions in writing, the said obligations shall be modified to such an extent as may be justified. Until the Engineer so confirms his instructions they shall be deemed not to have been given.

Contractor's Default

12.—If the Contractor shall neglect to perform the Contract with due diligence and expedition or shall refuse or neglect to comply with any reasonable orders given him in writing by the Engineer in connection with the performance of the Contract, or shall contravene the provisions thereof the Purchaser may give notice in writing to the Contractor to make good the neglect, refusal, or contravention complained of. Should the Contractor fail to comply with the notice within 14 days from the date of service thereof, in the case of a failure, neglect or contravention capable of being made good within that time, or otherwise within such time as may be reasonably necessary for making it good, then and in such case it shall be lawful for the Purchaser forthwith to terminate the Contract by notice in writing to the Contractor without prejudice to any rights which may have accrued thereunder to either party prior to such termination. The Purchaser shall be entitled to retain and apply any balance which may be otherwise due on the Contract by him to the Contractor, or such part thereof as may be necessary, to the payment of the cost of completing the Plant or of purchasing equivalent plant from another source and the supervision of erection thereof. If such cost shall exceed the amount that would otherwise become due to the Contractor in accordance with the Contract, the Contractor shall pay such excess. If the Purchaser pursuant to this clause completes the Plant or obtains equivalent plant from another source the Contractor's liability under Clause 21 (*Delay in Completion*) shall cease in respect of the Plant concerned with effect from the termination of the Contract under this clause, without prejudice to any such liability as shall have already accrued.

Bankruptcy.

13.—If the Contractor shall become bankrupt or insolvent, or have a receiving order made against him, or compound with his creditors, or being a corporation commence to be wound up, not being a members' voluntary winding up for the purpose of amalgamation or reconstruction, or carry on its business under a receiver for the benefit of its creditors or any of them, the Purchaser shall be at liberty either—

(*a*) to terminate the Contract forthwith by notice in writing to the Contractor or to the receiver or liquidator or to any person in whom the Contract may become vested, in which case the relevant provisions of Clause 12 (*Contractor's Default*) shall apply as though the Contract had been terminated pursuant thereto, or

(*b*) to give such receiver, liquidator or other person the option of carrying out the Contract subject to his providing a guarantee for the due and faithful performance of the Contract up to an amount to be agreed.

Inspection, Testing and Rejection.

14.—(i) The Engineer shall be entitled at all reasonable times during manufacture to inspect, examine and test on the Contractor's premises the materials, workmanship and performances of all Plant to be supplied under contract, and if part of the said Plant is being manufactured on other premises, the Contractor shall obtain for the Engineer permission to inspect, examine and test as if the said Plant were being manufactured on the Contractor's premises. Such inspection, examination or testing shall not release the Contractor from any obligaion under the Contract.

(ii) Where the Contract provides for tests of the Plant or any part thereof when completely manufactured such tests shall, in the absence of any arrangements to the contrary, take place on the premises of the Contractor.

(iii) The Contractor shall, after consulting the Engineer, give the Engineer reasonable notice in writing of the date on and the place at which any Plant will be ready for testing as provided in the Contract and unless the Engineer shall attend at the place so named on the date which the Contractor has stated in his notice the Contractor may proceed with the tests, which shall be deemed to have been made in the Engineer's presence, and shall forthwith forward to the Engineer duly certified copies of the test readings. The Engineer shall give the Contractor 24 hour's notice in writing of his intention to attend the tests.

(iv) Where the Contract provides for tests on the premises of the Contractor or of any Sub-contractor the Contractor, except where otherwise specified, shall provide free of charge such assistance, labour, materials, electricity, fuel, stores, apparatus and instruments as may be requisite and as may be reasonably demanded to carry out such tests efficiently.

(v) As and when the Engineer is satisfied that any Plant shall have passed the tests referred to in this clause he shall notify the Contractor in writing to that effect.

(vi) If after inspecting, examining or testing any Plant the Engineer shall decide that such Plant or any part thereof is defective or not in accordance with the Contract, he may reject the said Plant or part thereof by giving to the Contractor within a reasonable time notice in writing of such rejection, stating therein the grounds upon which the said decision is based.

(vii) The provisions of Sub-Clause (iii) of Clause 22 (*Tests on Completion and Taking Over*) shall relate also to inspections, examinations and tests carried out under this clause.

Delivery.

15.–(i) Save as varied by these General Conditions the obligations of the parties in relation to the delivery terms specified in the Contract and the rules determining the passing of property in and the risk of loss of or damage to the Plant shall be fixed in accordance with the International Rules for the Interpretation of Trade Terms (Incoterms) of the International Chamber of Commerce in force at the date of the Contract.

(ii) If the Contract provides for delivery FOB the Contractor shall not be required to give the Purchaser the notice relating to insurance mentioned in Section 32(3) of the Sale of Goods Act, 1979.

Customs and Import Duties.

(iii) The Purchaser shall in good time procure any necessary import permits and promptly pay all customs and import duties which may become payable upon the importation of the Plant into the country in which the Plant is to be erected.

Delayed Plant.

(iv) If by delay or failure on the part of the Purchaser or the Engineer to give any necessary instructions or from any cause for which the Purchaser or some other contractor employed by him is responsible, the Contractor shall be prevented, or at the request of the Purchaser refrains, from delivering any Plant at the time specified for delivery thereof or, if no time is specified, within a reasonable time, and shall have given notice in writing to the Purchaser that such Plant (hereinafter referred to as "the delayed Plant") is ready for delivery, and shall have suitably and sufficiently marked the delayed Plant as appropriated to the Contract, and shall have given to the Engineer an opportunity of inspecting the delayed Plant, then and in any such case the following provisions shall have effect–

(*a*) There shall be added to the Contract Price a reasonable sum for storing and taking reasonable measures to protect and preserve the delayed Plant from, and insuring it against loss, deterioration and damage however caused from the time when but for the said delay, failure or other cause the delayed Plant would have been delivered (hereinafter referred to as "the normal delivery date") until the Contractor shall no longer be prevented from delivering it or shall be relieved of responsibility therefor under paragraph (*b*) of this sub-clause, whichever shall first happen.

(*b*) If at the expiration of two months from the normal delivery date the Contractor shall still be prevented as aforesaid from delivering the delayed Plant he may by notice in writing expiring 30 days after receipt thereof by the Purchaser require the Purchaser to assume responsibility for storing, protecting, preserving and insuring the delayed Plant. Upon the expiration of the last-mentioned notice the Contractor shall be relieved of any responsibility for the delayed Plant either until the expiration of 30 days after the receipt of notice in writing from the Engineer that the delayed Plant may be delivered (hereinafter referred to as "the notice to deliver") or until the Contractor, having received the notice to deliver, has proceeded to fulfil the obligation imposed on him by paragraph (*c*) of this sub-clause, whichever shall first occur, provided always that if the notice to deliver shall be given within 30 days after the receipt of the last-mentioned notice given by the Contractor that notice shall not have effect.

(*c*) After the receipt of the notice to deliver, the Contractor, if he has been relieved of responsibility under the last preceding paragraph of this sub-clause, shall (and in any other case may) examine the delayed Plant and make good any deterioration or defect therein that may have developed or loss thereof that may have occurred after the normal delivery date.

(*d*) There shall be added to the Contract Price any reasonable expense to which the Contractor may be put in making the examination referred to in paragraph (*c*) of this sub-clause and in making good any deterioration, defect or loss as therein mentioned, except insofar as the same was caused by faulty workmanship or materials or by the Contractor's failure to take the measures referred to in paragraph (*a*) of this sub-clause. Any expense to which the Contractor may be put in delivering the delayed Plant or in performing his obligations under Clause 23 (*Defects after Delivery*) which would not have been incurred had the delivery of the delayed Plant not been prevented as aforesaid shall also be added to the Contract Price.

(*e*) Without prejudice to the provisions of Sub-Clause (viii) of Clause 23 (*Defects after Delivery*), the obligations of the Contractor under that clause with respect to delayed Plant shall not apply to any defect that may develop therein after the expiration of three years from the normal delivery date.

Damage before Erection.

16.– If for any cause for which the Contractor is not responsible the Plant after delivery and before erection shall suffer deterioration, damage or loss, the Contractor shall be relieved of his obligation to supervise the erection until the Plant has been put into a satisfactory condition or replaced at the cost of the Purchaser.

Supervision of Erection.

17.–(i) The Contractor shall provide the services of one or more competent engineers to supervise and to give instructions to the skilled and unskilled labour provided by the Purchaser so as to secure–

(*a*) the reception and unpacking of the Plant at the Site

(*b*) the erection of the Plant by the Purchaser, and

(*c*) if the Specification so provides, the checking for accuracy, testing and commissioning of the Plant.

(ii) The Engineer shall be at liberty by notice in writing to the Contractor to object to any representative employed by the Contractor in the execution of the Contract who shall, in the opinion of the Engineer, misconduct himself or be incompetent or negligent, and the Contractor shall remove such person from employment on the Contract.

Erection Equipment.

(iii) The Purchaser shall, at his own expense, provide all skilled and unskilled labour, lifting equipment, haulage and power necessary for the erection of the Plant, and all Erection Equipment except such Erection Equipment as is by the Specification to be provided by the Contractor.

Purchaser's Labour.

(iv) The skilled and unskilled labour provided by the Purchaser shall remain the servants and under the control of the Purchaser. The Contractor shall not be liable for any act or omission of such labour, but if in giving instructions to be carried out by such labour, or by omitting to give such instructions, the Contractor fails to use proper skill and care, he shall be deemed to have been negligent and shall, subject to the provisions of Clause 18 (*Contractor's Negligence*), be liable for the consequences of such negligence.

Fencing, Lighting, and Guarding.

(v) The Purchaser shall be responsible for the proper fencing, lighting, guarding and watching of all the Plant and Erection Equipment on the Site during erection and for the proper provision during that period of temporary roadways and footways so far as such roadways and footways may be rendered necessary by the Contract for the accommodation and protection of the owners and occupiers of adjacent property, the public and others.

Levels, etc.

(vi) The Contractor shall be responsible for ensuring that the positions, levels and dimensions of the Plant are correct according to the drawings, notwithstanding that he may have been assisted by the Engineer in setting out the said positions, levels and dimensions.

Contractor's Negligence

18.–(i) The Contractor shall, subject to Sub-Clauses (iii) and (iv) of this clause and Clause 19 (*Limitations on Contractors's Liability*), indemnify the Purchaser in respect of all damage or injury occurring before all the Plant shall have been taken over under Clause 22 (*Tests on Completion and Taking Over*) to any property or to any person and against all actions, suits, claims, demands, costs, charges and expenses arising in connection therewith which shall be occasioned by the negligence of or breach of statutory duty by the Contractor or any Sub-contractor, or by defective design (other than a design made, furnished or specified by the Purchaser and for which the Contractor has disclaimed responsibility in writing within a reasonable time after the receipt of the Purchaser's instructions), materials or workmanship in the manufacture of the Plant, but not otherwise.

Provided that the Contractor shall not be liable by virtue of this sub-clause in respect of damage or injury attributable to defects in any section or portion of the Plant taken over under Clause 22.

(ii) If there shall occur any loss of or damage to any property or injury to any person while the Contractor is on the Site for the purpose of making good a defect in any section or portion of the Plant pursuant to Clause 23 (*Defects after Delivery*) the Contractor shall be liable, subject to the provisions of Sub-Clauses (iii) and (iv) of this clause and Clause 19 (*Limitations on Contractor's Liability*) as follows:—

(*a*) In respect of loss of or damage to the said section or portion the Contractor's liability shall be as defined in Clause 23

(*b*) In respect of damage or injury to any other property or to any person and of any actions, claims, demands, costs, charges and expenses arising in connection therewith the Contractor shall be liable to the extent that such damage or injury was caused by the negligence or breach of statutory duty of the Contractor or a Sub-contractor while on the Site as aforesaid or by defective materials or workmanship used in making good the said defect but not otherwise.

The said section or portion of the Plant shall be defined by reference to the taking over certificate issued in respect thereof pursuant to Clause 22 (*Tests on Completion and Taking Over*).

(iii) The Contractor shall not be liable to the Purchaser for—

(*a*) any loss, damage or injury to the extent that it is caused by or arises from the acts or omissions of the Purchaser or of others (not being the Contractor's servant or Sub-contractors)

(*b*) any loss, damage or injury in circumstances over which the Contractor has no control.

(iv) Except in respect of personal injury or damage to property conferring on a person other than the Purchaser a good cause of action against the Contractor, the liability of the Contractor for any one act or default shall not exceed the Contract Price or £500,000, whichever is the greater.

(v) In the event of any claim being made against the Purchaser arising out of the matters referred to in and in respect of which the Contractor may be liable under this clause or Sub-Clause (iv) of Clause 17 (*Supervision of Erection*), the Contractor shall be promptly notified thereof, and may at his own expense conduct all negotiations for the settlement of the same and any litigation that may arise therefrom. The Purchaser shall not, unless and until the Contractor shall have failed to take over the conduct of the negotiations or litigation, make any admission which might be prejudicial thereto. The conduct by the Contractor of such negotiations or litigation shall be conditional upon the Contractor having first given to the Purchaser such reasonable security as shall from time to time be required by the Purchaser to cover the amount ascertained or agreed or estimated, as the case may be, of any compensation, damages, expenses, and costs for which the Purchaser may become liable. The Purchaser shall, at the request of the Contractor, afford all available assistance for any such purpose, and shall be repaid all reasonable expenses incurred in so doing.

(vi) The Contractor's liability under this clause shall be in lieu of any other liability for injury loss or damage occurring before all the works are taken over and caused by his breach of contract, or by the negligence or breach of statutory duty of the Contractor or any Sub-contractor or by defective design (other than a design made, furnished or specified by the Purchaser and for which the Contractor has disclaimed responsibility in writing within a reasonable time after receipt of the Purchaser's instructions), materials or workmanship.

Limitations on Contractor's Liability.

19.—Subject as provided in Clause 21 (*Delay in Completion*) for the deduction of liquidated damages for delay, the Contractor shall not be liable to the Purchaser, whether by way of indemnity or by reason of breach of contract or of negligence or of breach of statutory duty, for loss of use, whether complete or partial, of the Plant, or of profit or of any contract that may be suffered by the Purchaser.

Extension of Time for Completion.

20.—If, by reason of any act or omission on the part of the Purchaser or the Engineer or any industrial dispute or any cause beyond the reasonable control of the Contractor arising after the acceptance of the Tender, the Contractor shall have been delayed or impeded in the performance of his obligations under the Contract, except for his obligations under Clause 23 (*Defects after Delivery*), whether the delay or impediment occur before or after the time, if any, or extended time fixed for completion thereof, then provided that the Contractor shall without delay have given to the Purchaser or the Engineer notice in writing of his claim for an extension of time, the Engineer shall on receipt of such notice grant the Contractor from time to time in writing either prospectively or retrospectively such extension of time as may be reasonable. Any delay on the part of a Sub-contractor which prevents the Contractor from completing his obligations under the Contract within the time fixed therefor shall entitle the Contractor to an extension thereof provided such delay was due to any cause for which the Contractor himself would have been entitled to an extension of time under this clause.

Delay in Completion.

21.–If the Contractor fails to perform his obligations under the Contract (except for his obligations under Clause 23 (*Defects after Delivery*)) within the time fixed therefor by the Contract, or if no time be fixed, within a reasonable time, or within any extension of time granted by the Engineer, there shall be deducted from the Contract Price the percentage named in the Appendix of the Contract Value of such portion or portions only of the Plant as cannot in consequence of the said failure be put to the use intended for each week by which the putting of the Plant to the intended use has been delayed in consequence of such failure, but the amount so deducted shall not in any case exceed the maximum percentage named in the Appendix of the Contract Value, and such deduction shall be in full satisfaction of the Contractor's liability for the said failure.

22.–(i) Where the Contract provides for Tests on Completion, they shall be carried out by the Purchaser in the presence of the Contractor, who shall be given reasonable notice thereof.

(ii) The Tests on Completion (if any) shall be carried out promptly after the erection of the Plant has been completed (except in minor respects that do not affect the use of the Plant for the purpose for which it is intended).

Tests on Completion and Taking Over.

(iii) If for any reason for which the Contractor is responsible any portion of the Plant fails to pass the Tests on Completion, tests of the said portion shall, if required by the Engineer or by the Contractor, be repeated within a reasonable time upon the same terms and conditions, save that all reasonable expenses to which the Purchaser may be put by the repetition of the tests shall be deducted from the Contract Price.

(iv) As soon as the erection of the Plant has been completed (except as aforesaid) and the Plant has passed the Tests on Completion (if any), the Engineer shall issue a certificate (herein called a "taking-over certificate") in which he shall certify the date on which the erection of the Plant has been so completed and on which the Plant has passed the said tests and the Purchaser shall be deemed to have taken over the Plant on the date so certified.

(v) If the Plant is divided into two or more sections, Sub-Clause (iv) hereof shall apply to each section as it applied to the entire Plant. If by agreement between the Purchaser and the Contractor any portion of the Plant (other than a section or sections) shall be taken over before the remainder of the Plant, the Engineer shall issue a taking-over certificate in respect of that portion.

(vi) If by reason of any default on the part of the Contractor the issue of a taking-over certificate in respect of any portion of the Plant has been delayed and such portion is reasonably capable of being used without endangering the safety of the Plant or persons, then the Purchaser shall be at liberty to use such portion of the Plant, provided that the Contractor shall be afforded reasonable opportunity of taking such steps as may be necessary to permit the issue of the taking-over certificate.

(vii) If for any reason for which the Contractor is not responsible the Tests on Completion (if any) have not been carried out promptly after the erection of the Plant has been completed as provided in Sub-Clause (ii) of this clause, or have not been carried out successfully within three months after erection has been so completed, then the Purchaser shall be deemed to have taken over the Plant, and the Engineer shall issue a taking-over certificate accordingly. Any additional expense to which the Contractor may be put in attending any tests delayed in circumstances to which this sub-clause applies, shall be added to the Contract Price, and such allowances shall be made from the performances required to be attained in the said tests as may be reasonable having regard to any use of the Plant by the Purchaser prior to the Tests.

Defects after Delivery.

23.–(i) The Contractor shall be responsible for making good with all possible speed any defect in or damage to any portion of the Plant which may appear or occur during a period of 18 months after that portion has been delivered or 12 months after that portion shall have been taken over, whichever expires the earlier and which arises either

(*a*) from defective materials, workmanship or design (other than a design made, furnished or specified by the Purchaser and for which the Contractor has disclaimed responsibility in writing within a reasonable time after receipt of the Purchaser's instructions), or

(*b*) from any act or omission of the Contractor done or omitted during the said period of 18 months or 12 months as the case may be

provided always that the said period of 18 months shall be extended by the length of any period in respect of which a deduction is made from the Contract Price pursuant to Clause 21 (*Delay in Completion*) otherwise than for failure to deliver or for delay in delivery.

(ii) If any such defect shall appear or damage occur, the Engineer shall inform the Contractor thereof stating in writing the nature of the defect or damage. If the Contractor replaces or renews any part of the Plant, the provisions of this clause shall apply to the part of the Plant so replaced or renewed, except that the period during which the Contractor's responsibility pursuant to Sub-Clause (i) of this clause shall subsist shall be 12 months from the date of replacement or renewal.

(iii) The periods mentioned in Sub-Causes (i) and (ii) of this clause shall be extended by a period equal to the period during which the Plant or that portion thereof in which a defect to which this clause applies has appeared cannot be used by reason of that defect.

(iv) The supply to the Purchaser carriage paid of a defective or damaged part of the Plant properly repaired or of a part in replacement thereof shall constitute fulfilment by the Contractor of his obligation under Sub-Clause (i) of this clause in respect of that defective or damaged part. If it is reasonably practicable for a defective or damaged part to be returned to the Contractor and the Contractor shall call for its return the Purchaser shall cause it to be returned to the Contractor at the Contractor's expense.

(v) Where pursuant to this clause the Contractor supplies a part in replacement of a defective or damaged part the defective or damaged part shall become the property of the Contractor.

(vi) If any such defect or damage be not remedied within a reasonable time, the Purchaser may proceed to do the work at the Contractor's risk and expense.

(vii) The Contractor's liability under this clause shall be in lieu of any condition or warranty implied by law as to the quality or fitness for any particular purpose of any portion of the Plant delivered and save as in this clause expressed neither the Contractor nor his Sub-contractors, servants or agents shall be liable, whether in contract, tort or otherwise in respect of defects in or damage to such portion, or for any injury, damage or loss of whatsoever kind attributable to such defects or damage. For the purposes of this sub-clause the Contractor contracts on his own behalf and on behalf of and as trustee for his Sub-contractors, servants and agents.

Interim and Final Certificates

24.—(i) The Contractor may at the times and in the manner following apply for interim and final certificates.

Interim Certificates.

(ii) Applications for interim certificates may be made to the Engineer in respect of Plant ready and due for despatch or despatched and in respect of the fee for supervision from time to time as erection progresses. Each such application in respect of Plant shall identify the Plant in respect of which it is made and state the amount claimed and, where appropriate, shall be accompanied by such evidence of shipment and of payment of freight and insurance as the Engineer may reasonably require.

(iii) The Engineer shall issue to the Contractor an interim certificate within 14 days after receiving an application therefor which the Contractor was entitled to make. If the Engineer shall fail to issue an interim certificate as provided in this clause or if the Purchaser shall interfere with or obstruct the issue of any such certificate the Contractor may, without prejudice to any other remedy, either

(*a*) after giving to the Purchaser or the Engineer 14 days' notice of his intention so to do, cease the performance of the Contract until the said certificate be issued; in which case the expenses of the Contractor occasioned by such cessation and the subsequent resumption shall be added to the Contract Price, or

(*b*) after giving to the Purchaser or the Engineer one month's notice of his intention so to do, terminate the Contract, whether or not the Contractor shall have ceased performance of the Contract in accordance with paragraph (*a*) hereof or have given notice of his intention so to do.

(iv) Every interim certificate shall certify the total value of Plant ready and due for despatch or despatched as aforesaid and the amount of any fee payable in respect of supervision of erection up to the date named in the application for the certificate, less the total of any sums previously certified in interim certificates, provided that no sum shall be included in any interim certificate in respect of any Plant or work that, according to the decision of the Engineer, does not comply with the Contract.

(v) No interim certificate shall be relied on as conclusive evidence of any matter stated therein, or prejudice any right of the Purchaser or the Contractor against the other.

Final Certificate.

(vi) Application for a final certificate in respect of the Plant or any section thereof may be made to the Engineer at any time after the Contractor has ceased to be under any obligations in respect thereof under Clause 23 (*Defects after Delivery*). If in respect of any portion of the Plant replaced or renewed the obligations of the Contractor under Clause 23 continue after the period of eighteen months mentioned in Sub-Clause (i) of that Clause, the right of the Contractor to apply for a final certificate in respect of the Plant or a section thereof other than the portion so replaced or renewed shall not be affected. After the Contractor has ceased to be under any obligation under Clause 23 in respect of any portion replaced or renewed he may apply for a final certificate in respect of such portion.

(vii) The Engineer shall issue to the Contractor a final certificate within 14 days after receiving an application therefor which the Contractor was entitled to make.

(viii) A final certificate shall certify the total of all amounts comprised in interim certificates previously issued in respect of the Plant or the portion thereof to which the final certificate relates, subject to such additions thereto or deductions therefrom as may be authorised in Sub-Clause (x) of this clause.

(ix) A final certificate shall, save in the case of fraud or dishonesty relating to or affecting any matter dealt with in the certificate, be conclusive evidence as to the sufficiency of the Plant and of the value thereof unless any proceedings arising out of the Contract whether under Clause 30 (*Arbitration*) or otherwise shall have been commenced by either party before the final certificate has been issued or within one month thereafter.

Interim and Final Certificates.

(x) If any sum shall become payable to the Contractor under the Contract otherwise than for work executed or Plant delivered the amount thereof shall be included in the next certificate (interim or final) issued by the Engineer, and if any sum shall become payable under the Contract by the Contractor to the Purchaser, prior to the issue of the final certificate, whether by deduction from the Contract Price or otherwise, the amount thereof shall be deducted in the next certificate.

(xi) The Engineer may in any certificate give effect to any correction or modification that should properly be made in respect of any previous certificate.

Provisional Sums.

25.–(i) A provisional sum included in the Contract Price shall be expended or used as the Engineer may in writing direct and not otherwise. Insofar as a provisional sum is not expended or used it shall be deducted from the Contract Price.

Prime Cost Items.

(ii) All sums in respect of Prime Cost (P.C.) items included in the Contract Price shall be expended or used as the Engineer may in writing direct and not otherwise. To the net amount paid by the Contractor in respect of each P.C. item there shall be added the percentage named in the Appendix of the said amount. The sum by which the net amount so paid in respect of any P.C. item plus the said percentage thereof exceeds or is less than the sum included in the Contract Price in respect of that item shall be added to or deducted from the Contract Price as the case may be.

(iii) The Contractor shall have no responsibility for work done or Plant supplied by any other person in pursuance of directions given by the Engineer under this clause unless the Contractor shall have approved the person by whom such work is to be done or such Plant is to be supplied and the Plant, if any, to be supplied.

Payments due from the Contractor.

26.–Without prejudice to any other remedy which the Purchaser may have he shall be entitled to deduct from any moneys due, or becoming due to the Contractor under the Contract, all costs, damages or expenses for which under the Contract the Contractor is liable to the Purchaser.

Terms of Payment.

27.–(i) The Purchaser shall make payment to the Contractor as follows:

(*a*) Ten per cent of the Contract Price as an advance payment within fourteen days after the Contractor has furnished the Purchaser an irrevocable letter of guarantee from a guarantor or surety acceptable to the Purchaser with a value and currency equivalent to the advance payment. The letter of guarantee shall provide for its value to reduce by an amount equal to ten per cent of the sum certified in each interim certificate in respect of Plant ready and due for despatch or despatched and be in a form acceptable to the Purchaser.

(*b*) Within 14 days after presentation of each interim certificate, a sum equal to 85 per cent of the sum certified therein in respect of Plant, and the whole of any sum certified on account of the fee for supervision;

(*c*) Within one month after the presentation of the final certificate, the balance of the Contract Price adjusted to give effect to such additions thereto and such deductions therefrom as are provided for in these Conditions. Provided that if the Contractor shall have furnished to the Purchaser a guarantee acceptable to the Purchaser for the repayment on demand of such balance, he shall be entitled to payment thereof within one month from the date certified in the taking-over certificate.

(ii) If a final certificate shall be issued in respect of any portion of the Plant the expression "the balance of the Contract Price" as used in this clause shall be construed as meaning the balance of such part of the

Contract Price as shall be apportioned to such portion of the Plant by agreement or in default of agreement by the Engineer.

Statutory and other Regulations.

(iii) If the payment of any sum payable under sub-clause (i) of this clause shall be improperly delayed by the Purchaser or by reason of the failure of the Engineer to issue the relevant certificate within the time prescribed therefor interest at the rate of two per cent per annum over the Bank of England minimum lending rate from time to time in force on the amount of the delayed payment for the period of the delay shall be added to the Contract Price.

28.—If the cost to the Contractor of the performance of the Contract shall be increased or reduced by reason of the making after the date of his tender in the United Kingdom or elsewhere of any law or of any order, regulation or by-law having the force of law the amount of such increase or reduction shall be added to or deducted from the Contract Price as the case may be.

Local Taxation.

29.—Unless otherwise provided in the Contract, the Purchaser will neither be responsible for nor pay any Income Tax or other Taxation payable to properly constituted Authorities in the country in which the Site is located, to which the Contractor or his representatives may become legally liable whilst employed in that country for the purposes of the Contract.

Arbitration.

30.—(i) If at any time any question, dispute or difference shall arise between the Purchaser and the Contractor, either party shall, as soon as reasonably practicable, give to the other notice in writing of the existence of such question, dispute or difference specifying its nature and the point at issue, and the same shall be referred to the arbitration of a person to be agreed upon, or failing such agreement within six weeks, to some person appointed on the application of either of the parties hereto by the President for the time being of the Institution named in the Appendix, provided that a question, dispute or difference relating to a decision, instruction or order of the Engineer shall not be referred to arbitration unless notice has been given by the Contractor in accordance with Clause 10(*b*) (*Engineer's Decisions*).

(ii) Performance of the Contract shall continue during arbitration proceedings unless the Engineer or the Purchaser shall order the suspension thereof or of any part thereof, and if any such suspension shall be ordered the reasonable expenses of the Contractor occasioned by such suspension shall be added to the Contract Price. No payments due or payable by the Purchaser shall be withheld on account of pending reference to arbitration.

Construction of Contract.

31.—Unless otherwise agreed the Contract shall in all respects be construed and operate as an English contract and in conformity with English Law and the English courts shall have exclusive jurisdiction over any matter arising out of or in connection with the provisions for arbitration under Clause 30 (*Arbitration*). The marginal notes hereto shall not affect the construction hereof.

Variation in Costs.

32.*—If the cost to the Contractor of performing his obligations under the Contract shall be increased or reduced by reason of any rise or fall in the rates of wages payable to labour or in the cost of material or transport above or below such rates and costs ruling at the date of tender, the amount of such increase or reduction shall be added to or deducted from the Contract Price as the case may be, provided that no account shall be taken of any amount by which any cost incurred by the Contractor has been increased by the default or negligence of the Contractor. For the purpose of this clause 'the cost of material' shall be construed as including any duty or tax by whomsoever payable which is payable under or by virtue of any Act of Parliament on the import, purchase, sale appropriation, processing or use of such material.

Metrication.

33.—(i) If any Plant described in the Contract or the subject of a variation (hereinafter called "a variation order") under Clause 11 (*Variations*) is described by dimensions in the metric or imperial measure and the Contractor cannot procure such Plant in the measure specified in sufficient time to avoid delay in the performance of his other obligations under the Contract, but can obtain such Plant in the other measure to dimensions approximating to those described in the Contract or in the variation order, then the Contractor shall forthwith give notice to the Engineer of the facts stating the dimensions to which such Plant is procurable in the other measure.

(ii) The Engineer shall within 14 days after the receipt of a notice under the preceding sub-clause, give notice to the Contractor in writing pursuant to Clause 11 (*Variations*).

Such notice shall either:-

(*a*) direct the Contractor to supply such Plant to the dimensions stated in the said notice instead of to the dimensions described in the Contract or variation order as the case may be, or

*NOTE:— Unless this Clause is excluded any quoted price will be subject to adjustment in accordance with its terms.

(*b*) direct the Contractor to make some other variation whereby the need to supply such plant to the dimensions described in the Contract or variation order will be avoided.

The provisions of the Contract shall apply to directions given under this clause as though such directions had been included in the Specification.

APPENDIX

Clause

21. (Delay in Completion)

(*a*) .... Percentage to be deducted as damages.

(*b*) .... Maximum percentage which the deductions may not exceed.

25. .... Percentage on Prime Cost items.

30. .... The Institution of Mechanical Engineers*
The Institution of Electrical Engineers*

*Delete one line

AGREEMENT

This Agreement is made the day of 19

BETWEEN

(hereinafter referred to as "the Contractor") of the one part and

(hereinafter called "the Purchaser") of the other part **Whereas** the Purchaser desires to have certain Plant provided and the erection thereof supervised as mentioned, enumerated or referred to in certain General Conditions, Specification, Schedules, Drawings, Plans, Schedule of Prices, the Contractor's Tender and covering letter (if any), the acceptance of the Tender and any other relevant correspondence* which for the purpose of identification have been signed by

on behalf of the Contractor

and by
on behalf of the Purchaser all of which are deemed to form part of this Contract as though separately set out herein and are included in the expression "Contract" whenever herein used. **And Whereas** the Purchaser has accepted the tender of the Contractor for the provision of the said Plant for the sum of

(hereinafter called "the Contract Price") and for the supervision of the erection thereof upon the terms and subject to the Conditions hereinafter mentioned: **Now this Deed Witnesseth** and it is hereby agreed and declared as follows, that is to say, in consideration of the payments to be made to the Contractor by the Purchaser as hereinafter mentioned the Contractor hereby covenants with the Purchaser, his legal personal representatives, successors and assigns that the Contractor shall and will duly provide the said Plant and supervise the erection thereof and shall do and perform all other acts and things in the Contract mentioned or described or which are to be implied therefrom or may be reasonably necessary for the provision of the said Plant and the supervision of the erection thereof within and at the times and in the manner and subject to the terms conditions and stipulations mentioned in the Contract.

And in consideration of the due provision of the Plant and the supervision of the erection thereof as aforesaid the Purchaser does hereby for himself, his legal personal representatives, successors and assigns covenant with the Contractor that he, the Purchaser, his legal personal representatives, successors or assigns will pay the Contractor the Contract Price and such other sums as may become payable to the Contractor under the provisions of the Contract together with the fee payable pursuant to the Contract for supervision of erection, such payments to be made at such time and in such manner as is provided by the Contract.

In Witness whereof, etc.

*Delete as required

14 I.MECH.E./I.E.E. MODEL FORM OF GENERAL CONDITIONS OF CONTRACT 'B2/1981'

INDEX

Printed by A. McLay & Co. Ltd., Cardiff. 12/81.

MEMORANDUM ON THE USE OF THE GENERAL CONDITIONS OF CONTRACT (B2/1981)

A. Applicability

The General Conditions are intended to be suitable for the mechanical engineering and the electrical engineering industries and cover the main conditions for export contracts for the supply and supervision of erection of plant and machinery, the delivery thereof being in accordance with any of the methods detailed in the International Rules for the Interpretation of Trade Terms (INCOTERMS) published by the International Chamber of Commerce. INCOTERMS may be obtained in the United Kingdom from the British National Committee of the International Chamber of Commerce, London.

It is hoped that, in order to secure the advantage of uniformity and to avoid trouble and expense, the General Conditions may be found suitable in their entirety for most contracts. It is, however, inevitable that additions and/or amendments will have to be incorporated to meet the requirements of particular contracts or of the governmental, financial or other authorities of the countries from which plant is to be exported or of the countries into which it is to be imported. It is suggested, in order to facilitate checking, that any additions or alterations be made in such a way as to be clearly identifiable.

B. Contract documents

Contracts are commonly made in one of three ways–

(*a*) By a formal agreement, such as the model form supplied with these Conditions, incorporating in the contract the General Conditions, Specification, Schedules, Drawings, etc. In this case all the terms of the Contract are contained in the Agreement, the Conditions, Specification, and other incorporated documents on which tenders were invited, and agreed variations in or additions to the provisions of these documents are represented by suitable amendments.

(*b*) By a formal agreement, as in the method (*a*), but incorporating also the Contractor's Tender and covering letter, if any, the Purchaser's acceptance of the Tender, and all correspondence containing terms accepted by the Purchaser and the Contractor. When this method is adopted any modifications of the General Conditions, Specification, etc., have to be sought in the covering letter and other correspondence.

(*c*) Without a formal agreement, the Tender and its acceptance, together with other relevant correspondence, being relied on as constituting the Contract between the parties.

When either method (*a*) or (*b*) is adopted, a copy of the form of agreement to be used should be included with the documents supplied to or open to the inspection of intending Tenderers.

When method (*a*) is used, the references in the model form of agreement to the Tender, covering letter, acceptance of the Tender, and other correspondence should be deleted.

C. Delivery

It should be made clear in the contract documents whether delivery is to be f.o.b., c.i.f., f.o.r., or by any other method defined in INCOTERMS eg:

(*a*) f.o.b.–It should be specified in the contract documents at what port or ports the Plant is to be shipped.

(*b*) c.i.f.–The Purchaser should specify the port to which the Plant is to be shipped and the place where the documents (i.e. invoice, bill of lading and policy of insurance) are to be tendered, if other than the Purchaser's place of business. Any special stipulations with regard to these matters should also be specified. Delivery is not complete until the Plant has been shipped and the documents mentioned above have been tendered to the Purchaser. The Contractor should therefore not delay the tendering of the policy of insurance, bill of lading and invoice until the interim certificate is available.
The port of destination named in the Contract is often loosely referred to as the place of delivery. The parties should therefore, when referring to delivery in the contract documents be clear whether they intend to refer to the port of shipment or the port of destination.

ii I.MECH.E./I.E.E. MODEL FORM OF GENERAL CONDITIONS OF CONTRACT 'B2/1981'

(*c*) f.o.r. and f.o.t. – completion of delivery requires loading of the Plant on a railway or road vehicle at the place of delivery specified in the Contract. In case of deliveries by rail the place specified is generally on a railway conveniently near to the Contractor's works. Some Purchasers, however, often stipulate delivery f.o.r. either at the port of destination or at the Purchaser's own railhead. Neither of these methods is consistent with the INCOTERMS' definition. If either of these methods is chosen the parties' respective obligations will need to be specified in the Contract.

D. Advance Payments and Progress Payments During Manufacture

The General Conditions provide for an advance payment against a guarantee, but the Contractor does not become entitled to any further payments for Plant he has manufactured until he has shipped it. In the case of contracts involving considerable expenditure on manufacture before any shipment can be made, the General Conditions may be modified to provide for progress payments to the Contractor in respect of Plant in the course of manufacture, in which case the insurance arrangements may also need to be modified.

E. Customs and Imports Duties

The General Conditions do not deal with the responsibility for payment of customs and import duties and licenses nor with the responsibility for the obtaining of any export licenses or payment of export levies. Care should be taken to ensure that the form of delivery chosen by reference to INCOTERMS reflects the agreed intentions of the parties and if not to deal with the matter specifically in the Contract documents.

F. Contract Price Adjustment Formula

If the parties so wish they may operate Clause 32 by means of an agreed formula.

MEMORANDUM AS TO TENDERS

Tenderer may propose modifications.

The Tenderer is at liberty to add any details and conditions that he may deem desirable, and in the event of this doing so must print or type the same and annex the added matter to the Specification or General Conditions returned by him, but such additional details and conditions will not be binding on the Purchaser unless they are approved by him and incorporated in the Contract.

Where in special circumstances the Tenderer deems it advisable to provide for the nondisclosure of drawings and information of a confidential nature furnished with his Tender he should include in the Tender or covering letter a stipulation to this effect. If the Tender embodying such a stipulation is accepted the stipulation should be incorporated in any subsequent formal agreement.

If the Tenderer has any doubt as to the meaning of any portion of the General Conditions or of the Specification, he should, when submitting his Tender, set out in his covering letter the interpretation upon which he relies.

Tenders

One copy of the Specification supplied, together with the General Conditions and Drawings, enclosed in a plain sealed cover bearing the works "Tender for (*here follows the short title of the Plant*)

"

and not bearing any name or mark indicating the sender shall, subject to modifications mentioned above, be returned intact with the Tender Form and Schedules, if any, filled up and signed, addressed to

and must be received by him on or before

The Purchaser does not bind himself to accept the lowest or any Tender, nor will he be responsible for, or pay for, expenses or losses which may be incurred by any Tenderer in the preparation of this Tender.

Copies of Specification etc.

A duplicate copy of the Specification, General Conditions and Drawings will be supplied to intending Tenderers without charge, and extra copies of the Drawings at the cost of reproduction.

The sum deposited by the Tenderer on application for the Specification will be refunded to him within one month of the date of the adjudication upon the Tenders, if all copies of the Specification, Drawings, etc., have been returned by the Tenderer, unless the Tender was not made in good faith, in which case the deposit may be forfeited.

Agreement.

The Tenderer whose Tender is accepted may be required to enter into a formal agreement with the Purchaser.

NOTE

The advertisement inviting Tenders should state whether and if so where the Agreement, General Conditions, Specification and Drawings may be inspected, and should give such particulars of the class of plant and apparatus required under each Section as will enable contracting firms to decide, without obtaining the Specification, whether they desire to tender, and should also state the amount of the deposit to be paid for the General Conditions, Specification and Drawings.

B3

MODEL FORM OF

GENERAL CONDITIONS OF CONTRACT

INCLUDING FORMS OF AGREEMENT & GUARANTEE

RECOMMENDED BY

THE INSTITUTION OF MECHANICAL ENGINEERS
THE INSTITUTION OF ELECTRICAL ENGINEERS
AND
THE ASSOCIATION OF CONSULTING ENGINEERS

FOR USE IN CONNECTION WITH

EXPORT CONTRACTS (INCLUDING DELIVERY TO AND ERECTION ON SITE)

THIRD EDITION 1980

Published for the Joint Committee on Model Forms of General Conditions of Contract by
THE INSTITUTION OF ELECTRICAL ENGINEERS
and obtainable from
Publications Sales Department, Station House, 70 Nightingale Road, Hitchin, Herts, SG5 1RJ
or to callers, at the Reception Desk, Savoy Place
or from
THE INSTITUTION OF MECHANICAL ENGINEERS
Publications Sales Department, P.O. Box 24, Northgate Avenue, Bury St. Edmunds, Suffolk, IP32 6BW
or to callers, at the Reception Desk, Birdcage Walk, London
or at
THE ASSOCIATION OF CONSULTING ENGINEERS, Alliance House, 12, Caxton Street, London, SW1H 0QL.

HISTORICAL NOTE

A Model Form of General Conditions 'B2' (Export, including Complete Erection or Supervision of Erection) was first published by The Institution of Electrical Engineers in 1925. A revision was issued in 1928.

Following consultations, and by agreement between the Council of The Institution of Mechanical Engineers and the Council of The Institution of Electrical Engineers, a Model Form 'B3' was prepared for export contracts including delivery and erection of mechanical or electrical plant. On this basis the Model Form B3 was first issued jointly by the two Councils as—

1954 Edition

Model Form of General Conditions of Contract 'B3'— Export Contracts (including Delivery to and Erection on Site).

Following an approach in 1952 by The Association of Consulting Engineers, it was agreed that the Association should adopt and recommend to its members the use of this Model Form, and the name of the Association accordingly appears on the title page.

The 1971 Edition incorporated substantial revisions to take account of developments into practice and in the light of substantial changes made in other Model Forms published by the Institution since the 1954 Edition.

This 1980 Edition takes account of changes in legislation (the Unfair Contract Terms Act, 1977 and the Arbitration Act, 1979), advice received from Leading Counsel on certain aspects of the 1971 Edition, provision for an advance payment, an updated rate of interest payable by the Purchaser on delayed payment and other revisions to bring the Form up to date in the light of comments received and developments in practice.

ERRATUM to Model Form of General Conditions of Contract 'B3/1980'

Page 1 Line 15 for 'the' read 'in'
Page 10 Line 22 add the following sentence: 'Provided that the Contractor shall not be liable by virtue of this sub-clause in respect of damage or injury attributable to defects in any section or portion of the Works taken over under Clause 29.'
Page 14 Line 29 for '(vii)' read '(viii)'

This form may be cited as "Model Form B3/1980'

GENERAL CONDITIONS

Definition of Terms.

1.–In construing these General Conditions and the Specifications, the following words shall have the meanings herein assigned to them unless there is something in the subject matter or context inconsistent with such construction:

The "Contract" shall mean the agreement between the Purchaser and the Contractor for the execution of the Works howsoever made including therein all documents to which reference may properly be made in order to ascertain the rights and obligations of the parties under the said agreement.

The "Contractor" shall mean the tenderer whose tender has been accepted by the Purchaser and shall include the Contractor's legal personal representatives, successors and assigns.

"Contractor's Equipment" shall mean tools, tackle, stores and other things brought upon the Site by the Contractor and required thereon for the purposes of the Works but not for incorporation therein.

The "Contract Price" shall mean the sum named in the Contract as the contract price.

The "Contract Value" shall mean that part of the Contract Price, adjusted to give effect to such additions or deductions as are provided for the Clause 10 (*Variations*), which is properly apportionable to the Plant or work in question, having regard to the state, condition and location of the Plant, the amount of work done and all other relevant circumstances but disregarding any changes pursuant to Clause 40 in the cost of executing the Works.

The "Engineer" shall mean
or the person for the time being or from time to time notified in writing by the Purchaser to the Contractor as the Engineer for the Contract, or in default of any notification the Purchaser.

"Month" shall mean calendar month.

"Plant" shall mean machinery, apparatus, materials, articles, and things of all kinds other than Contractor's Equipment.

The "Purchaser" shall mean
and shall include the Purchaser's legal personal representatives, successors and assigns.

The "Site" shall mean the actual place to which Plant is to be delivered or where work is to be done by the Contractor, together with so much of the area surrounding the said place as the Contractor shall with the consent of the Engineer actually use in connection with the Works otherwise than merely for the purpose of access to the said place.

The "Specification" shall mean the specification annexed to or issued with these General Conditions.

"Sub-Contractor" shall mean any person (other than the Contractor) named in the Contract for any part of the Works or any person to whom any part of the Contract has been sub-let with the consent in writing of the Engineer, and the legal representatives, successors and assigns of such person.

"Tests on Completion" shall mean such tests to be made by the Contractor on completion of erection as are provided for in the Contract or otherwise agreed between the Purchaser and the Contractor.

The "Works" shall mean all Plant to be provided and work to be done by the Contractor under the Contract.

"Writing" shall include any manuscript, type-written, or printed statement, under seal or hand as the case may be.

Words importing persons shall include firms and corporations.

Words importing the singular only shall also include the plural and vice versa.

Contractor to inform himself fully.

2.–The Contractor shall be deemed to have examined the Site, if access thereto has been made available to him, and the General Conditions and Specifications, with such schedules, drawings and plans as are annexed thereto or referred to therein.

Security for due performance.

3.–(i) If required by the Purchaser, the Contractor shall provide a surety or sureties, or a grantor of an insurance or guarantee policy, who in either case shall be subject to the approval of the Purchaser, which approval shall not be unreasonably withheld, and who shall execute (if two or more jointly and severally) a guarantee, or grant an insurance or guarantee policy to an extent not exceeding 15 per cent of the Contract Price, by way of guarantee for the due and faithful performance of the Contract, such guarantee to be binding notwithstanding such variations, alterations, or extensions of time as may be made, given, conceded or agreed under these General Conditions. The instrument of guarantee shall be in the form annexed to these General Conditions with such modifications as may be necessary and the expenses of procuring, preparing, completing and stamping such instrument shall be paid by the Purchaser.

(ii) If the guarantee or other security for the due performance of the Contract required to be furnished pursuant to this clause shall not be duly furnished by the Contractor to the Purchaser within one month after the Contract has been entered into, the Purchaser may, at his option, without prejudice to any rights or claims he may have against the Contractor by reason of the Contractor's non-compliance with any of the provisions of this clause, and within seven days after the expiry of the said period, by notice in writing to the Contractor terminate the Contract forthwith, and the Purchaser shall thereupon not be liable for any claim or demand from the Contractor in respect of any thing then already done or furnished, or in respect of any other matter or thing whatsoever, in connection with the Contract, but the Purchaser shall be entitled to be repaid by the Contractor all out-of-pocket expenses properly incurred by the Purchaser incidental to the obtaining of new tenders.

Expenses of Agreement

(iii) The expenses of preparing, completing and stamping the agreement, if any, shall be paid by tne Purchaser, and an executed counterpart thereof properly stamped together with copies of all other documents comprising the Contract shall be furnished to the Contractor free of charge.

Drawings.

4–(i) The Contractor shall submit to the Engineer for approval:

(*a*) within the time specified in the Specification or, if no time is specified, then a reasonable time, such drawings, samples, patterns and models as may be called for therein, and

(*b*) during the progress of the Works, and within such reasonable times as the Engineer may require, such drawings of the general arrangement and of details of the Works or any part thereof as the Engineer may require, provided that the Contractor shall not be under any obligation to supply copies of shop drawings.

Within a reasonable period after receiving such drawings, samples, patterns and models the Engineer shall signify his approval or otherwise. If the Engineer shall not approve any drawing, sample, pattern or model thus submitted the same shall be forthwith modified to meet the reasonable requirements of the Engineer and shall be re-submitted. Copies of all drawings which require to be approved by the Engineer shall be provided in triplicate by the Contractor. Drawings, samples, patterns and models when approved shall, if required by either party be signed or identified by both parties and if required by the Contractor, one copy shall be returned to him. All dimensions marked on drawings shall be considered correct although measurements by scale may differ therefrom. Detailed drawings shall take precedence where they differ from general arrangement drawings.

(ii) Drawings, samples, patterns and models approved as above described shall not be departed from except as provided in Clause 10 (*Variations*).

(iii) The Engineer shall have the right at all reasonable times to inspect at the premises of the Contractor all drawings of any portion of the Works.

(iv) The Contractor shall, if so desired by the Purchaser, furnish to the Purchaser in writing at the commencement of the period referred to in Clause 31 (i) (*Defects after Taking Over*), or at such earlier times as may be named in the Specification, such information, accompanied by drawings, as may be necessary to enable the Purchaser to operate, maintain, dismantle, reassemble and adjust all parts of the Works.

(v) Drawings submitted in pursuance of paragraphs (*a*) and (*b*) of Sub-Clause (i) of this clause shall not, without the consent of the Contractor, be used by the Purchaser except for the purposes of the Contract, nor shall they without such consent be communicated to third parties save insofar as may be necessary for the proper execution of the Works.

Mistakes in information.

5.–(i) The Contractor shall be responsible for any discrepancies, errors or omissions in the drawings and information supplied by him, whether they have been approved by the Engineer or not, provided that such discrepancies, errors or omissions be not due to inaccurate drawings or information furnished to the Contractor by the Purchaser or the Engineer.

(ii) The Contractor shall at his own expense carry out any alterations or remedial work necessitated by reason of such discrepancies, errors or omissions and modify the drawings and information accordingly, or if the same be done by or on behalf of the Purchaser shall bear all costs reasonably incurred therein. The performance of his obligations under this sub-clause shall be in full satisfaction of the Contractor's liability under Sub-Clause (i) of this clause, but shall not relieve him of his liability under Clause 27 (*Delay in Completion*) insofar as that liability arises as a result of such discrepancies, errors or omissions.

(iii) The Purchaser shall be responsible for drawings and information supplied in writing by the Purchaser or the Engineer and for the details of special work specified by either of them. The Purchaser shall pay the extra cost reasonably incurred by the Contractor due to alterations of the work necessitated by reason of inaccurate drawings or information so supplied to the Contractor.

Assignment.

6.–(i) The Contractor shall not, without the consent in writing of the Purchaser, which shall not be unreasonably withheld, assign or transfer the Contract or the benefits or obligations thereof of any part thereof to any other person, provided that this shall not affect any right of the Contractor to assign, either absolutely or by way of charge, any moneys due or to become due to him, or which may become payable to him under the Contract.

Sub-letting.

(ii) The Contractor shall not, without the consent in writing of the Engineer, which shall not be unreasonably withheld, sub-let the Contract or any part thereof, or make any sub-contract with any person or persons for the execution of any portion of the Works but the restriction contained in this clause shall not apply to sub-contracts for materials, for minor details or for any part of the Works of which the makers are named in the Contract. Any such consent shall not relieve the Contractor from his obligations under the Contract.

Patent Rights, etc.

7.–(i) The Contractor shall indemnify the Purchaser against all actions, claims, demands, costs, charges and expenses arising from or incurred by reason of any infringement or alleged infringement of letters patent, registered design, copyright, trade mark or trade name protected in the United Kingdom, or in the country in which the Plant is to be erected, by the use of any Plant supplied by the Contractor, but such indemnity shall not cover any use of the Works otherwise than for the purpose indicated by or reasonably to be inferred from the Specification or to any infringement which is due to the use of any Plant in association or combination with any other plant not supplied by the Contractor.

(ii) In the event of any claim being made or action brought against the Purchaser arising out of the matters referred to in this clause, the Contractor shall be promptly notified thereof and may at his own expense conduct all negotiations for the settlement of the same, and any litigation that may arise therefrom. The Purchaser shall not, unless and until the Contractor shall have failed to take over the conduct of the negotiations or litigation, make any admission which might be prejudicial thereto. The conduct by the Contractor of such negotiations or litigation shall be conditional upon the Contractor having first given to the Purchaser such reasonable security as shall from time to time be required by the Purchaser to cover the amount ascertained or agreed or estimated, as the case may be, of any compensation, damages, expenses and costs for which the Purchaser may become liable. The Purchaser shall, at the request of the Contractor, afford all available assistance for the purpose of contesting any such claim or action, and shall be repaid all reasonable expenses incurred in so doing.

(iii) The Purchaser on his part warrants that any design or instructions furnished or given by him shall not be such as will cause the Contractor in the performance of the Contract to infringe any letters patent, registered design, copyright, trade mark or trade name in the country in which the Plant is used.

Manner of Execution.

8.–All Plant to be supplied and all work to be done under the Contract shall be manufactured and executed in the manner set out in the Specification or, where not so set out, to the reasonable satisfaction of the Engineer.

Contractor's Equipment, Labour, etc.

9.–(i) Unless specific arrangements be made to the contrary the Contractor shall, at his own expense, provide all Contractor's Equipment and all materials, labour, haulage and power necessary to execute and complete the Works.

(ii) The Contractor shall be responsible for the proper fencing, guarding, lighting and watching of all the Works on the Site until taken over under Clause 29 (*Taking Over*) and for the proper provision during a like period of temporary roadways, footways, guards and fences as far as the same may be rendered necessary by reason of the Works for the accommodation and protection of the owners and occupiers of adjacent property, the public and others. No naked light shall be used by the Contractor on the Site otherwise than in the open air without special permission in writing from the Engineer.

Electricity, Water and Gas.

(iii) The Contractor shall be entitled to use for the purposes of the Works such supplies of electricity, water and gas as may be available therefor on the Site and shall pay to the Purchaser for such use such sum as may be reasonable in the circumstances, and shall at his own expense provide any apparatus necessary for such use.

Lifting Equipment.

(iv) The Purchaser shall at the request of the Contractor and for the execution of the Works operate any suitable lifting equipment belonging to the Purchaser that may be available on the Site and the Contractor shall pay a reasonable sum therefor. The Purchaser shall during such operation retain control of and be responsible for the safe working of the lifting equipment but shall not be responsible for any negligence of the Contractor.

Variations

10.–(i) The Contractor shall not alter any of the Works, except as directed in writing by the Engineer; but the Engineer shall have full power, subject to the proviso hereinafter contained, from time to time during the execution of the Contract by notice in writing to direct the Contractor to alter, amend, omit, add to or otherwise vary any of the works, and the Contractor shall carry out such variations, and be bound by the same conditions, so far as applicable, as though the said variations were stated in the Specification. Provided that no such variation shall, except with the consent in writing of the Contractor, be such as will, with any variations already directed to be made, involve a net addition to or deduction from the Contract Price of more than 15 per cent thereof. In any case in which the Contractor has received any such direction from the Engineer which either then or later will, in the opinion of the Contractor, involve an addition to or deduction from the Contract Price, the Contractor shall, as soon as reasonably possible, advise the Engineer in writing to that effect. The amount to be added to or deducted from the Contract Price shall be ascertained and determined in accordance with the rates specified in the schedules of prices, so far as the same may be applicable, and where rates are not contained in the said schedules, or are not applicable, such amount shall be such sum as is reasonable in the circumstances. Due account shall be taken of any partial execution of the Works which is rendered useless by any such variation.

(ii) If the Engineer shall make any such variation in any part of the Works, such reasonable notice in writing shall be given to the Contractor as will enable him to make his arrangements accordingly. If in the opinion of the Contractor any such variation is likely to prevent or prejudice the Contractor from or in fulfilling any of his obligations under the Contract, he shall notify the Engineer thereof in writing, and the Engineer shall decide forthwith whether or not the same shall be carried out. If the Engineer confirms his instructions in writing, the said obligations shall be modified to such an extent as may be justified. Until the Engineer so confirms his instructions they shall be deemed not to have been given.

Underground Works.

11.–In the case of work underground or involving excavation where the actual conditions of the ground are not stated in the Contract and could not reasonably have been inferred from an inspection of the Site by the Contractor before he prepared his tender, if rock, rocky soil, solid chalk, water, running sand, slag, pipes, concrete or other obstructions are found, or if it should be necessary to leave in timber or provide support for existing works (such necessity not having been indicated in the Contract), the Contractor shall inform the Engineer as soon as reasonably practicable of the steps he proposes to take to deal with the hazard. If, as a consequence thereof, extra cost is incurred by the Contractor, a sum ascertained and determined in like manner to the valuation of variations under Clause 10 (*Variations*) shall be added to the Contract Price.

Contractor's Default.

12.–If the Contractor shall neglect to execute the Works with due diligence and expedition, or shall refuse or neglect to comply with any reasonable orders given him in writing by the Engineer in connection with the Works, or shall contravene the provisions of the Contract, the Purchaser may give notice in writing to the Contractor to make good the failure, neglect or contravention complained of. Should the

Contractor fail to comply with the notice within fourteen days from the date of service thereof in the case of a failure neglect or contravention capable of being made good within that time, or otherwise within such time as may be reasonably necessary for making it good, then and in such case the Purchaser shall be at liberty to employ other workmen and forthwith execute such part of the Works as the Contractor may have neglected to do, or, if the Purchaser shall think fit, it shall be lawful for him, without prejudice to any other rights he may have under the Contract, to take the Works wholly or in part out of the Contractor's hands and either by his own workmen or by re-contracting with any other person or persons to complete the Works or any part thereof, and in either event the Purchaser shall have the free use of all Contractor's Equipment that may be on the Site in Connection with the Works, without being responsible to the Contractor for fair wear and tear thereof, and to the exclusion of any right of the Contractor over the same, and the Purchaser shall be entitled to retain and apply any balance which may be otherwise due on the Contract by him to the Contractor, or such part thereof as may be necessary to the payment of the cost of executing the said part of the Works or of completing the Works as the case may be. If the cost of completing the Works or executing a part thereof as aforesaid shall exceed the amount that would otherwise become due to the Contractor in accordance with the Contract, the Contractor shall pay such excess. If the Purchaser pursuant to this clause takes the Works or part thereof out of the Contractor's hands, the Contractor's liability under Clause 27 (*Delay in Completion*) shall immediately cease in respect of the Works or part thereof, without prejudice to any such liability that shall have already accrued.

Bankruptcy

13.–If the Contractor shall become bankrupt or insolvent, or have a receiving order made against him, or compound with his creditors, or being a corporation commence to be wound up, not being a members' voluntary winding up for the purpose of reconstruction or amalgamation, or carry on its business under a receiver for the benefit of its creditors or any of them, the Purchaser shall be at liberty either–

(*a*) to terminate the Contract forthwith by notice in writing to the Contractor or to the receiver or liquidator or to any person in whom the Contract may become vested, and to act in the manner provided in Clause 12 (*Contractor's Default*) as though the last mentioned notice had been the notice referred to in such clause and the Works had been taken out of the Contractor's hands, or

(*b*) to give such receiver, liquidator or other person the option of carrying out the Contract subject to his providing a guarantee for the due and faithful performance of the Contract up to an amount to be agreed.

Inspection, Testing and Rejection of Plant.

14.–(i) The Engineer shall be entitled at all reasonable times during manufacture to inspect, examine and test on the Contractor's premises the materials and workmanship and performances of all Plant to be supplied under the Contract, and if part of the said Plant is being manufactured on other premises the Contractor shall obtain for the Engineer permission to inspect, examine and test as if the said Plant were being manufactured on the Contractor's premises. Such inspection, examination or testing shall not release the Contractor from any obligation under the Contract.

(ii) The Contractor shall, after consulting the Engineer, give the Engineer reasonable notice in writing of the date on and the place at which any Plant will be ready for testing as provided in the Contract and, unless the Engineer shall attend at the place so named on the date which the Contractor has stated in his notice, the Contractor may proceed with the tests, which shall be deemed to have been made in the Engineer's presence, and shall forthwith forward to the Engineer duly certified copies of the test readings. The Engineer shall give the Contractor 24 hours' notice in writing of his intention to attend the tests.

(iii) Where the Contract provides for tests on the premises of the Contractor or of any Sub-contractor the Contractor, except where otherwise specified, shall provide free of charge such assistance, labour, materials, electricity, fuel, stores, apparatus and instruments as may be requisite and as may be reasonably demanded to carry out such tests efficiently.

(iv) Where the Contract provides for tests on the Site, the Purchaser, except where otherwise specified, shall provide free of charge such labour, materials, electricity, fuel, water, stores and apparatus as may be requisite and as may be reasonably demanded to carry out such tests efficiently.

(v) As and when the Engineer is satisfied that any Plant shall have passed the tests referred to in this clause he shall forthwith notify the Contractor in writing to that effect.

(vi) If after inspecting, examining or testing any Plant the Engineer shall decide that such Plant or any part thereof is defective or not in accordance with the Contract, he may reject the said Plant or part thereof by giving to the Contractor within a reasonable time notice in writing of such rejection, stating therein the grounds upon which the said decision is based.

(vii) The provisions of Clause 28 (v) (*Tests on Completion*) shall relate also to inspections, examinations and tests carried out under this clause.

Delivery.

15.–(i) No Plant or Contractor's Equipment shall be delivered to the Site until an authorisation in writing has been applied for and obtained by the Contractor from the Engineer that delivery may be made. The Contractor shall be responsible for the reception on the Site of all Plant and Contractor's Equipment delivered for the purposes of the Contract.

(ii) For the purposes of this clause only–

"delayed Plant" means (*a*) Plant which, by delay or failure on the part of the Engineer to give such authorisation as is mentioned in Sub-Clause (i) of this clause or from any cause for which the Purchaser or some other contractor employed by him is responsible, the Contractor is prevented from delivering to the Site at the time specified for the delivery thereof or, if no time is specified, at the time when it is reasonable for it to be delivered having regard to the date by which the Works ought to be completed: and (*b*) Plant which has been delivered to the Site but which, by delay or failure on the part of the Engineer or from any cause for which the Purchaser or some other contractor employed by him is responsible, the Contractor is for the time being prevented from erecting.

"the normal delivery date" means the time when but for such delay, failure or other cause as aforesaid delayed Plant would have been delivered to the Site.

"notice to proceed" means notice in writing from the Engineer to the Contractor that delayed Plant may forthwith be delivered to the Site or (as the case may be) erected.

(iii) If delayed Plant is ready for delivery and has been suitably and sufficiently marked as appropriated to the Contract and the Contractor has given to the Engineer an opportunity of inspecting it or, if delayed Plant has been delivered to the Site, the Contractor may give notice in writing to the Purchaser requiring that the provisions of Sub-Clause (iv) of this clause shall have effect with respect to such delayed Plant.

(iv) Where notice has been given in accordance with Sub-Clause (iii) of this clause–

(*a*) There shall be added to the Contract Price a sum, ascertained and determined in like manner to the valuation of variations under Clause 10 (*Variations*), for storing and taking reasonable measures to protect and preserve the delayed Plant from and insuring it against loss, deterioration and damage however caused from the date of the said notice or the normal delivery date, if this shall be later, until the Contractor shall no longer be prevented from delivering the delayed Plant or (as the case may be) erecting it or shall be relieved of responsibility therefor under Sub-Clause (v) of this clause, whichever shall first happen.

(*b*) The Contractor shall after one month from the normal delivery date or from the date of the said notice (whichever shall be the later) be entitled to have the Contract Value of the delayed Plant included in an interim certificate.

(*c*) If at the expiration of six months from the normal delivery date or from the date of the said notice (whichever shall be the later) the Contractor shall still be prevented from delivering the delayed Plant to the Site or (as the case may be) from erecting it, the Engineer shall, on the application of the Contractor, certify accordingly and within one month from the presentation of such certificate the Contractor shall be entitled to be paid 97½ per cent of the Contract Value of the delayed Plant less any sum previously paid to him in respect thereof.

(*d*) If at the expiration of 12 months from the normal delivery date or from the date of the said notice (whichever shall be the later) the Contractor shall still be prevented from delivering the delayed Plant to the Site or (as the case may be) from erecting it, the Engineer shall, on the application of the Contractor, certify accordingly and within one month from the presentation of such certificate the Contractor shall be entitled to be paid 100% of the Contract Value of the delayed Plant less any sum previously paid to him in respect thereof provided always that, if notice to proceed shall be given to the Contractor prior to the expiration of the said period of 12 months, this paragraph of this sub-clause shall not operate.

(v) The Purchaser may at any time after receipt of the notice referred to in Sub-Clause (iii) of this clause assume responsibility for storing, protecting and preserving the delayed Plant. If at any time after the expiration of 12 months from the date of the said notice or at any time after the delayed Plant has been delivered to the Site the Purchaser shall not have assumed such responsibility, the Contractor may by a further notice in writing expiring 30 days after receipt thereof by the Purchaser require the Purchaser to assume the responsibility aforesaid and upon the expiration of the last mentioned notice the Purchaser shall assume such responsibility provided always that, if notice to proceed shall be given within 30 days after receipt by the Purchaser of the last mentioned notice given by the Contractor, the foregoing provisions of this Sub-Clause shall not operate. As and when the Purchaser assumes the responsibility aforesaid the Contractor shall be relieved of any responsibility for the delayed Plant until either the expiration of 30 days after the receipt of a notice to proceed or, the Contractor having received the notice to proceed, resumes possession of the said Plant, whichever shall first occur.

(vi) After the receipt of notice to proceed the Contractor, if he has been relieved of responsibility under Sub-Clause (v) of this clause, shall (and in any other case may) after due notice in writing to the Engineer and if required by the Engineer, examine in his presence the delayed Plant and any Plant on the Site that has been erected but not taken over under Clause 29 (*Taking Over*) by reason of delay in the delivery or erection of the delayed Plant, and make good any deterioration or defect therein or loss thereof that may have appeared or occurred after the normal delivery date or (if later) the date when the Contractor was by such delay, failure or other cause as before-mentioned first prevented from erecting the delayed Plant.

(vii) There shall be added to the Contract Price a reasonable sum for making the examination referred to in Sub-Clause (vi) of this clause or in making good any deterioration, defect or loss as therein mentioned except insofar as the same was caused by faulty workmanship or materials or by the Contractor's failure to take the measures referred to in paragraph (*a*) of Sub-Clause (iv) of this clause or in Sub-Clause (i) of Clause 22 (*Liability for Accidents and Damage*). If the Contractor incurs additional expense in delivering the delayed Plant to the Site or in erecting the same or any other Plant or in carrying out the tests on completion or in performing his obligations under Clause 31 (*Defects after Taking Over*) which would not have been incurred had the delivery or erection of the delayed Plant not been prevented as aforesaid the Purchaser shall pay a reasonable sum in respect thereof which shall be added to the Contract Price.

(viii) Without prejudice to the provisions of Sub-Clause (vii) of Clause 31 (*Defects after Taking Over*), the obligations of the Contractor under that clause with respect to delayed Plant shall not apply to any defect that may appear or occur therein after the expiration of three years from the date of the said notice referred to in Sub-Clause (iii) of this clause or the normal delivery date if this shall be later.

Customs and Import Duties.

16.—The Purchaser shall in good time procure any necessary import permits and promptly pay all customs and import duties which may become payable upon the importation of the Plant into the country in which the Plant is to be erected.

Access to and possession of the Site.

17.—(i) Subject to Sub-Clause (v) of this clause, access to and possession of the Site shall be afforded to the Contractor by the Purchaser in reasonable time and, except in so far as the Specification may provide to the contrary, the Purchaser shall provide a road or railway suitable for the transport of all Plant and Contractor's Equipment necessary for the execution of the Works from a convenient point on a public thoroughfare suitable for such transport or on a railway available to the Contractor to the point on the Site where it is to be delivered or used.

(ii) The Purchaser shall, before the time specified for the delivery of any Plant to the Site, or if no time be specified, before it is reasonable for the Plant to be delivered having regard to the date fixed by the Contract for completion of the Works, obtain all consents, wayleaves and approvals required in connection with the regulations and by-laws of local or other authority which shall be applicable to the Works.

(iii) If a building, structure, foundation or approach is by the Contract to be provided by the Purchaser, such building, structure, foundation or approach shall be in a condition suitable for the efficient transport, reception, erection and maintenance of the Works.

(iv) In the execution of the Works, the Contractor shall not authorize or purport to authorize any person other than his employees and Sub-contractors and their employees to come upon the Site, except by the written permission of the Engineer, but facilities to inspect the Works at all times shall be afforded to the Engineer and his representatives and authorized representatives of the Purchaser.

(v) The access to and possession of the Site referred to in Sub-Clause (i) hereof will not be exclusive to the Contractor but only such as shall enable him to execute the Works. The Contractor shall afford to the Purchaser and to other contractors whose names shall have been previously communicated in writing to the Contractor by the Engineer every reasonable facility for the execution of work concurrently with his own.

(vi) Unless otherwise provided in the Specification, the Purchaser shall give the Contractor facilities for carrying out the Works on the Site continuously during the normal working hours generally recognized in the district. The Engineer may, after consulting with the Contractor, direct that work shall be done at other times if it shall be practicable in the circumstances for work to be so done, and a sum for work so done shall be added to the Contract Price unless such work has, by the default of the Contractor, become necessary for the completion of the Works within the time fixed by the Contract or, if no time be fixed within a reasonable time. Such sum shall be ascertained and determined in like manner to the valuation of variations under Clause 10 (*Variations*).

Vesting of Plant and Contractor's Equipment.

18.—(i) Plant supplied pursuant to the Contract shall become the property of the Purchaser at whichever is the earlier of the following times, namely—

(*a*) when the Plant is delivered pursuant to the Contract,

(*b*) when by virtue of Clause 15 (*Delivery*) or Clause 30 (*Suspension of Works*) the Contractor becomes entitled to require that the Contract Value of the Plant be included in an interim certificate.

(ii) All Contractor's Equipment owned by the Contractor or by any firm or corporation in which the Contractor has a controlling interest shall when brought upon the Site vest in and become the property of the Purchaser, and shall be used solely for the purpose of the Works and shall not be removed by the Contractor without the permission in writing of the Engineer, which permission shall not be unreasonably withheld in the case of Contractor's Equipment not currently required for the purpose of the Works. The Contractor shall be liable for the loss or destruction of such Contractor's Equipment or for damage thereto which may happen otherwise than through the fault of the Purchaser. If there shall be due, owing or accruing to the Purchaser from the Contractor any moneys under or in respect of the Contract, of which the Purchaser shall be unable to obtain payment, the Purchaser shall be at liberty at the cost of the Contractor to sell and dispose as he shall think fit of such Contractor's Equipment and to apply the proceeds in or towards the satisfaction of such moneys as aforesaid. Subject to the foregoing and to the right of the Purchaser under Clause 12 (*Contractor's Default*) and Clause 13 (*Bankruptcy*) the property in such Contractor's Equipment and in any Plant which is no longer required for completion of the Works shall revert to the Contractor on their proper removal from the Site or on the completion of the Works or on the termination of the Contract, whichever may be the earliest.

(iii) If the Purchaser shall become bankrupt or insolvent, or have a receiving order made against him, or compound with his creditors, or being a corporation commence to be wound up, not being a members' voluntary winding up for the purpose of reconstruction or amalgamation, or carry on its business under a receiver for the benefit of its creditors or any of them, then notwithstanding the provisions of Sub-Clause (ii) of this clause the property in Contractor's Equipment shall forthwith revert to the Contractor.

Engineer's Supervision

19.—(i) All instructions and orders to the Contractor shall, except as herein otherwise provided, be given by the Engineer.

(ii) The Contractor shall be responsible for ensuring that the positions, levels and dimensions of the Works are correct according to the drawings, notwithstanding that he may have been assisted by the Engineer in setting out the said positions, levels and dimensions.

(iii) All the Works shall be carried out under the direction and to the reasonable satisfaction of the Engineer.

Engineer's Representatives

(iv) The Engineer may from time to time delegate any of the powers, discretions, functions and authorities vested in him and may at any time revoke any such delegation. Any such delegation or revocation shall be in writing signed by the Engineer and, in the case of a delegation, shall specify the powers, discretion, functions and authorities thereby delegated and the person or persons to whom the same are delegated. No such delegation or revocation shall have effect until a copy thereof has been delivered to the Contractor.

Clerk of Works.

(v) If a Clerk of the Works be appointed to watch the carrying out of the Contract, the Contractor shall afford him every reasonable facility for so doing, but the Clerk of the Works shall not be authorized to relieve the Contractor in any way of his duties or obligations under the Contract. Any written notice from the Clerk of the Works condemning any Plant or workmanship shall have the effect of a similar notice given by the Engineer under Clause 25 (*Defects Prior to Taking Over*) except that the Contractor may appeal to the Engineer for his decision in the matter.

Engineer's Decisions.

20.—The Contractor shall proceed with the Works in accordance with decisions, instructions and orders given by the Engineer in accordance with these Conditions, provided always that—

(*a*) if the Contractor shall, without undue delay after being given any decision, instruction or order otherwise than in writing, require it to be confirmed in writing, such decision, instruction or order shall not be effective until written confirmation thereof has been received by the Contractor, and

(*b*) if the Contractor shall, by written notice to the Engineer within 14 days after receiving any decision, instruction or order of the Engineer in writing or written confirmation thereof, intimate that he disputes or questions the decision, instruction or order, giving his reasons for so doing, either party to the Contract shall be at liberty to refer the matter to arbitration pursuant to Clause 38 (*Arbitration*), but such an intimation shall not relieve the Contractor of his obligation to proceed with the Works in accordance with the decision, instruction or order in respect of which the intimation has been given. The Contractor shall be at liberty in any such arbitration to rely on reasons additional to the reasons stated in the said intimation.

Contractor's Representatives and Workmen

21. (i) The Contractor shall employ one or more competent representatives, whose name or names shall have been communicated previously in writing to the Engineer by the Contractor, to superintend the carrying out of the Works on the Site. The said representative, or if more than one shall be employed, then one of such representatives, shall be present on the Site during working hours, and any orders or instructions which the Engineer may give to the said representative of the Contractor shall be deemed to have been given to the Contractor.

(ii) The Engineer shall be at liberty by notice in writing to the Contractor to object to any representative or person employed by the Contractor in the execution of or otherwise about the Works who shall, in the opinion of the Engineer, misconduct himself or be incompetent or negligent, and the Contractor shall remove such person from the Works.

(iii) The Purchaser will neither be responsible for nor pay any Income Tax or other Taxation payable to properly constituted Authorities in the country in which the Site is located, for which the Contractor's representatives or workmen may become legally liable whilst employed in that country for the purposes of the Contract.

Liability for Accidents and Damage.

22.—(i) The Contractor shall properly cover up and protect until taken over under Clause 29 (*Taking Over*) any section or portion of the Works liable to suffer damage by exposure to the weather, and shall take every reasonable precaution to protect any section or portion of the Works not taken over against loss or damage from any cause.

(ii) In the case of loss or damage to the Works on the Site arising from or occasioned by causes for which the Contractor is not responsible under the Contract the same shall, if required by the Purchaser, be made good by the Contractor but at the cost of the Purchaser at a price to be agreed between the Contractor and the Purchaser or in default of agreement to be settled by arbitration and such cost shall be added to the Contract Price.

(iii) Subject to Sub-Clauses (vi) and (vii) of this clause and Clause 23 (*Limitations on Contractor's Liability*), all losses of and damage to any section or portion of the Works that shall not have been taken over under Clause 29 (*Taking Over*), which shall arise from or be occasioned by defective design (other than a design made furnished or specified by the Purchaser and for which the Contractor has disclaimed responsibility in writing within a reasonable time after receipt of the Purchasers' instructions), materials or workmanship or by any act of the Contractor or any Sub-contractor or by a failure of the Contractor to comply with any obligation imposed on him by Sub-Clause (i) of this clause, shall be made good by and at the sole cost of the Contractor and to the reasonable satisfaction of the Engineer.

(iv) The Contractor shall, subject to Sub-Clauses (vi) and (vii) of this clause and Clause 23 (*Limitations on Contractor's Liability*), indemnify the Purchaser in respect of all damage or injury occurring before all the Works shall have been taken over under Clause 29 (*Taking Over*) to any person or to any property (other than property forming part of the Works not yet taken over) and against all actions, suits, claims, demands, costs, charges and expenses arising in connection therewith which shall be occasioned by the negligence of, or breach of statutory duty by, the Contractor or any Sub-contractor, or by defective design (other than a design made, furnished or specified by the Purchaser and for which the Contractor has disclaimed responsibility in writing within a reasonable time after the receipt of the Purchaser's instructions), materials or workmanship, but not otherwise.

(v) If there shall occur any loss of or damage to any property or injury to any person while the Contractor is on the Site for the purpose of making good a defect in any section or portion of the Works pursuant to Clause 31 (*Defects after Taking Over*), or for the purpose of carrying out tests on completion as provided in Clause 29(iv) (*Interference with Tests*) during the period referred to in Clause 31(i), the Contractor shall be liable, subject to the provisions of Sub-Clauses (vi) and (vii) of this clause and Clause 23 (*Limitations on Contractor's Liability*), as follows–

(*a*) In respect of loss of or damage to the said section or portion the Contractor's liability shall be as defined in Clause 31.

(*b*) In respect of damage or injury to any other property or to any person and of any actions, claims, demands, costs, charges and expenses arising in connection therewith the Contractor shall be liable, subject to the provisions of Sub-Clauses (i) and (iii) of this clause, to the extent that such damage or injury was caused by the negligence or breach of statutory duty of the Contractor or a Sub-contractor while on the Site as aforesaid or by defective materials or workmanship used in making good the said defect but not otherwise.

The said section or portion of the Works shall be defined by reference to the taking over certificate issued in respect thereof pursuant to Clause 29 (*Taking Over*).

(vi) The Contractor shall not be liable to the Purchaser for–

(*a*) any damage or injury to the extent that it is caused by or arises from the acts or omissions of the Purchaser or of others (not being the Contractor's servants or Sub-contractors),

(*b*) any loss or damage in circumstances over which the Contractor has no control.

(vii) Except in respect of personal injury or damage to property conferring on a person other than the Purchaser a good cause of action against the Contractor, the liability of the Contractor for any one act or default shall not exceed the Contract Price or £100,000, whichever is the greater.

(viii) In the event of any claim being made against the Purchaser arising out of the matters referred to in and in respect of which the Contractor may be liable under this clause, the Contractor shall be promptly notified thereof, and may at his own expense conduct all negotiations for the settlement of the same and any litigation that may arise therefrom. The Purchaser shall not, unless and until the Contractor shall have failed to take over the conduct of the negotiations or litigation, make any admission which might be prejudicial thereto. The conduct by the Contractor of such negotiations or litigation shall be conditional upon the Contractor having first given to the Purchaser such reasonable security as shall from time to time be required by the Purchaser to cover the amount ascertained or agreed or estimated, as the case may be, of any compensation, damages, expenses and costs for which the Purchaser may become liable. The Purchaser shall, at the request of the Contractor, afford all available assistance for any such purpose, and shall be repaid all reasonable expenses incurred in so doing.

(ix) The Contractor's liability under this clause shall be in lieu of any other liability for injury loss or damage occurring before all the works are taken over and caused by his breach of contract, or by the negligence or breach of statutory duty of the Contractor or any Sub-contractor or by defective design (other than a design made, furnished or specified by the Purchaser and for which the Contractor has disclaimed responsibility in writing within a reasonable time after receipt of the Purchaser's instructions), materials or workmanship.

Limitations on Contractor's Liability

23.–Subject as provided in Clause 27 (*Delay in Completion*) for the deduction of liquidated damages for delay, the Contractor shall not be liable to the Purchaser, whether by way of indemnity or by reason of breach of contract or of negligence or of breach of statutory duty, for loss of use, whether complete or partial, of the Works, or of profit or of any contract that may be suffered by the Purchaser.

Insurance of Works.

24.–Unless the Purchaser shall have approved in writing other arrangements, the Contractor shall in the joint names of the Contractor and the Purchaser insure the Works and keep each portion thereof insured for the Contract Price thereof or each other sum as may be agreed between the parties against loss, damage or destruction by fire, explosion, lightning, earthquake, theft, flood, storm, tempest, perils of the sea, and aircraft and other aerial devices or articles dropped therefrom, against malicious damage and such other risks, if any, as are specified in the Appendix from the date of shipment or the date on which it becomes the property of the Purchaser, whichever is the earlier, until a taking-over certificate has been issued under Clause 29 (*Taking Over*) and shall from time to time, when so required by the Engineer, produce evidence of satisfactory insurance cover. All moneys received under any such policy shall be applied in or towards the replacement and repair of the Works lost, damaged or destroyed, but this provision shall not affect the Contractor's liabilities under the Contract.

Defects prior to Taking Over.

25.–If, in respect of any section or portion of the Works not yet taken over, the Engineer shall at any time–

(*a*) decide that any work done or Plant supplied or materials used by the Contractor or any Sub-contractor is or are defective or not in accordance with the Contract or that such section or portion of the Works is defective or does not fulfil the requirements of the Contract (all such matters being hereinafter in this clause called "defects"), and

(*b*) as soon as reasonably practicable give to the Contractor notice in writing of the said decision specifying particulars of the defects alleged and of where the same are alleged to exist or to have occurred, and

(*c*) so far as may be necessary place the Plant at the Contractor's disposal,

then the Contractor shall with all speed and, except as provided in Sub-Clause (vii) of Clause 15 (*Delivery*), at his own expense make good the defects so specified. In case the Contractor shall fail so to do the Purchaser may, provided he does so without undue delay, take at the cost of the Contractor, such steps as may in all the circumstances be reasonable to make good such defects. All Plant provided by the Purchaser to replace defective Plant shall comply with the Contract and shall be obtained at reasonable prices and where reasonably practicable under competitive conditions. The Contractor shall be entitled to remove and retain all Plant that the Purchaser may have replaced at the Contractor's cost.

Nothing contained in this clause shall affect any claim by the Purchaser under Clause 27 (*Delay in Completion*).

Extension of time for Completion

26.–If by reason of any act or omission on the part of the Purchaser or the Engineer or by reason of any industrial dispute or any cause beyond the reasonable control of the Contractor arising after the acceptance of the tender, the Contractor shall have been delayed or impeded in the completion of the Works, whether such delay or impediment occur before or after the time (if any) or extended time fixed for completion, then provided that the Contractor shall without delay have given to the Purchaser or the Engineer notice in writing of his claim for an extension of time, the Engineer shall on receipt of such notice grant the Contractor from time to time in writing either prospectively or retrospectively such extension of the time fixed by the Contract for the completion of the Works as may be reasonable. Any delay on the part of a Sub-contractor which prevents the Contractor from completing the Works within the time for completion shall entitle the Contractor to an extension thereof provided such delay was due to any cause for which the Contractor himself would have been entitled to an extension of time under this clause.

Delay in Completion.

27.–If the Contractor fail to complete the Works in accordance with the Contract (save as regards his obligations under Clause 31 (*Defects after Taking Over*) and such tests as are to be made in accordance with Clause 28 (*Tests on Completion*) within the time fixed by the Contract for the completion of the Works or any extension of such time, or if no time be fixed, within a reasonable time, there shall be deducted from the Contract Price as liquidated damages the percentage named in the Appendix of the Contract Value of such portion or portions only of the Works as cannot in consequence of the said failure be put to the use intended for each week between the time for completion of the Works as aforesaid and the actual date of completion, but the amount so deducted shall not in any case exceed the maximum percentage named in the Appendix of the Contract Value of such portion or portions of the Works, and such deduction shall be in full satisfaction of the Contractor's liability for the said failure.

Tests on Completion.

28.–(i)The Contractor shall give to the Engineer in writing 21 days' notice of the date after which he will be ready to make the Tests on Completion. Unless otherwise agreed, the tests shall take place within 10 days after the said date, on such day or days as the Engineer shall in writing notify the Contractor.

(ii) If the Engineer fail to appoint a time after having been asked to do so, or to attend at any time or place duly appointed for making the said tests, the Contractor shall be entitled to proceed in his absence, and the said tests shall be deemed to have been made in the presence of the Engineer.

(iii) If, in the opinion of the Engineer, the tests are being unduly delayed, he may by notice in writing call upon the Contractor to make such tests within 10 days from the receipt of the said notice and the Contractor shall make the said tests on such day within the said 10 days as the Contractor may fix and of which he shall give notice to the Engineer. If the Contractor fail to make such tests within the time aforesaid the Engineer may himself proceed to make the tests. All tests so made by the Engineer shall be at the risk and expense of the Contractor unless the Contractor shall establish that the tests were not being unduly delayed, in which case tests so made shall be at the risk and expense of the Purchaser.

(iv) The Purchaser, except where otherwise specified, shall provide free of charge, subject to the provisions of Sub-Clause (v) of this clause, such labour, materials, electricity, fuel, water, stores and apparatus, as may be requisite and as may be reasonably demanded to carry out such tests efficiently.

(v) If the Works or any portion thereof fail to pass the tests, tests of the Works or the said portion shall, if required by the Engineer or by the Contractor, be repeated within a reasonable time upon the same terms and conditions, save that all reasonable expenses to which the Purchaser may be put by the repetition of the tests shall be deducted from the Contract Price.

Taking Over.

29.–(i) As soon as the Works have been completed in accordance with the Contract (except in minor respects that do not affect their use for the purpose for which they are intended and save for the obligations of the Contractor under Clause 31 (*Defects after Taking Over*) and have passed the Tests on Completion, the Engineer shall issue a certificate (herein called a "taking-over certificate") in which he shall certify the date on which Works have been so completed and have passed the said tests and the Purchaser shall be deemed to have taken over the Works on the date so certified, but the issue of the taking-over certificate shall not operate as an admission that the Works have been completed in every respect. Save as provided in Sub-Clause (iii) of this clause the Purchaser shall not use the Works or any section or portion thereof until a taking-over certificate has been issued in respect thereof. If nevertheless the Purchaser does so use the Works or any section or portion thereof the Works or section or portion shall be deemed to have been taken over.

(ii) If the Works are divided into two or more sections, Sub-Clause (i) hereof shall apply to each section as it applies to the Works. If by agreement between the Purchaser and the Contractor any portion of the Works (other than a section or sections) shall be taken over before the remainder of the Works, the Engineer shall issue a taking-over certificate in respect of that portion.

(iii) If by reason of any default on the part of the Contractor a taking-over certificate has not been issued in respect of every portion of the Works within one month after the date fixed by the Contract for the completion of the Works, or, if no date be fixed, within a reasonable time, the Purchaser shall be at liberty to use the Works or any portion thereof in respect of which a taking-over certificate has not been issued if and so long as the Works or the portion so used as aforesaid shall be reasonably capable of being used, provided that the Contractor shall be afforded a reasonable opportunity of taking such steps as may be necessary to permit the issue of the taking-over certificate. The provisions of Sub-Clause (i) of Clause 22 (*Liability for Accidents and Damage*) shall not apply to any portion of the Works while being so used by the Purchaser.

Interference with Tests.

(iv) If by reason of any act or omission of the Purchaser, or the Engineer, or some other contractor employed by the Purchaser, the Contractor shall be prevented from carrying out the Tests on Completion as provided in Sub-Clause (i) of Clause 28 (*Tests on Completion*) then, unless in the meantime the Works shall have been proved not to be substantially in accordance with the Contract, the Purchaser shall be deemed to have taken over the Works, and the Engineer shall issue a taking-over certificate accordingly; nevertheless the Contractor shall make the said tests during the period referred to in Sub-Clause (i) of Clause 31 (*Defects after Taking Over*) as and when required by the Engineer by 14 days notice in writing, and Sub-Clauses (ii), (iii), (iv) and (v) of Clause 28 and Sub-Clause (vi) of Clause 14 (*Inspection, Testing and Rejection of Plant*) shall apply. Such allowances shall be made from the performances required to be attained in the said tests as may be reasonable having regard to any use of the Works by the Purchaser prior to the tests and if the Contractor incurs extra expense in making the said tests pursuant to this sub-clause the Purchaser shall pay to the Contractor a sum in respect thereof which shall be ascertained and determined in like manner to the valuation of variations under Clause 10 (*Variations*).

Suspension of Works.

30.–(i) If by reason of the suspension of the Works by the Purchaser or the Engineer (otherwise than in consequence of some default on the part of the Contractor) or by reason of the Contractor being prevented from or delayed in proceeding with the Works by the Purchaser, the Engineer, or some other contractor employed by the Purchaser, the Contractor shall incur additional expense, there shall be added to the Contract Price a sum in respect thereof, such sum to be ascertained and determined in like manner to the valuation of variations under Clause 10 (*Variations*). Provided that no claim shall be made under this clause unless the Contractor has, within a reasonable time after the event giving rise to the claim, given notice in writing to the Engineer of his intention to make such claim.

(ii) If work on the Plant or any portion thereof is suspended as aforesaid by the Purchaser or the Engineer before the Plant or such portion thereof is delivered to the Site and the suspension exceeds three months and the Contractor has suitably and sufficiently marked the Plant or such portion thereof as the Purchaser's property and insured it as provided by Clause 24 (*Insurance of Works*) (the provisions of which clause shall thereafter until actual delivery to the Site apply as if the Plant or such portion thereof were for the time being upon the Site) then without prejudice to the provisions of Sub-Clause (iv)(*b*) of Clause 15 (*Delivery*) the Contractor shall be entitled to have the Contract Value thereof as at the commencement of the suspension included in an interim certificate on the expiration of the said three months or (if later) at the time when, but for such suspension, the Plant or such portion thereof would have been delivered: provided that the Contract Value of any Plant that according to the decision of the Engineer, is defective or not in accordance with the Contract shall not be included in any such certificate.

Defects after Taking Over.

31.–(i) The Contractor shall be responsible for making good with all possible speed any defect in or damage to any portion of the Works which may appear or occur during a period of 12 months after that portion shall have been taken over and which arises either–

(*a*) from defective materials, workmanship or design (other than a design made, furnished or specified by the Purchaser and for which the Contractor has disclaimed responsibility in writing within a reasonable time after receipt of the Purchaser's instructions), or

(*b*) from any act or omission of the Contractor done or omitted during the said period.

(ii) If any such defect shall appear or damage occur the Engineer shall inform the Contractor thereof stating in writing the nature of the defect or damage. If the Contractor replaces or renews any portion of the Works, the provisions of this clause shall apply to the portion of the Works so replaced or renewed until the expiration of 12 months from the date of such replacement or renewal.

(iii) The period of 12 months mentioned in Sub-Clauses (i) and (ii) of this clause shall be extended by a period equal to the period during which the Works or that portion thereof in which a defect or damage to which this clause applies has appeared or occurred cannot be used by reason of that defect or damage.

(iv) If any such defect or damage be not remedied within a reasonable time, the Purchaser may proceed to do the work at the Contractor's risk and expense.

(v) If the replacements or renewals are of such a character as may affect the efficiency of the Works or any portion thereof, the Purchaser may within one month of such replacement or renewal give to the Contractor notice in writing requiring that Tests on Completion be made, in which case such tests shall be carried out as provided in Clause 28 (*Tests on Completion*).

(vi) These General Conditions shall apply to all inspections, adjustments, replacement, and renewals and to all tests occasioned thereby, carried out by the Contractor pursuant to this clause.

(vii) The Contractor's liability under this clause shall be in lieu of any condition or warranty implied by law as to the quality or fitness for any particular purpose of any portion of the Works taken over under Clause 29 (*Taking Over*) and save as in this clause expressed neither the Contractor nor his Sub-contractors, servants or agents shall be liable, whether in contract, tort or otherwise, in respect of defects in or damage to such portion, or for any injury, damage or loss of whatsoever kind attributable to such defects or damage. For the purposes of this sub-clause the Contractor contracts on his own behalf and on behalf of and as trustee for his Sub-contractors, servants and agents. Nothing in this clause shall affect the liability of the Contractor under Sub-Clauses (i) and (iii) of Clause 22 (*Liability for Accidents and Damage*) in respect of any portion of the Works not yet taken over.

(vii) Until the final certificate shall have been issued, the Contractor shall have the right of access, at all reasonable working hours, at his own risk and expense, by himself or his duly authorized representatives, whose names shall have previously been communicated in writing to the Engineer, to all parts of the Works for the purpose of inspecting the working thereof and to records of the working and performance thereof for the purpose of inspecting the same and taking notes therefrom. Subject to the Engineer's approval, which shall not be unreasonably withheld, the Contractor may at his own risk and expense make any tests which he considers desirable.

Interim and Final Certificates.

32.–(i) The Contractor may at the times and in the manner following apply for interim and final certificates, as referred to in Clause 35 (*Terms of Payment*), for Plant shipped and *en route* to the Site and for work executed on the Site.

Interim Certificates.

(ii) Applications for interim certificates may be made to the Engineer in respect of each shipment of Plant and from time to time as work on the Site progresses. Each such application in respect of shipment shall identify the Plant shipped, state the amount claimed, and be accompanied by such evidence of shipment and of payment of freight and insurance as the Engineer may reasonably require.

Each other such application shall state the amount claimed and shall set forth in detail in the order of the schedule of prices particulars of the work executed on the Site and of the Plant delivered to the Site pursuant to the Contract to a date named in the application and since the period covered by the last preceding certificate, if any, which includes work on Site.

(iii) The Engineer shall issue to the Contractor an interim certificate within 28 days after receiving an application therefor which the Contractor was entitled to make. If the Engineer shall fail to issue an interim certificate as provided in this clause or if the Purchaser shall interfere with or obstruct the issue of any such certificate the Contractor may, without prejudice to any other remedy, either

(*a*) after giving to the Purchaser or the Engineer 14 days' notice of his intention so to do, stop the Works or any part thereof until the said certificate be issued; in which case the expenses of the Contractor occasioned by such stoppage and the subsequent resumption of work shall be added to the Contract Price, or

(*b*) after giving to the Purchaser or the Engineer one month's notice of his intention so to do, terminate the Contract, whether or not the Contractor shall have stopped the Works in accordance with paragraph (*a*) hereof or have given notice of his intention so to do.

(iv) Every interim certificate shall certify the total value of Plant shipped or, as the case may be, the total value of the work duly executed on the Site and of the Plant delivered to the Site for use in the Works pursuant to the Contract up to the date named in the application for the certificate, less the total of any sums previously certified in interim certificates, provided that no sum shall be included in any interim certificate in respect of any Plant that, according to the decision of the Engineer, does not comply with the Contract or has been brought and is at the date of the certificate prematurely upon the Site.

(v) No interim certificate shall be relied on as conclusive evidence of any matter stated therein, or prejudice any right of the Purchaser or the Contractor against the other.

Final Certificate.

(vi) Application for the final certificate may be made to the Engineer at any time after the Contractor has ceased to be under any obligation under Clause 31 (*Defects after Taking Over*), provided that, if a taking-over certificate has been issued in respect of any section or portion of the Works, the Contractor may apply for a separate final certificate in respect of each such section or portion at any time after the said obligation has ceased in relation to such section or portion, and provided also that, if by reason of the fact that it has been necessary for the Contractor to replace or renew any portion of the Works the obligations of the Contractor under Clause 31 shall continue after the period of 12 months first therein mentioned, the right of the Contractor to apply for a final certificate in respect of the Works, section or portion thereof other than the portions so replaced or renewed shall not be affected by that fact, and after the Contractor has ceased to be under any obligation under Clause 31 in respect of the portions replaced or renewed he may apply for a final certificate in respect thereof.

(vii) The Engineer shall issue to the Contractor a final certificate within 28 days after receiving an application therefor which the Contractor was entitled to make.

(viii) A final certificate shall certify the total of all amounts comprised in interim certificates previously issued in respect of the Works or the portion thereof to which the final certificate relates, subject to such additions thereto or deductions therefrom as may be authorized in Sub-Clause (x) of this clause.

(ix) A final certificate shall, save in the case of fraud or dishonesty relating to or affecting any matter dealt with in the certificate, be conclusive evidence as to the sufficiency of the Works and of the value thereof unless any proceedings arising out of the Contract whether under Clause 38 (*Arbitration*) or otherwise shall have been commenced by either party before the final certificate has been issued or within one month thereafter.

Interim and Final Certificates.

(x) If any sum shall become payable to the Contractor under the Contract otherwise than for work executed or Plant delivered the amount thereof shall be included in the next certificate (interim or final) issued by the Engineer, and if any sum shall become payable under the Contract by the Contractor to the Purchaser, prior to the issue of the final certificate, whether by deduction from the Contract Price or otherwise, the amount thereof shall be deducted in the next certificate.

(xi) The Engineer may in any certificate give effect to any correction or modification that should properly be made in respect of any previous certificate.

Provisional Sums and P.C. Items.

33.–(i) A provisional sum included in the Contract Price shall be expended or used as the Engineer may in writing direct and not otherwise. In so far as a provisional sum is not expended or used it shall be deducted from the Contract Price.

(ii) All P.C. (prime cost) items included in the Contract Price shall be expended or used as the Engineer may in writing direct and not otherwise. To the net amount paid by the Contractor in respect of each P.C. item there shall be added the percentage named in the Appendix of the said amount. The sum by which the net amount so paid in respect of any P.C. item plus the said percentage thereon exceeds or is less than the sum included in the Contract Price in respect of that item shall be added to or deducted from the Contract Price as the case may be.

(iii) The Contractor shall have no responsibility for work done or Plant supplied by any other person in pursuance of directions given by the Engineer under this clause unless the Contractor shall have approved the person by whom such work is to be done or such Plant is to be supplied and the Plant, if any, to be supplied.

Payments due from the Contractor.

34.–Without prejudice to any other remedy which the Purchaser may have he shall be entitled to deduct from any moneys due, or becoming due to the Contractor under the Contract, all costs, damages or expenses for which under the Contract the Contractor is liable to the Purchaser.

Terms of Payment.

35.–(i) The Purchaser shall pay to the Contractor in the following manner the Contract Price adjusted to give effect to such additions thereto and such deductions therefrom as are provided for in these Conditions:

(*a*) 10% of the FOB portion (as specified in the Contract) of the Contract Price as an advance payment within 14 days after the Contractor has furnished to the Purchaser an irrevocable letter of guarantee, from a guarantor or surety acceptable to the Purchaser, with a value equivalent to the advance payment. The letter of guarantee shall provide for its value to reduce by an amount equal to 10% of the sum certified in each Interim Certificate issued in respect of Plant shipped and be in a form acceptable to the Purchaser.

(*b*) within 14 days after presentation to the Purchaser of each Interim Certificate a sum equal to

(i) 85% of the sum certified therein in respect of Plant shipped and

(ii) 95% of the sum certified therein in respect of work done on site and of freight and insurance paid in respect of Plant shipped.

(*c*) 97½% of the Contract Price adjusted as aforesaid within 14 days after presentation to the Purchaser of the Taking-Over Certificate.

(*d*) The balance of the Contract Price adjusted as aforesaid within 14 days after presentation to the Purchaser of the Final Certificate. Provided that if the Contractor shall have furnished to the Purchaser a guarantee acceptable to the Purchaser for the payment on demand of such balance, he shall be entitled to payment thereof with or at any time after the payment provided for by paragraph (*c*) hereof.

If any section or portion of the Works shall be taken over separately under Clause 29 (*Taking Over*) the payments herein provided for on or after taking over shall be made in respect of the section or portion taken over and reference to the Contract Price shall mean such part of the Contract Price as shall, in the absence of agreement, be apportioned thereto by the Engineer.

In determining the amount of any payment under this clause in respect of any section or portion of the Works due account shall be taken of all payments previously made in respect of the same section or portion whether under this clause or under Clause 15 (*Delivery*).

(ii) If at any time at which any payment would fall to be made under paragraph (*c*) of Sub-Clause (i) of this clause there shall be any defect in any portion of the Works in respect of which such payment is proposed, the Purchaser may retain the whole of such payment provided that in the event of the said defect being of a minor character and not such as to affect the use of the Works or the said portion thereof for the purpose intended the Purchaser shall not retain a greater sum than represents the cost of making good the said minor defect. Any sum retained by the Purchaser pursuant to the provisions of this sub-clause shall be paid to the Contractor upon the said defect being made good.

(iii) If the payment of any sum payable under Sub-Clause (i) of this clause shall be improperly delayed by the Purchaser or the Engineer interest at the rate of two percent per annum over the Bank of England minimum lending rate from time to time in force on the amount of the delayed payment for the period of the delay shall be added to the Contract Price.

(iv) If the Purchaser shall fail to make any payment as provided in this clause the Contractor shall have the like remedies, without prejudice to any other, as are provided in Clause 32 (iii) (*Interim and Final Certificates*).

Statutory and other Regulations.

36.–If the cost to the Contractor of performing his obligations under the Contract shall be increased or reduced by reason of the making after the date of his tender in the United Kingdom or elsewhere of any law or of any order, regulation or by-law having the force of law in the United Kingdom or elsewhere that shall affect the Contractor in the performance of his obligations under the Contract, the amount of such increase or reduction (to the extent that it arises directly in respect of the Works) shall be added to or deducted from the Contract Price as the case may be.

Metrication

37.–(i) If any Plant described in the Contract or the subject of a variation (hereinafter called “a variation order”) under Clause 10 (*Variations*) is described by dimensions in the metric or imperial measure and the Contractor cannot procure such Plant in the measure specified in sufficient time to avoid delay in the performance of his other obligations under the Contract, but can obtain such Plant in the other measure to dimensions approximating to those described in the Contract or in the variation order, then the Contractor shall forthwith give notice to the Engineer of the facts stating the dimensions to which such Plant is procurable in the other measure.

(ii) The Engineer shall within 14 days after the receipt of a notice under the preceding sub-clause give notice to the Contractor in writing pursuant to Clause 10 (*Variations*). Such notice shall either:-

(*a*) direct the Contractor to supply such Plant to the dimensions stated in the said notice instead of to the dimensions described in the Contract or variation order as the case may be, or,

(*b*) direct the Contractor to make some other variation whereby the need to supply such plant to the dimensions described in the Contract or variation order will be avoided.

The provisions of the Contract shall apply to directions given under this clause as though such directions had been included in the Specification.

Arbitration

38.–(i) If at any time any question, dispute, or difference shall arise between the Purchaser and the Contractor, either party shall, as soon as reasonably practicable, give to the other notice in writing of the existence of such question, dispute or difference specifying its nature and the point at issue, and the same shall be referred to the arbitration of a person to be agreed upon, or failing such agreement within six weeks, to some person appointed on the application of either of the parties hereto by the President for the time being of the Institution named in the Appendix, provided that a question, dispute or difference relating to a decision, instruction, or order of the Engineer shall not be referred to arbitration unless notice have been given by the Contractor in accordance with Clause 20 (*b*) (*Engineer's Decisions*).

(ii) Performance of the Contract shall continue during arbitration proceedings unless the Engineer or the Purchaser shall order the suspension thereof or of any part thereof, and if any such suspension shall be ordered the reasonable expenses of the Contractor occasioned by such suspension shall be added to the Contract Price. No payments due or payable by the Purchaser shall be withheld on account of pending reference to arbitration.

Construction of Contract.

39.–Unless otherwise agreed the Contract shall in all respects be construed and operate as an English contract and in conformity with English Law and the English courts shall have exclusive jurisdiction over any matter arising out of or in connection with the provisions for arbitration under Clause 38 (*Arbitration*). The marginal notes hereto shall not affect the construction hereof.

Variation in Costs.

40.*–If the cost to the Contractor of performing his obligations under the Contract shall be increased or reduced by reason of any rise or fall in the rates of wages payable to labour or in the cost of material or transport above or below such rates and costs ruling at the date of tender, the amount of such increase or reduction shall be added to or deducted from the Contract Price as the case may be, provided that no account shall be taken of any amount by which any cost incurred by the Contractor has been increased by the default or negligence of the Contractor. For the purpose of this clause 'the cost of material' shall be construed as including any duty or tax by whomsoever payable which is payable under or by virtue of any Act of Parliament on the import, purchase, sale, appropriation, processing or use of such material.

***NOTE:– Unless this Clause is excluded any quoted price will be subject to adjustment in accordance with its terms.**

APPENDIX

Clause	
24	Additional risks to be covered by insurance.
27	Delay in Completion:
(*a*)	(*a*) percentage of Contract Value to be deducted as damages.
(*b*)	(*b*) maximum percentage of Contract Value which the deductions may not exceed.
33	Percentage on Prime Cost items
38	The Institution of Mechanical Engineers.* The Institution of Electrical Engineers.*

*Delete one line

FORM OF GUARANTEE

(when required)

Whereas by an Agreement dated .... and

made between .... (hereinafter called

"the Purchaser") and ....

.... (hereinafter called "the Contractor")

the parties thereto entered into a contract as therein stated:

NOW we hereby jointly and severally guarantee to the Purchaser punctual true and faithful performance and observance by the Contractor of the covenant on his part contained in the said Agreement and undertake to be responsible to the Purchaser his legal personal representatives successors or assigns as Sureties for the Contractor for the payment by him of all sums of money losses damages costs charges and expenses that may become due or payable to the Purchaser his legal personal representatives successors or assigns by or from the Contractor by reason or in consequence of the default of the Contractor in the performance or observance of his said covenant but so nevertheless that the total amount to be demanded or recovered by the Purchaser his legal personal representatives successors or assigns of or from us as Sureties shall not exceed 15 per cent of the Contract Price.

This Guarantee shall not be revocable by notice or by reason of the death of us or either of us and our liability as Sureties hereunder shall not be impaired or discharged by any extensions of time or variations or alterations made given conceded or agreed (with or without our knowledge or consent) under the General Conditions referred to in the said Agreement or (where the Purchaser or the Contractor is a firm) by any change in the constitution of the Purchaser's or the Contractor's respective firms.

FORM OF AGREEMENT

This Agreement is made the day of 19

BETWEEN

(hereinafter referred to as the "Contractor") of the one part and

(hereinafter called the "Purchaser") of the other part Whereas the Purchaser desires to have provided and executed certain Works mentioned, enumerated, or referred to in certain General Conditions, Specifications, Schedules, Drawings, Plans, Schedule of Prices, the Contractor's Tender and covering letter (if any), the acceptance of the Tender, and any other relevant correspondence* which for the purpose of identification have been signed by

on behalf of the Contractor

and by

of (the Engineer of the Purchaser) on behalf of the Purchaser all of which are deemed to form part of this Contract as though separately set out herein and are included in the expression "Contract" whenever herein used. And Whereas the Purchaser has accepted the Tender of the Contractor for the provision and execution of the said Works for the sum of (hereinafter called "the Contract Price") upon the terms and subject to the Conditions hereinafter mentioned: Now this Deed Witnesseth and it is hereby agreed and declared as follows, that is to say, in consideration of the payments to be made to the Contractor by the Purchaser as hereinafter mentioned the Contractor hereby covenants with the Purchaser, his legal personal representatives, successors, and assigns that the Contractor shall and will duly provide, execute, complete and maintain the said Works and shall do and perform all other acts and things in the Contract mentioned or described or which are to be implied therefrom or may be reasonably necessary for the completion of the said Works within and at the times and in the manner and subject to the terms conditions and stipulations mentioned in the Contract.

And in consideration of the due provision, execution and completion of the Works and the maintenance thereof as aforesaid the Purchaser does hereby for himself his legal personal representatives, successors, and assigns covenant with the Contractor that he, the Purchaser, his legal personal representatives, successors, or assigns, will pay to the Contractor the Contract Price or such other sum as may become payable to the Contractor under the provisions of the Contract, such payments to be made at such time and in such manner as is provided by the Contract.

In Witness whereof, etc.

*Delete as required

INDEX

I.MECH.E./I.E.E. MODEL FORM OF GENERAL CONDITIONS 'B/3 1980' i

MEMORANDUM ON THE USE OF THE GENERAL CONDITIONS OF CONTRACT

A. Applicability

The General Conditions are intended to be suitable for the mechanical engineering and the electrical engineering industries and to cover the main conditions for export contracts for the supply of plant and materials and the erection thereof on Site.

It is hoped that, in order to secure the advantage of uniformity and to avoid trouble and expense, they may be found suitable in their entirety for most contracts. It is, however, inevitable that additions and/or amendments will have to be incorporated to meet the requirements of particular contracts or of the governmental, financial or other authorities of the countries from which plant is to be exported or of the countries into which it is to be imported. It is suggested, in order to facilitate checking, that when reprinted elsewhere any additions or alterations be made so as to attract attention, e.g. by the use of different type or underlining.

B. Contract documents

Contracts are commonly made in one of three ways:

(*a*) By a formal agreement, such as the model form supplied with these Conditions, incorporating in the contract the General Conditions, Specification, Schedules, Drawings, etc. In this case all the terms of the contract are contained in the Agreement, the Conditions, Specification, and other incorporated documents on which Tenders were invited, and agreed variations in or additions to the provisions of these documents are represented by suitable amendments.

(*b*) By a formal agreement, as in the method (*a*), but incorporating also the Contractor's Tender and covering letter, if any, the Purchaser's acceptance of the Tender, and all correspondence containing terms accepted by the Purchaser and the Contractor. When this method is adopted any modifications of the General Conditions, Specifications, etc., have to be sought in the covering letter and other correspondence.

(*c*) Without a formal agreement, the Tender and its acceptance, together with other relevant correspondence, being relied on as constituting the Contract between the parties.

When either method (*a*) or (*b*) is adopted, a copy of the form of Agreement to be used should be included with the documents supplied to or open to the inspection of intending Tenderers.

When method (*a*) is used, the references in the model form of Agreement to the Tender, covering letter, acceptance of the Tender, and other correspondence should be deleted.

C. Progress payments during manufacture

The General Conditions provide for an advance payment against a guarantee, but the Contractor does not become entitled to any further payments for Plant he has manufactured until he has shipped it. In the case of contracts involving considerable expenditure on manufacture before any shipment can be made, the General Conditions may be modified to provide for progress payments to the Contractor in respect of Plant in the course of manufacture, in which case the insurance arrangements may also need to be modified.

D. Customs and import duties

Under Clause 16 of the General Conditions, the Purchaser is responsible for paying all customs and import duties and the Contract Price should not therefore include any sum to cover such duties. Should it be desired to adapt the General Conditions to contracts under which the Contractor is to be responsible for paying such duties, Clause 16 will require modification.

E. Variation in costs

If the parties so wish they may operate Clause 40 by means of an agreed formula.

ii I.MECH.E./I.E.E. MODEL FORM OF GENERAL CONDITIONS 'B3/1980'

MEMORANDUM AS TO TENDERS

Tenderer may propose modifications.

The Tenderer is at liberty to add any details and conditions that he may deem desirable, and in the event of his doing so must print or type the same and annex the added matter to the Specification or General Conditions returned by him, but such additional details and conditions will not be binding on the Purchaser unless they are approved by him and incorporated in the Contract.

Where in special circumstances the Tenderer deems it advisable to provide for the nondisclosure of drawings and information of a confidential nature furnished with his Tender he should include in the Tender or covering letter a stipulation to this effect. If the Tender embodying such a stipulation is accepted the stipulation should be incorporated in any subsequent formal agreement.

If the Tenderer has any doubt as to the meaning of any portion of the General Conditions or of the Specification, he should, when submitting his Tender, set out in his covering letter the interpretation upon which he relies.

Tenders.

One copy of the Specification supplied, together with the General Conditions and Drawings, enclosed in a plain sealed cover bearing the words "Tender for (*here follows the short title of the Works*)"

and not bearing any name or mark indicating the sender shall, subject to modifications mentioned above, be returned intact with the Tender Form and Schedules, if any, filled up and signed, addressed to

and must be received by him on or before

The Purchaser does not bind himself to accept the lowest or any Tender, nor will he be responsible for, or pay for, expenses or losses which may be incurred by any Tenderer in the preparation of this Tender.

Copies of Specification etc.

A duplicate copy of the Specification, General Conditions and Drawings will be supplied to intending Tenderers without charge, and extra copies of the Drawings at the cost of reproduction.

The sum deposited by the Tenderer on application for the Specification will be refunded to him within one month of the date of the adjudication upon the Tenders, if all copies of the Specification, Drawings, etc., have been returned by the Tenderer, unless the Tender was not made in good faith, in which case the deposit may be forfeited.

Agreement.

The tenderer whose Tender is accepted may be required to enter into a formal agreement with the Purchaser.

NOTE

The advertisement inviting Tenders should state whether and if so where the Agreement, General Conditions, Specification, and Drawings may be inspected, and should give such particulars of the class of plant and apparatus required under each Section as will enable contracting firms to decide, without obtaining the Specification, whether they desire to tender, and should also state the amount of the deposit to be paid for the General Conditions, Specification, and Drawings.

C

MODEL FORM OF

GENERAL CONDITIONS OF CONTRACT

RECOMMENDED BY

THE INSTITUTION OF MECHANICAL ENGINEERS

AND

THE INSTITUTION OF ELECTRICAL ENGINEERS

FOR THE SUPPLY OF

ELECTRICAL AND MECHANICAL GOODS, OTHER THAN ELECTRIC CABLES

(HOME – WITHOUT ERECTION)

1975 EDITION

Published for the Joint Committee on Model Forms of General Conditions of Contract by
THE INSTITUTION OF ELECTRICAL ENGINEERS
and obtainable from
Publication Sales Department, Station House, 70 Nightingale Road, Hitchin, Hertfordshire SG5 1RJ
or to callers, at the Reception Desk, Savoy Place, or at
THE INSTITUTION OF MECHANICAL ENGINEERS
1 Birdcage Walk, London SW1H 9JJ

FOREWORD

These General Conditions are intended to cover the main conditions for Home Contracts for the supply of electrical and mechanical goods, other than electric cables, without erection, and it is hoped that in order to secure the advantage of uniformity and to avoid trouble and expense they may be found suitable in their entirety for the majority of such Contracts. They may, however, be added to or amended as may be required to suit each particular case. In order to facilitate checking, it is recommended that if the Conditions are reproduced elsewhere with additions or alterations, these should be so made as to attract attention.

Attention is directed to the blank which requires to be completed in Clause 15.

HISTORICAL NOTE

A Model Form of General Conditions 'C' – Home (without erection) was first published by The Institution of Electrical Engineers in 1924. A revision was issued in 1940.

Following consultations, and by agreement between the Council of the Institution of Mechanical Engineers and the Council of the Institution of Electrical Engineers, the scope of the Model Form was enlarged to make it suitable for both the electrical and mechanical engineering industries, and on this basis the Model Form was first issued jointly by the two Councils as:

1956 Edition

Model Form of General Conditions of Contract 'C' for the sale of Electrical and Mechanical Goods (other than Electric Cables) (Home – without erection).

This 1975 Edition is issued to bring these General Conditions up to date.

This form may be cited as 'Model Form C/1975, Home – without erection'.

I.Mech.E./I.E.E. MODEL FORM OF GENERAL CONDITIONS (C/1975)

GENERAL CONDITIONS

1. These General Conditions shall have effect subject to any express stipulation or condition at variance with these Conditions that may be contained in the Specification or may otherwise be incorporated in the Contract.

Information

2. The Purchaser shall within a reasonable time furnish all such further information, beyond that which is contained in the Specification or has been otherwise given to the Vendor, as the Vendor may reasonably call for to execute the Contract. The Purchaser shall pay all reasonable extra costs caused to the Vendor by unreasonable delay in so doing or by the supply of inaccurate information.

Drawings

3. (i) Drawings, illustrations, descriptions, price lists, and catalogues issued by the Vendor shall not form part of the Contract unless incorporated therein by reference or otherwise.

(ii) The Vendor shall within a reasonable time supply to the Purchaser:

(*a*) the particulars and drawings (if any) called for in the Contract, and

(*b*) such drawings (other than shop drawings) and other particulars of the goods as may be reasonably necessary for the purposes of installation and maintenance (including such dismantling and re-assembling as maintenance may involve).

Tests

4. (i) Before delivering any goods the Vendor shall inspect and test the same for compliance with the Contract and, if so requested, shall supply to the Purchaser a certificate of the results of the test.

(ii) Where the Contract provides that the goods shall pass any prescribed tests or shall give a specified performance they shall be tested by the Vendor before delivery for compliance with the prescribed tests or for performance or for both as the case may be, the Vendor providing free of charge what may be requisite for the purpose. The Vendor shall give the Purchaser seven days' notice in writing of the date on and the place at which any of the goods will be ready for testing as provided in this sub-clause. If the Purchaser shall fail to give the Vendor 24 hours' notice appointing a day within seven days after the date which the Vendor has stated in his notice, or shall fail to attend on the day he has appointed, the Vendor may proceed with the tests, which shall be deemed to have been made in the Purchaser's presence. The Vendor shall forthwith forward to the Purchaser a certificate of the results of the tests.

(iii) If on a test made pursuant to Sub-Clause (ii) of this clause the goods or any part thereof fail to pass the prescribed tests or to give the specified performance such goods or part thereof shall, if the Vendor so desires, be tested again or the Vendor may submit for test other goods in their place. If the goods or the said other goods shall fail to pass the test or to give the specified performance, the Purchaser shall be entitled by notice in writing to reject the goods or such part thereof as shall have failed as aforesaid.

Rejection and Replacement

5. (i) The Purchaser shall be entitled, by notice in writing given within a reasonable time after delivery, to reject goods delivered which are not in accordance with the Contract.

(ii) When goods have been rejected, either under Clause 4 (*Tests*) or Sub-Clause (i) of this clause, the Purchaser shall be entitled, provided he does so without undue delay, to replace the goods so rejected. There shall be deducted from the Contract Price that part thereof which is properly apportionable to the goods rejected. The Vendor shall pay to the Purchaser any sum by which the expenditure reasonably incurred by the Purchaser in replacing the rejected goods exceeds the sum deducted. All goods obtained by the Purchaser to replace rejected goods shall comply with the Contract and shall be obtained at reasonable prices and, when reasonably practicable, under competitive conditions. Where goods have been rejected as aforesaid the Vendor shall not be under any liability to the Purchaser except as provided in this clause and as may arise under Clause 7 (*Time for Delivery*).

Place of Delivery

6. The Vendor shall deliver the goods at the place (if any) named in the Contract or, if none be named, at the Vendor's works. Where delivery is to be made otherwise than at the Vendor's works the Vendor shall convey the goods to the point nearest to the place of delivery to which there is suitable access and the Purchaser shall be responsible for offloading the goods. Where delivery is to be made at the Vendor's works the Vendor shall, if required, load the goods on the Purchaser's vehicle. The property in the goods shall pass to the Purchaser on delivery.

Time for Delivery

7. (i) Any time fixed by the Contract for delivery shall run from the acceptance of the Vendor's tender (or, if there be no tender, of the Purchaser's order) or from the date on which the Vendor is placed by the Purchaser in possession of such information and drawings as may be necessary to enable him to put the work in hand, whichever may be the later. Any time described as an estimate shall not be construed as a time fixed by the Contract.

(ii) If for any cause beyond the reasonable control of the Vendor or by reason of any industrial dispute delivery of the goods shall be delayed the above-mentioned time for delivery shall be extended by such period as may be reasonable.

(iii) If the Purchaser shall have suffered any loss by the failure of the Vendor to deliver goods in accordance with the Contract within the time fixed thereby or any extension thereof or, if no time be fixed, within a reasonable time, the Purchaser shall be entitled to recover liquidated damages from the Vendor. Such damages shall be a sum equal to the percentage specified in the Contract of that part of the Contract Price which is properly apportionable to such portion of the goods as cannot in consequence of such failure be put to the use intended for each week until the Vendor has delivered goods in accordance with the Contract, or goods in replacement have been provided by the Purchaser pursuant to Sub-Clause (ii) of Clause 5 (*Rejection and Replacement*). Provided always that the amount so recoverable shall not exceed the maximum percentage specified in the Contract. In default of specification in the Contract the percentages above mentioned shall be one-half per cent and five per cent respectively.

(iv) When the sum recoverable by the Purchaser as liquidated damages has amounted to the maximum above provided the Purchaser shall be entitled by notice in writing to the Vendor to require him to deliver the goods within such time (not being less than 28 days) as the Purchaser may specify in the notice. If the Vendor shall fail to deliver the goods within the time so specified the Purchaser shall, without prejudice to his rights under Sub-Clause (iii) of this clause, be entitled, after having informed the Vendor in writing of his intention so to do, to obtain goods in place of those which the Vendor has failed to deliver and there shall be deducted from the Contract Price that part thereof which is properly apportionable to the undelivered goods. The Vendor shall pay to the Purchaser any sum by which the expenditure reasonably incurred by the Purchaser in obtaining goods in place of undelivered goods exceeds the sum deducted. All goods obtained by the Purchaser in place of undelivered goods shall comply with the Contract and shall be obtained at reasonable prices and when practicable under competitive conditions.

(v) The Purchaser's remedies under Sub-Clauses (iii) and (iv) of this clause shall be in lieu of any other remedy in respect of the Vendor's failure to deliver goods in accordance with the Contract within the time fixed thereby or any extension thereof or, if no time be fixed, within a reasonable time.

Storage

8. If by reason of instructions or lack of instructions from the Purchaser the delivery of goods in accordance with the Contract is delayed for 14 days after the Vendor has given notice in writing to the Purchaser that the said goods are ready for delivery, then provided they are suitably and sufficiently marked as appropriated to the Contract and the Vendor has given to the Purchaser an opportunity of inspecting them, the said goods shall be deemed to have been delivered in accordance with the Contract and subject as provided in Clause 9 (*Damage or Loss in Transit*) the goods shall at the expiration of the said 14 days become the property of the Purchaser and shall be at the risk of the Purchaser. Nevertheless the Vendor shall use his best endeavours to deliver the said goods in accordance with such instructions as may be given to him by the Purchaser and in the meantime shall store, protect, preserve and, if required by the Purchaser, insure them to the extent so required. The Purchaser shall repay to the Vendor the reasonable cost of so storing, protecting, preserving and insuring the said goods.

Damage or Loss in Transit

9. (i) The Vendor shall repair or replace free of charge goods damaged in transit to the place of delivery, and in the event of such damage delivery shall not be deemed to have taken place until repaired or replacement goods have been delivered. Provided always that, if the Vendor has given to the Purchaser notice of the date of despatch and has with that notice required the Purchaser to give him within a stated period notice of any damage suffered and the Purchaser has failed to do so, the Vendor shall not be liable to repair or replace the damaged goods, and delivery of the damaged goods shall be deemed to be delivery for the purpose of the Contract.

For the purposes of this sub-clause:

(*a*) where the goods are delivered by a carrier employed by the Vendor under a contract of carriage which frees the carrier from liability for damage in transit unless notice of damage is given to the carrier within a specified time, the stated period shall be such period (not being less than 24 hours) after the receipt of the goods as will allow the Vendor at least 24 hours after receiving notice from the Purchaser within which to give notice to the carrier in compliance with the terms of the carrier's said contract with the Vendor, and

(b) where the goods are delivered by a carrier employed by the Vendor but not under such a contract as aforesaid or are delivered by the Vendor's own transport, the stated period shall be such period as is reasonable.

(ii) The Vendor shall replace goods lost in transit to the place of delivery provided always that, if the Vendor has given the Purchaser notice of the date of despatch and method of transport and has with that notice required the Purchaser to give him notice of non-delivery within a stated period and the Purchaser has failed to do so, the goods shall notwithstanding their non-receipt within that period be deemed to have been delivered at the expiry of that period. In such last mentioned event the Vendor shall at the request and expense of the Purchaser pursue for the benefit of the Purchaser such rights (if any) as the Vendor may have against the carrier.

(iii) The liability imposed on the Vendor in this clause shall be accepted by the Purchaser in substitution for all or any other liability on the part of the Vendor arising from the delivery of goods damaged in transit or the non-delivery of goods in consequence of loss in transit.

Terms of Payment

10. Unless otherwise agreed, payment for the goods shall become due on completion of delivery in accordance with the Contract.

Defects after Delivery

11. (i) If within 12 months after delivery there shall appear in the goods any defect which shall arise under proper use from faulty materials, workmanship or design (other than a design made, furnished, or specified by the Purchaser for which the Vendor has in writing disclaimed responsibility), and the Purchaser shall give notice thereof in writing to the Vendor, the Vendor shall, provided that the defective goods or defective parts thereof have been returned to the Vendor if he shall have so required, make good the defects either by repair or, at the option of the Vendor, by the supply of a replacement. The Vendor shall refund the cost of carriage on the return of the defective goods or parts and shall deliver any repaired or replacement goods or parts as if Clause 6 (*Place of Delivery*) applied.

(ii) The Vendor's liability under this clause or under Clause 5 (*Rejection and Replacement*) shall be accepted by the Purchaser in lieu of any liability implied by law as to the quality or fitness for any particular purpose of the goods and save as provided in this clause the Vendor shall not be under any liability to the Purchaser (whether in contract, tort or otherwise) for any defects in the goods or for any damage, injury or loss resulting from such defects or from any work done in connection therewith.

Patents and Design

12. (i) The Vendor shall fully indemnify the Purchaser against all actions, claims, demands, costs, charges and expenses arising from or incurred by reason of any infringement or alleged infringement of letters patent, registered design, copyright, trade mark or trade name protected in the United Kingdom by the use of any goods supplied by the Vendor, but such indemnity shall not cover any use of the goods otherwise than for the purpose indicated by or reasonably to be inferred from the Specification or to any infringement which is due to the use of any goods in association or combination with any other goods not supplied by the Vendor.

(ii) In the event of any claim being made or action brought against the Purchaser arising out of the matters referred to in this clause, the Vendor shall be promptly notified thereof and may at his own expense conduct all negotiations for the settlement of the same, and any litigation that may arise therefrom. The Purchaser shall not, unless and until the Vendor shall have failed to take over the conduct of the negotiations or litigation, make any admission which might be prejudicial thereto. The conduct by the Vendor of such negotiations or litigation shall be conditional upon the Vendor having first given to the Purchaser such reasonable security as shall from time to time be required by the Purchaser to cover the amount ascertained or agreed or estimated, as the case may be, of any compensation, damages, expenses, and costs for which the Purchaser may become liable. The Purchaser shall, at the request of the Vendor, afford all available assistance for the purpose of contesting any such claim or action, and shall be repaid all reasonable expenses incurred in so doing.

(iii) The Purchaser on his part warrants that any design or instructions furnished or given by him shall not be such as will cause the Vendor to infringe any letters patent, registered design, trade mark, or copyright in the performance of the Contract.

Metrication

13. (i) If any goods are described in the Contract by dimensions in the metric or imperial measure and the Vendor cannot procure such goods in the measure specified in sufficient time to avoid delay in the performance of his other obligations under the Contract, but can obtain such goods in the other measure to dimensions approximating to those described in the Contract, then the Vendor shall forthwith give notice in writing to the Purchaser of the facts stating the dimensions to which such goods are procurable in the other measure.

(ii) The Purchaser shall within 14 days after the receipt of a notice under the preceding sub-clause, give notice to the Vendor in writing which shall either:

(*a*) direct the Vendor to supply such goods to the dimensions stated in the said notice instead of to the dimensions described in the Contract, or

(*b*) direct the Vendor to make some other variation whereby the need to supply such goods to the dimensions described in the Contract will be avoided, or

(*c*) direct the Vendor to supply such goods to the dimensions stated in the Contract.

In the event that the Purchaser shall give notice under paragraphs (*b*) or (*c*) of this sub-clause the Purchaser shall grant such extension of time for delivery as may be reasonable, and the Contract Price shall be adjusted by such amount as may be reasonable in the circumstances.

The provisions of the Contract shall apply to directions given under this clause as though such directions had been included in the Specification.

Statutory and other Regulations

14. (i) If the cost to the Vendor of performing his obligations under the Contract shall be increased or reduced by reason of the making or amendment after the date of the Vendor's tender of any law or any order, regulation, or by-law having the force of law in the United Kingdom that shall affect the Vendor in the performance of his obligations under the Contract, the amount of such increase or reduction (to the extent that it arises directly on or in respect of the goods) shall be added to or deducted from the Contract Price as the case may be.

Value Added Tax

(ii) Unless otherwise stated in the Tender the Contract Price is deemed to exclude Value Added Tax. To the extent that the Tax is properly chargeable on the supply to the Purchaser of any goods or services provided by the Vendor under the Contract, the Purchaser shall pay such Tax as an addition to payments otherwise due to the Vendor under the Contract.

Arbitration

15. If at any time any question, dispute, or difference whatsoever shall arise between the Vendor and the Purchaser upon, in relation to, or in connection with the Contract, either of them shall give to the other notice in writing of the existence of such question, dispute, or difference, and the same shall be referred to the arbitration of a person to be agreed upon or failing agreement within 14 days after the date of such notice, or some person to be appointed, on the application of either party, by the President for the time being of the Institution of* Engineers.

Marginal Notes

16. The marginal notes hereto shall not affect the construction hereof.

**Insert Mechanical or Electrical as preferred.*

Supplementary Clause

Variations in Costs

If, by reason of any rise or fall in the rates of wages payable to labour or in the cost of material or transport above or below such rates and costs ruling at the date of the Vendor's tender, the cost to the Vendor of performing his obligations under the Contract shall be increased or reduced, the amount of such increase or reduction shall be added to or deducted from the Contract Price as the case may be, provided that no account shall be taken of any amount by which any cost incurred by the Vendor has been increased by the default or negligence of the Vendor. For the purposes of this clause 'the cost of material' shall be construed as including any duty or tax by whomsoever payable which is payable under or by virtue of any Act of Parliament on the import, purchase, sale, appropriation, processing or use of such material.

SEPTEMBER 1978 AMENDMENTS TO IMECHE/IEE MODEL FORMS OF GENERAL CONDITIONS OF CONTRACT A/1976, B3/1971, C/1975 and E/1973

Amendments in connection with the Unfair Contract Terms Act 1977.

MODEL FORM A 1976

Amend the last sentence of Sub-Clause (vii) of Clause 30 to read:

"Nothing in this clause shall affect either the liability of the Contractor under Sub-Clauses (i) and (iii) of Clause 21 *(Liability for accidents and Damage)* in respect of any portion of the Works not yet taken over or his liability for death or personal injury caused by negligence on his part as defined in Section 1 of the Unfair Contract Terms Act 1977".

MODEL FORM C 1975

Amend Sub-Clause (ii) of Clause 11 to read:

"The Vendor's liability under this clause or under Clause 5 *(Rejection and Replacement)* shall be accepted by the Purchaser in lieu of any warranty or condition implied by law as to the quality or fitness for any particular purpose of the goods and save as provided in this clause the Vendor shall not be under any liability to the Purchaser (whether in contract, tort or otherwise) for any defects in the goods or for any damage, loss, death or injury (other than death or personal injury caused by the negligence of the Vendor as defined in Section 1 of the Unfair Contract Terms Act, 1977) resulting from such defects or from any work done in connection therewith."

MODEL FORM E 1973 INCLUDING 1976 AMENDMENTS

Amend the last sentence of Sub-Clause (vii) of Clause 34 to read:

"Nothing in this clause shall affect either the liability of the Contractor under Sub-Clauses (i) and (iii) of Clause 25 *(Liability for Accidents and Damage)* in respect of any portion of the Works not yet taken over or his liability for death or personal injury caused by negligence on his part as defined in Section 1 of the Unfair Contract Terms Act 1977".

General Amendments and Editorial Changes

MODEL FORM A 1976

Amend the marginal heading relating to Clause 14 to read:

"Inspection, Testing and Rejection of Plant"

Amend Sub-Clause (vii) of Clause 14 to read:

"The provisions of Clause 27(v) *(Tests on Completion)* shall relate also to inspections, examinations, and tests carried out under this clause".

Amend line 35 in Sub-Clause (iv) of Clause 28 to read:

"(iii), (iv) and (v) of Clause 27 and and Sub-Clause (vi) of Clause 14 *(Inspection, Testing and Rejection of Plant)* shall"

Amend lines 15 and 16 of Sub-Clause (ii) of Clause 34 to read:

"If at any time at which any payment would fall to be made under paragraph (b) of Sub-Clause (i) of this clause there shall be any defect in any portion of the Works in respect of which"

Amend line 1 of Sub-Clause (iii) of Clause 34 to read:

"If the payment of any sum payable under Sub-Clause (i) of this clause"

MODEL FORM B3 1971

Amend lines 1 and 2 of Sub-Clause (ii) of Clause 35 to read:

"If at any time at which any payment would fall to be made under paragraph (b) of Sub-Clause (i) of this clause there shall be any defect in any portion of the Works in respect of"

MODEL FORM E INCLUDING 1973 AMENDMENTS

Amend the marginal heading relating to Clause 14 to read:

"Inspection, Testing and Rejection of Plant"

Amend line 45 of Sub-Clause (iv) of Clause 32 to read:

"Clause 31 and Sub-Clause (vi) of Caluse 14 *(Inspection, Testing and Rejection of Plant)*"

Amend lines 1 and 2 of Sub-Clause (ii) of Clause 40 to read:

"If at any time at which any payment would fall to be made under paragraph (b) of Sub-Clause (i) of this clause there shall be any defect in any portion of the Works in respect of"

Amend line 1 of Sub-Clause (iii) of Clause 40 to read:

"If the payment of any sum payable under Sub-Clause (i) of this clause"

E

MODEL FORM OF

GENERAL CONDITIONS OF CONTRACT

INCLUDING FORMS OF AGREEMENT AND GUARANTEE

RECOMMENDED BY

THE INSTITUTION OF ELECTRICAL ENGINEERS

FOR USE IN CONNECTION WITH

HOME CABLE CONTRACTS— WITH INSTALLATION

SECOND EDITION, 1973

Published by
THE INSTITUTION OF ELECTRICAL ENGINEERS
and obtainable from
The IEE Publication Sales Department, Station House, 70 Nightingale Road, Hitchin, Herts, SG5 1RJ
or
by calling at the Reception Desk at Savoy Place, London.

HISTORICAL NOTE

A Model Form of General Conditions 'E' (Home Cable Contracts—with Installation) was first published by The Institution of Electrical Engineers in 1954.

This Model Form was revised and up-dated during 1972 and 1973 and is now issued as Model Form 'E/1973'—(Home Cable Contracts—with Installation) Second Edition.

This 1973 Edition incorporates substantial revisions to take account of developments in practice.

These may be cited as: Model Form 'E/1973'.

MEMORANDUM ON THE USE OF THE GENERAL CONDITIONS OF CONTRACT

A. Applicability

The General Conditions are intended to be suitable for the electrical industry and others, and to cover the main conditions for Home Cable contracts for the supply of plant and materials and their installation on Site.

It is hoped that, in order to secure the advantages of uniformity and to avoid trouble and expense, they may be found suitable in their entirety for most Home Cable contracts. They may, however, be added to or amended as may be required to suit each particular case. It is suggested, in order to facilitate checking, that when reprinted elsewhere any additions or alterations be made so as to attract attention, e.g. by the use of different type or underlining.

B. Contract documents

Contracts are commonly made in one of three ways:

(*a*) By a formal agreement, such as the model form supplied with these Conditions, incorporating in the contract the General Conditions, specification, schedules, drawings, etc. In this case all the terms of the contract are contained in the agreement, the Conditions, specification, and other incorporated documents on which tenders were invited, and agreed variations in or additions to the provisions of these documents are represented by suitable amendments.

(*b*) By a formal agreement, as in the method (*a*), but incorporating contractor's tender and covering letter, if any, the purchaser's acceptance of the tender, and all correspondence containing terms accepted by the purchaser and the contractor. When this method is adopted any modifications of the General Conditions, specifications, etc., have to be sought in the covering letter and other correspondence.

(*c*) Without a formal agreement, the tender and its acceptance, together with other relevant correspondence, being relied on as constituting the contract between the parties.

When either method (*a*) or (*b*) is adopted, a copy of the form of agreement to be used should be included with the documents supplied to or open to the inspection of intending tenderers.

When method (*a*) is used, the references in the model form to the tender, covering letter, acceptance of the tender, and other correspondence should be deleted.

C. Progress Payments during Manufacture

Under the General Conditions the contractor does not become entitled to any payments for plant he has manufactured until it has been delivered to the site. In the case of contracts involving considerable expenditure on manufacture before any delivery can be made, the General Conditions may be modified to provide for progress payments to the contractor in respect of plant in course of manufacture.

D. Deductions for delay in completion

It is recommended that the percentages to be inserted in the Appendix under Clause 30 should not exceed one per cent in the case of (*a*) as per Appendix and 25 per cent in the case of (*b*) as per Appendix.

MEMORANDUM AS TO TENDERS

Tenderer may propose modifications.

The tenderer is at liberty to add any details and conditions that he may deem desirable, and in the event of his doing so must print or type the same and annex the added matter to the specification or General Conditions returned by him, but such additional details and conditions will not be binding on the purchaser unless they are approved by him and incorporated in the contract.

Where in special circumstances the tenderer deems it advisable to provide for the non-disclosure of drawings and information of a confidential nature furnished with his tender, he should include in the tender or covering letter a stipulation to this effect. If the tender embodying such a stipulation is accepted, the stipulation should be incorporated in any subsequent formal agreement.

Tenders.

If the tenderer has any doubt as to the meaning of any portion of the General Conditions or of the specification, he should, when submitting his tender, set out in his covering letter the interpretation upon which he relies.

One copy of the specification supplied, together with the General conditions and drawings, enclosed in a plain sealed cover bearing the words "Tender for (*here follows the short title of the Works*)
"

and not bearing any name or mark indicating the sender shall, subject to modifications mentioned above, be returned intact with the tender form and schedules, if any, filled up and signed, addressed to

and must be received by him on or before

The purchaser does not bind himself to accept the lowest or any tender, nor will he be responsible for, or pay for, expenses or losses which may be incurred by any tenderer in the preparation of his tender.

Copies of Specification etc.

A duplicate copy of the specification, General Conditions and drawings will be supplied to intending tenderers without charge, and extra copies of the drawings at the cost of reproduction.

The sum, if any, deposited by the tenderer on application for the specification will be refunded to him within one month of the date of the adjudication upon the tenders, if all copies of the specification, drawings, etc., have been returned by the tenderer, unless the tender was not made in good faith, in which case the deposit may be forfeited.

Agreement.

The tenderer whose tender is accepted may be required to enter into a formal agreement with the purchaser.

NOTE

Any advertisement inviting tenders should state whether and if so where the agreement, General Conditions, specification, and drawings may be inspected, and should give such particulars of the class of plant and apparatus required under each section as will enable contracting firms to decide, without obtaining the specifications, whether they desire to tender, and should also state the amount of the deposit, if any, to be paid for the General Conditions, specification, and drawings.

GENERAL CONDITIONS

Definition of Terms.

1.—In construing these General Conditions and the Specification, the following words shall have the meanings herein assigned to them unless there is something in the subject matter or context inconsistent with such construction:

The "Contract" shall mean the agreement between the Purchaser and the Contractor for the execution of the Works howsoever made, including therein all documents to which reference may properly be made in order to ascertain the rights and obligations of the parties under the said agreement.

The "Contractor" shall mean the tenderer whose tender has been accepted by the Purchaser, and shall include the Contractor's legal personal representatives, successors, and assigns.

"Contractor's Equipment" shall mean tools, tackle, stores and other things brought upon the Site by the Contractor and required thereon for the purposes of the Works but not for incorporation therein.

The "Contract Price" shall mean the sum payable in accordance with the Contract in respect of the Works, without any additions thereto, or deductions therefrom, made pursuant to these Conditions.

The "Contract Value" shall mean that part of the Contract Price which is properly apportionable to the Plant or work in question having regard to the state, condition, and location of the Plant, the amount of work done, and all other relevant circumstances, and disregarding any changes that may have occurred since the date of the Contract in the cost of executing the Works.

The "Engineer" shall mean

or the person for the time being or from time to time notified in writing by the Purchaser to the Contractor as the Engineer for the Contract, or in default of any notification the Purchaser.

"Month" shall mean calendar month.

"Plant" shall mean cables, machinery, apparatus, materials, articles, and things of all kinds other than Contractor's Equipment.

The "Purchaser" shall mean

and shall include the Purchaser's legal personal representatives, successors, and assigns.

The "Site" shall mean the actual place or places to which plant is to be delivered or where work is to be done by the Contractor, together with so much of the area surrounding the said place or places as the Contractor shall with the consent of the Engineer actually use in connection with the Works otherwise than merely for the purposes of access to the said place or places.

The "Specification" shall mean the specification annexed to or issued with these General Conditions.

"Sub-contractor" shall mean any person (other than the Contractor) named in the Contract for any part of the Works or any person to whom any part of the Contract has been sub-let with the consent in writing of the Engineer, and the legal representatives, successors and assigns of such person.

"Tests on Completion" shall mean such tests to be made by the Contractor on completion of installation as are provided for in the Contract or otherwise agreed between the Purchaser and the Contractor.

The "Works" shall mean all Plant to be provided and work to be done by the Contractor under the Contract.

"Writing" shall include any manuscript, type-written, or printed statement, under seal or hand as the case may be.

Words importing persons shall include firms and corporations.

Words importing the singular only shall also include the plural and *vice versa.*

Contractor to inform himself fully.

2.—(i) The Contractor shall be deemed to have inspected the Site if access has been available to him and satisfied himself as far as can reasonably be done as to the condition of and all circumstances affecting the Site, and to have allowed in his tender for all such matters but not for the removal, alteration or diversion of mains, sewers, drains, conduits, and the like. Any such removal, alteration, or diversion which the Engineer shall direct the Contractor to undertake shall be deemed to be an addition to the Works and shall be dealt with in accordance with Clause 11 (*Variations and Omissions*).

(ii) If during the progress of the Works any unfavourable physical condition (other than weather or conditions due to weather) or artificial obstruction is encountered which could not have reasonably been foreseen and in consequence thereof additional cost will be or has been incurred by the Contractor, a sum ascertained as provided in Clause 38 (*Measurement for Payment*) shall be added to the Contract Price provided that the Contractor has given to the Purchaser as soon as reasonably practicable notice in writing that he intends to make a claim in respect of such additional cost.

Security for due performance.

3.—(i) If required by the Purchaser, the Contractor shall provide a surety or sureties, or a grantor of an insurance or guarantee policy, who in either case shall be subject to the approval of the Purchaser, which approval shall not be unreasonably withheld, and who shall execute (if two or more jointly and severally) a guarantee, or grant an insurance or guarantee policy to an extent not exceeding 10 per cent of the Contract Price, by way of guarantee for the due and faithful performance of the Contract, such guarantee to be binding notwithstanding such variations, alterations, or extensions of time as may be made, given, conceded, or agreed under these General Conditions. The instrument of guarantee shall be in the form annexed to these General Conditions with such modifications as may be necessary, and the expenses of procuring, preparing, completing and stamping such instrument shall be paid by the Purchaser.

(ii) If the guarantee or other security for the due performance of the Contract required to be furnished pursuant to this clause shall not be duly furnished by the Contractor to the Purchaser within one month after the Contract has been entered into, the Purchaser may, at his option, without prejudice to any rights or claims he may have against the Contractor by reason of the Contractor's non-compliance with any of the provisions of this clause, and within seven days after the expiry of the said period, by notice in writing to the Contractor terminate the Contract forthwith, and the Purchaser shall thereupon not be liable for any claim or demand from the Contractor in respect of any thing then already done or furnished, or in respect of any other matter or thing whatsoever in connection with the Contract, but the Purchaser shall be entitled to be repaid by the Contractor all out-of-pocket expenses properly incurred by the Purchaser incidental to the obtaining of new tenders.

Expenses of Agreement.

(iii) The expenses of preparing, completing, and stamping the Agreement (if any) shall be paid by the Purchaser, and an executed counterpart thereof properly stamped together with copies of all other documents comprising the Contract shall be furnished to the Contractor free of charge.

Drawings.

4.—(i) The Contractor shall submit to the Engineer for approval:

(*a*) within the time specified in the Specification or if no time is specified then a reasonable time such drawings as may be called for therein, and

(*b*) during the progress of the Works, and within such reasonable times as the Engineer may require, such drawings of the general arrangement and of details of the Works or any part thereof as the Engineer may reasonably require, provided that the Contractor shall not be under any obligation to supply copies of shop drawings.

Within a reasonable period after receiving such drawings the Engineer shall signify his approval or otherwise. If the Engineer shall not approve any drawing thus submitted the same shall be forthwith modified to meet the reasonable requirements of the Engineer and shall be re-submitted. Copies of all drawings which require to be approved by the Engineer shall be provided in triplicate by the Contractor. Drawings when approved shall if required by either party be signed or identified by both parties and if required by the Contractor one copy shall be returned to him. All dimensions marked on drawings shall be considered correct although measurements by scale may differ therefrom. Detailed drawings shall take precedence where they differ from general arrangement drawings.

(ii) Drawings approved as above described shall not be departed from except as provided in Clause 11 (*Variations and Omissions*).

(iii) The Engineer shall have the right at all reasonable times to inspect at the premises of the Contractor all drawings of any portion of the Works.

(iv) The Contractor shall, if so desired by the Purchaser, furnish to the Purchaser in writing at the commencement of the maintenance period, or at such earlier times as may be named in the Specification, such information, accompanied by drawings, as may be necessary to enable the Purchaser to operate, maintain, dismantle, reassemble, and adjust all parts of the Works.

(v) Drawings submitted in pursuance of paragraphs (*a*) and (*b*) of Sub-Clause (i) of this clause shall not, without the consent of the Contractor, be used by the Purchaser except for purposes of the Contract, nor shall they without such consent be communicated to third parties save insofar as may be necessary for the proper execution of the Works.

Records.

5.—The Contractor shall during the progress of the Works keep at his office nearest to the Site such particulars of the progress of the Works as will enable him to furnish to the Purchaser such records and information as may be required by the Specification to be furnished, and shall at all reasonable times produce such particulars to the Engineer for his inspection.

Mistakes in Information.

6.— (i) The Contractor shall be responsible for any discrepancies, errors, or omissions in the drawings and information supplied by him, whether they have been approved by the Engineer or not, provided that such discrepancies, errors, or omissions be not due to defective drawings or information furnished to the Contractor by the Purchaser or the Engineer.

(ii) The Contractor shall at his own expense carry out any alterations or remedial work necessitated by reason of such discrepancies, errors or omissions and modify the drawings and information accordingly, or if the same be done by or on behalf of the Purchaser shall bear all costs reasonably incurred therein. The performance of his obligations under this sub-clause shall be in full satisfaction of the Contractor's liability under Sub-Clause (i) of this clause, but shall not relieve him of his liability under Clause 30 (*Delay in Completion*) insofar as that liability is the result of a failure of the Contractor to perform his obligations under that sub-clause.

(iii) The Purchaser shall be responsible for drawings and information supplied in writing by the Purchaser or the Engineer for the details of special work specified by either of them. The Purchaser shall pay the extra cost reasonably incurred by the Contractor due to alterations of the work necessitated by reason of defective drawings or information so supplied to the Contractor.

Assignment.

7.—(i) The Contractor shall not, without the consent in writing of the Purchaser, which shall not be unreasonably withheld, assign or transfer the Contract or the benefits or obligations thereof or any part thereof to any other person, provided that this shall not affect any right of the Contractor to assign, either absolutely or by way of charge, any moneys due or to become due to him, or which may become payable to him under the Contract.

Sub-letting.

(ii) The Contractor shall not, without the consent in writing of the Engineer, which shall not be unreasonably withheld, sub-let the Contract or any part thereof, or make any sub-contract with any person or persons for the execution of any portion of the Works but the restriction contained in this clause shall not apply to sub-contracts for materials, for minor details, or for any part of the Works of which the makers are named in the Contract. Any such consent shall not relieve the Contractor from his obligations under the Contract.

Patent Rights, etc.

8.—(i) The Contractor shall fully indemnify the Purchaser against all actions, claims, demands, costs, charges, and expenses arising from or incurred by reason of any infringement or alleged infringement of letters patent, design, or copyright protected in the United Kingdom or in the country in which the Plant is to be erected by the use of any Plant supplied by the Contractor, but such indemnity shall not cover any use of the Works otherwise than for the purpose indicated by or reasonably to be inferred from the Specification or to any infringement which is due to the use of any Plant in association or combination with any other plant not supplied by the Contractor.

(ii) In the event of any claim being made or action brought against the Purchaser arising out of the matters referred to in this clause, the Contractor shall be promptly notified thereof and may at his own expense conduct all negotiations for the settlement of the same, and any litigation that may arise therefrom. The Purchaser shall not, unless and until the Contractor shall have failed to take over the conduct of the negotiations or litigation, make any admission which might be prejudicial thereto. The conduct by the Contractor of such negotiations or litigation shall be conditional upon the Contractor having first given to the Purchaser such reasonable security as shall from time to time be required by the Purchaser to cover the amount ascertained or agreed or estimated, as the case may be, of any compensation, damages, expenses and costs for which the Purchaser may become liable. The Purchaser shall, at the request of the Contractor, afford all available assistance for the purpose of contesting any such claim or action and shall be repaid all reasonable expenses incurred in so doing.

(iii) The Purchaser on his part warrants that any design or instructions furnished or given by him shall not be such as will cause the Contractor in the performance of the Contract to infringe any letters patent, registered design, trade mark, or copyright in the country in which the Plant is used.

Manner of Execution.

9.—All Plant to be supplied and all work to be done under the Contract shall be manufactured and executed in the manner set out in the Specification or, where not so set out, to the reasonable satisfaction of the Engineer.

Contractor's Equipment, Labour, etc.

10.—(i) Unless specific arrangements be made to the contrary the Contractor shall, at his own expense, provide all Contractor's Equipment and all materials, labour, haulage, and power necessary to execute and complete the Works.

Fencing, Guarding and Lighting.

(ii) The Contractor shall be responsible for the proper fencing, guarding, lighting, and watching of all the Works on the Site until taken over under Clause 32 (*Taking Over*) and for the proper provision during a like period of temporary roadways, footways, guards and fences as far as the same may be rendered necessary by reason of the Works for the accommodation and protection of the owners and occupiers of adjacent property, the public, and others. No naked light shall be used by the Contractor on the Site otherwise than in the open air without special permission in writing from the Engineer.

Electricity, Water and Gas.

(iii) The Contractor shall be entitled to use for the purposes of the Works such supplies of electricity, water, and gas as may be available therefor on the Site and shall pay to the Purchaser for such use such sum as may be reasonable in the circumstances, and shall at his own expense provide apparatus necessary for such use.

Lifting Equipment.

(iv) The Purchaser shall at the request of the Contractor and for the execution of the Works operate any suitable lifting equipment belonging to the Purchaser that may be available on the Site and the Contractor shall pay a reasonable sum therefor. The Purchaser shall during such operation retain control of and be responsible for the safe working of the lifting equipment but shall not be responsible for any negligence of the Contractor.

Variations and Omissions.

11.—(i) The Contractor shall not alter any of the Works except as directed in writing by the Engineer; but the Engineer shall have power, subject to the proviso hereinafter contained, from time to time during the execution of the Contract by notice in writing to direct the Contractor to alter, amend, omit, add to, or otherwise vary any of the Works, and the Contractor shall carry out such variations, and be bound by the same conditions, so far as applicable, as though the said variations were stated in the Specification. Provided that no such variation shall, except with the consent in writing of the Contractor be such as will, with variations already directed to be made, involve a net addition to or deduction from the Contract Price of more than 15 per cent thereof where the Contract Price is known, and in any other case 15 per cent of the estimated Contract Price named in the Contract. In any case in which the Contractor has received any such direction from the Engineer which either then or later will, in the opinion of the Contractor, involve any addition to or deduction from the Contract Price, the Contractor shall, as soon as reasonably possible, advise the Engineer in writing to that effect. The difference of cost (if any) occasioned by such variations shall be added to or deducted from the Contract Price as the case may require. The amount of such difference shall be ascertained in the manner provided in Clause 38 (*Measurement for Payment*).

(ii) If the Engineer shall make any such variation in any part of the Works, such reasonable notice in writing shall be given to the Contractor as will enable him to make his arrangements accordingly, and in cases where Plant is already manufactured or in course of manufacture, or any matter done or drawings or patterns made that require to be altered, a reasonable sum in respect thereof shall be allowed by the Engineer.

(iii) If in the opinion of the Contractor any such variation is likely to prevent or prejudice the Contractor from or in fulfilling any of his obligations under the Contract, or to involve an increased risk of loss or damage, he shall notify the Engineer thereof in writing, and the Engineer shall decide forthwith whether or not the same shall be carried out. If the Engineer confirms his instructions in writing the said obligations shall be modified to such an extent as may be just and the Purchaser shall indemnify the Contractor against all claims, demands, costs, charges and expenses arising from or incurred by the Contractor in consequence of any such loss or damage, provided always that this indemnity shall not apply to any loss or damage which would not have occurred if the Contractor had taken such precautions as would be reasonable in the circumstances. If the Engineer does not so confirm his instructions they shall be deemed to be withdrawn.

Contractor's Default.

12.—If the Contractor shall neglect to execute the Works with due diligence and expedition, or shall refuse or neglect to comply with any reasonable orders given him in writing by the Engineer in connection with the Works, or shall contravene the provisions of the Contract, the Purchaser may give notice in writing to the Contractor to make good the failure, neglect, or contravention complained of. Should the Contractor fail to comply with the notice within seven days from the date of service thereof in the case of a failure, neglect, or contravention capable of being made good within that time, or otherwise within such time as may be reasonably necessary for making it good, then and in such case the Purchaser shall be at liberty to employ other workmen, and forthwith execute such part of the Works as the Contractor may have neglected to do, or, if the Purchaser shall think fit, it shall be lawful for him, without prejudice to any other rights he may have under the Contract, to take the Works wholly or in part out of the Contractor's hands and either by his own workmen or by re-contracting with any other person or persons to complete the Works or any part thereof, and in either event the Purchaser shall have the free use of all Contractor's Equipment that may be on the Site in connection with the Works, without being responsible to the Contractor for fair wear and tear thereof, and to the exclusion of any right of the Contractor over the same, and the Purchaser shall be entitled to retain and apply any balance which may be otherwise due on the Contract by him to the Contractor, or such part thereof as may be necessary to the payment of the cost of executing the said part of the Works or of completing the Works as the case may be. If the cost of completing the Works or executing a part thereof as aforesaid shall exceed the balance due to the Contractor, the Contractor shall pay such excess.

Bankruptcy.

13.—If the Contractor shall become bankrupt or insolvent, or have a receiving order made against him, or compound with his creditors, or being a corporation commence to be wound up, not being a members' voluntary winding up for the purpose of reconstruction or amalgamation, or carry on its business under a receiver for the benefit of its creditors or any of them, the Purchaser shall be at liberty either—

(*a*) to terminate the Contract forthwith by notice in writing to the Contractor or to the receiver or liquidator or to any person in whom the Contract may become vested, and to act in the manner provided in Clause 12 (*Contractor's Default*) as though the last mentioned notice had been the notice referred to in such clause and the Works had been taken out of the Contractor's hands, or

(*b*) to give such receiver, liquidator, or other person the option of carrying out the Contract subject to his providing a guarantee for the due and faithful performance of the Contract up to an amount to be agreed.

Inspection, Testing and Rejection.

14.—(i) The Engineer shall be entitled at all reasonable times during manufacture to inspect, examine, and test on the Contractor's premises the materials and workmanship and performances of all Plant to be supplied under the Contract, and if part of the said Plant is being manufactured on other premises the Contractor shall obtain for the Engineer permission to inspect, examine, and test as if the said Plant were being manufactured on the Contractor's premises. Such inspection, examination, or testing, shall not release the Contractor from any obligation under the Contract.

(ii) The Contractor shall, after consulting the Engineer, give the Engineer reasonable notice in writing of the date on and the place at which any Plant will be ready for testing as provided in the Contract and unless the Engineer shall attend at the place so named on the date which the Contractor has stated in his notice the Contractor may proceed with the tests, which shall be deemed to have been made in the Engineer's presence, and shall forthwith forward to the Engineer duly certified copies of the test readings. The Engineer shall give the Contractor 24 hours' notice in writing of his intention to attend the tests.

(iii) Where the Contract provides for tests on the premises of the Contractor or of any Sub-contractor or on the Site, the Contractor except where otherwise specified, shall provide all assistance, labour, materials, electricity, fuel, stores, apparatus, and instruments as may be requisite and as may be reasonably demanded to carry out such tests efficiently.

(iv) As and when the Engineer is satisfied that any Plant shall have passed the tests referred to in this clause he shall notify the Contractor in writing to that effect.

(v) If after inspecting, examining, or testing any Plant the Engineer shall decide that such Plant or any part thereof is defective or not in accordance with the Contract, he may reject the said Plant or part thereof by giving to the Contractor within a reasonable time notice in writing of such rejection, stating therein the grounds upon which his decision is based.

(vi) The provisions of Clause 31 (iv) (*Tests on Completion*) shall relate also to inspections, examinations, and tests carried out under this clause.

Delivery.

15.—(i) No Plant or Contractor's Equipment shall be delivered to the Site until an intimation in writing has been applied for and obtained by the Contractor from the Engineer that delivery may be made. The Contractor shall be responsible for the reception on the Site of all Plant and Contractor's Equipment delivered for the purposes of the Contract.

(ii) For the purposes of this clause only—"delayed Plant" means:

(*a*) Plant which by delay or failure on the part of the Engineer to give such intimation as is mentioned in Sub-Clause (i) of this clause or from any cause for which the Purchaser or some other contractor employed by him is responsible the Contractor is prevented from delivering to the Site at the time specified for the delivery thereof, or if no time is specified, at the time when it is reasonable for it to be delivered having regard to the date by which the Works ought to be completed: and

(*b*) Plant which has been delivered to the Site but which by delay or failure on the part of the Engineer or from any cause for which the Purchaser or some other contractor employed by him is responsible the Contractor is for the time being prevented from erecting or installing it.

"the normal delivery date" means the time when but for such delay, failure or other cause as aforesaid delayed Plant would have been delivered to the Site.

"notice to proceed" means notice in writing from the Engineer to the Contractor that delayed Plant may forthwith be delivered to the Site or (as the case may be) erected or installed.

(iii) If delayed Plant is ready for delivery and has been suitably and sufficiently marked as appropriated to the Contract and the Contractor has given to the Engineer an opportunity of inspecting it or if delayed Plant has been delivered to the Site the Contractor may give notice in writing to the Purchaser requiring that the provisions of Sub-Clause (iv) of this clause shall have effect with respect to such delayed Plant.

(iv) Where notice has been given in accordance with Sub-Clause (iii) of this clause—

(*a*) There shall be added to the Contract Price the reasonable additional expense incurred in storing and taking reasonable measures to protect and preserve the delayed Plant from and insuring it against loss, deterioration and damage however caused from the date of the said notice or the normal delivery date if this shall be later until the Contractor shall no longer be prevented from delivering the delayed Plant or (as the case may be) erecting or installing it or shall be relieved of responsibility therefor under Sub-Clause (v) of this clause whichever shall first happen.

(*b*) The Contractor shall after one month from the normal delivery date or from the date of the said notice (whichever shall be the later) be entitled to have the Contract Value of the delayed Plant included in an interim certificate.

(*c*) If at the expiration of six months from the normal delivery date or from the date of the said notice (whichever shall be the later) the Contractor shall still be prevented from delivering the delayed Plant to the Site or (as the case may be) from erecting or installing it the Engineer shall, on the application of the Contractor, certify accordingly and within one month from the presentation of such certificate the Contractor shall be entitled to be paid 95 per cent of the Contract Value of the delayed plant less any sum previously paid to him in respect thereof.

(*d*) If at any time after the expiration of 12 months from the date of the said notice or at any time after the delayed Plant has been delivered to the Site the Purchaser shall not have assumed responsibility for the storage of the delayed Plant the Contractor may by a further notice in writing expiring 30 days after receipt thereof by the Purchaser require the Purchaser to assume responsibility for storing, protecting, and preserving the delayed Plant and upon the expiration of the last mentioned notice the Purchaser shall assume responsibility for storing the delayed Plant provided always that if notice to proceed shall be given within 30 days after receipt of the last mentioned notice given by the Contractor this paragraph of this sub-clause shall not operate.

(*e*) Without prejudice to the provisions of Sub-Clause (vi) of Clause 34 (*Defects after Taking Over*), the obligations of the Contractor under that clause with respect to delayed Plant shall not apply to any defect that may develop therein after the expiration of three years from the date of the said notice referred to in Sub-Clause (iii) of this clause or the normal delivery date if this shall be later.

(v) If at any time the Purchaser assumes responsibility for storing delayed Plant whether pursuant to paragraph (*d*) of the last preceeding sub-clause or otherwise the Contractor shall thereupon be relieved of any responsibility for the delayed Plant until either the expiration of 30 days after the receipt of a notice to proceed or the Contractor having received the notice to proceed resumes possession of the said Plant whichever shall first occur.

(vi) After the receipt of notice to proceed the Contractor, if he has been relieved of responsibility under Sub-Clause (v) of this clause, shall (and in any other case may) after due notice in writing to the Engineer and if required by the Engineer in his presence examine the delayed Plant and any Plant on the Site that has been erected or installed but not taken over under Clause 32 (*Taking Over*) by reason of delay in the delivery, erection or installation of the delayed Plant, and make good any deterioration or defect therein that may have developed or loss thereof that may have occurred after the normal delivery date or (if later) the date when the Contractor was by such delay, failure or other cause as before-mentioned first prevented from erecting or installing the delayed Plant.

(vii) There shall be added to the Contract Price any reasonable expense to which the Contractor may be put in making the examination referred to in Sub-Clause (vi) of this clause or in making good any deterioration, defect or loss as therein mentioned except insofar as the same was caused by faulty workmanship or materials or by the Contractor's failure to take the measures referred to in paragraph (*a*) of Sub-Clause (iv) of this clause or in Sub-Clause (i) of Clause 25 (*Liability for Accidents and Damage*). Any reasonable expense to which the Contractor may be put in delivering the delayed Plant to the Site or in erecting or installing the same or any other Plant or in carrying out the Tests on Completion or in performing his obligations under Clause 34 (*Defects after Taking Over*) which would not have been incurred had the delivery, erection or installation of the delayed Plant not been prevented as aforesaid shall also be added to the Contract Price.

Access to and Possession of the Site.

16.—(i) Subject to Sub-Clause (iv) of this clause, access to and possession of the Site shall be afforded to the Contractor by the Purchaser in reasonable time. The Purchaser shall provide such roads and other means of access to the Site as may be stated in the Specification, subject to such limitations as may there be imposed. The Contractor shall provide all means of access that he may require in addition to such as will be provided by the Purchaser as above-mentioned or may otherwise be available. The Contractor shall make good all damage done by him or Sub-contractors by the improper use of roads or other means of access provided by the Purchaser and shall indemnify the Purchaser against claims in respect of damage done by the Contractor or Sub-contractors to public roads and other means of access not provided by the Purchaser.

(ii) If a building, structure, foundation, approach, or other provision for the execution of the Works is by the Contract to be provided by the Purchaser, it shall be in a condition suitable for the efficient transport, reception, installation, and maintenance of the Works.

(iii) During the execution of the Works, persons other than the Contractor, Sub-contractors, and his and their employees and persons authorized in writing by the Engineer shall be excluded from the Site so far as they may be lawfully excluded, but facilities to inspect the Works at all times shall be afforded to the Engineer and his representatives and other authorized representatives of the Purchaser.

(iv) The access to and possession of the Site referred to in Sub-Clause (i) hereof shall not be exclusive to the Contractor but only such as shall enable him to execute the Works. The Contractor shall afford to the Purchaser and to other contractors whose names shall have been previously communicated in writing to the Contractor by the Engineer every reasonable facility for the execution of work concurrently with his own.

(v) Unless otherwise provided in the Specification, the Purchaser shall give the Contractor facilities for carrying out the Works on the Site continuously during the normal working hours generally recognized in the district. The Engineer may, after consulting with the Contractor, direct that work shall be done at other times if it shall be practicable in the circumstances for work to be so done, and the extra cost of work so done shall be added to the Contract Price unless such work has, by the default of the Contractor, become necessary for the completion of the Works within the time fixed by the Contract, or, if no time be fixed, within a reasonable time.

Statutory and Other Requirements.

17.—The Contractor shall in all matters arising in the performance of the Contract observe and perform such of the obligations and restrictions imposed on the Purchaser in the capacity of undertaker by the Public Utilities Street Works Act, 1950, or any other statute or by any order, regulation, or by-law made with statutory authority as apply to or affect the Contractor or anything done or omitted by the Contractor in the execution of the Works, provided always that (*a*) Where the Contractor is not required by the Contract to undertake or the Purchaser ceases to be under the obligation to undertake permanent reinstatement and making good of the street or controlled land at upper levels, the obligation of the Contractor hereunder to execute interim restoration (as defined in the said Act) at any place shall not continue beyond the expiration of three months from the time when the Purchaser is to be treated for the purposes of paragraph 5(2) of the Third Schedule to the said Act as having completed the undertakers' works there and such of the permanent reinstatement and making good there as the undertakers are under an obligation to execute, and

(*b*) Where any work is required to be executed by any government municipal or other public body, the Purchaser shall procure that such work is duly executed, and

(*c*) The Contractor shall not be responsible for the payment of any moneys payable by the undertakers pursuant to the said Act or the cost of any work executed by any public body other than moneys payable or work required in consequence of any default on the part of the Contractor.

Reinstatement.

18.—The Contractor shall not be responsible for the cost of new materials for the replacement of any flagstones, bricks, setts, or other surface materials or any pipes, conduits, or other underground structures agreed by the Engineer before lifting or when uncovered to be in a damaged or cracked condition, or so worn as to be incapable of removal without damage or of satisfactory replacement.

Wayleaves, Contents, Notices, and Restrictions.

19.—(i) The Purchaser shall be responsible for obtaining all wayleaves and consents necessary for the execution of the Works and shall inform the Contractor of any conditions attached to such consents in so far as such conditions affect the execution of the Works. The Purchaser shall also be responsible for giving all notices required to be given by any statute or any order, regulation, or by-law made with statutory authority.

(ii) The Contractor shall not enter upon or disturb any land or buildings without the previous written consent of the Engineer.

(iii) In so far as they relate to the execution or maintenance of the Works, the Contractor shall, subject to Sub-Clause (ii) of this clause, observe and perform all restrictions and obligations contained in agreements, contracts, conveyances, leases, and other instruments with or to local or other authorities or the owners of any interest in land on or adjoining which the Works are to be executed or their predecessors in title so far as such restrictions and obligations shall have been brought to his notice provided always that the Contractor shall not be under any obligation with respect to damage caused in the execution of the Works to the property of third parties which could not have been avoided by the exercise of reasonable care and is incapable of being made good.

Clearing of Site.

20.—The Contractor shall when required by the Engineer or any person or body having the right to impose such requirement carry away excavated and other material arising from the execution of the Works and carry back so much thereof as may be required for the completion of the Works. The Contractor shall from time to time as work on the Site is completed remove all surplus material, debris, rubbish, unused materials, temporary erections, and Contractor's Equipment, and shall leave the Site clear and tidy to the reasonable satisfaction of the Engineer.

Vesting of Plant and Contractor's Equipment.

21.—(i) Plant supplied pursuant to the Contract shall become the property of the Purchaser at whichever is the earlier of the following times, namely—

(*a*) when the Plant is delivered pursuant to the Contract,

(*b*) when by virtue of Clause 15 (*Delivery*) or Clause 33 (*Suspension of Works*) the Contractor becomes entitled to require that the Contract Value of the Plant be included in an interim certificate.

(ii) All Contractor's Equipment owned by the Contractor or by any firm or corporation in which the Contractor has a controlling interest shall when brought upon the Site vest in and become the property of the Purchaser, and shall be used solely for the purpose of the Works and shall not be removed by the Contractor without the permission in writing of the Engineer, which permission shall not be unreasonably withheld in the case of Contractor's Equipment not currently required for the purpose of the Works. The Contractor shall be liable for the loss or destruction of such Contractor's Equipment or for damage thereto which may happen otherwise than through the fault of the Purchaser. If there shall be due, owing, or accruing to the Purchaser from the Contractor any moneys under or in respect of the Contract, of which the Purchaser shall be unable to obtain payment, the Purchaser shall be at liberty at the cost of the Contractor to sell and dispose as he shall think fit of such

Contractor's Equipment and to apply the proceeds in or towards the satisfaction of such moneys as aforesaid. Subject to the foregoing and to the right of the Purchaser under Clause 12 (*Contractor's Default*) and Clause 13 (*Bankruptcy*) the property in such Contractor's Equipment and in any Plant which is no longer required for completion of the Works shall revert to the Contractor on their proper removal from the Site or on the completion of the Works or on the termination of the Contract, whichever may be the earliest.

(iii) If the Purchaser shall become bankrupt, or insolvent, or have a receiving order made against him, or compound with his creditors, or being a corporation commence to be wound up, then notwithstanding the provisions of Sub-Clause (ii) the Contractor's Equipment shall revert to and become the property of the Contractor.

Engineer's Supervision.

22.—(i) After the tender has been accepted by the Purchaser, all instructions and orders to the Contractor shall, except as herein otherwise provided, be given by the Engineer.

(ii) The Contractor shall be responsible for ensuring that the positions, level, dimensions, and routes of the Works are correct according to the drawings notwithstanding that he may have been assisted by the Engineer in setting out the said positions, levels, dimensions and routes.

Engineer's Representation.

(iii) The Engineer may from time to time delegate any of the powers, discretions, functions, and authorities vested in him and may at any time revoke any such delegation. Any such delegation or revocation shall be in writing signed by the Engineer and, in the case of a delegation, shall specify the powers, discretions, functions, and authorities thereby delegated and the person or persons to whom the same are delegated. No such delegation or revocation shall have effect until a copy thereof has been delivered to the Contractor.

Engineer's Decisions.

23.—The Contractor shall proceed with the Works in accordance with decisions, instructions, and orders given by the Engineer in accordance with these Conditions, provided always that—

(*a*) if the Contractor shall, without undue delay after being given any decision, instruction, or order otherwise than in writing, require it to be confirmed in writing, such decision, instruction, or order shall not be effective until written confirmation thereof has been received by the Contractor, and

(*b*) if the Contractor shall, by written notice to the Engineer within 14 days after receiving any decision, instruction, or order of the Engineer in writing or written confirmation thereof, intimate that he disputes or questions the decision, instruction, or order, giving his reasons for so doing, either party to the Contract shall be at liberty to refer the matter to arbitration pursuant to Clause 41 (*Arbitration*), but such an intimation shall not relieve the Contractor of his obligation to proceed with the Works in accordance with the decision, instruction, or order in respect of which the intimation has been given. The Contractor shall be at liberty in any such arbitration to rely on reasons additional to the reasons stated in the said intimation.

Contractor's Representatives and Workmen.

24.—(i) The Contractor shall employ one or more competent representatives, whose name or names shall have previously been communicated in writing to the Engineer by the Contractor, to superintend the carrying out of the Works on the Site. The said representative, or if more than one shall be employed, then one of such representatives, shall be present on the Site during working hours, and any orders or instructions which the Engineer may give to the said representative of the Contractor shall be deemed to have been given to the Contractor.

(ii) The Engineer shall be at liberty by notice in writing to the Contractor to object to any representative or person employed by the Contractor in the execution of or otherwise about the Works who shall, in the opinion of the Engineer, misconduct himself or be incompetent or negligent, and the Contractor shall remove such person from the Works.

Liability for Accidents and Damage.

25.—(i) The Contractor shall properly cover up and protect until taken over under Clause 32 (*Taking Over*) any section or portion of the Works liable to injury by exposure to the weather, and shall take every reasonable precaution to protect any section or portion of the Works not taken over against loss or damage from any cause.

(ii) In the case of loss or damage to the Works on the Site arising from or occasioned by causes for which the Contractor is not responsible under the Contract the same shall, if required by the Purchaser, be made good by the Contractor but at the cost of the Purchaser at a price to be agreed between the Contractor and the Purchaser or in default of agreement to be settled by arbitration and such cost shall be added to the Contract Price.

(iii) Subject to Clause 26 (*Limitations on Contractor's Liability*), all losses of and damage to any section or portion of the Works that shall not have been taken over under Clause 32 (*Taking Over*), which shall arise from or be occasioned by any act of the Contractor or any Sub-contractor or by a failure of the Contractor to comply with any obligation imposed on him by Sub-Clause (i) of this clause, shall be made good by and at the sole cost of the Contractor and to the reasonable satisfaction of the Engineer.

(iv) The Contractor shall, subject to Clause 26, indemnify the Purchaser in respect of all damage or injury occurring before all the Works shall have been taken over under Clause 32 to any person or to any property (other than property forming part of the Works not yet taken over) and against all actions, suits, claims, demands, costs, charges, and expenses arising in connection therewith which shall be occasioned by the negligence of or breach of statutory duty by the Contractor or any Sub-contractor, or by defective design (other than a design made, furnished, or specified by the Purchaser and for which the Contractor has disclaimed responsibility in writing within a reasonable time after the receipt of the Purchaser's instructions), materials, or workmanship, but not otherwise. Provided that the Contractor shall not be liable by virtue of this sub-clause in respect of damage or injury attributable to defects in any section or portion of the Works taken over under Clause 32.

(v) If there shall occur any loss of or damage to any property or person while the Contractor is on the Site for the purpose of making good a defect in any section or portion of the Works pursuant to Clause 34 (*Defects After Taking Over*), or for the purpose of carrying out tests on completion during the period referred to in Clause 34 (i) as provided in Clause 32 (iv) (*Interference with Tests*), the Contractor shall be liable, subject to the provisions of Clause 26 as follows:—

(*a*) In respect of loss of or damage to the said section or portion the Contractor's liability shall be as defined in Clause 34.

(*b*) In respect of damage or injury to any other person and of any actions, claims, demands, costs, charges and expenses arising in connection therewith the Contractor shall be liable, subject to the provisions of Sub-Clauses (i) and (iii) of this clause, to the extent that such damage or injury was caused by the negligence or breach of statutory duty of the Contractor or a Sub-contractor while on the Site as aforesaid or by defective materials or workmanship used in making good the said defect but not otherwise.

The said section or portion of the Works shall be defined by reference to the taking over certificate issued in respect thereof pursuant to Clause 32 (*Taking Over*).

(vi) In the event of any claim being made against the Purchaser arising out of the matters referred to in and in respect of which the Contractor may be liable under this clause, the Contractor shall be promptly notified thereof, and may at his own expense conduct all negotiations for the settlement of the same and any litigation that may arise therefrom. The Purchaser shall not, unless and until the Contractor shall have failed to take over the conduct of the negotiations or litigation, make any admission which might be prejudicial thereto. The conduct by the Contractor of such negotiations or litigation shall be conditional upon the Contractor having first given to the Purchaser such reasonable security as shall from time to time be required by the Purchaser to cover the amount ascertained or agreed or estimated, as the case may be, of any compensation, damages, expenses, and costs for which the Purchaser may become liable. The Purchaser shall, at the request of the Contractor, afford all available assistance for any such purpose, and shall be repaid all reasonable expenses incurred in so doing.

Limitations on Contractor's Liability.

26.—(i) The Contractor shall not be liable to the Purchaser by way of indemnity or by reason of any breach of the Contract for—

(*a*) any loss of use (whether complete or partial) of the Works or of profit or of any contract that may be suffered by the Purchaser except as provided in Clause 30 (*Delay in Completion*) by the deduction of liquidated damages for delay;

(*b*) any loss, damage or injury to the extent that it is caused by or arises from the acts or omissions of the Purchaser or of others (not being the Contractor's servants or Sub-contractors);

(*c*) any loss, damage or injury in circumstances over which the Contractor has no control.

(ii) Except in respect of personal injury or damage to property conferring on a person other than the Purchaser a good cause of action against the Contractor, the liability of the Contractor for any one act or default shall not exceed the Contract Price or £100,000, whichever is the greater.

Insurance of Works.

27.—Unless the Purchaser shall have approved in writing other arrangements for the insurance hereinafter mentioned the Contractor shall in the joint names of the Purchaser and the Contractor and without prejudice to his liability under Clause 25 (*Liability for Accidents and Damage*) insure and keep insured until a taking over certificate has been issued under Clause 32 (*Taking Over*) such Works as may for the time being be upon the site for the full replacement value thereof against all loss or damage from whatever cause arising (other than the excepted risks). The Contractor shall from time to time, when so required by the Engineer, produce evidence of satisfactory insurance cover and of the payment of the premium for inspection by the Purchaser. All moneys received under any such policy shall be applied in or towards the replacement or repair of the Works destroyed or damaged, but the provisions of this Clause shall not affect the Contractor's liabilities under the Contract.

Defects prior to Taking Over.

28.—If at any time before the Works are taken over under Clause 32 (*Taking Over*) the Engineer shall—

(*a*) decide that any work done or Plant supplied or materials used by the Contractor or any Sub-contractor is or are defective or not in accordance with the Contract or that the Works or any portion thereof are defective or do not fulfil the requirements of the Contract (all such matters being hereinafter in this clause called "defects"), and

(*b*) as soon as reasonably practicable give to the Contractor notice in writing of the said decision specifying particulars of the defects alleged and of where the same are alleged to exist or to have occurred, and

(*c*) so far as may be necessary place the Plant at the Contractor's disposal,

then the Contractor shall at his own expense and with all speed make good the defects so specified. In case the Contractor shall fail so to do the Purchaser may, provided he does so without undue delay, take, at the cost of the Contractor, such steps as may in all the circumstances be reasonable to make good such defects. All Plant provided by the Purchaser to replace defective Plant shall comply with the Contract and shall be obtained at reasonable prices and where reasonably practicable under competitive conditions. The Contractor shall be entitled to remove and retain all Plant that the Purchaser may have replaced at the Contractor's cost.

Nothing contained in this clause shall affect any claim by the Purchaser under Clause 30 (*Delay in Completion*).

Extension of Time for Completion.

29.—If, by reason of any industrial dispute or any cause beyond the reasonable control of the Contractor arising after the acceptance of the tender, the Contractor shall have been delayed or impeded in the completion of the Works, whether such delay or impediment occur before or after the time (if any) or extended time fixed for completion, provided that the Contractor shall without delay have given to the Purchaser or the Engineer notice in writing of his claim for an extension of time, the Engineer shall on the receipt of such notice grant the Contractor from time to time in writing either prospectively or retrospectively such extension of the time fixed by the Contract for the completion of the Works as may be reasonable.

Delay in Completion.

30.—If the Contractor fails to complete the Works in accordance with the Contract (save as regards his obligations under Clause 34 (*Defects after Taking Over*) and such tests as are to be made in accordance with Clause 34 (*Tests on Completion*) within the time fixed by the Contract for the completion of the Works or any extension of such time, or if no time be fixed, within a reasonable time, and the Purchaser shall have suffered any loss from such failure, there shall be deducted from the Contract Price the percentage named in the Appendix of the Contract Value of such portion or portions only of the Works as cannot in consequence of the said failure be put to the use intended for each week between the time for completion of the Works as aforesaid and the actual date of completion, but the amount so deducted shall not in any case exceed the maximum percentage named in the Appendix of the Contract Value of such portion or portions of the Works, and such deduction shall be in full satisfaction of the Contractor's liability for the said failure.

Tests on Completion.

31.—(i) The Contractor shall give to the Engineer in writing 21 days' notice of the date after which he will be ready to make the Tests on Completion. Unless otherwise agreed, the tests shall take place within 10 days after the said date, on such day or days as the Engineer shall in writing notify the Contractor.

(ii) If the Engineer fail to appoint a time after having been asked to do so, or to attend at any time or place duly appointed for making the said tests the Contractor shall be entitled to proceed in his absence, and the said tests shall be deemed to have been made in the presence of the Engineer.

(iii) If, in the opinion of the Engineer, the tests are being unduly delayed, he may by notice in writing call upon the Contractor to make such tests within 10 days from the receipt of the said notice and the Contractor shall make the said tests on such day within the said 10 days as the Contractor may fix and of which he shall give notice to the Engineer. If the Contractor fail to make such tests within the time aforesaid the Engineer may himself proceed to make the tests. All tests so made by the Engineer shall be at the risk and expense of the Contractor unless the Contractor shall establish that the tests were not being unduly delayed in which case tests so made shall be at the risk and expense of the Purchaser.

(iv) If any portion of the Works fails to pass the tests, tests of the said portion shall, if required by the Engineer or by the Contractor, be repeated within a reasonable time upon the same terms and conditions, save that all reasonable expenses to which the Purchaser may be put by the repetition of the tests shall be deducted from the Contract Price.

Taking Over.

32.—(i) As soon as the Works have been completed in accordance with the Contract (except in minor respects that do not affect their use for the purpose for which they are intended and except for the maintenance thereof as provided in Clause 34 (*Defects after Taking Over*) and have passed the Tests on Completion the Engineer shall issue a certificate (herein called a "taking-over certificate") in which he shall certify the date on which the Works have been so completed and have passed the said tests and the Purchaser shall be deemed to have taken over the Works on the date so certified, but the issue of a taking-over certificate shall not operate as an admission that the Works have been completed in every respect.

The Purchaser shall not use the Works or any section or portion thereof until a taking-over certificate has been issued in respect thereof. If the Purchaser does so use the Works or any section or portion thereof the Contractor's liability in respect thereof shall be the same as if the Works or any section or portion thereof had been taken over.

(ii) If the Works are divided into two or more sections, Sub-Clause (i) hereof shall apply to each section as it applies to the Works. If by agreement between the Purchaser and the Contractor any portion of the Works (other than a section or sections) shall be taken over before the remainder of the Works, the Engineer shall issue a taking-over certificae in respect of that portion.

(iii) If by reason of any default on the part of the Contractor a taking-over certificate has not been issued in respect of every portion of the Works within one month after the date fixed by the Contract for the completion of the Works, or, if no date be fixed, within a reasonable time, the Purchaser shall be at liberty to use the Works or any portion thereof in respect of which a taking-over certificate has not been issued if and so long as the Works or the portion so used as aforesaid shall be reasonably capable of being used provided that the Contractor shall be afforded reasonable opportunity of taking such steps as may be necessary to permit of the issue of the taking-over certificate. The provisions of Clause 25 (i) (*Liability for Accidents and Damage*) shall not apply to any portion of the Works while being so used by the Purchaser.

Interference with Tests.

(iv) If by reason of any act or omission of the Purchaser, or the Engineer, or some other contractor employed by the Purchaser, the Contractor shall be prevented from carrying out the Tests on Completion as provided in Clause 31 (i) (*Tests on Completion*) then, unless in the meantime the Works shall have been proved not to be substantially in accordance with the Contract, the Purchaser shall be deemed to have taken over the Works, and the Engineer shall issue a taking-over certificate accordingly; nevertheless the Contractor shall make the said tests during the period referred to in Clause 34 (i) as and when required by the Engineer by 14 days' notice in writing, and Sub-Clauses (ii), (iii) and (iv) of Clause 31 and Sub-Clause (vi) of Clause 14 (*Inspection, Testing and Rejection during Manufacture*) shall apply. Any additional expense to which the Contractor may be put in making the said tests pursuant to this sub-clause shall be added to the Contract Price, and such allowances shall be made from the performances required to be attained in the said tests as may be reasonable having regard to any use of the Works by the Purchaser prior to the tests.

Suspension of Works.

33.—(i) All reasonable expenses incurred by the Contractor by reason of the suspension of the Works by the Purchaser or the Engineer (otherwise than in consequence of some default on the part of the Contractor) or by reason of the Contractor being prevented from or delayed in proceeding with the Works by the Purchaser, the Engineer, or some other contractor employed by the Purchaser, shall be added to the Contract Price, provided that no claim shall be made under this clause unless the Contractor has, within a reasonable time after the event giving rise to the claim, given notice in writing to the Engineer of his intention to make such claim.

(ii) If work on the Plant or any portion thereof is suspended as aforesaid by the Purchaser or the Engineer before the Plant or such portion thereof is delivered to the Site and the suspension exceeds three months and the Contractor has suitably and sufficiently marked the Plant or such portion thereof as the Purchaser's property and insured it as provided by Clause 27 (*Insurance of Works*) (the provisions of which clause shall thereafter until actual delivery to the Site apply as if the Plant or such portion thereof were for the time being upon the Site) then the Contractor shall be entitled to have the Contract Value of the Plant or such portion thereof as at the commencement of the suspension included in an interim certificate on the expiration of the said three months or (if later) at the time when, but for such suspension, the Plant or such portion thereof would have been delivered: provided that the Contract Value of any Plant that, according to the decision of the Engineer, is defective or not in accordance with the Contract shall not be included in any such certificate.

Defects after Taking Over.

34.—(i) The Contractor shall be responsible for making good with all possible speed any defect in or damage to any portion of the Works which may appear or occur during a period of 12 months after that portion shall have been taken over and which arises either:—

(*a*) from defective materials, workmanship or design (other than a design made, furnished, or specified by the Purchaser and for which the Contractor has disclaimed in writing responsibility within a reasonable time after receipt of the Purchaser's instructions), or

(*b*) from any act or omission of the Contractor done or omitted during the said period.

(ii) If any such defect shall appear or damage occur the Engineer shall inform the Contractor thereof stating in writing the nature of the defect or damage. If the Contractor replaces or renews any portion of the Works, the provisions of this clause shall apply to the portion of the Works so replaced or renewed until the expiration of 12 months from the date of such replacement or renewal.

(iii) The period of 12 months mentioned in Sub-Clauses (i) and (ii) of this clause shall be extended by a period equal to the period during which the Works or that portion thereof in which a defect to which this clause applies has developed cannot be used by reasons of that defect.

(iv) If any such defect or damage be not remedied within a reasonable time, the Purchaser may proceed to do the work at the Contractor's risk and expense.

(v) If the replacements or renewals are of such a character as may affect the efficiency of the Works or any portion thereof, the Purchaser may within one month of such replacement or renewal give to the Contractor notice in writing requiring that Tests on Completion be made, in which case such tests shall be carried out as provided in Clause 31 (*Tests on Completion*).

(vi) These General Conditions shall apply to all inspections, adjustments, replacements, and renewals and to all tests occasioned thereby, carried out by the Contractor pursuant to this clause.

(vii) The Contractor's liability under this clause shall be in lieu of any condition or warranty implied by law as to the quality or fitness for any particular purpose of any portion of the Works taken over under Clause 32 (*Taking Over*) and save as in this clause expressed neither the Contractor nor his Sub-contractors, servants or agents shall be liable, whether in contract, tort or otherwise, in respect of defects in or damage to such portion, or for any injury, damage or loss of whatsoever kind attributable to such defects or damage. For the purposes of this sub-clause the Contractor contracts on his own behalf and on behalf of and as trustee for his Sub-contractors, servants and agents. Nothing in this clause shall affect the liability of the Contractor under Sub-Clauses (i) and (iii) of Clause 25 (*Liability for Accidents and Damage*) in respect of any portion of the Works not yet taken over.

(viii) Until the final certificate shall have been issued, the Contractor shall have the right of access, at all reasonable working hours, at his own risk and expense, by himself or his duly authorized representatives, whose names shall have previously been communicated in writing to the Engineer, to all parts of the Works for the purpose of inspecting the working thereof and to records of the working and performance thereof for the purpose of inspecting the same and taking notes therefrom. Subject to the Engineer's approval, which shall not be unreasonably withheld, the Contractor may at his own risk and expense make any tests which he considers desirable.

Interim, Taking-over (Payment), and Final Certificates.

35.—(i) The Contractor may at the times and in the manner following apply for interim and final certificates, as referred to in Clause 40 (*Terms of Payment*), for Plant delivered to, and work executed on the Site.

Interim Certificates.

(ii) Applications for interim certificates may be made to the Engineer from time to time during the progress of the Works. Each such application shall state the amount claimed and shall set forth in detail in the order of the schedule of prices particulars of the works executed on the Site and of the Plant delivered to the Site pursuant to the Contract to a date named in the application and since the period covered by the last preceding certificate, if any.

(iii) The Engineer shall issue to the Contractor an interim certificate within 28 days after receiving an application therefor which the Contractor was entitled to make. If the Engineer shall fail to issue an interim certificate as provided in this clause the Contractor shall be at liberty, after giving to the Purchaser or the Engineer 14 days' notice in writing of his intention so to do, to stop the Works or any part thereof until the said certificate be issued, and the expenses of the Contractor occasioned by the stoppage and the subsequent resumption of work shall be added to the Contract Price.

(iv) Every interim certificate shall certify the total value of the work duly executed on the Site and of the Plant delivered to the Site for use in the Works pursuant to the Contract up to the date named in the application for the certificate, less the said total value so certified in the last previous certificate (if any), provided that the value of any Plant that, according to the decision of the Engineer, does not comply with the Contract or had been brought and is at the date of the certificate prematurely upon the Site shall not be included in any such certificate.

(v) No interim certificate shall be relied on as conclusive evidence of any matter stated therein, or prejudice any right of the Purchaser or the Contractor against the other.

Taking-over (Payment) Certificate.

(vi) As soon as a taking-over certificate has been issued under Clause 32 (*Taking Over*) in respect of the Works or any portion thereof application may be made for a taking-over (payment) certificate for the Works or that portion. Each such application shall be accompanied by a statement of final account setting forth the amount which the Contractor claims will be payable to him after the issue of a final certificate.

(vii) The Engineer shall in a taking-over (payment) certificate certify the sum which, subject to the due performance by the Contractor of his obligations under Clause 34 (*Defects after Taking Over*) and to the matters referred to in Sub-clauses (xiii) and (xiv) of this clause, will be certified in a final certificate.

(viii) The Engineer shall issue to the Contractor a taking-over (payment) certificate as soon as reasonably practicable after receiving an application therefor which the Contractor was entitled to make.

Final Certificate.

(ix) Application for the final certificate may be made to the Engineer at any time after the Contractor has ceased to be under any obligation under Clause 34 (*Defects after Taking Over*), provided that if a taking-over certificate has been issued in respect of any portion of the Works, the Contractor may apply for a separate final certificate in respect of each such portion at any time after the said obligation has ceased in relation to such portion, and provided also that, if by reason of the fact that it has become necessary for the Contractor to replace or renew any portion of the Works the obligations of the Contractor under Clause 34 shall continue after the period of 12 months first therein mentioned, the right of the Contractor to apply for a final certificate in respect of the Works or portion thereof other than the portions so replaced or renewed shall not be affected by that fact, and after the Contractor has ceased to be under any obligation under Clause 34 in respect of the portions replaced or renewed he may apply for a final certificate in respect thereof.

(x) The Engineer shall issue to the Contractor a final certificate within one month after receiving an application therefor which the Contractor was entitled to make, or a taking-over (payment) certificate has been issued in respect of the Works whichever shall be the later.

(xi) A final certificate shall certify the amount certified in the taking-over (payment) certificate or the total of the amounts certified in all taking-over (payment) certificates, if more than one has been issued, or, where a final certificate is applied for in respect of a portion of the Works, the amount certified in the taking-over (payment) certificate issued in respect of that portion, subject to such additions thereto or deductions therefrom as may be authorized in Sub-Clause (xiii) of this clause.

(xii) A final certificate shall, save in the case of fraud or dishonesty relating to or affecting any matter dealt with in the certificate, be conclusive evidence as to the sufficiency of the Works and of the value thereof.

Interim and Final Certificates.

(xiii) If any sum shall become payable to the Contractor under the Contract otherwise than for work executed or Plant delivered the amount thereof shall be included in the next certificate (interim, taking-over (payment), or final) issued by the Engineer, and if any sum shall become payable under the Contract by the Contractor to the Purchaser, prior to the issue of the final certificate, whether by deduction from the Contract Price or otherwise, the amount thereof shall be deducted in the next certificate.

(xiv) The Engineer may in any certificate give effect to any correction or modification that should properly be made in respect of any previous certificate.

Provisional Sums.

36.—(i) A provisional sum included in the Contract Price shall be expended or used as the Engineer may in writing direct and not otherwise. In so far as a provisional sum is not expended or used it shall be deducted from the Contract Price.

Prime Cost Items.

(ii) All sums in respect of Prime Cost (P.C.) items included in the Contract Price shall be expended or used as the Engineer may in writing direct and not otherwise. To the net amount paid by the Contractor in respect of each P.C. item there shall be added the percentage named in the Appendix of the said amount. The sum by which the net amount so paid in respect of any P.C. item plus the said percentage thereof exceeds or is less than the sum included in the Contract Price in respect of that item shall be added to or deducted from the Contract Price as the case may be.

(iii) The Contractor shall have no responsibility for work done or Plant supplied by any other person in pursuance of directions given by the Engineer under this clause unless the Contractor shall have approved the person by whom such work is to be done or such Plant is to be supplied and the Plant, if any, to be supplied.

Statutory and other Regulations.

37.—(i) If the cost to the Contractor of the performance of the Contract shall be increased or reduced by reason of the making in the United Kingdom after the date of the tender of any law or of any order, regulation, or by-law having the force of law that shall be applicable to the Works or shall be increased by observing or performing any such condition, restriction, or obligation as is referred to in Clause 19 (*Wayleaves, Consents, Notices, and Restrictions*), the existence and nature of which was not fully disclosed to the Contractor before the date of his tender the amount of such increase or reduction shall be added to or deducted from the Contract Price as the case may be.

Value Added Tax.

(ii) Unless otherwise stated in the tender the Contract Price is deemed to exclude Value Added Tax. To the extent that the Tax is properly chargeable on the supply to the Purchaser of any goods or services provided by the Contractor under the Contract, the Purchaser shall pay such Tax as an addition to payments otherwise due to the Contractor under the Contract.

Measurement for Payment.

38.—Where any part of the Works is to be paid for according to the quantity of Plant supplied or work done the following provisions shall apply:—

(*a*) Such part of the Works shall be measured from time to time by the Engineer or by the Contractor. No such measurement shall be made by either of them without the other being afforded a reasonable opportunity of attending and agreeing the measurements. The Contractor shall at the request of the Engineer open up any part of the Works which may have been covered up without his having been afforded a reasonable opportunity of measuring or agreeing the measurements thereof and the Contractor shall restore the same at his own cost. All measurements shall be made in accordance with the provisions of the Specification respecting methods of measurement.

(*b*) The sum payable in respect of such part of the Works shall be ascertained according to the price or rate appropriate thereto as specified in the Contract. If no appropriate price or rate has been specified the price or rate shall be a fair and reasonable price or rate taking into account any prices or rates that may be specified in the Contract for similar Plant or work.

Payment due from the Contractor.

39.—Without prejudice to any other remedy which the Purchaser may have he shall be entitled to deduct from any moneys due, or becoming due to the Contractor under the Contract, all costs damages or expenses for which under the Contract the Contractor is liable to the Purchaser.

Terms of Payment.

40.—(i) The Purchaser shall pay to the Contractor in the following manner the Contract Price adjusted to give effect to such additions thereto and such deductions therefrom as are provided for in these Conditions:

(*a*) Within 14 days from the presentation of each interim certificate a sum equal to 90 per cent of the sum certified therein.

(*b*) 95 per cent of the Contract Price adjusted as aforesaid within 14 days after the presentation of a taking-over (payment) certificate.

(*c*) The balance of the Contract Price adjusted as aforesaid within one month after the presentation of the final certificate provided that if the Contractor shall have furnished to the Purchaser a guarantee acceptable to the Purchaser for the repayment on demand of such balance he shall be entitled to payment thereof with or at any time after the payment provided for by paragraph (*b*) hereof.

If any section or portion of the Works shall be taken over separately under Clause 32 (*Taking Over*) the payments herein provided for on or after taking over shall be made in respect of the section or portion taken over and reference to the Contract Price shall mean such part of the Contract Price as shall, in the absence of agreement, be apportioned thereto by the Engineer.

In determining the amount of any payment under this clause in respect of any portion of the Works due account shall be taken of all payments previously made in respect of the same portion whether under this clause or under Clause 15 (*Delivery*).

(ii) If at any time at which any payment would fall to be made under paragraph (*b*) or paragraph (*c*) of Sub-Clause (i) of this clause there shall be any defect in any portion of the Works in respect of which such payment is proposed, the Purchaser may retain the whole of such payment provided that in the event of the said defect being of a minor character and not such as to affect the use of the Works or the said portion thereof for the purpose intended without serious risk the Purchaser shall not retain a greater sum than represents the cost of making good the said minor defect. Any sum retained by the Purchaser pursuant to the provisions of this sub-clause shall be paid to the Contractor upon the said defect being made good.

(iii) If the payment of any sum payable under paragraphs (*b*) or (*c*) of Sub-Clause (i) of this clause shall be improperly delayed by the Purchaser or the Engineer interest at the rate of one and one half per cent per annum over the Bank of England minimum lending rate from time to time in force on the amount of the delayed payment for the period of the delay shall be added to the Contract Price.

(iv) If the Purchaser shall fail to make any payment as provided in this clause the Contractor shall be at liberty, without prejudice to any other remedy, after giving to the Purchaser 14 days' notice in writing of his intention so to do, to stop the Works or any part thereof until the said payment be made, and the expenses of the Contractor occasioned by the stoppage and the subsequent resumption of work shall be added to the Contract Price.

Arbitration.

41.—(i) If at any time any question, dispute or difference shall arise between the Purchaser and the Contractor, either party shall, as soon as reasonably practicable, give to the other notice in writing of the existence of such question, dispute or difference, specifying its nature and the point at issue, and the same shall be referred to the arbitration of a person to be agreed upon, or failing such agreement, to some person appointed on the application of either of the parties hereto by the President for the time being of The Institution of Electrical Engineers provided that a question, dispute or difference relating to a decision, instruction, or order of the Engineer shall not be referred to arbitration unless notice has been given by the Contractor in accordance with Clause 23 (*b*) (*Engineer's Decisions*).

(ii) Performance of the Contract shall continue during arbitration proceedings unless the Engineer shall order the suspension thereof or of any part thereof, and if any such suspension shall be ordered the reasonable expenses of the Contractor occasioned by such suspension shall be added to the Contract Price. No payments due or payable by the Purchaser shall be withheld on account of a pending reference to arbitration.

Metrication.

42.—(i) If any Plant described in the Contract or the subject of a variation (hereinafter called "a variation order") under Clause 11 (*Variations and Omissions*) is described by dimensions in the metric or imperial measure and the Contractor cannot procure such Plant in the measure specified in sufficient time to avoid delay in the performance of his other obligations under the Contract, but can obtain such Plant in the other measure to dimensions approximating to those described in the Contract or in the variation order, then the Contractor shall forthwith give notice to the Engineer of the facts stating the dimensions to which such Plant is procurable in the other measure.

(ii) The Engineer shall within 14 days after the receipt of a notice under the preceding sub-clause give notice to the Contractor in writing pursuant to Clause 11 (*Variations and Omissions*). Such notice shall either:—

(*a*) direct the Contractor to supply such Plant to the dimensions stated in the said notice instead of to the dimensions described in the Contract or variation order as the case may be, or,

(*b*) direct the Contractor to make some other variation whereby the need to supply such plant to the dimensions described in the Contract or variation order will be avoided.

The provisions of the Contract shall apply to directions given under this clause as though such directions had been included in the Specification.

Construction of Contract.

43.—The Contract shall in all respects be construed and operate as an English contract and in conformity with English law, and all payments thereunder shall be made in sterling money. The marginal notes hereto shall not affect the construction hereof.

Variation Costs.

44.—* If, by reason of any rise or fall in the rates of wages payable to labour or in the cost of material or transport or of conforming to such laws, orders, regulations and by-laws as are applicable to the Works above or below such rates and costs ruling at the date of the tender, the cost to the Contractor of performing his obligations under this Contract shall be increased or reduced, the amount of such increase or reduction shall be added to or deducted from the Contract Price as the case may be, provided that no account shall be taken of any amount by which any cost incurred by the Contractor has been increased by the default or negligence of the Contractor. For the purpose of this clause "the cost of material" shall be construed as including any duty or tax by whomsoever payable which is payable under or by virtue of any Act of Parliament on the import, purchase, sale, appropriation, processing or use of such material.

* **NOTE:—Unless this Clause is excluded any quoted price will be subject to adjustment in accordance with its terms.**

AGREEMENT

This Agreement is made the day of 19

BETWEEN

(hereinafter referred to as the "Contractor") of the one part and

(hereinafter called the "Purchaser") of the other part **Whereas** the Purchaser desires to have provided and executed certain Works mentioned, enumerated, or referred to in certain General Conditions, Specification, Schedules, Drawings, Plans, Schedule of Prices, the Contractor's Tender and covering letter (if any), the acceptance of the Tender, and any other relevant correspondence* which for the purpose of identification have been signed by on behalf of the Contractor and by of (the Engineer of the Purchaser) on behalf of the Purchaser all of which are deemed to form part of this Contract as though separately set out herein and are included in the expression "Contract" whenever herein used. **And Whereas** the Purchaser has accepted the Tender of the Contractor for the provision and execution of the said Works for the sum of

(hereinafter called "the Contract Price) upon the terms and subject to the Conditions hereinafter mentioned: **Now this Deed Witnesseth** and it is hereby agreed and declared as follows, that is to say, in consideration of the payments to be made to the Contractor by the Purchaser as hereinafter mentioned the Contractor hereby covenants with the Purchaser, his legal personal representatives, successors, and assigns that the Contractor shall and will duly provide, execute, complete, and maintain the said Works and shall do and perform all other acts and things in the Contract mentioned or described or which are to be implied therefrom or may be reasonably necessary for the completion of the said Works within and at the times and in the manner and subject to the terms conditions and stipulations mentioned in the Contract.

And in consideration of the due provision, execution and completion of the Works and the maintenance thereof as aforesaid the Purchaser does hereby for himself his legal personal representatives, successors, and assigns covenant with the Contractor that he, the Purchaser, his legal personal representatives, successors, or assigns will pay to the Contractor the Contract Price or such other sum as may become payable to the Contractor under the provisions of the Contract, such payments to be made at such time and in such manner as is provided by the Contract.

In Witness whereof, etc.

*Delete as required; see Memorandum on the use of the General Conditions, paragraph B.

APPENDIX

Clause.		
30		Delay in Completion:
	(a)	*(a)* percentage of Contract Value to be deducted as damages.
	(b)	*(b)* maximum percentage of Contract Value which the deductions may not exceed.
36	....	Percentage on Prime Cost items.

FORM OF GUARANTEE

(When required, and in cases where an Insurance Policy is not used.)

Whereas by an Agreement dated .... and made between .... (hereinafter called "the Purchaser") and .... (hereinafter called "the Contractor") the parties thereto entered into a contract as therein stated: NOW we hereby jointly and severally guarantee to the Purchaser punctual true and faithful performance and observance by the Contractor of the covenant on his part contained in the said Agreement and undertake to be responsible to the Purchaser his legal personal representatives successors or assigns as Sureties for the Contractor for the payment by him of all sums of money losses damages costs charges and expenses that may become due or payable to the Purchaser his legal personal representatives successors or assigns by or from the Contractor by reason or in consequence of the default of the Contractor in the performance or observance of his said covenant but so nevertheless that the total amount to be demanded or recovered by the Purchaser his legal personal representatives successors or assigns of or from us as Sureties shall not exceed £ *

This Guarantee shall not be revocable by notice or by reason of the death of us or either of us and our liability as Sureties hereunder shall not be impaired or discharged by any extensions of time or variations or alterations made given conceded or agreed (with or without our knowledge or consent) under the General Conditions referred to in the said Agreement or (where the Purchaser or the Contractor is a firm) by any change in the constitution of the Purchaser's or the Contractor's respective firms.

*See Clause 3 (Security for due performance).

INDEX

JULY 1976 AMENDMENTS TO IEE MODEL FORM OF GENERAL CONDITIONS OF CONTRACT E/1973 SECOND EDITION.

1. Delete Clauses 25, 26 and 27 in toto and substitute the following Clauses 25, 26 and 27.

Liability for Accidents and Damage

25.—(i) The Contractor shall properly cover up and protect until taken over under Clause 32 (*Taking Over*) any section or portion of the Works liable to injury by exposure to the weather, and shall take every reasonable precaution to protect any section or portion of the Works not taken over against loss or damage from any cause.

(ii) In the case of loss of or damage to the Works on the Site arising from or occasioned by causes for which the Contractor is not responsible under the Contract the same shall, if required by the Purchaser, be made good by the Contractor but at the cost of the Purchaser at a price to be agreed between the Contractor and the Purchaser or in default of agreement to be settled by arbitration and such cost shall be added to the Contract Price.

(iii) Subject to Sub-Clauses (vi) and (vii) of this clause and Clause 26 (*Limitation on Contractor's Liability*), all losses of and damage to any section or portion of the Works that shall not have been taken over under Clause 32 (*Taking Over*), which shall arise from or be occasioned by any act of the Contractor or any Sub-contractor or by a failure of the Contractor to comply with any obligation imposed on him by Sub-Clause (i) of this clause, shall be made good by and at the sole cost of the Contractor and to the reasonable satisfaction of the Engineer.

(iv) The Contractor shall, subject to Sub-Clauses (vi) and (vii) of this clause and Clause 26 (*Limitation on Contractor's Liability*), indemnify the Purchaser in respect of all damage or injury occurring before all the Works shall have been taken over under Clause 32 (*Taking Over*) to any person or to any property (other than property forming part of the Works not yet taken over) and against all actions, suits, claims, demands, costs, charges, and expenses arising in connection therewith which shall be occasioned by the negligence of or breach of statutory duty by the Contractor or any Sub-contractor, or by defective design (other than a design made, furnished, or specified by the Purchaser and for which the Contractor has disclaimed responsibility in writing within a reasonable time after the receipt of the Purchaser's instructions), materials, or workmanship, but not otherwise. Provided that the Contractor shall not be liable by virtue of this sub-clause in respect of damage or injury attributable to defects in any section or portion of the Works taken over under Clause 32.

(v) If there shall occur any loss of or damage to any property or injury to any person while the Contractor is on the Site for the purpose of making good a defect in any section or portion of the Works pursuant to Clause 34 (*Defects after Taking Over*), or for the purpose of carrying out tests on completion during the period referred to in Clause 34 as provided in Clause 32 (iv) (*Interference with Tests*), the Contractor shall be liable, subject to the provisions of Sub-Clauses (vi) and (vii) of this clause and Clause 26 (*Limitation on Contractor's Liability*) as follows:

(*a*) In respect of loss of or damage to the said section or portion the Contractor's liability shall be as defined in Clause 34 (*Defects after Taking Over*).

(*b*) In respect of damage to any other property or injury to any person and of any actions, claims, demands, costs, charges and expenses arising in connection therewith the Contractor shall be liable, subject to the provisions of Sub-Clauses (i) and (iii) of this clause, to the extent that such damage or injury was caused by the negligence or breach of statutory duty of the Contractor or a Sub-contractor while on the Site as aforesaid or by defective materials or workmanship used in making good the said defect but not otherwise.

The said section or portion of the Works shall be defined by reference to the taking-over certificate issued in respect thereof pursuant to Clause 32 (*Taking Over*).

(vi) The Contractor shall not be liable to the Purchaser for:

(*a*) any damage or injury to the extent that it is caused by or arises from the acts or omission of the Purchaser or of others (not being the Contractor's servants or Sub-contractors),

(*b*) any loss or damage in circumstances over which the Contractor has no control.

(vii) Except in respect of personal injury or damage to property conferring on a person other than the Purchaser a good cause of action against the Contractor, the liability of the Contractor to the Purchaser under this clause for any one act or default shall not exceed the Contract Price or £100,000 whichever is the greater.

(viii) In the event of any claim being made against the Purchaser arising out of the matters referred to in and in respect of which the Contractor may be liable under this clause, the Contractor shall be promptly notified thereof, and may at his own expense conduct all negotiations for the settlement of the same and any litigation that may arise therefrom. The Purchaser shall not, unless and until the Contractor shall have failed to take over the conduct of the negotiations or litigation, make any admission which might be prejudicial thereto. The conduct by the Contractor of such negotiations or litigation shall be conditional upon the Contractor having first given to the Purchaser such reasonable security as shall from time to time be required by the Purchaser to cover the amount ascertained or agreed or estimated, as the case may be, of any compensation, damages, expenses, and costs for which the Purchaser may become liable. The Purchaser shall, at the request of the Contractor, afford all available assistance for any such purpose, and shall be repaid all reasonable expenses incurred in so doing.

Limitation on Contractor's Liability

26.—Subject as provided in Clause 30 (*Delay in Completion*) for the deduction of liquidated damages for delay, the Contractor shall not be liable to the Purchaser by way of indemnity or by reason of any breach of the Contract for loss of use (whether complete or partial) of the Works or of profit or of any contract that may be suffered by the Purchaser.

Insurance of Works.

27.—Unless the Purchaser shall have approved in writing other arrangements for the insurance hereinafter mentioned the Contractor shall in the joint names of the Contractor and the Purchaser insure, and keep insured until a taking-over certificate has been issued under Clause 32 (*Taking Over*), such Works as may for the time being be upon the Site against loss, damage or destruction by fire, explosion, lightning, earthquake, malicious damage, theft, flood, storm, tempest, and aircraft and other aerial devices or articles dropped therefrom for the full replacement value thereof, and shall from time to time, when so required by the Engineer, produce the policy and receipts for the premiums. All moneys received under any such policies shall be applied in or towards the replacement and repair of the Works lost, damaged or destroyed but this provision shall not affect the Contractor's liabilities under the Contract.

2. In Sub-Clause 40 (iii) delete 'one and one half per cent' and substitute 'two per cent'.

A

Edition June, 1979

Conditions of Sale (A) for machinery and equipment (Exclusive of Erection) United Kingdom

1. GENERAL.—The acceptance of our tender includes the acceptance of the following terms and conditions :—

2. VALIDITY.—Unless previously withdrawn, our tender is open for acceptance within the period stated therein or, when no period is so stated, within thirty days only after its date.

3. ACCEPTANCE.—The acceptance of our tender must be accompanied by sufficient information to enable us to proceed with the order forthwith, otherwise we shall be at liberty to amend the tender prices to cover any increase in cost which has taken place after acceptance. Any samples submitted to you and not returned to our works within one month from date of receipt shall be paid for by you.

4. PACKING.—Unless otherwise specified in our tender, all packing cases, skids, drums and other packing materials must be returned to our works at your expense and in good condition within one month from date of receipt. If not so returned they will be charged for.

5. LIMITS OF CONTRACT.—Our tender includes only such goods, accessories and work as are specified therein.

6. DRAWINGS, ETC.—All specifications, drawings, and particulars of weights and dimensions submitted with our tender are approximate only, and the descriptions and illustrations contained in our catalogues, price lists and other advertisement matter are intended merely to present a general idea of the goods described therein, and none of these shall form part of the contract. After acceptance of our tender a set of certified outline drawings will be supplied free of charge on request.

7. INSPECTION AND TESTS.—Our products are carefully inspected and, where practicable, submitted to our standard tests at our works before despatch. If tests other than those specified in our tender or tests in the presence of you or your representative are required, these will be charged for. In the event of any delay on your part in attending such tests or in carrying out any inspection required by you after seven days' notice that we are ready, the tests will proceed in your absence and shall be deemed to have been made in your presence.

8. PERFORMANCE.—We will accept no liability for failure to attain any performance figures quoted by us unless we have specifically guaranteed them, subject to any tolerances specified or agreed to by us, in an agreed sum as liquidated damages.

If the performance figures obtained on any test provided for in the contract are outside the acceptance limits specified therein, you will be entitled to reject the goods.

Before you become entitled to claim liquidated damages or to reject the goods we are to be given reasonable time and opportunity to rectify their performance. If you become entitled to reject goods, we will repay to you any sum paid by you to us on account of the contract price thereof and any sum that may have accrued due to you in respect of delay in despatch under Clause 9 up to the date of such rejection.

You assume responsibility that goods stipulated by you are sufficient and suitable for your purpose save in so far as your stipulations are in accordance with our advice.

9. LIABILITY FOR DELAY.—Any times quoted for despatch or delivery are to date from receipt by us of a written order to proceed and of all necessary information and drawings to enable us to put the work in hand. The time for despatch or delivery shall be extended by a reasonable period if delay in despatch or delivery is caused by instructions or lack of instructions from you or by industrial dispute or by any cause beyond our reasonable control.

If a fixed time be quoted for despatch or delivery, and we fail to despatch or deliver within that time or within any extension thereof provided by this clause, and if as a result you shall have suffered loss, we undertake to pay for each week or part of a week of delay, liquidated damages at the rate of.............. per cent. up to a maximum of.............. per cent. of that portion of the price named in the contract which is referable to such portion only of the contract goods as cannot in consequence of the delay be used commercially and effectively. Such payment shall be in full satisfaction of our liability for delay.

Any time described as an estimate shall not be construed as a fixed time quoted for the purpose of this clause.

10. VARIATIONS.—In the event of variation or suspension of work by your instructions or lack of instructions the contract price shall be adjusted accordingly.

11. DELIVERY.—Unless otherwise specified in our tender, the price quoted includes delivery by any method of transport at our option.

Unless otherwise specified, we shall not be responsible for offloading.

12. LOSS OR DAMAGE IN TRANSIT.—When the price quoted includes delivery other than at our works, we will repair or at our option replace free of charge goods lost or damaged in transit; Provided that we are given written notification of such loss or damage within such time as will enable us to comply with the carrier's conditions of carriage as affecting loss or damage in transit or, where delivery is made by our own transport, within a reasonable time after receipt of the Advice Note.

13. TERMS OF PAYMENT.—Unless otherwise agreed, payment in full shall be due for goods on notification by us that they are ready for despatch.

14. STORAGE.—If we do not receive forwarding instructions sufficient to enable us to despatch the goods within 14 days after the date of notification that they are ready for despatch, you shall take delivery or arrange for storage. If you do not take delivery or arrange for storage, we shall be entitled to arrange storage either at our own works or elsewhere on your behalf and all charges for storage, for insurance or for demurrage shall be payable by you.

15. DEFECTS AFTER DELIVERY.—We will make good, by repair or at our option by the supply of a replacement, defects which, under proper use, appear in the goods within a period of twelve calendar months after the goods have been delivered and arise solely from faulty design (other than a design made, furnished or specified by you for which we have disclaimed responsibility in writing), materials or workmanship: Provided always that defective parts have been returned to us if we shall have so required. We shall refund the cost of carriage on such returned parts and the repaired or new parts will be delivered by us free of charge as provided in Clause 11 (Delivery).

Our liability under this clause shall be in lieu of any warranty or condition implied by law as to the quality or fitness for any particular purpose of the goods, and save as provided in this clause we shall not be under any liability, whether in contract, tort or otherwise, in respect of defects in goods delivered or for any injury (other than personal injury caused by our negligence as defined in Section 1 of the Unfair Contract Terms Act, 1977), damage or loss resulting from such defects or from any work done in connection therewith.

Continued overleaf

16. PATENTS.—We will indemnify you against any claim for infringement of Letters Patent, Registered Design, Trade Mark or Copyright (published at the date of the contract) by the use or sale of any article or material supplied by us to you and against all costs and damages which you may incur in any action for such infringement or for which you may become liable in any such action. Provided always that this indemnity shall not apply to any infringement which is due to our having followed a design or instruction furnished or given by you or to the use of such article or material in a manner or for a purpose or in a foreign country not specified by or disclosed to us, or to any infringement which is due to the use of such article or material in association or combination with any other article or material not supplied by us. And provided also that this indemnity is conditional on your giving to us the earliest possible notice in writing of any claim being made or action threatened or brought against you and on your permitting us at our own expense to conduct any litigation that may ensue and all negotiations for a settlement of the claim. You on your part warrant that any design or instruction furnished or given by you shall not be such as will cause us to infringe any Letters Patent, Registered Design, Trade Mark or Copyright in the execution of your order.

17. LIABILITY FOR ACCIDENTS AND DAMAGE.—If we, our agents or sub-contractors are on site for the purposes of the contract then, notwithstanding the provisions of Clause 15 we will indemnify you against direct damage or injury to your property or person or that of others occurring while we are working on site to the extent caused by the negligence of ourselves, our sub-contractors or agents, but not otherwise, by making good such damage to property or compensating personal injury. Provided that :

(a) our total liability for damage to your property (including damage caused by our breach of contract, tort or breach of statutory duty) shall not exceed £100,000 or the contract price, whichever sum is the greater, and—

(b) we shall not be liable to you for any loss of profit or of contracts or, save as aforesaid, for any loss or damage of any kind wh. tsoever and whether caused by our breach of contract, tort, breach of statutory duty or otherwise howsoever.

Save as provided in Clause 15, we shall not be liable for any damage or injury occurring after our completion of work on site.

18. FAIR WAGES CLAUSE.—We undertake to be bound by a fair wages clause in the terms of the House of Commons Resolution of 14th October, 1946. We also undertake that when the installation of machinery and equipment is carried out by men sent from the Contractor's or Sub-Contractor's establishment they shall receive the time rate payable in terms of the Fair Wages Clause to such workpeople in such establishment, and in addition shall receive the outworking allowances recognised for outworkers sent from such establishment.

19. ARBITRATION.—If at any time any question, dispute or difference whatsoever shall arise between you and ourselves upon, in relation to, or in connection with the contract, either of us may give to the other notice in writing of the existence of such question, dispute, or difference, and the same shall be referred to the arbitration of a person to be mutually agreed upon, or failing agreement within 30 days of receipt of such notice, of some person appointed by the President for the time being of the Institution of Electrical Engineers.

20. LEGAL CONSTRUCTION.—Unless otherwise agreed in writing the contract shall in all respects be construed and operate as an English contract and in conformity with English law.

21. STATUTORY AND OTHER REGULATIONS.—If the cost to us of performing our obligations under the contract shall be increased or reduced by reason of the making or amendment after the date of tender of any law or of any order, regulation, or bye-law having the force of law that shall affect the performance of our obligations under the contract, the amount of such increase or reduction shall be added to or deducted from the contract price as the case may be.

Published by: The British Electrical and Allied Manufacturers' Association Limited, 8 Leicester Street, Leicester Square, London, WC2H 7BN

AE

Edition December, 1980

Conditions of Sale (AE) for machinery and equipment (Exclusive of Erection) Export F.O.B., F.O.R. and F.O.T.

1. **DEFINITIONS.**–The expressions F.O.B., F.O.R., and F.O.T. shall bear the meanings assigned to them in **INCOTERMS 1980** save insofar as the same may have been varied by these Conditions.

2. **GENERAL.**–The acceptance of our tender includes the acceptance of the following terms and conditions:–

3. **VALIDITY.**–Unless previously withdrawn, our tender is open for acceptance for the period stated therein or, when no period is so stated, within thirty days only after its date.

4. **ACCEPTANCE.**–The acceptance of our tender must be accompanied by sufficient information to enable us to proceed with the order forthwith, otherwise we shall be at liberty to amend the tender prices to cover any increase in cost which has taken place after acceptance. Any samples submitted to you and not returned to our works within one month from date of receipt shall be paid for by you.

5. **PACKING.**–Unless otherwise specified in our tender, packing in accordance with our standard export practice is included.

6. **LIMITS OF CONTRACT.**–Our tender includes only such goods, accessories, and work as are specified therein.

7. **DRAWINGS, ETC.**–All descriptive and shipping specifications, drawings, and particulars of weights and dimensions submitted with our tender are approximate only, and the descriptions and illustrations contained in our catalogues, price lists, and other advertisement matter are intended merely to present a general idea of the goods described therein, and none of these shall form part of the contract. After acceptance of our tender, a set of certified outline drawings will be supplied free of charge.

8. **INSPECTION AND TESTS.**–Our products are carefully inspected and, where practicable, submitted to our standard tests at our works before despatch. If tests other than those specified in our tender or tests in the presence of you or your representative are required, these will be charged for. In the event of any delay on your part in attending such tests or in carrying out any inspection required by you after seven days' notice that we are ready, the tests will proceed in your absence and shall be deemed to have been made in your presence, and the inspection will be deemed to have been made by you.

9. **PERFORMANCE.**–Any performance figures given by us are based upon our experience and are such as we expect to obtain on test in our works. We shall be under no liability for damages for failure to attain such figures unless we have specifically guaranteed performance figures subject to the recognised tolerances applicable to such figures in an agreed sum as liquidated damages with provision for a corresponding bonus.

If the performance figures obtained on test in our works are outside the acceptance limits specified in the contract you will be entitled to reject the goods.

Before you become entitled to claim liquidated damages or to reject the goods, we are to be given reasonable time and opportunity to rectify their performance. If you become entitled to reject goods, we will pay any sum that may have accured due to you up to the date of such rejection.

You assume responsibility that goods stipulated by you are sufficient and suitable for your purpose save in so far as your stipulations are in accordance with our advice.

10. **DELIVERY.**–

(i) If the contract provides for delivery F.O.B. we shall deliver the goods on board a vessel named by you or on your behalf at the port stated in our tender and we shall not be required to give you the notice relating to insurance mentioned in Section 32(3) of the Sale of Goods Act, 1979

(ii) If the contract provides for delivery F.O.R. or F.O.T. we shall at our own cost (except as mentioned in this sub-clause) do what may be necessary to load the goods on a railway vehicle, container or truck suitable for any further carriage by rail or road (as the case may be) and to place the goods at your disposal at the place of delivery specified in our tender, and shall promptly give you notice in writing when the goods are loaded or placed in the custody of the railway or carrier. The railway or carriers' conditions of carriage for international deliveries shall apply. You will be responsible for paying promptly all customs and import duties and transit dues which may become payable in order to convey the goods to the place of delivery specified in the contract.

11. **LIABILITY FOR DELAY.**–Any times quoted for despatch or delivery are to date from receipt by us of a written order to proceed and of all necessary information and drawings to enable us to put the work in hand. The time for despatch or delivery shall be extended by a reasonable period if delay in despatch or delivery is caused by instructions or lack of instructions from you or by industrial dispute or by any cause beyond our reasonable control.

If a fixed time be quoted for despatch or delivery, and we fail to despatch or deliver within that time or within any extension thereof provided by this clause, and if as a result you shall have suffered loss, we undertake to pay for each week or part of a week of delay, liquidated damages at the rate of ....per cent.,up to a maximum of ....per cent.,of that portion of the price named in the contract which is referable to such portion only of the contract goods as cannot in consequence of the delay be used commercially and effectively. Such payment shall be in full satisfaction of our liability for delay.

Any time described as an estimate shall not be construed as a fixed time quoted for the purpose of this clause.

12. **VARIATIONS.**–In the event of variation or suspension of the work by your instructions or lack of instructions, the contract price shall be adjusted accordingly.

13. **TERMS OF PAYMENT.**–The prices quoted are strictly net and payment in full shall be due as follows:–

(i) In the case of goods delivered F.O.B., on presentation of shipping documents and invoices in the United Kingdom.

(ii) In the case of goods delivered F.O.R. or F.O.T., upon receipt of notice in writing that the goods have been loaded or placed in the custody of the railway or carrier.

Or, if we are unable by reason of your instructions or lack of instructions to deliver goods when ready, payment in full shall be due upon presentation of invoices and notification from us that the goods are ready for despatch.

14. **STORAGE.**–If we do not receive forwarding instructions sufficient to enable us to despatch within fourteen days after notification that the goods have been tested under Clause 8 or that they are ready for despatch, you shall take delivery or arrange for storage. If you do not take delivery or arrange for storage, we shall be entitled to arrange storage either at our own works or elsewhere on your behalf and all charges for storage, for insurance or for demurrage shall be payable by you.

Continued overleaf

15. **DEFECTS AFTER DELIVERY.**–We will made good, by repair or at our option by the supply of a replacement, defects which, under proper use, appear in the goods within a period of twelve calendar months after the goods have been delivered or, if delivery is delayed by reason of your instructions or lack of instructions, within a period of 18 months after the goods have been notified as ready for despatch (whichever period expires the earlier) and arise solely from faulty design (other than a design made, furnished or specified by you for which we have disclaimed responsibility in writing), materials or workmanship: Provided always that defective parts have been returned to us if we shall have so required. We shall refund the cost of carriage on such returned parts and the repaired or new parts will be delivered by us free of charge as provided in Clause 10 (Delivery).

Our liability under this clause shall be in lieu of any warranty or condition implied by law as to the quality or fitness for any particular purpose of the goods, and save as provided in this clause we shall not be under any liability, whether in contract, tort or otherwise, in respect of defects in goods delivered or for any injury,damage or loss resulting from such defects or from any work done in connection therewith.

16. **PATENTS.**–We will indemnify you against any claim of infringement of Letters Patent, Registered Design, Trade Mark or Copyright (published at the date of the Contract) by the use or sale of any article or material supplied by us to you and against all costs and damages which you may incur in any action for such infringement or for which you may become liable in any such action. Provided always that this indemnity shall not apply to any infringement which is due to our having followed a design or instruction furnished or given by you or to the use of such article or material in a manner or for a purpose or in a foreign country not specified by or disclosed to us, or to any infringement which is due to the use of such article or material in association or combination with any other article or material not supplied by us. And provided also that this indemnity is conditional on your giving to us the earliest possible notice in writing of any claim being made or action threatened or brought against you and on your permitting us at our own expense to conduct any litigation that may ensue and all negotiations for a settlement of the claim. You on your part warrant that any design or instruction furnished or given by you shall not be such as will cause us to infringe any Letters Patent, Registered Design, Trade Mark or Copyright in the execution of your order.

17. **LIABILITY FOR ACCIDENTS AND DAMAGE.**–If we, our agents or sub-contractors are on site for the purposes of the contract then, notwithstanding the provisions of Clause 15, we will indemnify you against direct damage or injury to your property or person or that of others occurring while we are working on site to the extent caused by the negligence of ourselves, our sub-contractors or agents, but not otherwise, by making good such damage to property or compensating personal injury.

Provided that:

(a) our total liability for damage to your property (including damage caused by our breach of contract, tort or breach of statutory duty) shall not exceed £100,000 (United Kingdom currency) or the contract price, whichever is the greater, and–

(b) we shall not be liable to you for any loss of profit or of contracts or, save as aforesaid, for any loss, damage or injury of any kind whatsoever and whether caused by our breach of contract, tort, breach of statutory duty or otherwise howsoever.

Save as provided in Clause 15, we shall not be liable for any damage or injury occurring after our completion of work on site.

18. **ARBITRATION.**–If at any time any question, dispute, or difference whatsoever shall arise between you and ourselves upon, in relation to, or in connection with the contract, either of us may give to the other notice in writing of the existence of such question, dispute, or difference, and the same shall be referred to the arbitration of a person to be mutually agreed upon, or failing agreement within thirty days of the receipt of such notice, of some person appointed by the President for the time being of the Institution of Electrical Engineers.

19. **LEGAL CONSTRUCTION.**–Unless otherwise agreed in writing the contract shall in all respects be construed and operate as an English contract and in conformity with English law and the English courts shall have exclusive jurisdiction over any matter arising out of the provisions of Clause 18 (Arbitration).

20. **STATUTORY AND OTHER REGULATIONS.** — If the cost to us of performing our obligations under the contract shall be increased or reduced by reason of the making or amendment after the date of tender of any law or of any order, regulation, or bye-law having the force of law that shall affect the performance of our obligations under the contract, the amount of such increase or reduction shall be added to or deducted from the contract price as the case may be.

Published by: The British Electrical and Allied Manufacturers Association Limited, 8 Leicester Street, Leicester Square, London, WC2H 7BN

AEC

Edition December, 1980

Conditions of Sale (AEC) for machinery and equipment
(Exclusive of Erection)
Export C.I.F., and C. & F.

1. **DEFINITIONS.**—The expressions C.I.F. and C. & F. shall bear the meanings assigned to them in INCOTERMS 1980 save insofar as the same may have been varied by these Conditions.

2. **GENERAL.**—The acceptance of our tender includes the acceptance of the following terms and conditions:—

3. **VALIDITY.**—Unless previously withdrawn, our tender is open for acceptance for the period stated therein or, when no period is so stated, within thirty days only after its date.

4. **ACCEPTANCE.**–The acceptance of our tender must be accompanied by sufficient information to enable us to proceed with the order forthwith, otherwise we shall be at liberty to amend the tender prices to cover any increase in cost which has taken place after acceptance. Any samples submitted to you and not returned to our works within one month from date of receipt shall be paid for by you.

5. **PACKING.**—Unless otherwise specified in our tender, packing in accordance with our standard export practice is included.

6. **LIMITS OF CONTRACT.** Our tender includes only such goods, accessories, and work as are specified therein.

7. **DRAWINGS, ETC.**—All descriptive and shipping specifications, drawings, and particulars of weights and dimensions submitted with our tender are approximate only, and the descriptions and illustrations contained in our catalogues, price lists, and other advertisement matter are intended merely to present a general idea of the goods described therein, and none of these shall form part of the contract. After acceptance of our tender, a set of certified outline drawings will be supplied free of charge.

8. **INSPECTION AND TESTS.**–Our products are carefully inspected and, where practicable, submitted to our standard tests at our works before despatch. If tests other than those specified in our tender or tests in the presence of you or your representative are required, these will be charged for. In the event of any delay on your part in attending such tests or in carrying out any inspection required by you after seven days' notice that we are ready, the tests will proceed in your absence and shall be deemed to have been made in your presence, and the inspection will be deemed to have been made by you.

9. **PERFORMANCE.**—Any performance figures given by us are based upon our experience and are such as we expect to obtain on test in our works. We shall be under no liability for damages for failure to attain such figures unless we have specifically guaranteed performance figures subject to the recognised tolerances applicable to such figures in an agreed sum as liquidated damages with provision for a corresponding bonus.

If the performance figures obtained on test in our works are outside the acceptance limits specified in the contract you will be entitled to reject the goods.

Before you become entitled to claim liquidated damages or to reject the goods, we are to be given reasonable time and opportunity to rectify their performance. If you become entitled to reject goods, we will pay any sum that may have accrued due to you up to the date of such rejection.

You assume responsibility that goods stipulated by you are sufficient and suitable for your purpose save in so far as your stipulations are in accordance with our advice.

10. **DELIVERY.**—Delivery will be made C.I.F. or C. & F. (as the case may be) at the port stated in our tender. No lighterage, landing charges, dock, wharf or customs dues are included. Freight and insurance charges (where applicable) are based on the rates obtainable at the date of our tender. If these rates have increased or decreased from any cause between the date of tender and the date on which the goods are shipped the contract price will be increased or decreased by the net amount of the increase or decrease due to the variation of such rates. In the case of C.I.F. contracts a document certifying in proper form that insurance has been effected (and whether or not other goods are included in or covered by such insurance) and endorsed by us may, at our option, be tendered instead of a policy of insurance in respect of any goods shipped, you shall accept such documents in lieu of any policy together with invoice or invoices and bill or bills of lading, as complete tender of shipping documents by us.

In the case of C. & F. contracts, marine insurance having been omitted, we shall, if so requested in good time by you, be prepared to give you such notice as will enable you to insure the goods during sea transit. In the absence of such request we shall not be liable for failure to give you such notice under Section 32 (3) of the Sale of Goods Act, 1979.

11. **LIABILITY FOR DELAY.**—Any times quoted for despatch or delivery are to date from receipt by us of a written order to proceed and of all necessary information and drawings to enable us to put the work in hand. The time for despatch or delivery shall be extended by a reasonable period if delay in despatch or delivery is caused by instructions or lack of instructions from you or by industrial dispute or by any cause beyond our reasonable control.

If a fixed time be quoted for despatch or delivery, and we fail to despatch or deliver within that time or within any extension thereof provided by this clause, and if as a result you shall have suffered loss, we undertake to pay for each week or part of a week of delay, liquidated damages at the rate of............per cent. up to a maximum of ....per cent. of that portion of the price named in the contract which is referable to such portion only of the contract goods as cannot in consequence of the delay be used commercially and effectively. Such payment shall be in full satisfaction of our liability for delay.

Any time described as an estimate shall not be construed as a fixed time quoted for the purpose of this clause.

12. **VARIATIONS.**—In the event of variation or suspension of the work by your instructions or lack of instructions, the contract price shall be adjusted accordingly.

13. **TERMS OF PAYMENT.**—The prices quoted are strictly net and payment in full shall be due for any goods shipped upon presentation of shipping documents and invoices in the United Kingdom, or, if we are unable by reason of your instructions or lack of instructions to ship goods when ready, upon presentation of invoices and notification from us that the goods are ready for despatch.

14. **STORAGE.**—If we do not receive forwarding instructions sufficient to enable us to despatch the goods within 14 days after notification that the goods have been tested under Clause 8 or that they are ready for despatch, you shall take delivery or arrange for storage. If you do not take delivery or arrange for storage, we shall be entitled to arrange storage either at our own works or elsewhere on your behalf and all charges for storage, for insurance or for demurrage shall be payable by you.

continued overleaf

15. **DEFECTS AFTER DELIVERY.**—We will make good, by repair or at our option by the supply of a replacement, defects which, under proper use, appear in the goods within a period of twelve calendar months after the goods have been shipped or, if shipment is delayed by reason of your instructions or lack of instructions, within a period of 18 months after the goods have been notified as ready for despatch (whichever period expires the earlier) and arise solely from faulty design (other than a design made, furnished or specified by you for which we have disclaimed responsibility in writing), materials or workmanship: Provided always that defective parts have been returned to us if we shall have so required. We shall refund the cost of carriage on such returned parts and the repaired or new parts will be delivered by us free of charge as provided in Clause 10 (Delivery).

Our liability under this clause shall be in lieu of any warranty or condition implied by law as to the quality or fitness for any particular purpose of the goods, and save as provided in this clause we shall not be under any liability, whether in contract, tort or otherwise, in respect of defects in goods delivered or for any injury, damage or loss resulting from such defects, or from any work done in connection therewith.

16. **PATENTS.**—We will indemnify you against any claim of infringement of Letters Patent, Registered Design, Trade Mark or Copyright (published at the date of the Contract) by the use or sale of any article or material supplied by us to you and against all costs and damages which you may incur in any action for such infringement or for which you may become liable in any such action. Provided always that this indemnity shall not apply to any infringement which is due to our having followed a design or instruction furnished or given by you or to the use of such article or material in a manner or for a purpose or in a foreign country not specified by or disclosed to us, or to any infringement which is due to the use of such article or material in association or combination with any other article or material not supplied by us. And provided also that this indemnity is conditional on your giving to us the earliest possible notice in writing of any claim being made or action threatened or brought against you and on your permitting us at our own expense to conduct any litigation that may ensue and all negotiations for a settlement of the claim. You on your part warrant that any design or instruction furnished or given by you shall not be such as will cause us to infringe any Letters Patent, Registered Design, Trade Mark, or Copyright in the execution of your order.

17. **LIABILITY FOR ACCIDENTS AND DAMAGE.**—If we, our agents or sub-contractors are on site for the purposes of the contract then, notwithstanding the provisions of Clause 15, we will indemnify you against direct damage or injury to your property or person or that of others occurring while we are working on site to the extent caused by the negligence of ourselves, our sub-contractors or agents, but not otherwise, by making good such damage to property or compensating personal injury.

Provided that:

(a) our total liability for damage to your property (including damage caused by our breach of contract, tort or breach of statutory duty) shall not exceed £100,000 (United Kingdom currency) or the contract price, whichever is the greater, and–

(b) we shall not be liable to you for any loss of profit or of contracts or, save as aforesaid, for any loss, damage or injury of any kind whatsoever and whether caused by our breach of contract, tort, breach of statutory duty or otherwise howsoever.

Save as provided in Clause 15, we shall not be liable for any damage or injury occurring after our completion of work on site.

18. **ARBITRATION.**—If at any time any question, dispute, or difference whatsoever shall arise between you and ourselves upon, in relation to, or in connection with the contract, either of us may give to the other notice in writing of the existence of such question, dispute, or difference, and the same shall be referred to the arbitration of a person to be mutually agreed upon, or failing agreement within thirty days of the receipt of such notice, of some person appointed by the President for the time being of the Institution of Electrical Engineers.

19. **LEGAL CONSTRUCTION.**–Unless otherwise agreed in writing the contract shall in all respects be construed and operate as an English contract and in conformity with English law and the English courts shall have exclusive jurisdiction over any matter arising out of the provisions of Clause 18 (Arbitration).

20. **STATUTORY AND OTHER REGULATIONS.**—If the cost to us of performing our obligations under the contract shall be increased or reduced by reason of the making or amendment after the date of tender of any law or of any order, regulation, or bye-law having the force of law that shall affect the performance of our obligations under the contract, the amount of such increase or reduction shall be added to or deducted from the contract price as the case may be.

Published by: The British Electrical and Allied Manufacturers Association Limited, 8 Leicester Street, Leicester Square, London WC2H 7BN

B

Edition June, 1979

Conditions of Sale (B) for machinery and equipment including supervision of erection United Kingdom

1. GENERAL–The acceptance of our tender includes the acceptance of the following terms and conditions:–

2. VALIDITY–Unless previously withdrawn, our tender is open for acceptance within the period stated therein or, when no period is so stated, within thirty days only after its date.

3. ACCEPTANCE–The acceptance of our tender must be accompanied by sufficient information to enable us to proceed with the order forthwith, otherwise we shall be at liberty to amend the tender prices to cover any increase in cost which has taken place after acceptance. Any samples submitted to you and not returned to our works within one month from date of receipt shall be paid for by you.

4. PACKING–Unless otherwise specified in our tender, all packing cases, skids, drums and other packing materials must be returned to our works at your expense and in good condition within one month from date of receipt. If not so returned they will be charged for.

5. LIMITS OF CONTRACT–Our tender includes only such plant, accessories and work as are specified therein.

6. DRAWINGS, ETC–All specifications, drawings and particulars of weights and dimensions submitted with our tender are approximate only, and the descriptions and illustrations contained in our catalogues, price lists, and other advertisement matter are intended merely to present a general idea of the plant described therein, and none of these shall form part of the contract. After acceptance of our tender, a set of certified outline drawings will be supplied free of charge on request.

7. INSPECTION AND TESTS–Our products are carefully inspected and, where practicable, submitted to our standard tests at our works before despatch. If tests other than those specified in our tender or tests in the presence of you or your representative are required, these will be charged for. In the event of any delay on your part in attending such tests or in carrying out any inspection required by you after seven days' notice that we are ready, the tests will proceed in your absence and shall be deemed to have been made in your presence, and the inspection will be deemed to have been made by you.

In the case of tests being carried out on site, they shall be carried out within one month after completion of erection and due notice in writing shall be given to us so that our representative may have reasonable opportunity of witnessing these tests if we so desire. Should the result of the tests not come within the margin specified the tests shall, if required by us, be repeated within one month after the date when the plant is ready for re-test, and we shall repay to you all reasonable expenses to which you may be put by such re-tests.

8. PERFORMANCE–We will accept no liability for failure to attain any performance figures quoted by us unless we have specifically guaranteed them, subject to any tolerances specified or agreed to by us, in an agreed sum as liquidated damages.

If the performance figures obtained on any test provided for in the contract are outside the acceptance limits specified therein, you will be entitled to reject the plant.

Before you become entitled to claim liquidated damages or to reject the plant we are to be given reasonable time and opportunity to rectify its performance. If you become entitled to reject plant, we will repay you any sum paid by you to us on account of the contract price thereof and any sum that may have accrued due to you in respect of delay in delivery or completion under Clause 10 up to the date of such rejection.

You assume responsibility that plant stipulated by you is sufficient and suitable for your purpose save in so far as your stipulations are in accordance with our advice.

9. DELIVERY–Unless otherwise specified in our tender, the price quoted includes delivery by any method of transport at our option.

Unless otherwise specified, we shall not be responsible for offloading.

10. LIABILITY FOR DELAY–Any times quoted for delivery or completion are to date from receipt by us of a written order to proceed and of all necessary information and drawings to enable us to put the work in hand. The time for delivery or completion shall be extended by a reasonable period if delay in delivery or completion is caused by instructions or lack of instructions from you or by industrial dispute or by any cause beyond our reasonable control.

If a fixed time be quoted for delivery or completion, and we fail to deliver or complete within that time or within any extension thereof provided by this clause, and if as a result you shall have suffered loss, we undertake to pay for each week or part of a week of delay, liquidated damages at the rate of .... per cent. up to a maximum of .... per cent. of that portion of the price named in the contract which is referable to such portion only of the contract plant as cannot in consequence of the delay be used commercially and effectively. Such payment shall be in full satisfaction of our liability for delay.

Any time described as an estimate shall not be construed as a fixed time quoted for the purpose of this clause.

11. ERECTION–Unless otherwise stated, our tender includes the supervision of erection only.

You shall provide suitable access to and possession of the site, proper foundations ready to receive the plant as and when delivered, adequate lifting facilities and scaffolding, all skilled and unskilled labour, masons', joiners' and builders' work, suitable protection for the plant from time of delivery, any lighting and heating necessary on the site during erection, and all necessary facilities and adequate assistance. All of these to be supplied at your expense to enable the work to be expeditiously and continuously carried out.

12. EXTRA COST–Should we incur extra cost owing to suspension of the work by your instructions or lack of instructions, interruptions, delays, overtime, unusual hours, mistakes, or work for which we are not responsible, a reasonable sum in respect of such extra cost, as well as the cost incurred by keeping any of our men on the site after completion of erection, shall be added to the contract price and paid for accordingly.

13. TIME OF TAKING OVER–The plant shall be deemed to have been taken over by you when erection has been completed and the plant has passed tests on site when these are included, or one calendar month after it shall have been put into commercial use (whichever may be the earlier): Provided that in any case the plant shall be deemed to have been taken over at the expiration of two calendar months after we shall have given you written notice that erection is complete, unless in the meantime tests shall have been made showing that it does not comply with the terms of the contract.

The time of taking over shall not be delayed on account of additions, minor omissions or defects which do not materially affect the commercial use of the plant.

Continued overleaf

14. TERMS OF PAYMENT–Unless otherwise agreed, payment shall be made by you as follows:–

(a) Ninety five per cent. of the contract value of plant as and when delivered to site from time to time and of work done on site respectively.

(b) Two and one-half per cent. of the contract price as and when plant has been taken over, or has been deemed to have been taken over, by you.

(c) The balance of the contract price one calendar month after payment of the above two and one-half per cent. has become due.

Minor defects in the plant, not of such importance as to affect materially its commercial use, shall entitle you to retain from the payment mentioned in (c) only such sum as represents the value of such incomplete or defective details, and any sum so retained shall be paid upon such omissions or defects being remedied, which will be done by us at the earliest opportunity. In the event of a portion of the plant being rejected under Clause 8, any sums paid to us shall be applied to payments due for the accepted portions of the plant, and the balance shall be refunded; should the whole plant be rejected all sums paid shall be refunded. Any liability on our part is subject to the terms of payment and all your other obligations to us under the contract being strictly observed.

If we are unable by reason of your instruction or lack of instructions or from causes beyond our control, to deliver all or any of the plant when ready, or to proceed with the supervision of erection of such plant as we have already delivered, you shall take delivery or arrange for storage. If you do not take delivery or arrange for storage, we shall be entitled to arrange storage either at our own works or elsewhere on your behalf and all charges for storage, for insurance or for demurrage shall be payable by you. In any case, you will make the payments of ninety five per cent. as if delivery had been made; further, you will make the first abovementioned payment of two and one-half per cent. within one calendar month from the date of notification that the plant is ready for delivery, and the final payment of two and one-half per cent. within three calendar months from date of such notification.

Subject to Clause 15 you shall pay any balance of the contract price outstanding in respect of plant lost or damaged after delivery.

15. LIABILITY FOR ACCIDENTS AND DAMAGE.–We will indemnify you against damage or injury to your property or person or that of others occurring before the plant is taken over to the extent directly caused by the negligence of ourselves, our sub-contractors or agents, or by defective design (other than a design made, furnished or specified by you for which we have disclaimed responsibility in writing), workmanship or materials, but not otherwise, by making good such damage to property or compensating personal injury.

Provided that:–

(a) our total liability for damage to your property (including damage caused by our breach of contract, tort or breach of statutory duty) shall not exceed £100,000 or the contract price, whichever sum is the greater, and

(b) we shall not be liable to you for any loss of profit or of contracts or, save as aforesaid, for any loss or damage of any kind whatsoever and whether caused by our breach of contract, tort, breach of statutory duty or otherwise howsoever.

Save as provided in the next following paragraph and in Clause 16, we shall not be liable for any damage or injury occurring after the plant has been taken over.

If we, our agents or sub-contractors are on site after taking over for the purpose of remedying a defect pursuant to Clause 16 or for any other purpose of the contract, the provisions of Clause 15 shall apply as though the plant had not yet been taken over. Save as provided in Clause 16, we shall not be liable for any damage or injury after the completion of such work on site as aforesaid.

16. DEFECTS AFTER TAKING OVER–We will make good, by repair or at our opinion by the supply of a replacement, defects which, under proper use, appear in the plant within a period of twelve calendar months after the plant has been taken over and arise solely from faulty design (other than a design made, furnished or specified by you for which we have disclaimed responsibility in writing), materials or workmanship: Provided always that defective parts have been returned to us if we shall have so required. We shall refund the cost of carriage on such returned parts and the repaired or new parts will be delivered by us free of charge as provided in Clause 9 (Delivery).

Our liability under this clause shall be in lieu of any warranty or condition implied by law as to the quality or fitness for any particular purpose of the plant, and save as provided in this clause we shall not be under any liability, whether in contract, tort or otherwise, in respect of defects in plant taken over or for any injury (other than personal injury caused by our negligence as defined in Section 1 of the Unfair Contract Terms Act, 1977), damage or loss resulting from such defects or from any work done in connection therewith.

17. PATENTS–We will indemnify you against any claim of infringement of Letters Patent, Registered Design, Trade Mark or Copyright (published at the date of the contract) by the use or sale of any article or material supplied by us to you and against all costs and damages which you may incur in any action for such infringement or for which you may become liable in any such action. Provided always that this indemnity shall not apply to any infringement which is due to our having followed a design or instruction furnished or given by you or to the use of such article or material in a manner or for a purpose or in a foreign country not specified by or disclosed to us, or to any infringement which is due to the use of such article or material in association or combination with any other article or material not supplied by us. And provided also that this indemnity is conditional on your giving to us the earliest possible notice in writing of any claim being made or action threatened or brought against you and on your permitting us at our own expense to conduct any litigation that may ensue and all negotiations for a settlement of the claim. You on your part warrant that any design or instructions furnished or given by you shall not be such as will cause us to infringe any Letters Patent, Registered Design, Trade Mark or Copyright in the execution of your order.

18. FAIR WAGES CLAUSE–We undertake to be bound by a fair wages clause in the terms of the House of Commons Resolution of 14th October, 1946. We also undertake that when the installation of machinery and equipment is carried out by men sent from the Contractor's or Sub-Contractor's establishment they shall receive the time rate payable in terms of the Fair Wages Clause to such work-people in such establishment, and in addition shall receive the outworking allowances recognised for outworkers sent from such establishment.

19. ARBITRATION–If at any time any question, dispute or difference whatsoever shall arise between you and ourselves upon, in relation to, or in connection with the contract, either of us may give to the other notice in writing of the existence of such question, dispute, or difference, and the same shall be referred to the arbitration of a person to be mutually agreed upon, or failing agreement within 30 days of receipt of such notice, of some person appointed by the President for the time being of the Institution of Electrical Engineers.

20. LEGAL CONSTRUCTION–Unless otherwise agreed in writing the contract shall in all respects be construed and operate as an English contract and in conformity with English law.

21. STATUTORY AND OTHER REGULATIONS–If the cost to us of performing our obligations under the contract shall be increased or reduced by reason of the making or amendment after the date of tender of any law or of any order, regulation, or bye-law having the force of law that shall affect the performance of our obligations under the contract, the amount of such increase or reduction shall be added to or deducted from the contract price as the case may be.

Published by: The British Electrical and Allied Manufacturers' Association Limited, 8 Leicester Street, Leicester Square, London WC2H 7BN

BE

Edition June, 1979

Conditions of Sale (BE) for machinery and equipment, including supervision of erection Export F.O.B.

1. **DEFINITIONS**–The expression FOB shall bear the meaning assigned to it in INCOTERMS 1953 save insofar as the same may have been varied by these Conditions.

2. **GENERAL**–The acceptance of our tender includes the acceptance of the following terms and conditions:–

3. **VALIDITY**–Unless previously withdrawn, our tender is open for acceptance for the period stated therein or, when no period is so stated, within thirty days only after its date.

4. **ACCEPTANCE**–The acceptance of our tender must be accompanied by sufficient information to enable us to proceed with the order forthwith, otherwise we shall be at liberty to amend the tender prices to cover any increase in cost which has taken place after acceptance. Any samples submitted to you and not returned to our works within one month from date of receipt shall be paid for by you.

5. **PACKING**–Unless otherwise specified in our tender, packing in accordance with our standard export practice is included.

6. **LIMITS OF CONTRACT**–Our tender includes only such plant, accessories and work as are specified therein.

7. **DRAWINGS, ETC.**–All descriptive and shipping specifications, drawings, and particulars of weights and dimensions submitted with our tender are approximate only, and the descriptions and illustrations contained in our catalogues, price lists, and other advertisement matter are intended merely to present a general idea of the plant described therein, and none of these shall form part of the contract. After acceptance of our tender, a set of certified outline drawings will be supplied free of charge.

8. **INSPECTION AND TESTS**–Our products are carefully inspected and, where practicable, submitted to our standard tests at our works before despatch. If tests other than those specified in our tender or tests in the presence of you or your representative are required, these will be charged for. In the event of any delay on your part in attending such tests or in carrying out any inspection required by you after seven days' notice that we are ready, the tests will proceed in your absence and shall be deemed to have been made in your presence, and the inspection will be deemed to have been made by you.

In the case of tests being carried out on site, they shall be carried out within one month after completion of erection and due notice in writing shall be given to us so that our representative may have reasonable opportunity of witnessing these tests if we so desire. Should the result of these tests not come within the margin specified the tests shall, if required by us, be repeated within one month after the date when the plant is ready for re-test, and we shall repay to you all reasonable expenses to which you may be put by such re-tests.

9. **PERFORMANCE**–We will accept no liability for failure to attain any performance figures quoted by us unless we have specifically guaranteed them, subject to any tolerances specified or agreed to by us, in an agreed sum as liquidated damages.

If the performance figures obtained on any test provided for in the contract are outside the acceptance limits specified therein, you will be entitled to reject the plant.

Before you become entitled to claim liquidated damages or to reject the plant we are to be given reasonable time and opportunity to rectify its performance. If you become entitled to reject plant, we will repay you any sum paid by you to us on account of the contract price thereof and any sum that may have accrued due to you in respect of delay in delivery or completion under Clause 11 up to the date of such rejection.

You assume responsibility that plant stipulated by you is sufficient and suitable for your purpose save insofar as your stipulations are in accordance with our advice.

10. **DELIVERY**–Unless otherwise agreed, delivery will be made f.o.b. at the port stated in our tender.

We shall not be required to give you the notice relating to insurance mentioned in Section 32(3) of the Sale of Goods Act, 1893.

11. **LIABILITY FOR DELAY**–Any times quoted for delivery or completion are to date from receipt by us of a written order to proceed and of all necessary information and drawings to enable us to put the work in hand. The time for delivery or completion shall be extended by a reasonable period if delay in delivery or completion is caused by instructions or lack of instructions from you or by industrial dispute or by any cause beyond our reasonable control.

If a fixed time be quoted for delivery or completion, and we fail to deliver or complete within that time or within any extension thereof provided by this clause, and if as a result you shall have suffered loss, we undertake to pay for each week or part of a week of delay, liquidated damages at the rate of .... per cent. up to a maximum of .... per cent. of that portion of the price named in the contract which is referable to such portion only of the contract plant as cannot in consequence of the delay be used commercially and effectively. Such payment shall be in full satisfaction of our liability for delay.

Any time described as an estimate shall not be construed as a fixed time quoted for the purpose of this clause.

12. **ERECTION**–Our tender includes supervision of erection only.

You shall provide suitable access to and possession of the site, proper foundations ready to receive the plant as and when delivered, adequate lifting tackle and scaffolding, all skilled and unskilled labour, masons', joiners', and builders' work, suitable protection for the plant from time of delivery, any lighting and heating necessary on the site during erection, and all necessary facilities and adequate assistance. All of these shall be supplied at your expense to enable erection to be expeditiously and continuously carried out under our supervision.

13. **EXTRA COST**–Should we incur extra cost owing to suspension of the work by your instructions or lack of instructions, interruptions, delays, overtime, unusual hours, mistakes, or work for which we are not responsible, a reasonable sum in respect of such extra cost, as well as the cost incurred by keeping any of our men on the site after completion of erection, shall be added to the contract price and paid for accordingly.

14. **TIME OF TAKING OVER**–The plant shall be deemed to have been taken over by you when erection has been completed and the plant has passed tests on site when these are included, or one calendar month after it shall have been put into commercial use (whichever may be the earlier) : Provided that in any case the plant shall be deemed to have been taken over at the expiration of two calendar months after we shall have given you written notice that erection is complete, unless in the meanwhile tests shall have been made showing that the plant does not comply with the terms of the contract.

The time of taking over shall not be delayed on account of additions, minor omissions or defects, which do not materially affect the commercial use of the plant.

Continued overleaf

15. TERMS OF PAYMENT–Payments shall be made by you as follows:–

(a) Ninety five per cent. of the contract value of plant as and when shipped.

(b) The balance of the Contract Price as and when plant has been taken over, or has been deemed to have been taken over, by you.

(c) For supervision of erection, monthly at the agreed rate.

Minor defects in the plant, not of such importance as to affect materially its commercial use, shall entitle you to retain from the payment mentioned in paragraph *(b)* only such sum as represents the value of such incomplete or defective details, and any sum so retained shall be paid upon such omissions or defects being remedied, which shall be done by us at the earliest opportunity. In the event of any portion of the plant being rejected under Clause 9, any sums paid to us shall be applied to payments due for the accepted portions of the plant, and the balance shall be refunded; should the whole plant be rejected all sums paid shall be refunded. Any liability on our part is subject to the terms of payment and all your other obligations to us under the contract being strictly observed.

If we are unable, by reason of your instructions or lack of instructions, to ship part or all of the plant when ready, or if, having shipped the plant or part thereof where delivery under the contract is beyond the port of shipment we are unable for causes beyond our control to deliver it in accordance with the contract, you shall find and pay the cost of suitable storage and efficient protection, including fire insurance. If you do not take delivery or arrange for storage, we shall be entitled to arrange storage either at our own works or elsewhere on your behalf and all charges for storage, for insurance or for demurrage shall be payable by you. In any such case, you shall make payment under paragraph *(a)* as if shipment had been made and payment under paragraph *(b)* within three calendar months after the date of notification that the plant is ready for shipment.

Subject to Clause 16 you shall pay any balance of the contract price outstanding in respect of plant lost or damaged after delivery.

16. LIABILITY FOR ACCIDENTS AND DAMAGE–We will indemnify you against damage or injury to your property or person or that of others occurring before the plant is taken over to the extent directly caused by the negligence of ourselves, our sub-contractors or agents, or by defective design (other than a design made, furnished or specified by you for which we have disclaimed responsibility in writing), workmanship or materials, but not otherwise, by making good such damage to property or compensating personal injury.

Provided that:–

(a) our total liability for damage to your property (including damage caused by our breach of contract, tort or breach of statutory duty) shall not exceed £100,000 (United Kingdom currency) or the contract price, whichever is the greater, and

(b) we shall not be liable to you for any loss of profit or of contracts or, save as aforesaid, for any loss, damage or injury of any kind whatsoever and whether caused by our breach of contract, tort, breach of statutory duty or otherwise howsoever.

Save as provided in the next following paragraph and in Clause 17, we shall not be liable for any damage or injury occurring after the plant has been taken over.

If we, our agents or sub-contractors are on site after taking over for the purpose of remedying a defect pursuant to Clause 17 or for any other purpose of the contract, the provisions of Clause 16 shall apply as though the plant had not yet been taken over. Save as provided in Clause 17, we shall not be liable for any damage or injury after the completion of such work on site as aforesaid.

17. DEFECTS AFTER TAKING OVER–We will make good, by repair or at our option by the supply of a replacement, defects which, under proper use, appear in the plant within a period of twelve calendar months after the plant has been taken over and arise solely from faulty design (other than a design made, furnished or specified by you for which we have disclaimed responsibility in writing), materials or workmanship : Provided always that defective parts have been returned to us if we shall have so required. We shall refund the cost of carriage on such returned parts and the repaired or new parts will be delivered by us free of charge as provided in Clause 10 (Delivery).

Our liability under this clause shall be in lieu of any warranty or condition implied by law as to the quality or fitness for any particular purpose of the plant, and save as provided in this clause we shall not be under any liability, whether in contract, tort or otherwise, in respect of defects in plant taken over or for any injury, damage or loss resulting from such defects or from any work done in connection therewith.

18. PATENTS–We will indemnify you against any claim of infringement of Letters Patent, Registered Design, Trade Mark or Copyright (published at the date of the Contract) by the use or sale of any article or material supplied by us to you and against all costs and damages which you may incur in any action for such infringement or for which you may become liable in any such action. Provided always that this indemnity shall not apply to any infringement which is due to our having followed a design or instruction furnished or given by you or to the use of such article or material in a manner or for a purpose or in a foreign country not specified by or disclosed to us, or to any infringement which is due to the use of such article or material in association or combination with any other article or material not supplied by us. And provided also that this indemnity is conditional on your giving to us the earliest possible notice in writing of any claim being made or action threatened or brought against you and on your permitting us at our own expense to conduct any litigation that may ensue and all negotiations for a settlement of the claim. You on your part warrant that any design or instruction furnished or given by you shall not be such as will cause us to infringe any Letters Patent, Registered Design, Trade Mark or Copyright in the execution of your order.

19. ARBITRATION–If at any time any question, dispute, or difference whatsoever shall arise between you and ourselves upon, in relation to, or in connection with the contract, either of us may give to the other notice in writing of the existence of such question, dispute, or difference, and the same shall be referred to the arbitration of a person to be mutually agreed upon, or failing agreement within thirty days of the receipt of such notice, of some person appointed by the President for the time being of the Institution of Electrical Engineers.

20. LEGAL CONSTRUCTION–Unless otherwise agreed in writing the contract shall in all respects be construed and operate as an English contract and in conformity with English law and the English courts shall have exclusive jurisdiction over any matter arising out of the provisions of Clause 19 (Arbitration).

21. STATUTORY AND OTHER REGULATIONS–If the cost to us of performing our obligations under the contract shall be increased or reduced by reason of the making or amendment after the date of tender of any law or of any order, regulation, or bye-law having the force of law that shall affect the performance of our obligations under the contract, the amount of such increase or reduction shall be added to or deducted from the contract price as the case may be.

Published by: The British Electrical and Allied Manufacturers' Association Limited, 8 Leicester Street, Leicester Square, London WC2N 7BN

C

Edition June, 1979

Conditions of Sale (C) for Electronic Equipment including Installation United Kingdom

1. **GENERAL**–The acceptance of our tender includes the acceptance of the following terms and conditions and of the special conditions (if any) stated in or referred to in the tender:–

2. **VALIDITY**–Unless previously withdrawn, our tender is open for acceptance within the period stated therein or, when no period is so stated, within thirty days only after its date.

3. **ACCEPTANCE**–The acceptance of our tender must be accompanied by sufficient information to enable us to proceed with the order forthwith, otherwise we shall be at liberty to amend the tender prices to cover any increase in cost which has taken place after acceptance. Any samples submitted to you and not returned to our works within one month from date of receipt shall be paid for by you.

4. **PACKING**–Unless otherwise specified in our tender, all packing cases, skids, drums and other packing materials must be returned to our works at your expense and in good condition. If not so returned they will be charged for.

5. **LIMITS OF CONTRACT**–Our tender includes only such equipment, accessories and work as are specified therein.

6. **DRAWINGS, ETC.**–Unless otherwise specified in our tender all specifications, drawings and particulars of weights and dimensions submitted therewith are approximate only, and the descriptions and illustrations contained in our catalogues, price lists, and other advertisement matter are intended merely to present a general idea of the equipment described therein, and none of these shall form part of the contract. After acceptance of our tender, a set of certified outline drawings will be supplied free of charge on request, sufficient to enable you to make provision for installation.

All specifications, drawings and technical description submitted with or in connection with our tender are our copyright.

All such copyright material, and all information and "know-how", whenever supplied, shall at all times be treated by you as confidential and shall not without our consent be used by you except for purposes of (i) adjudicating the tender, (ii) the contract (if any) placed with us, and (iii) the operation of the equipment supplied thereunder, nor shall they without our consent be communicated to third parties save insofar as may be necessary for the permitted purposes.

7. **INSPECTION AND TESTS**–The equipment, where practicable, is submitted to our standard tests before despatch. If tests other than those specified in our tender or tests in the presence of you or your representative are required, these will be charged for. In the event of any delay on your part in attending such tests or in carrying out any inspection required by you after seven days' notice that we are ready, the tests will proceed in your absence and shall be deemed to have been made in your presence, and the inspection shall be deemed to have been made by you.

If site tests are specified, we shall give you 21 days' notice in writing of the date after which we will be ready to make such tests. Unless otherwise agreed, such tests shall take place within 10 days after the said date on such day or days as you shall in writing notify us. If you fail to appoint a date after having been asked to do so, we shall be entitled to proceed in your absence, and the tests shall be deemed to have been made in your presence. We will supply you with full data of the results of such tests.

If any portion of the equipment supplied fails to pass the requisite tests, tests of the said portion shall, if required by you or by us, be repeated within a reasonable time upon the same terms and conditions, save that all reasonable expenses incurred by the party NOT responsible for the failure shall be reimbursed by the other.

Save where otherwise agreed the equipment shall be deemed to have passed all the requisite tests when its ability to accept and/or provide the previously agreed signals at the interface or simulated interface has been demonstrated.

8. **ENVIRONMENT**–Where the environmental conditions specified in our tender do not exist at the site at the time when delivery is due, we shall not be required to deliver the equipment until we are satisfied that the specified environmental conditions do exist.

9. **PERFORMANCE**–We will accept no liability for failure to attain any performance figures quoted by us unless (i) we have specifically guaranteed them, subject to any tolerance specified or agreed to by us, in an agreed sum as liquidated damages, and (ii) the environmental conditions specified in our tender are maintained.

If the results obtained in the tests specified in the contract fail to achieve the performance figures (subject to the tolerances, if any) specifically guaranteed by us, you will be entitled to reject the equipment.

Before you become entitled to claim liquidated damages or to reject the equipment we are to be given reasonable time and opportunity to rectify its performance. If you exercise your right to reject equipment, we will repay you any sum paid by you to us on account of the contract price thereof and any sum that may have accrued due to you in respect of delay in delivery or completion under Clause 11 up to the date of such rejection. Such liquidated damages or payments on rejection shall be in full satisfaction of our liability under this clause.

You assume responsibility that performance data or equipment stipulated by you are sufficient and suitable for your purpose save insofar as your stipulations are in accordance with our advice.

10. **TRANSPORT**–Unless otherwise specified in our tender, the price quoted includes delivery by any method of transport at our option.

Unless otherwise specified, we shall not be responsible for offloading.

11. **LIABILITY FOR DELAY**–Any times quoted for delivery or completion are to date from receipt by us of a written order to proceed and of all necessary information and drawings to enable us to put the work in hand. The time for delivery or completion shall be extended by a reasonable period if delay in delivery or completion is caused by instructions or lack of instructions from you or by industrial dispute or by any cause beyond our reasonable control.

If a fixed time be quoted for delivery or completion, and we fail to deliver or complete within that time or within any extension thereof provided by this clause, and if as a result you shall have suffered loss, we undertake to pay for each week or part of a week of delay, liquidated damages at the rate of   per cent up to a maximum of   per cent of that portion of the price named in the contract which is referable to such portion only of the contract equipment as cannot in consequence of the delay be used commercially and effectively. Such payment shall be in full satisfaction of our liability for delay.

Any time described as an estimate shall not be construed as a fixed time quoted for the purpose of this clause.

12. **INSTALLATION**–Unless otherwise specified you shall provide suitable access to and possession of the site, proper foundations ready to receive the equipment as and when delivered, adequate lifting facilities and scaffolding, all unskilled labour, masons', joiners', and builders' work, suitable guarding and protection for the equipment from time of delivery, any electric power, lighting and heating necessary on the site during and after installation, and all necessary facilities and adequate assistance. All of these to be supplied at your expense to enable the work to be expeditiously and continuously carried out.

13. **EXTRA COST**–Should we incur extra cost owing to variation or suspension of the work by your instructions or lack of instructions, interruptions, delays, overtime, unusual hours, mistakes, or work, for which we are not responsible, a reasonable sum in respect of such extra cost, as well as the cost incurred by keeping any of our men on the site after completion of installation, shall be added to the contract price and paid for accordingly.

14. **TIME OF TAKING OVER**–The equipment shall be deemed to have been taken over at the earlier of the following times:

(i) when installation has been completed and the equipment has passed or is deemed to have passed all the tests provided for in the contract:

(ii) one calendar month after the equipment shall have been put into use.

Continued Overleaf

Provided that in any case the equipment shall be deemed to have been taken over at the expiration of two calendar months after we shall have given you written notice that installation is complete, unless in the meantime tests shall have been made showing that it does not comply with the terms of the contract.

If by agreement between us any portion shall have been put into use before the remainder, the preceding paragraph shall apply in respect of that portion.

The time of taking over shall not be delayed on account of additions, minor omissions or defects which do not materially affect the use of the equipment.

15. **TERMS OF PAYMENT**–Unless otherwise agreed, payment shall be made by you as follows:

(i) per cent of the contract price at the time of placing the order.

(ii) per cent of the contract value of equipment as and when delivered from time to time on site and of work done on site respectively.

(iii) 2½ per cent of the contract value of equipment as and when this has been taken over, or has been deemed to have been taken over by you.

(iv) The balance of the contract price one calendar month after payment of the above 2½ per cent has become due.

Minor defects or omissions in the equipment not of such importance as to affect materially their use, shall entitle you to retain from the payment mentioned in (iv) only such sum as represents the value of such incomplete or defective portion of the equipment, and any sum so retained shall be paid upon such omissions or defects being remedied, which will be done by us at the earliest opportunity. In the event of a portion of the equipment being rejected under Clause 9, any sums paid to us shall be applied to payments due for the accepted portions of the equipment, and the balance shall be refunded; should the whole equipment be rejected all sums paid shall be refunded. Any liability on our part is subject to the terms of payment and all your other obligations to us under the contract being strictly observed. If we are unable by reason of your instructions or lack of instructions or from causes beyond our control, to deliver or if, pursuant to Clause 8, we cannot be required to deliver, all or any of the equipment when ready, or to proceed with the installation of such equipment as we have already delivered, we shall be entitled to arrange storage either at our own works or elsewhere on your behalf and all charges for packing and storage, for insurance, for demurrage, for carriage and for any retesting and necessary refurbishing shall be payable by you. In any case, you will make the payments of per cent mentioned in (ii) as if delivery had been made; further, you will make the payment of two and one half per cent mentioned in (iii) within one calendar month from the date of notification that the equipment is ready for delivery, and the balance mentioned in (iv) within three calendar months from the date of such notification.

Subject to Clause 16 you shall pay any balance of the contract price outstanding in respect of equipment lost or damaged after delivery.

16. **LIABILITY FOR ACCIDENTS AND DAMAGE**–We will indemnify you against damage or injury to your property or person or that of others occurring before the equipment is taken over or is deemed to have been taken over to the extent directly caused by the negligence of ourselves, our sub-contractors or agents, or by defective design (other than a design made, furnished or specified by you for which we have disclaimed responsibility in writing), workmanship or materials, but not otherwise, by making good such damage to property or compensating personal injury.

Provided that:–

(a) our total liability for damage to your property (including damage caused by our breach of contract, tort or breach of statutory duty) shall not exceed £100,000 or the contract price, whichever is the greater, and

(b) we shall not be liable to you for any loss of profit or of contracts or, save as aforesaid, for any loss or damage of any kind whatsoever and whether caused by our breach of contract, tort, breach of statutory duty or otherwise howsoever.

Save as provided in Clause 17 and in the next following paragraph we shall not be liable for any damage or injury occurring after the equipment has been taken over, or is deemed to have been taken over.

If we, our agents or sub-contractors are on site after taking over for the purpose of remedying a defect pursuant to Clause 17 or for any other purpose of the contract, the preceding provisions of this clause shall apply as though the equipment had not yet been taken over. Save as provided in Clause 17 we shall not be liable for any damage or injury after the completion of such work on site as aforesaid.

17. **DEFECTS AFTER TAKING OVER**–We will make good, by repair or at our option by the supply of a replacement, defects which, under proper use, appear in the equipment within a period of twelve calendar months after the equipment has been taken over or has been deemed to have been taken over and arise solely from faulty design (other than a design made, furnished or specified by you for which we have disclaimed responsibility in writing), materials or workmanship: Provided that in respect of the items listed in Appendix A there shall be substituted for the period of twelve months the periods set against such items respectively, and provided always that defective parts have been returned to us if we shall have so required. We shall refund the cost of carriage on such returned parts and the repaired or new parts will be delivered by us free of charge as provided in Clause 10 (Transport).

Our liability under this clause shall be in lieu of any warranty or condition implied by law as to the quality or fitness for any particular purpose of the equipment, and save as provided in this clause we shall not be under any liability, whether in contract, tort or otherwise, in respect of defects in equipment taken over or for any injury (other than personal injury caused by our negligence as defined in Section 1 of the Unfair Contract Terms Act, 1977), damage or loss resulting from such defects or from any work done in connection therewith.

18. **PATENTS**–We will indemnify you against any claim of infringement of Letters Patent, Registered Design, Trade Mark or Copyright (published at the date of the contract) by the use or sale of any equipment supplied by us to you and against all costs and damages which you may incur in any action for such infringement or for which you may become liable in any such action. Provided always that this indemnity shall not apply to any infringement which is due to our having followed a design or instruction furnished or given by you or to the use of such equipment in a manner or for a purpose or in a foreign country not specified by or disclosed to us, or to any infringement which is due to the use of such equipment in association or combination with any other equipment not supplied by us. And provided also that this indemnity is conditional on your giving to us the earliest possible notice in writing of any claim being made or action threatened or brought against you on your permitting us at our own expense to conduct any litigation that may ensue and all negotiations for a settlement of the claim. You on your part warrant that any design or instructions furnished or given by you shall not be such as will cause us to infringe any Letters Patent, Registered Design, Trade Mark or Copyright in the execution of your order.

19. **FAIR WAGES CLAUSE**–We undertake to be bound by a fair wages clause in the terms of the House of Commons Resolution of 14th October, 1946. We also undertake that when the installation of equipment is carried out by men sent from our or our sub-contractor's establishment they shall receive the time rate payable in terms of the Fair Wages Clause to such workpeople in such establishment, and in addition shall receive the outworking allowances recognised for outworkers sent from such establishment.

20. **ARBITRATION**–If at any time any question, dispute or difference whatsoever shall arise between you and ourselves upon, in relation to, or in connection with the contract, either of us may give to the other notice in writing of the existence of such question, dispute, or difference, and the same shall be referred to the arbitration of a person to be mutually agreed upon, or failing agreement within 30 days of receipt of such notice, of some person appointed by the President for the time being of the Institution of Electrical Engineers.

21. **LEGAL CONSTRUCTION**–Unless otherwise agreed in writing the contract shall in all respects be construed and operate as an English contract and in conformity with English law.

22. **STATUTORY AND OTHER REGULATIONS**–If the cost to us of performing our obligations under the contract shall be increased or reduced by reason of the making or amendment after the date of tender of any law or of any order, regulation, or bye-law having the force of law that shall affect the performance of our obligations under the contract, the amount of such increase or reduction shall be added to or deducted from the contract price as the case may be.

APPENDIX A

(Guarantees on bought-out components)

Published by: The British Electrical and Allied Manufacturers' Association (Ltd.), 8 Leicester Street, Leicester Square, London WC2N 7BN

E

Edition November, 1980

Conditions of Contract (E) for erection of electrical plant and machinery (Home or Export)

DEFINITION OF TERMS – "the Contract" shall mean the tender and the acceptance thereof together with all documents to which reference may properly be made in order to ascertain the rights and obligations of the parties.

- "the Contract Price" shall mean the sum named in the Contract as the Contract Price.
- "erection" shall mean the fixing in position of the various items of plant, their inter-connection and the connection to the plant of supplies and services all as detailed in the Contract.
- "plant" means all machinery, apparatus, materials and articles (whether manufactured or supplied by us under a separate contract or not) and which are to be erected by us in accordance with the provisions of the Contract.
- "Site" means the place where the plant is to be erected together with so much of the surrounding area as may be used by us in connection with the Work.
- "Work" means the erection of the plant and all other work to be done by us under the Contract.

1. **GENERAL.** – The acceptance of our tender includes acceptance of the following terms and conditions:–

2. **VALIDITY.** – Unless previously withdrawn, our tender is open for acceptance within the period stated therein, or when no period is so stated, within thirty days only after its date.

3. **ACCEPTANCE OF TENDER.** – The acceptance of our tender must be accompanied by all necessary information which we may reasonably require to enable us to proceed with the erection without hindrance from the date or dates specified.

4. **LIMITS OF CONTRACT.** – Our tender includes only such work as is specified therein.

5. **DRAWINGS ETC.** – You shall provide us free of charge before commencement of the Work with any information, plans or drawings required for erection. If the information, plans or drawings required for erection have not been furnished to us in reasonable time or if they do not contain the necessary details, we shall be entitled either to suspend performance of the Contract until such time as drawings are furnished or until the necessary details have been provided or to prepare such plans or drawings ourselves at your cost and in either event the time for performance of the Contract shall be extended by a reasonable period.

6. **SITE FACILITIES AND WORKING CONDITIONS.** – Our tender is based on our estimate of the extent of the Work in the light of information (including your Site Safety Regulations) produced to us and unless otherwise specified assumes the following conditions:–

(a) You will provide suitable access to and possession of the Site in reasonable time.

(b) The erection will not be carried out in unhealthy or unsafe conditions.

(c) Our employees will be able to obtain suitable and convenient board and lodging in the neighbourhood of the Site and have access to adequate medical services.

(d) Such equipment craneage, consumable stores, water and power, scaffolding, lighting, heating and unskilled labour as are specified in the Contract will be available to us on Site and in reasonable time, and except as otherwise agreed, free of charge to us.

(e) You will provide us, free of charge, with closed or guarded premises on or near the Site as a protection against the theft and deterioration of the plant to be erected and of our tools and equipment.

(f) We shall not be required to undertake any construction or demolition work.

If these conditions are not satisfied we shall be entitled to charge extra.

7. **HOURS OF WORK.** – Our employees will work normal hours applicable to the engineering industry, Monday to Friday inclusive, local public holidays excepted. Unless stated to the contrary in our tender or otherwise agreed, night work, overtime and holiday working are specifically excluded.

8. **EXTRA COST.** – Should we incur extra cost owing to variation or suspension of the Work by your instructions, or your lack of instructions, or to interruptions, delays, overtime, unusual hours, mistakes or work for which we are not responsible, or to any specified site conditions not being maintained by you, the Contract Price will be adjusted in accordance with our rates ruling at the times such extra costs are incurred.

9. **PREPARATORY WORK.** – The plant must be on the Site in reasonable time. You will provide us in reasonable time with suitable access to the Site and plant and to all necessary equipment to be provided by you and furnish us with all information required for making all necessary connections to the plant. If you are responsible for preparatory work such as foundation or other work specified in the Contract as a pre-requisite to commencement of the Work, it must be completed in reasonable time. If we are responsible for foundation work you will provide us in reasonable time with all necessary information relating to the work necessary for preparing suitable foundations. Any extra cost incurred by us which results from an error or omission in the information furnished by you shall be added to the Contract Price and paid for accordingly.

10. **LIABILITY FOR DELAY.** – Any times quoted by us for completion of erection are to date from receipt by us of full and final instructions, sufficient and suitable availability of and access to the plant to allow us to proceed. The time for completion shall be extended by a reasonable period if delay in completion is caused by instructions or lack of instructions from you or by industrial dispute or by any cause beyond our reasonable control. If a fixed time be quoted for completion of erection and we fail to complete within that time or within any extension thereof under this clause or under Clause 5 (Drawings etc.), and if as a result you shall have suffered loss, we undertake to pay for each week or part of a week of delay liquidated damages at the rate of   per cent. up to a maximum of   per cent. of the Contract Price. Such payment shall be in full satisfaction of our liability for delay.

Any time described as an estimate shall not be construed as a fixed time quoted for the purposes of this clause.

11. **TAKING OVER AND ACCEPTANCE OF ERECTION.** – We shall give you notice in writing of the date when the Work is ready for acceptance in sufficient time to enable you to make any necessary arrangements for inspection. If the Contract provides for tests on acceptance these shall take place in the presence of both parties. If in the course of the acceptance it is found that the Work or any part thereof is defective as a result of defective erection, assembly or connection of the plant we shall with all speed and at our expense make good the defect and thereafter if you so require offer the Work for re-inspection and acceptance at our expense.

As soon as the Work has been completed in accordance with the Contract and has been accepted, you shall be deemed to have taken over the Work and the defects liablility period shall start to run. You will thereupon issue a certificate (herein called "taking over certificate") in which you will certify the date on which the Work was completed and passed the tests (if any). If you do not take the steps necessary for such acceptance or testing, the Work shall be deemed to have been taken over and the defects liability period shall start to run on receipt by you of written notice from us to that effect.

Continued overleaf

12. DEFECTS LIABILITY PERIOD. – We undertake to remedy at our expense and with all reasonable speed, any defect in the Work which may appear during the period of twelve months after taking over by making good any such defect. If in consequence of any such defect it is necessary to repair or replace any part of the plant any such repair or replacement shall in all respects be at our expense, but not so as to impose upon us any liability greater than the Contract Price.

Our liability under this clause is in lieu of any condition or warranty implied by law as to the quality, workmanship, or fitness for purpose of the Work and save as provided in this clause and in Clause 13 (Liability for Accidents and Damage) we shall not be liable to you whether in contract, tort or otherwise for any loss, damage or injury (other than personal injury caused by our negligence as defined in the Unfair Contract Terms Act 1977) resulting from any defects in work done or materials provided under the Contract or from any services or advice rendered in connection therewith.

13. LIABILITY FOR ACCIDENTS AND DAMAGE. – We will indemnify you against direct damage or injury to your property or person or that of others occurring while we are working on the Site for the purpose of the Contract to the extent caused by the negligence of ourselves, our sub-contractors, or agents, but not otherwise, by making good such damage to property or compensating personal injury. Provided that:

(a) our total liability for damage to the plant (including damage caused by our breach of contract, tort or breach of statutory duty) is limited in accordance with Clause 12 (Defects Liability Period) to the Contract Price

(b) our total liability for damage to your property (other than the plant) including damage caused by our breach of contract, tort or breach of statutory duty shall not exceed £250,000 or the Contract Price, whichever shall be the greater, and

(c) we shall not be liable to you for any loss of profit or contracts or, save as aforesaid, for any loss or damage of any kind whatsoever and whether caused by our breach of contract, tort, breach of statutory duty or otherwise howsoever.

Save as provided in Clause 12 (Defects Liability Period) we shall not be liable for any injury (other than personal injury caused by our negligence as defined in the Unfair Contract Terms Act, 1977) or damage occurring after completion of work on site.

14. TERMS OF PAYMENT. – Unless otherwise agreed, payment in full shall be made by you on issue of the taking over certificate under Clause 11 (Taking Over and Acceptance of Erection).

15. PATENTS. – We will indemnify you against any claim for infringement of Patents, Registered Design, Copyright or Trade Mark (published at the date of the Contract) by the use of any machinery, apparatus, work, material or method in the erection of the plant and against all costs and damages which you may incur in any action for such infringement or for which you may become liable in any such action. Provided always that this indemnity shall not apply to any infringement which is due to our having followed any instruction furnished or given by you as to the method of erection or as to the machinery, apparatus, work, or material to be used in connection therewith. And Provided also that this indemnity is conditional on your giving to us the earliest possible notice in writing of any claim being made or action threatened or brought against you and on your permitting us at our own expense to conduct any litigation that may ensue and all negotiations for a settlement of the claim. You on your part warrant that any instructions furnished or given by you will not be such as will cause us to infringe any Letters Patent, Registered Design, Trade Mark or Copyright in the execution of the Contract.

16. ARBITRATION. – If at any time any question, dispute or difference whatsoever shall arise between you and ourselves upon, in relation to or in connection with the Contract, either of us may give to the other notice in writing of the existence of such question, dispute or difference, and the same shall be referred to the arbitration of a person to be mutually agreed upon, or failing agreement within 30 days of receipt of such notice, of some person appointed by the President for the time being of the Institution of Electrical Engineers.

17. LEGAL CONSTRUCTION. – Unless otherwise agreed in writing the Contract shall in all respects be construed and operate as an English contract and in conformity with English law and the English courts shall have exclusive jurisdiction over any matter arising out of the provisions of Clause 16 (Arbitration).

18. STATUTORY AND OTHER REGULATIONS. – If the cost to us of performing our obligations under the Contract shall be increased or reduced by reason of the making or amendment after the date of tender, of any law or of any order, regulation or bye-law, having the force of law that shall affect the performance of our obligations under the Contract, the amount of such increase or reduction shall be added to or deducted from the Contract Price as the case may be.

19. VARIATIONS IN COSTS. – If the cost to us of performing our obligations under the Contract shall be increased or reduced by reason of any rise or fall in the cost of labour or in the cost of materials or transport above or below such rates and costs ruling at the date of tender the amount of such increase or reduction shall be added to or deducted from the Contract Price as the case may be, provided that no account shall be taken of any amount by which any cost incurred by us has been increased by our default or negligence.

Published by: The British Electrical and Allied Manufacturers Association Limited, 8 Leicester Street, Leicester Square, London WC2H 7BN

R

Edition June, 1979

Conditions (R) for the repair of machinery and equipment United Kingdom

1. **GENERAL**–The acceptance of our tender includes the acceptance of the following terms and conditions:–

2. **VALIDITY**–Unless previously withdrawn, our tender is open for acceptance for the period stated therein or, when no period is so stated, within thirty days only after its date.

3. **PRICES**–Unless otherwise stated, prices are approximate only.

4. **LIMITS OF CONTRACT**–Our tender is made on the assumption that the repairs are reasonably capable of being carried out. If, on inspection, this is found not to be the case, we will advise you as soon as reasonably practicable; no liability shall attach to us for any loss occasioned by the repairs not being carried out and the cost of such inspection shall be borne by you.

Plant sent to us for repair shall be delivered to our works free of all cost.

5. **LIABILITY FOR DELAY**–Any time quoted by us for delivery of plant repaired in our works or for completion of repairs on site shall not begin to run until we have received the plant to be repaired or, where repairs are to be carried out on site, have obtained access to the plant and in either case until we have received a written order to proceed and all necessary information to enable us to put the work in hand. Any such time is to be treated as an estimate only not involving us in any liability for failure to deliver or complete within such time unless you have suffered loss and the amount payable in respect thereof shall have been agreed in writing as liquidated damages, in which case our liability shall be limited to the amount so agreed to be paid. In all cases, whether a time for delivery or completion be quoted or not, the time therefor shall be extended by a reasonable period if delay in delivery or completion is caused by instructions, or lack of instructions, from you or by industrial dispute or by any cause whatsoever beyond our reasonable control.

6. **DELIVERY**–Unless otherwise specified in our tender, the price quoted includes delivery by any method of transport at our option.

Unless otherwise specified, we shall not be responsible for offloading.

7. **LOSS OR DAMAGE IN TRANSIT**–When the price quoted includes delivery other than at our works, we will repair or at our option replace free of charge plant lost or damaged in transit: Provided that we are given written notification of such loss or damage within such time as will enable us to comply with the carrier's conditions of carriage as affecting loss or damage in transit or, where delivery is made by our own transport, within a reasonable time after receipt of the Advice Note.

8. **PACKING**–Unless otherwise specified in our tender, all packing cases, skids, drums and other packing materials must be returned to our works at your expense and in good condition within one month from date of receipt. If not so returned they will be charged for.

9. **STORAGE**–If we do not receive forwarding instructions sufficient to enable us to despatch repaired plant within 14 days after the date of notification that the plant is ready for despatch, you shall take delivery or arrange for storage. If you do not take delivery or arrange for storage, we shall be entitled to arrange storage either at our own works or elsewhere on your behalf and all charges for storage, for insurance or for demurrage shall be payable by you.

10. **PAYMENT**–Unless otherwise agreed, payment in full shall be due on notification to you that the plant is ready for despatch, or on completion when the repairs are effected on site.

11. **EXTRA COST**–In the event of suspension of the work by your instructions or lack of instructions, any price quoted shall be increased by a reasonable sum to cover any extra expense thereby incurred by us.

12. **LIABILITY FOR ACCIDENTS AND DAMAGE**–We will indemnify you against direct damage or injury to your property or person or that of others to the extent directly caused by the negligence of ourselves, our sub-contractors or agents, by making good such damage to property or compensating personal injury.

Provided that:–

(a) our total liability for damage to your property (including damage caused by our breach of contract, tort or breach of statutory duty) shall not exceed £100,000 or the contract price, whichever sum is the greater, and

(b) we shall not be liable to you for any loss of profit or of contracts or, save as aforesaid, for any loss or damage of any kind whatsoever and whether caused by our breach of contract, tort, breach of statutory duty or otherwise howsoever.

Save as provided in Clause 7 and Clause 13 we shall not be liable for damage or injury occurring after repairs have been completed, where repairs are carried out on site, or after despatch of plant repaired at our works.

You assume all liability for accidents and damage whether on site or in our works caused by or arising out of the condition or nature of the plant not disclosed to us and not apparent on reasonable examination by us.

13. **DEFECTS AFTER COMPLETION**–We will remedy any defects which, under proper use, appear in repaired plant within 6 months after the completion of repairs or, where plant is repaired at our works, after delivery of the repaired plant, provided that such defects are due solely to faulty materials or workmanship used in carrying out the repairs.

Save as aforesaid and as provided in Clause 7, we shall not be under liability in contract, tort or otherwise in respect of defects in plant repaired, or for any injury (other than personal injury caused by our negligence as defined in Section 1 of the Unfair Contract Terms Act, 1977), damage or loss resulting from such defects.

14. **WORK ON SITE**–When the contract includes dismantling, repair or re-erection work on site we will provide all necessary skilled labour.

When our offer is limited to supervision of work on site we will provide a competent supervisor.

In either case you shall provide, at your expense, all other labour, suitable access to and possession of the site, proper foundations ready to receive the plant as and when delivered, adequate crane, lifting tackle and scaffolding, masons', joiners' and builders' work, suitable protection for the plant from time of delivery and all necessary facilities and adequate assistance.

Continued overleaf

15. PATENTS–We will indemnify you against any claim of infringement of Letters Patent, Registered Design, Trade Mark or Copyright (published at the date of the Contract) by the use or sale of any article or material supplied by us to you and against all costs and damages which you may incur in any action for such infringement or for which you may become liable in any such action. Provided always that this indemnity shall not apply to any infringement which is due to our having followed a design or instruction furnished or given by you or to the use of such article or material in a manner or for a purpose or in a foreign country not specified by or disclosed to us, or to any infringement which is due to the use of such article or material in association or combination with any other article or material not supplied by us. And provided also that this indemnity is conditional on your giving to us the earliest possible notice in writing of any claim being made or action threatened or brought against you and on your permitting us at our own expense to conduct any litigation that may ensue and all negotiations for a settlement of the claim. You on your part warrant that any design or instructions furnished or given by you shall not be such as will cause us to infringe any Letters Patent, Registered Design, Trade Mark or Copyright in the execution of your order.

16. FAIR WAGES CLAUSE–We undertake to be bound by a fair wages clause in the terms of the House of Commons Resolution of 14th October, 1946. We also undertake that when the installation of machinery and equipment is carried out by men sent from the Contractor's or Sub-Contractor's establishment they shall receive the time rate payable in terms of the Fair Wages Clause to such workpeople in such establishment, and in addition shall receive the outworking allowances recognised for outworkers sent from such establishment.

17. ARBITRATION–If at any time any question, dispute or difference whatsoever shall arise between you and ourselves upon, in relation to, or in connection with the contract, either of us may give to the other notice in writing of the existence of such question, dispute or difference, and the same shall be referred to the arbitration of a person to be mutually agreed upon, or failing agreement within thirty days of receipt of such notice, of some person appointed by the President for the time being of the Institution of Electrical Engineers.

18. LEGAL CONSTRUCTION–Unless otherwise agreed in writing the contract shall in all respects be construed and operate as an English contract and in conformity with English law.

19. STATUTORY AND OTHER REGULATIONS–If the cost to us of performing our obligations under the contract shall be increased or reduced by reason of the making or amendment after the date of tender of any law or of any order, regulation, or bye-law having the force of law that shall affect the performance of our obligations under the contract, the amount of such increase or reduction shall be added to or deducted from the contract price as the case may be.

Published by: The British Electrical and Allied Manufacturers' Association Limited, 8 Leicester Street, Leicester Square, London WC2H 7BN

RE

Edition June, 1979

Conditions (RE) for the repair of machinery and equipment Export F.O.B.

1. **DEFINITIONS**–The expression FOB shall bear the meaning assigned to it in INCOTERMS 1953 save insofar as the same may have been varied by these Conditions.

2. **GENERAL**–The acceptance of our tender includes the acceptance of the following terms and conditions:–

3. **VALIDITY**–Unless previously withdrawn, our tender is open for acceptance for the period stated therein or, when no period is so stated, within thirty days only after its date.

4. **PRICES**–Unless otherwise stated, prices are approximate only.

5. **PACKING**–Unless otherwise specified in our tender, packing in accordance with our standard export practice is included.

6. **LIMITS OF CONTRACT**–Our tender is made on the assumption that the repairs are reasonably capable of being carried out. If, on inspection, this is found not to be the case, we will advise you as soon as reasonably practicable; no liability shall attach to us for any loss occasioned by the repairs not being carried out and the cost of such inspection shall be borne by you.

Plant sent to us for repair shall be delivered to our works free of all cost.

7. **LIABILITY FOR DELAY**–Any time quoted by us for delivery of plant repaired in our works or for completion of repairs on site shall not begin to run until we have received the plant to be repaired or, where repairs are to be carried out on site, have obtained access to the plant and in either case until we have received a written order to proceed and all necessary information to enable us to put the work in hand. Any such time is to be treated as an estimate only not involving us in any liability for failure to deliver or complete within such time unless you have suffered loss and the amount payable in respect thereof shall have been agreed in writing as liquidated damages, in which case our liability shall be limited to the amount so agreed to be paid. In all cases, whether a time for delivery or completion be quoted or not, the time therefor shall be extended by a reasonable period if delay in delivery or completion is caused by instructions, or lack of instructions, from you or by industrial dispute or by any cause whatsoever beyond our reasonable control.

8. **DELIVERY**–Unless otherwise arranged, delivery of repaired plant will be f.o.b. at the port stated in our tender.

We shall not be required to give you the notice relating to insurance mentioned in Section 32(3) of the Sale of Goods Act, 1893.

9. **STORAGE**–If we do not receive forwarding instructions sufficient to enable us to despatch repaired plant within 14 days after the date of notification that the plant is ready for despatch, you shall take delivery or arrange for storage. If you do not take delivery or arrange for storage, we shall be entitled to arrange storage either at our own works or elsewhere on your behalf and all charges for such storage, for insurance or for demurrage shall be payable by you.

10. **PAYMENT**–Unless otherwise agreed, payment in full shall be due on notification to you that the plant is ready for despatch, or on completion when the repairs are effected on site.

11. **EXTRA COST**–In the event of suspension of the work by your instructions or lack of instructions, any price quoted shall be increased by a reasonable sum to cover any extra expense thereby incurred by us.

12. **LIABILITY FOR ACCIDENTS AND DAMAGE**–We will indemnify you against damage or injury to your property or person or that of others to the extent directly caused by the negligence of ourselves, our sub-contractors or agents, by making good such damage to property or compensating personal injury.

Provided that:

(*a*) Our total liability for such damage to your property (including damage caused by our breach of contract, tort or breach of statutory duty) shall not exceed £100,000 (United Kingdom currency) or the contract price, whichever is the greater, and

(*b*) We shall not be liable to you for any loss of profit or contracts or, save as aforesaid, for loss, damage or injury of any kind whatsoever and whether caused by our breach of contract, tort, breach of statutory duty or otherwise howsoever.

Save as provided in Clause 13 we shall not be liable for any damage or injury occurring after repairs have been completed, where repairs are carried out on site, or after delivery of plant repaired at our works.

You assume all liability for accidents and damage whether on site or in our works caused by or arising out of the condition or nature of the plant not disclosed to us and not apparent on reasonable examination by us.

13. **DEFECTS AFTER COMPLETION**–We will remedy any defects which, under proper use, appear in repaired plant within 6 months after the completion of repairs or, where plant is repaired at our works, within 6 months after the delivery of the repaired plant, provided that such defects are due solely to faulty materials or workmanship used in carrying out the repairs.

Save as aforesaid we shall not be under any liability whether in contract, tort or otherwise in respect of defects in plant repaired, or for any injury, damage or loss resulting from such defects.

14. **WORK ON SITE**–When the contract includes dismantling, repair or re-erection work on site we will provide all necessary skilled labour.

When our offer is limited to supervision of work on site we will provide a competent supervisor.

In either case you shall provide, at your expense, all other labour, suitable access to and possession of the site, proper foundations ready to receive the plant as and when delivered, adequate crane, lifting tackle and scaffolding, masons', joiners', and builders' work, suitable protection for the plant from time of delivery and all necessary facilities and adequate assistance.

15. **PATENTS**–We will indemnify you against any claim of infringement of Letters Patent, Registered Design, Trade Mark or Copyright (published at the date of the Contract) by the use or sale of any article or material supplied by us to you and against all costs and damages which you may incur in any action for such infringement or for which you may become liable in any such action. Provided always that this indemnity shall not apply to any infringement which is due to our having followed a design or instruction furnished or given by you or to the use of such article or material in a manner or for a purpose or in a foreign country not specified by or disclosed to us, or to any infringement which is due to the use of such article or material in association or combination with any other article or material not supplied by us. And provided also that this indemnity is conditional on your giving to us the earliest possible notice in writing of any claim being made or action threatened or brought against you and on your permitting us at our own expense to conduct any litigation that may ensue and all negotiations for a settlement of the claim. You on your part warrant that any design or instructions furnished or given by you shall not be such as will cause us to infringe any Letters Patent, Registered Design, Trade Mark or Copyright in the execution of your order.

16. **ARBITRATION**–If at any time any question, dispute or difference whatsoever shall arise between you and ourselves upon, in relation to, or in connection with the contract, either of us may give to the other notice in writing of the existence of such question, dispute or difference, and the same shall be referred to the arbitration of a person to be mutually agreed upon, or failing agreement within thirty days of the receipt of such notice, of some person appointed by the President for the time being of the Institution of Electrical Engineers.

17. **LEGAL CONSTRUCTION**–Unless otherwise agreed in writing the contract shall in all respects be construed and operate as an English contract and in conformity with English law and the English courts shall have exclusive jurisdiction over any matter arising out of the provisions of Clause 16 (Arbitration).

18. **STATUTORY AND OTHER REGULATIONS**–If the cost to us of performing our obligations under the contract shall be increased or reduced by reason of the making or amendment after the date of tender of any law or of any order, regulation, or bye-law having the force of law that shall affect the performance of our obligations under the contract, the amount of such increase or reduction shall be added to or deducted from the contract price as the case may be.

Published by: The British Electrical and Allied Manufacturers' Association Limited, 8 Leicester Street, Leicester Square, London WC2N 7BN

Edition 1978, first published
Summer 1979, with revised
Clause 21 (see text on p.8.).

RC

Edition, 1978

Conditions (RC) for the reconstruction, modification or repair of plant and equipment in the United Kingdom involving work on site.

1. **DEFINITIONS.**–In construing these General Conditions and the Specification, the following words shall have the meanings herein assigned to them unless there is something in the subject matter or context inconsistent with such construction:–

The "Owner" shall mean and shall include the Owner's legal personal representatives and assigns.

The "Contract" means the agreement howsoever made between the Owner and the Contractor for the execution of the Works, including therein all documents to which reference may properly be made in order to ascertain the rights and obligations of the parties under the said agreement.

The "Contractor" shall mean the tenderer whose tender has been accepted by the Owner, and shall include the Contractor's legal personal representatives, successors and assigns.

"Sub-Contractor" shall mean any person (other than the Contractor) named in the Contract for any part of the Works or any person to whom any part of the Contract has been sub-let with the consent in writing of the Engineer, and the legal representatives, successors, and assigns of such persons.

"Contractor's Equipment" shall mean tools, tackle, stores and other things brought upon the Site by the Contractor and required thereon for the purposes of the Works but not for incorporation therein.

The "Contract Price" shall mean the sum named in the Contract as the Contract Price.

The "Contract Value" shall mean that part of the Contract Price adjusted to give effect to such additions or deductions as are provided for in these Conditions which is properly apportionable to the Plant or work in question having regard to the state, condition, and location of the Plant, the amount of work done, and all other relevant circumstances, but disregarding any changes pursuant to Clause 36 (Variations in Costs) in the cost of executing the Works.

The "Engineer" shall mean or the person for the time being or from time to time notified in writing by the Owner to the Contractor as the Engineer for the Contract, or in default of any notification the Owner.

"Plant" shall mean and include machinery, apparatus, materials, articles and things of all kinds other than Contractor's Equipment.

The "Subject Matter" shall mean the piece or pieces of plant for the reconstruction, alteration, or repair of which the Contract provides.

"The Specification" shall mean the Specification annexed to or issued with these General Conditions.

"The Site" shall mean the actual place or places where the Subject Matter is situated, or to which the Subject Matter is to be delivered, together with so much of the area surrounding the said place or places as the Contractor shall, with the consent of the Engineer, actually use in connection with the Works otherwise than merely for the purpose of access to the said place or places.

"Tests on Completion" shall mean such tests to be made by the Contractor before the Works are taken over by the Owner as are provided for in the Contract and such other tests as may be agreed between the Owner and the Contractor.

"The Works" shall mean the work to be done by the Contractor to the Subject Matter pursuant to the Contract and shall include all Plant to be supplied by the Contractor for that purpose.

"Month" shall mean calendar month.

"Writing" shall include any manuscript, typewritten or printed statement, under seal or hand as the case may be.

Words importing persons shall include firms and corporations.

Words importing the singular only shall also include the plural and vice versa.

2. **CONTRACTOR TO INFORM HIMSELF FULLY.**–The Contractor shall be deemed to have examined the Site, if access thereto has been available to him, and the General Conditions and Specification, with such schedules, drawings and plans as are annexed thereto or referred to therein.

3. **DRAWINGS.**–(i) The Contractor shall submit to the Engineer for approval:–

(*a*) Within the time specified in the Specification or if no time is specified then a reasonable time such drawings, samples, patterns, information and models as may be called for therein, and

(*b*) During the progress of the Works, and within such reasonable times as the Engineer may require, such drawings of the general arrangement and of details of the Works or any part thereof as the Engineer may reasonably require.

Within a reasonable period after receiving such drawings, samples, patterns, information and models the Engineer shall signify his approval or otherwise. If the Engineer shall not approve any drawing, sample, pattern or model thus submitted the same shall be forthwith modified to meet the reasonable requirements of the Engineer and shall be resubmitted. The approval of drawings shall not relieve the Contractor from his obligation to execute the Works in accordance with the Contract. Copies of all drawings which require to be approved by the Engineer shall be provided by the Contractor in triplicate or such other numbers as may be provided by the Specification.

(ii) Drawings, samples, patterns and models approved as above described shall not be departed from except as follows:–

(*a*) If the Contractor, in order to comply with his obligations under the Contract, shall wish to modify or correct any such drawings he shall submit for the approval of the Engineer a revised drawing or drawings and the Engineer shall signify his approval otherwise as herein provided, or

(*b*) As provided in Sub-clauses (i) and (ii) of Clause 6 (Variations and Omissions).

(iii) The Engineer shall have the right at all reasonable times to inspect at the premises of the Contractor all drawings of any portion of the Works.

(iv) The Contractor shall, if so requested by the Owner in writing furnish to the Owner in writing at the commencement of the defects liability period referred to in Clause 29 (Defects after Taking Over), or at such earlier times as may be named in the Specification, such information, accompanied by drawings, as may be necessary to enable the Owner to operate, maintain, dismantle, re-assemble, and adjust all parts of the Works.

(v) Drawings submitted in pursuance of paragraphs (a) and (b) of Sub-Clause (i) of this Clause shall not, without the consent of the Contractor, be used by the Owner except for the purposes of the Contract, nor shall they without such consent be communicated to third parties save insofar as may be necessary for the proper execution of the Works.

(vi) Nothing in this Clause shall oblige the Contractor to supply copies of any shop drawings.

4. MANNER OF EXECUTION.–(i) All Plant to be supplied and all work to be done under the Contract shall be manufactured and executed in the manner set out in the Specification, if any, and to the reasonable satisfaction of the Engineer, but unless a contrary intention shall be manifest in the Specification or elsewhere in the Contract the Contractor shall not be responsible for the adequacy, sufficiency, suitability or desirability of the Works described in the Specification in relation to the objects of the Owner, whether known to the Contractor or not, or of any design which the Contractor is required by the Owner to use.

(ii) The Contractor shall not in the performance of the Contract in any manner endanger the safety or unlawfully interfere with the convenience of the public.

5. MISTAKES IN INFORMATION.–(i) The Contractor shall be responsible for any discrepancies, errors, or omissions in the drawings and information supplied by him, whether they have been approved by the Engineer or not, provided that such discrepancies, errors, or omissions be not due to defective drawings or information furnished to the Contractor by the Owner or the Engineer.

(ii) The Contractor shall at his own expense carry out any alterations or remedial work necessitated by reason of such discrepancies, errors, or omissions and modify the drawings and information accordingly, or if the same be done by or on behalf of the Owner shall bear all costs reasonably incurred therein. The performance of his obligations under this sub-clause shall be in full satisfaction of the Contractor's liability under Sub-Clause (i) of this clause, but shall not relieve him of his liability under Clause 28 (Delay in Completion) insofar as that liability arises as a result of such discrepancies errors or omissions.

(iii) The Owner shall be responsible for drawings and information supplied by the Owner or the Engineer and for the details of special work specified by either of them. The Owner shall pay the reasonable extra cost incurred by the Contractor due to alterations of the work necessitated by reason of defective drawings or information so supplied to the Contractor.

6. VARIATIONS AND OMISSIONS.–(i) The Contractor shall not alter any of the Works, except as directed in writing by the Engineer; but the Engineer shall have full power, subject to the proviso hereinafter contained, from time to time during the execution of the Works by notice in writing to direct the Contractor to add to, alter, omit, or otherwise vary any of the Works without prejudice to the Contract, and the Contractor shall carry out such variations, and be bound by the same conditions, so far as applicable, as though the said variations were stated in the Specification. Provided that no such variation shall, except with the consent in writing of the Contractor, be such as will, with any variations already directed to be made involve a net addition to or deduction from the Contract Price of more than 15% thereof.

(ii) In any case in which the Contractor has received any such direction from the Engineer which either then or later will, in the opinion of the Contractor involve an addition to or deduction from the Contract Price, the Contractor shall advise the Engineer in writing to that effect and the amount of such addition or deduction shall be determined under Sub-Clause (iii) of this clause before the work is put in hand unless the Contractor shall be otherwise instructed in writing by the Engineer.

(iii) The amount of such addition or deduction shall be agreed between the Engineer and the Contractor or, failing agreement, be determined under Sub-Clause (iv) of this clause by arbitration. Due account shall be taken of any partial execution of the Works which is rendered useless by any such variation.

(iv) If during the execution of the Works it appears to the Contractor as a result of the dismantling or closer examination of the Subject Matter that some addition, alteration, or omission should be made to or from the Works in order that the objects of the Owner as known to the Contractor may be better attained he shall inform the Engineer.

7. UNFORESEEN DIFFICULTIES.–(i) If during the progress of the Works it shall appear to the Contractor that for reasons which were not known to him at the date of his tender the execution of the Works is not reasonably practicable, or that after the completion of the Works the Subject Matter is unlikely to be satisfactory either as a piece of electrical plant or to serve the objects of the Owner, the Contractor shall give the Engineer notice in writing to that effect with full particulars of the reasons and the Engineer shall within a reasonable time thereafter inform the Contractor in writing whether he accepts or rejects the said notice.

(ii) If the Engineer shall reject a notice given pursuant to Sub-Clause (i) of this clause a difference shall be deemed to have arisen which shall be determined by arbitration in accordance with these Conditions.

(iii) If the Engineer shall accept a notice given pursuant to Sub-Clause (i) of this clause or if the Arbitrator shall determine in favour of the Contractor's contentions as stated in such notice then either:–

(*a*) The Engineer shall within a reasonable time by the exercise of the power conferred on him by Sub-Clause (i) of Clause 6 Variations and Omissions) or otherwise arrange for the rectification of the matters specified in the notice; or

(*b*) The Contract shall be deemed to be terminated and the Contractor shall be entitled to be paid by the Owner the Contract Value of work executed and on plant provided or intended to be provided for the purposes of the Contract, whether by the Contractor or Sub-Contractors, according to its state at the date of the acceptance by the Engineer of the said notice or the award of the Arbitrator as the case may be.

8. ASSIGNMENT.–(i) The Contractor shall not, without the consent in writing of the Owner, which shall not be unreasonably withheld, assign or transfer the Contract or the benefits or obligations thereof or any part thereof to any other person.

SUB-LETTING.–(ii) The Contractor shall not, without the consent in writing of the Engineer, which shall not be unreasonably withheld, sub-let the Contract or any part thereof, or make any sub-contract with any person or persons for the execution of any portion of the Works but the restriction contained in this clause shall not apply to sub-contracts for materials, for minor details or for any part of the Works of which the makers are named in the Contract. Any such consent shall not relieve the Contractor from his obligations under the Contract.

9. PATENT RIGHTS.–(i) The Contractor shall indemnify the Owner against all actions, claims, demands, costs, charges and expenses arising from or incurred by reason of any infringement or alleged infringement of letters patent, registered design, or copyright, trade mark or trade name protected in the United Kingdom by the use, sale or possession of any Plant supplied by the Contractor, but such indemnity shall not cover any use of the Works otherwise than for the purpose indicated by or reasonably to be inferred from the Specification. Provided always that this indemnity shall not apply to any infringement which is due to the Contractor having followed a design or instruction furnished or given by the Owner or to the use of such articles or material in a manner for a purpose not specified by or disclosed to the Contractor, or to any infringement which is due to the use of such article or material in association or combination with any other article or material not supplied by the Contractor.

(ii) In the event of any claim being made or action brought against the Owner arising out of the matters referred to in this clause, the Contractor shall be promptly notified thereof and may at his own expense conduct all negotiations for the settlement of the same, and any litigation that may arise therefrom. The Owner shall not, unless and until the Contractor shall have failed to take over the conduct of the negotiations or litigation, make any admission which might be prejudicial thereto. The conduct by the Contractor of such negotiations or litigation shall be conditional upon the Contractor having first given to the Owner such reasonable security as shall from time to time be required by the Owner to cover the amount ascertained or agreed or estimated, as the case may be, of any compensation, damages, expenses, and costs for which the Owner may become liable. The Owner shall, at the request of the Contractor, afford all available assistance for the purpose of contesting any such claim or action, and shall be repaid all reasonable expenses incurred in so doing.

(iii) If the Contractor shall be prevented from carrying out his obligations under the Contract due to any infringement or alleged infringement of letters patent, registered design, copyright, trade mark or trade name, protected in the United Kingdom, the Owner may treat such inability as a default by the Contractor and exercise the powers and remedies available to him under Clause 11 (Contractor's Default).

(iv) The Owner on his part warrants that any design or instruction furnished or given by him shall not be such as will cause the Contractor to infringe any letters patent, registered design, or copyright, trade mark or trade name in the performance of the Contract.

10. SERVICES AND FACILITIES.–Unless specific arrangements are made to the contrary the Owner shall, at his own expense, provide all unskilled labour, services and facilities necessary for the execution of the Works on the Site.

11. CONTRACTOR'S DEFAULT.–(i) If the Contractor shall neglect to execute the Works with due diligence and expedition, or shall refuse or neglect to comply with any reasonable orders given to him in writing by the Engineer in connection with the Works, or shall contravene the provisions of the Contract, the Owner may give seven days' notice in writing to the Contractor to make good the failure, neglect or contravention complained of, and should the Contractor fail to comply with the notice within seven days from the date of service thereof in the case of a failure, neglect, or contravention capable of being made good within that time, or otherwise within such time as may be reasonably necessary for making it good, then and in such case the Owner shall be at liberty to employ other workmen and forthwith execute such part of the Works as the Contractor may have neglected to do, or, if the Owner shall think fit, it shall be lawful for him without prejudice to any other rights he may have under the Contract, to take the Works wholly or in part out of the Contractor's hands and re-contract with any other person or persons to complete the Works or any part thereof, and in that event the Owner shall, without being responsible to the Contractor for fair wear and tear of the same, have the free use of all the materials, tools, tackle and other things that may be at any time on the Site in connection with the Works to the exclusion of any right of the Contractor over the same, and the Owner shall be entitled to retain and apply any balance which may be otherwise due on the Contract by him to the Contractor, or such part thereof as may be necessary, to the payment of the cost of executing the said part of the Works or of completing the Works as the case may be.

(ii) If the cost of completing the Works or executing a part thereof as aforesaid shall exceed the amount that would otherwise have become due to the Contractor in accordance with the Contract, the Contractor shall pay such excess.

(iii) If the Owner pursuant to this Clause takes the Works or part thereof out of the Contractor's hands, the Contractor's liability under Clause 25 (Delay in Completion) shall immediately cease in respect of the Works or part thereof, without prejudice to any such liability that shall have already accrued.

12. BANKRUPTCY.–If the Contractor shall have become bankrupt or insolvent, or have a receiving order made against him, or compound with his creditors, or being a corporation commence to be wound-up, not being a members' voluntary winding-up for the purposes of reconstruction or amalgamation, or carry on its business under a receiver for the benefit of its creditors or any of them, the Owner shall be at liberty either:–

(*a*) To terminate the Contract forthwith by notice in writing to the Contractor or to any such receiver or liquidator or to any person in whom the Contract may become vested, and to act in the manner provided in Clause 11 (Contractor's Default) as though the last mentioned notice had been the notice referred to in Sub-Clause (i) thereof and the Works had been taken out of the Contractor's hands, or

(*b*) to give the receiver, liquidator, or other person the option of carrying out the Contract subject to his providing a guarantee for the due and faithful performance of the Contract up to an amount to be agreed.

13. DELIVERY OF PLANT.–If, from any cause for which the Owner or some other contractor employed by him is responsible, the Contractor shall be prevented from delivering any Plant to the Site at the time when it is reasonable for it to be delivered having regard to the date by which the Works ought to be completed, and shall have given notice in writing to the Owner that such Plant is ready for delivery, and shall have suitably and sufficiently marked the said Plant as allocated to the Contract, and shall have given the Engineer an opportunity of inspecting the said Plant, then and in any such case, without prejudice to any other rights the Contractor may have, the reasonable extra cost of storing and adequately protecting and preserving such plant against loss, deterioration, or damage and of insuring it against fire and special perils from the time when the said Plant would have been delivered until the Contractor shall be no longer prevented from delivering it shall be added to the Contract Price.

The Contractor shall be entitled one month after the date of the said notice to have the Contract Value of the Plant to which the notice relates included in an interim Certificate.

14. ACCESS TO AND POSSESSION OF THE SITE.–(i) Subject to Sub-Clause (iv) of this clause, access to and possession of the Site shall be afforded to the Contractor by the Owner in reasonable time and, except insofar as the Specification may provide to the contrary the Owner shall provide a road or railway suitable for the transport of all Plant and Contractor's Equipment necessary for the execution of the Works from a convenient point on a public thoroughfare suitable for such transport or on a railway available to the Contractor to the point on the Site where it is to be delivered or is required for use.

(ii) If a building, structure, foundation, or approach is by the Contract to be provided by the Owner such building, structure, foundation or approach shall be in a condition suitable for the efficient transport, reception, installation and maintenance of the Works.

(iii) In the execution of the Works, the Contractor shall not authorise or purport to authorise any person other than his employees and Sub-Contractors and their employees to come upon the Site, except by the written permission of the Engineer, but facilities to inspect the Subject Matter and the Works at all times shall be afforded to the Engineer and his representatives, and other authorised officials or representatives of the Owner. The Contractor shall afford to the Owner and to other contractors whose names shall have been previously communicated in writing to the Contractor by the Engineer every reasonable facility for the execution of the Works concurrently with his own.

(iv) Unless otherwise provided in the Specification the Owner shall give the Contractor facilities for carrying out the Works on the Site continuously during the normal working hours generally recognised in the district. The Engineer may, after consulting with the Contractor, direct that work shall be done at other times if it shall be practicable in the circumstances for work to be so done, and the extra cost of work so done shall be added to the Contract Price unless such work has, by the default of the Contractor become necessary for the completion of the Works within the time fixed by the Contract, or if no time be fixed, within a reasonable time.

15. VESTING OF PLANT.–(i) Plant supplied or intended to be supplied pursuant to the Contract shall become the property of the Owner at whichever is the earlier of the following times:–

(*a*) when the Plant is brought upon the Site pursuant to the Contract;

(*b*) when pursuant to the Contract the Plant is fixed to or otherwise made part of the Subject Matter.

(ii) When instructed by the Engineer so to do the Contractor shall suitably and sufficiently mark as the property of the Owner Plant the property in which has passed to the Owner as aforesaid.

(iii) In the event of any Plant becoming the property of the Owner pursuant to the Contract and subsequently being rejected by the Engineer pursuant to Clause 18 (Inspection, Testing and Rejection), such Plant shall forthwith upon such rejection cease to be the property of the Owner and become the property of the Contractor.

16. ENGINEER'S SUPERVISION.–(i) After the acceptance of the tender by the Owner all instructions and all orders to the Contractor shall, except as herein otherwise provided, be given by the Engineer on behalf of the Owner, and all the Works shall be carried out under the direction and to the reasonable satisfaction of the Engineer.

DELEGATION.–(ii) The Engineer may from time to time delegate any of the powers, discretions, functions, and authorities vested in him and may at any time revoke any such delegation. Any such delegation or revocation shall be in writing signed by the Engineer and, in the case of a delegation, shall specify the powers, discretions, functions and authorities thereby delegated and the person or persons to whom the same are delegated. No such delegation or revocation shall have effect until a copy thereof has been delivered to the Contractor.

17. ENGINEER'S DECISIONS.–The Contractor shall proceed with the Works in accordance with decisions, instructions and orders given by the Engineer in accordance with these Conditions, provided always that:–

(*a*) if the Contractor shall, without undue delay after being given any decision, instructions, or order otherwise than in writing, require it to be confirmed in writing, such decision, instruction, or order shall not be effective until written confirmation thereof has been received by the Contractor, and

(*b*) if the Contractor shall, by written notice to the Engineer within 14 days after receiving any decision, instruction or order of the Engineer in writing or written confirmation thereof intimate either:–

(i) that he considers that any such decision, instruction or order constitutes a variation of the Works, or

(ii) that he disputes or questions the decision, instruction or order on other grounds and gives his reasons for so doing

then unless the intimation is given under (i) above and the Engineer has issued an instruction in writing to vary the Works pursuant to Clause 6 (Variations and Omissions) within 14 days of the receipt by the Engineer of the above-mentioned written notice then either party to the Contract shall be at liberty to refer the matter to arbitration pursuant to Clause 35 (Arbitration) but such intimation shall not relieve the Contractor of his obligation to proceed with the Works in accordance with the decision, instruction or order in respect of which any such intimation has been given. The Contractor shall be at liberty in any such arbitration to rely on reasons additional to the reasons stated in any such intimation.

18. INSPECTION, TESTING AND REJECTION.–(i) The Engineer shall be entitled at all reasonable times during manufacture to inspect, examine, and test on the Contractor's premises the materials and workmanship and performances of all Plant to be supplied under the Contract, and if part of the said Plant is being manufactured on other premises the Contractor shall obtain for the Engineer permission to inspect, examine, and test as if the said Plant were being manufactured on the Contractor's premises. Such inspection, examination, or testing, shall not release the Contractor from any obligation under the Contract.

(ii) The Contractor shall, after consulting the Engineer, give the Engineer reasonable notice in writing of the date on and place at which any Plant will be ready for testing as provided in the Contract and unless the Engineer shall attend at the place so named on the date which the Contractor has stated in his notice the Contractor may proceed with the tests, which shall be deemed to have been made in the Engineer's presence, and shall forthwith forward to the Engineer duly certified copies of the test reading. The Engineer shall give the Contractor 24 hours' notice in writing of his intention to attend the tests.

(iii) Where the Contract provides for tests on the premises of the Contractor or of any Sub-Contractor the Contractor, except where otherwise specified, shall provide free of charge such assistance, labour, materials, electricity, fuel, stores, apparatus, and instruments as may be requisite and as may be reasonably demanded to carry out such tests efficiently.

(iv) Where the Contract provides for tests on the Site, the Owner, except where otherwise specified, shall provide free of charge, such labour, materials, electricity, fuel, stores, and apparatus, as may be requisite and as may be reasonably demanded to carry out such tests efficiently.

(v) As and when the Engineer is satisfied that any Plant shall have passed the tests referred to in this clause he shall, if so required, notify the Contractor in writing to that effect.

(vi) If after inspecting, examining, or testing any Plant the Engineer shall decide that such Plant or any part thereof is defective or not in accordance with the Contract, he may reject the said Plant or part thereof by giving to the Contractor within a reasonable time notice in writing of such rejection, stating therein the grounds upon which the said decision is based.

(vii) The provisions of Clause 26 (Tests on Completion) shall relate also to inspections, examinations, and tests carried out under this Clause.

19. CONTRACTOR'S REPRESENTATIVES AND WORKMEN.–(i) The Contractor shall employ one or more competent representatives, whose name or names shall have previously been communicated in writing to the Engineer by the Contractor, to superintend the carrying out of the Works on the Site. The said representative or, if more than one shall be employed, then one of such representatives shall, if required by the Engineer, be present on the Site during working hours, and any orders or instructions which the Engineer may give to the said representative of the Contractor shall be deemed to have been given to the Contractor.

(ii) The Engineer shall be at liberty by notice in writing to the Contractor to object to any representative or person employed by the Contractor in the execution of or otherwise about the Works who shall, in the opinion of the Engineer, misconduct himself or be incompetent or negligent and the Contractor shall remove such person from the Works.

20. LIABILITY FOR ACCIDENTS AND DAMAGE.–(i) The Contractor shall properly cover up and protect until taken over under Clause 27 (Taking Over) any part of the Works or of the Subject Matter liable to suffer damage by exposure to the weather, and shall take every reasonable precaution to protect the Works or the Subject Matter not taken over against loss or damage from any cause.

(ii) In the case of loss of or damage to the Works or to the Subject Matter on the Site arising from or occasioned by causes for which the Contractor is not responsible under the Contract the same shall, if required by the Owner, be made good by the Contractor but at the cost of the Owner at a price to be agreed between the Contractor and the Owner or in default of agreement to be settled by arbitration and such cost shall be added to the Contract Price.

(iii) Subject to Clause 21 (Limitations on Contractor's Liability), all losses of or damage to any part of the Works or the Subject Matter, occurring before taking over which shall arise from or be occasioned by any act of the Contractor or any Sub-Contractor or by a failure of the Contractor to comply with any obligations imposed on him by Sub-Clause (i) of this clause, shall be made good by and at the sole cost of the Contractor and to the reasonable satisfaction of the Engineer.

(iv) The Contractor shall, subject to Clause 21 (Limitations on Contractor's Liability) indemnify the Owner in respect of all damage or injury occurring before all the Works shall have been taken over under Clause 27 (Taking Over) to any person or to any property and against all actions, suits, claims, demands, costs, charges and expenses arising in connection therewith which shall be occasioned by the negligence of or breach of statutory duty by the Contractor or any Sub-Contractor, or by defective design (other than a design made, furnished or specified by the Owner and for which the Contractor has disclaimed responsibility in writing within a reasonable time after the receipt of the Owner's instructions), materials or workmanship but not otherwise.

(v) If there shall occur any loss of or damage or injury to any property or person while the Contractor is on the Site for the purpose of making good a defect in the Works pursuant to Clause 29 (Defects after Taking Over) or for the purpose of carrying out tests on completion during the defects liability period as provided in Sub-Clause (iii) of Clause 27 (Taking Over), the Contractor shall be liable, subject to the provisions of Clause 21 (Limitations on Contractor's Liability) as follows:–

(*a*) in respect of any defect in or damage to the Works the Contractor's liability shall be as defined in Clause 29 (Defects after Taking Over);

(*b*) in respect of damage or injury to any other property or to any person and of any actions, claims, demands, costs, charges and expenses arising in connection therewith the Contractor shall be liable, subject to the provisions of Sub-Clauses (i) and (iii) of this clause, to the extent that such damage or injury was caused by the negligence or breach of statutory duty of the Contractor or a sub-Contractor while on the Site as aforesaid or by defective materials or workmanship used in making good the said defect but not otherwise.

(vi) In the event of any claim being made against the Owner arising out of the matters referred to in and in respect of which the Contractor may be liable under this clause, the Contractor shall be promptly notified thereof, and may at his own expense conduct all negotiations for the settlement of the same and any litigation that may arise therefrom. The Owner shall not, unless and until the Contractor shall have failed to take over the conduct of the negotiations or litigation, make any admission which might be prejudicial thereto. The conduct by the Contractor of such negotiations or litigation shall be conditional upon the Contractor having first given to the Owner such reasonable security as shall from time to time be required by the Owner to recover the amount ascertained or agreed or estimated, as the case may be, of any compensation, damages expenses and costs for which the Owner may become liable. The Owner shall, at the request of the Contractor, afford all available assistance for any such purpose, and shall be repaid any out-of-pocket expenses incurred in so doing.

(vii) The Contractor shall hold the Owner indemnified against all actions, suits, claims, demands, costs, charges or expenses arising from the death of or bodily injury to any person or damage to property (except where such death or injury arises out of or in the course of employment of such person by the Owner) as a result of an accident caused by or through or in connection with any motor vehicle brought on to the Owner's premises by an employee of the Contractor or any Sub-Contractor engaged on the Works.

(viii) The Contractor shall not be responsible for, and the Owner shall indemnify the Contractor against all loss or damage which shall be caused by the negligence of the Owner or any other Contractor employed by the Owner or by any defect or fault in the plant of the Owner other than the Subject Matter, or by any defect or fault in the Subject Matter if such defect or fault is one which ought to have been made known to the Contractor, or by a defective design, made, furnished or specified by the Owner and for which the Contractor has disclaimed responsibility in writing within a reasonable time after the receipt of the Owner's instructions, and against all claims against the Contractor in respect of loss or damage sustained by any employee of the Contractor by reason of any of the said causes.

21. LIMITATIONS ON CONTRACTOR'S LIABILITY.–(i) The Contractor shall not be liable to the Owner under Clause 20 (Liability for Accidents and Damage) for:–

(*a*) any damage or injury to the extent that it is caused by or arises from the acts or omissions of the Owner or of others (not being the Contractor's servants or sub-contractors);

(*b*) any loss or damage in circumstances over which the Contractor has no control.

(ii) Except in respect of personal injury or damage to property conferring on a person other than the Owner a good cause of action against the Contractor, the liability of the Contractor under Clause 20 (Liability for Accidents and Damage) for any one act or default shall not exceed the Contract Price, or £100,000, whichever is the greater.

(iii) Subject as provided in Clause 25 (Delay in Completion) for the deduction of liquidated damages for delay, the Contractor shall not be liable to the Owner by way of indemnity or by reason of any breach of the Contract for loss of use (whether complete or partial) of the Works or of profit or of any contract that may be suffered by the Owner.

22. INSURANCE OF WORKS.–Unless the Owner shall have approved in writing other arrangements for the insurance hereinafter mentioned the Contractor shall, in the joint names of the Contractor and the Owner, insure all parts of the Subject Matter which the Contractor may remove from the Site with a view to the same being returned to the Site after being worked on, and all parts of the Works which, after becoming the property of the Owner, may be similarly removed, and shall keep the same insured to the full value thereof against destruction or damage by fire, lightning or explosion and against such other risks as may be specified in the Appendix and damage in transit until the same be returned to the Site or other plant is substituted therefor, and shall from time to time, when so required by the Engineer, produce the policy and receipts for the premiums. All monies received under any such policies shall be applied in or towards the replacement or repair of the plant destroyed or damaged, but this provision shall not affect the Contractor's liabilities under the Contract.

23. THIRD PARTY AND OTHER INSURANCE.–The Contractor (but without limiting his obligations and responsibilities under Clause 20 (Liability for Accidents and Damage)) shall insure against any damage, loss or injury which may occur to any property or to any person for which he may be responsible or liable to indemnify the Purchaser under Clause 20 (Liability for Accidents and Damage) hereof and Clause 21 (Limitations on Contractor's Liability) insofar as insurance is not covered by Clause 22 (Insurance of Works) and shall from time to time when so required by the Engineer, produce the policies and receipts for the premiums.

24. EXTENSION OF TIME FOR COMPLETION.–If by reason of any act or omission of the Owner, a direction by the Engineer under Clause 6 (Variations and Omissions), industrial dispute or any cause beyond the reasonable control of the Contractor arising after acceptance of the tender, the Contractor shall have been delayed or impeded in the completion of the Works, whether such delay or impediment occur before or after the time, if any, or extended time fixed for completion, provided that the Contractor shall have given to the Owner or the Engineer the earliest possible notice in writing of his claim for an extension of time, the Engineer shall, on receipt of such notice grant the Contractor from time to time in writing either prospectively or retrospectively such extension of the time fixed by the Contract for the completion of the Works as may be reasonable. Any delay on the part of a Sub-Contractor which prevents the Contractor from completing the Works within the time for completion shall entitle the Contractor to an extension thereof provided such delay was due to any cause for which the Contractor himself would have been entitled to an extension of time under this Clause.

25. DELAY IN COMPLETION.–If the Contractor fails to complete the Works in accordance with the Contract (except for the obligations of the Contractor under Clause 29 (Defects after Taking Over) and such tests as are to be made in accordance with Clause 26 (Tests on Completion)) within the time, if any, fixed by the Contract for the completion of the Works or any extension thereof, and the Owner shall have suffered any loss from such failure, there shall be deducted from the Contract Price the percentage thereof named in the Appendix for each week between the fixed or extended time as the case may be and the actual date for completion, but the amount so deducted shall not in any case exceed the maximum percentage named in the Appendix of the Contract Price, and such deduction shall be in full satisfaction of the Contractor's liability for the said failure.

26. TESTS ON COMPLETION.–(i) Where the Contract provides for Tests on Completion, the Contractor shall give to the Engineer in writing 21 days notice of the date after which he will be ready to make the Tests on Completion. Unless otherwise agreed, the tests shall take place within 10 days after the said date, on such day or days as the Engineer shall in writing notify the Contractor.

(ii) If the Engineer fails to appoint a time after having been asked to do so, or to attend at any time or place duly appointed for making the said tests the Contractor shall be entitled to proceed in his absence, and the said tests shall be deemed to have been made in the presence of the Engineer.

(iii) If, in the opinion of the Engineer, the tests are being unduly delayed, he may by notice in writing call upon the Contractor to make such tests within 10 days from the receipt of the said notice and the Contractor shall make the said tests on such day within the said 10 days as the Contractor may fix and of which he shall give notice to the Engineer. If the Contractor fails to make such tests within the time aforesaid the Engineer may himself proceed to make the tests. All tests so made by the Engineer shall be at the risk and expense of the Contractor unless the Contractor shall establish that the said tests were not being unduly delayed in which case tests so made shall be at the risk and expense of the Owner.

(iv) If the Works fail to pass the tests the Contractor shall be afforded all reasonable opportunities of repeating the tests upon the same terms and conditions, and all reasonable expenses to which the Owner may be put by any repetition of the tests necessitated by some defect in the Works shall be deducted from the Contract Price.

(v) The Contractor shall be relieved from his obligation to make the Tests on Completion or to cause the Works to pass the said tests if and insofar as the condition of the Subject Matter or any other part of the Owner's plant shall, notwithstanding the proper execution of the Works by the Contractor and the observance of his obligations hereunder, be such as to render it unsafe or impracticable to make the said tests, and shall not within a reasonable time be put into such condition as will make it safe and practicable to make the said tests.

27. TAKING OVER.–(i) As soon as the Works have been completed in accordance with the Contract (except in minor respects that do not affect their use for the purposes for which they are intended and except for the obligations of the Contractor under Clause 29 (Defects after Taking Over)) and subject to Sub-Clause (v) of Clause 26 (Tests on Completion) have passed the tests on completion, if any, the Engineer shall issue a certificate (herein called a "taking over certificate") in which he shall certify the date on which the Works have been so completed and have passed the said tests and the Owner shall be deemed to have taken over the Works on the date so certified, but the issue of a taking over certificate shall not operate as an admission that the Works have been completed in every respect. Save as provided in Sub-Clause (ii) of this clause the Owner shall not use the Works until a taking over certificate has been issued. If nevertheless the Owner does so use the Works the Works shall be deemed to have been taken over.

(ii) If by reason of any default on the part of the Contractor a taking over certificate has not been issued in respect of the Works within one month after the date fixed by the Contract for the completion of the Works or, if no time be fixed, within a reasonable time, the Owner shall be at liberty to use the Works, provided that the Works shall be reasonably capable of being used and that the Contractor shall be afforded reasonable opportunity of taking such steps as may be necessary to permit the issue of the taking over certificate.

INTERFERENCE WITH TESTS.–(iii) If by reason of any act or omission of the Owner or the Engineer the Contractor shall be prevented from carrying out the Tests on Completion as provided in Sub-Clause (i) of Clause 26 (Tests on Completion) then, unless in the meantime the Works shall have been proved not to be substantially in accordance with the Contract the Owner shall be deemed to have taken over the Works, and the Engineer shall issue a taking over certificate accordingly; nevertheless the Contractor shall make the said tests during the period referred to under Clause 29 as and when required by the Engineer by 14 days notice in writing, and Sub-Clauses (ii), (iii) and (iv) of Clause 26 (Tests on Completion) and Sub-Clause (vi) of Clause 18 (Inspection, Testing and Rejection) shall apply. The reasonable extra cost to which the Contractor may be put in making the said tests during the period referred to in Clause 29 (Defects after Taking Over) pursuant to this Sub-Clause shall be added to the Contract Price, and such allowances shall be made for the performances required to be obtained in the said tests as may be reasonable having regard to any use of the Works by the Owner prior to the tests.

(iv) If, after a taking over certificate has been applied for by the Contractor, the Engineer shall, without proper cause, fail to issue it or to certify therein a date not later than the date which ought to have been certified there shall be added to the Contract Price interest at the rate of 2% per annum over Bank of England Minimum Lending Rate for the time being in force on all sums the payment of which has been delayed by the said failure for the period or periods for which the said payments have been delayed.

(v) If the Works are divided into two or more sections the provisions of this Clause shall apply to each section as it applies to the Works.

28. RECOVERY OF ADDITIONAL COSTS.–(i) All reasonable extra costs incurred by the Contractor:–

(*a*) by reason of the Contractor being prevented from or delayed in proceeding in the Works by the Owner, the Engineer or some other contractor employed by the Owner, or

(*b*) by reason of the suspension of the Works by the Owner or the Engineer (otherwise than in consequence of some default on the part of the Contractor), or

(*c*) as the result of the granting under Clause 24 (Extension of Time for Completion) of an extension of time fixed for completion of the Works or any section thereof in consequence of a variation made under the provisions of Clause 6 (Variations and Omissions)

shall be added to the Contract Price, provided that no such addition shall be made, under this clause unless the Contractor has, within a reasonable time after the event giving rise to a claim given notice in writing to the Engineer of his intention to make such claim, and has supplied to the Engineer such information in support of such claim as the Engineer may reasonably require.

(ii) If the work on the Plant or any portion thereof is suspended as aforesaid by the Owner or the Engineer before the Plant or such portion thereof is delivered to the Site and the suspension exceeds three months and the Contractor has suitably and sufficiently marked the Plant or such portion thereof as the Owner's property and insured it as provided by Clause 22 (Insurance of Works) (the provisions of which clause shall thereafter and until actual delivery to the Site apply as if the Plant or such portion thereof were for the time being upon the Site) then without prejudice to the provisions of Sub-Clause (ii) of Clause 13 (Delivery of Plant) the Contractor shall be entitled to have the Contract Value of the Plant or such portion thereof as at the commencement of the suspension included in an interim certificate on the expiration of the said three months or (if later) at the time when, but for such suspension, the Plant or such portion thereof would have been delivered: provided that the Contract Value of any Plant that, according to the decision of the Engineer, is defective or not in accordance with the Contract shall not be included in any such certificate.

(iii) Where under the clauses stated in the Appendix the Contractor is entitled to be paid reasonable extra costs there shall be added thereto the percentage stated in the Appendix.

29. DEFECTS AFTER TAKING OVER.–(i) The Contractor shall make good with all possible speed any defect in or damage to the Works which may develop during a period of 12 months after the Works shall have been taken over (herein referred to as "the Defects Liability Period") and which, arises either:–

(*a*) from defective materials, workmanship or design (other than a design furnished or specified by the Owner and for which the Contractor has disclaimed in writing responsibility within a reasonable time after receipt of the Owner's instructions), or

(*b*) from any act or omission of the Contractor done or omitted during the defects liability period.

(ii) If any such defect or damage shall occur the Engineer shall inform the Contractor thereof stating in writing the nature of the defect or damage. If the Contractor replaces or renews any portion of the Works, the provisions of this clause shall apply to the portion of the Works so replaced or renewed until the expiration of 12 months from the date of such replacement or renewal.

(iii) If any such defect or damage be not remedied within a reasonable time the Owner may proceed to do the Work at the Contractor's risk and expense, but without prejudice to any other rights which the Owner may have against the Contractor in respect of the failure of the Contractor to remedy such defect or damage.

(iv) If the replacements or renewals are of such a character as may affect the use of the Subject Matter for the purposes of which it is intended the Owner may within one month of such replacement or renewal give to the Contractor notice in writing requiring that Tests on Completion be made, in which case such tests shall be carried out as provided in Clause 26 (Tests on Completion).

(v) These General Conditions shall apply to all inspections, adjustments, replacements, and renewals and to all tests occasioned thereby, carried out by the Contractor during the defects liability period.

(vi) The Contractor's liability under this clause shall be in lieu of any condition or warranty implied by law as to the quality or fitness for any particular purpose of the Works taken over under Clause 27 (Taking Over) and save as in this clause expressed neither the Contractor nor his Sub-Contractors, servants or agents shall be liable whether in contract, tort or otherwise, in respect of defects in or damage to the Works, or for any damage, loss or injury (other than personal injury caused by negligence on the Contractor's part as defined in Section 1 of the Unfair Contract Terms Act, 1977) attributable to such defects or damage. For the purpose of this Sub-Clause the Contractor contracts on his own behalf and on behalf of and as trustee for his Sub-Contractors, servants and agents. Nothing in this clause shall affect the liability of the Contractor under Sub-Clauses (i) and (iii) of Clause 20 (Liability for Accidents and Damage) in respect of any portion of the Works not taken over.

(vii) Until the Contractor has ceased to be under any obligation under this clause, the Contractor shall have the right of access, at all reasonable working hours, at his own risk and expense, by himself or his duly authorised representatives whose names shall have previously been communicated in writing to the Engineer, to all parts of the Works and of the Subject Matter for the purpose of inspecting the working thereof and to the records of the working and performance thereof for the purpose of inspecting the same and taking notes therefrom.

30. PROVISIONAL SUMS.–A provisional sum included in the Contract Price shall be expended or used as the Engineer may in writing direct and not otherwise. Insofar as a provisional sum is not expended or used it shall be deducted from the Contract Price.

31. PAYMENTS DUE FROM THE CONTRACTOR.–All costs, damages or expenses for which under the Contract the Contractor is liable to the Owner may be deducted by the Owner from any monies due or becoming due to the Contractor under the Contract, or may be recovered by action at law or otherwise from the Contractor.

32. TERMS OF PAYMENT.–(i) Subject to the provisions of Clause 31 (Payments Due from the Contractor) and of this clause, the Owner shall pay to the Contractor in the following manner the Contract Price together with such additions thereto and subject to such deductions therefrom as are herein provided for, namely; as to 95% thereof within one month after the taking over of the Works, and as to the remaining 5% thereof within one month after the Contractor has ceased to be under any obligation under Clause 29 (Defects after Taking Over).

(ii) If for any cause for which the Owner, the Engineer or some other Contractor employed by the Owner is responsible the Contractor shall be prevented from delivering any plant to the Site or from installing the same at the time when it is reasonable for it to be delivered or installed as the case may be having regard to the date by which the Works ought to be completed, and shall have given notice in writing to the Owner that he is so prevented, the Contractor shall, if he is still so prevented at the end of one month after the giving of such notice, thereupon be entitled in addition to any other rights under the Contract to payment of 95% of the Contract Value of such Plant including the work done thereon.

(iii) If the Works are taken over in sections the provisions of this Clause shall apply to each such section taken over as it applies to the Works.

33. STATUTORY AND OTHER REGULATIONS.–If the cost to the Contractor of performing his obligations under the Contract shall be increased or reduced by reason of the making after the date of his tender of any law or of any order, regulation, or bye-law having the force of law in the United Kingdom the amount of such increase or reduction shall be added to or deducted from the Contract Price as the case may be.

The Owner shall when desired afford all reasonable assistance to the Contractor in obtaining information as to local conditions.

34. CONTINGENCIES.–(i) The Contractor shall be relieved of liabilities incurred under the Contract wherever and to the extent to which the fulfilment of the Contract is prevented, frustrated or impeded as a consequence of conforming to any statute or rule, regulation, order or requisition made thereunder which is passed or made subsequent to the date of the tender provided that the Contractor shall without delay have given to the Owner or the Engineer notice in writing that the fulfilment of the Contract is prevented, frustrated or impeded as aforesaid.

(ii) Any statute or any rule, regulation, order or requisition made thereunder which,

(*a*) though passed or made on or before the date of the tender is not to come into force until a future date, and at the date of the tender such date has not been prescribed; or

(*b*) at the date of the tender is in force but is not applicable thereto subsequently

shall for the purpose of this clause be deemed to be passed or made subsequent to the date of the tender.

35. ARBITRATION.–(i) If at any time any question, dispute or difference shall arise between the Owner and the Contractor, either party shall, as soon as reasonably practicable, give to the other notice in writing of the existence of such question, dispute, or difference specifying its nature and the point of issue, and the same shall be referred to the arbitration of a person to be agreed upon, or failing such agreement, to some person appointed on the application of either of the parties hereto by the President for the time being of the Institution of Electrical Engineers, provided that a question, dispute, or difference relating to a decision, instruction, or order of the Engineer shall not be referred to arbitration unless notice has been given by the Contractor in accordance with paragraph (b) of Clause 17 (Engineer's Decisions).

(ii) Work under the Contract shall continue during arbitration proceedings unless the Engineer shall order the suspension thereof or of any part thereof, and if any such suspension shall be ordered the reasonable extra cost of the Contractor occasioned by such suspension shall be added to the Contract Price. No payments due or payable by the Owner shall be withheld on account of a pending reference to arbitration.

36. VARIATION IN COSTS AND TAX FLUCTUATIONS.–(i) If, by reason of any rise or fall in the cost of labour or in the cost of material or transport or of conforming to such laws, orders, regulations and bye-laws as are applicable to the Works above or below such rates and costs ruling at the date of the tender, the cost to the Contractor of performing his obligations under the Contract shall be increased or reduced, the amount of such increase or reduction shall be added to or deducted from the Contract Price as the case may be, provided that no account shall be taken of any amount by which any cost incurred by the Contractor has been increased by the default or negligence of the Contractor. For the purposes of this clause "the cost of material" shall be construed as including any duty or tax by whomsoever payable which is payable under or by virtue of any Act of Parliament on the import, purchase, sale, appropriation, processing or use of such material.

(ii) In arriving at the amount of such increase no account shall be taken of Value Added Tax or any tax of a like nature.

37. CONSTRUCTION OF CONTRACT.–The contract shall in all respects be construed and operate as an English contract and in conformity with English law.

APPENDIX

Clause **Additional risks to be insured**

Clause 25 Delay in Completion
(*a*) ....(*a*) percentage of Contract Price to be deducted as damages.
(*b*) ....(*b*) maximum percentage of Contract Price which the deductions may not exceed.

Recovery of Additional Costs

Percentage to be added .... per cent.

Clauses to which this percentage is applicable:

Clause 5 (Mistakes in Information) Sub-Clause (iii).
Clause 13 (Delivery of Plant).
Clause 14 (Access to and Possession of Site) Sub-Clause (iv).
Clause 27 (Taking Over) Sub-Clause (iii).
Clause 28 (Recovery of Additional Costs) Sub-Clause (iii).
Clause 35 (Arbitration) Sub-Clause (iii).

AGREEMENT

THIS AGREEMENT is made the day of 19 BETWEEN
(hereinafter referred to as the "Contractor") of the one part and
(hereinafter called the "Owner" of the other part.

WHEREAS the Owner desires to have provided and executed certain Works mentioned, enumerated, or referred to in certain General Conditions, Specification, Schedules, Drawings, Plans, Schedule of Prices, the Contractor's tender and covering letter (if any), the acceptance of the tender, and any other relevant correspondence which for the purpose of identification have been signed by on behalf of the Contractor and by of on behalf of the Owner all of which are deemed to form part of this Contract as though separately set out herein are included in the expression "Contract" whenever herein used AND WHEREAS the Owner has accepted the Tender of the Contractor for the provision and execution of the said Works for the sum of (hereinafter called "the Contract Price") upon the terms and subject to the Conditions hereinafter mentioned: NOW THIS DEED WITNESSETH AND IT IS HEREBY AGREED AND DECLARED as follows, that is to say, in consideration of the payments to be made to the Contractor by the Owner as hereinafter mentioned the Contractor hereby covenants with the Owner, his legal personal representatives, successors, and assigns that the Contractor shall and will duly provide, execute, complete and maintain the said Works and shall do and perform all other acts and things in the Contract mentioned or described or which are to be implied therefrom or may be reasonably necessary for the completion of the said Works within and at the times and in the manner and subject to the terms, conditions, and stipulations mentioned in the Contract.

AND in consideration of the due provision, execution, and completion of the Works and the maintenance thereof as aforesaid the Owner does hereby for himself, his legal personal representatives, successors, or assigns will pay to the Contractor the Contract Price or such other sum as may become payable to the Contractor under the provisions of the Contract, such payments to be made at such time and in such manner as is provided by the Contract.

IN WITNESS whereof, etc.

21. LIMITATIONS ON CONTRACTOR'S LIABILITY.–(i) The Contractor shall not be liable to the Owner under Clause 20 (Liability for Accidents and Damage) for:–

(a) any damage or injury to the extent that it is caused by or arises from the acts or omissions of the Owner or of others (not being the Contractor's servants or sub-contractors);

(b) any loss or damage in circumstances over which the Contractor has no control.

(ii) Except in respect of personal injury or damage to property conferring on a person other than the Owner a good cause of action against the Contractor, the liability of the Contractor under Clause 20 (Liability for Accidents and Damage) for any one act or default shall not exceed the Contract Price, or £100,000, whichever is the greater.

(iii) The Contractor's liability under Clause 20 (Liability for Accidents and Damage), as limited by Sub-Clause (iii) of this Clause, shall be in lieu of any other liability for damage or injury occurring before all the Works are taken over and caused by bis breach of contract or by the negligence or breach of statutory duty by the Contractor or any Sub-Contractor or by defective design (other than a design made, furnished or specified by the Owner and for which the Contractor has disclaimed responsibility in writing within a reasonable time after the receipt of the Owner's instructions), materials or workmanship but not otherwise.

(iv) Subject as provided in Clause 25 (Delay in Completion) for the deduction of liquidated damages for delay, the Contractor shall not be liable to the Owner for loss of use (whether complete or partial) of the Works or of profit or of any contract that may be suffered by the Owner whether by way of indemnity and whether caused by the Contractor's breach of contract, tort, breach of statutory duty or otherwise howsoever.

(1979 text)

Published by: The British Electrical and Allied Manufacturers Association Limited, 8 Leicester Street, London, WC2H 7BN

SA

Conditions of Sale (SA) for stock and catalogue articles United Kingdom

Edition March, 1978

1. **GENERAL.**—All quotations are made and all orders are accepted subject to the following terms and conditions and no addition thereto or variation therein shall be made unless agreed in writing by the parties.

2. **VALIDITY OF QUOTATIONS.**—We reserve the right to refuse your acceptance of a quotation unless such quotation is stated to be open for a specific period and is not withdrawn within such period.

3. **CATALOGUES.**—Catalogues, price lists and other advertising matter are only an indication of the type of goods offered and no prices or other particulars contained therein shall be binding on us.

4. **DESPATCH.**—Any times quoted for despatch are to be treated as estimates only and we shall not be liable for failure to despatch within such time unless you have suffered loss thereby and the amount payable in respect thereof shall have been agreed in writing as liquidated damages, in which case our liability shall be limited to the amount so agreed to be paid. In all cases, whether a time for despatch be quoted or not, the time for despatch shall be extended by a reasonable period if delay in despatch is caused by instructions or lack of instructions from you, or by industrial dispute, or by any cause whatsoever beyond our reasonable control.

5. **STORAGE.**—If by reason of instructions or lack of instructions from you despatch in accordance with the contract is delayed for 14 days after you have been notified that the goods are ready for despatch, the property in the goods shall pass to yourselves who shall take delivery or arrange for storage and for purposes of Clause 10 (Payment) the goods shall thereupon be deemed to have been delivered. If and for so long as our storage facilities permit, we may store the goods and you shall pay a reasonable charge therefor.

6. DELIVERY.–Unless otherwise specified in our tender, the price quoted includes delivery by any method of transport at our option.

On orders below £..................................in value carriage will be charged except where delivery is made by our own transport.

7. **LOSS OR DAMAGE IN TRANSIT.**—When the price quoted includes delivery, we shall repair or replace free of charge goods damaged in transit or not delivered in accordance with the Advice Note ; Provided that we are given written notification of such damage or non-delivery within such time as will enable us to comply with the carrier's conditions of carriage as affecting loss or damage in transit, or, where delivery is made by our own transport, within a reasonable time after receipt of the Advice Note.

8. **PACKING.**—Packing cases, skids, drums and other packing materials, if charged for, will be credited in full if returned in good condition carriage paid to us within one month of delivery of the goods.

9. **REJECTION.**—Unless otherwise agreed, goods rejected by you as not complying with the contract must be so rejected within 7 days of receipt by you.

10. **PAYMENT.**—Unless otherwise agreed in writing, payment in full is due in respect of any goods delivered.

11. **DEFECTS AFTER DELIVERY.**–We will make good, by repair or at our option by the supply of a replacement, defects which, under proper use, appear in the goods within a period of twelve calendar months after the goods have been delivered and arise solely from faulty design, materials or workmanship; Provided always that defective parts have been returned to us if we shall have so required. We shall refund the cost of carriage on such returned parts and the repaired or new parts will be delivered by us free of charge as provided in Clause 6 (Delivery).

Our liability under this clause shall be in lieu of any warranty or condition implied by law as to the quality or fitness for any particular purpose of the goods, and save as provided in this clause we shall not be under any liability, whether in contract, tort or otherwise, in respect of defects in goods delivered or for any injury (other than personal injury caused by our negligence as defined in Section 1 of the Unfair Contract Terms Act, 1977), damage or loss resulting from such defects or from any work done in connection therewith. Provided however that nothing in this clause shall operate to exclude any warranty or condition implied by law as to the quality of the goods in the event that the goods when sold by you or when sold by any person or persons to whom you may sell the goods shall become the subject of a consumer sale as defined in the Supply of Goods (Implied Terms) Act, 1973 except to the extent that any claim under such warranty or condition shall have arisen from any act or omission by you or by any other person or persons selling the goods by way of a consumer sale.

12. **PATENTS.**—We will indemnify you against any claim of infringement of Letters Patent, Registered Design, Trade Mark or Copyright (published at the date of the Contract) by the use or sale of any article or material supplied by us to you and against all costs and damages which you may incur in any action for such infringement or for which you may become liable in any such action ; Provided always that this indemnity shall not apply to any infringement which is due to our having followed a design or instruction furnished or given by you or to the use of such article or material in a manner or for a purpose or in a foreign country not specified by or disclosed to us, or to any infringement which is due to the use of such article or material in association or combination with any other article or material not supplied by us. And provided also that this indemnity is conditional on your making no admission in respect of such alleged infringement and giving us the earliest possible notice in writing of any claim being made or action threatened or brought against you and on your permitting us at our own expense to conduct any litigation that may ensue and all negotiations for a settlement of the claim. You on your part warrant that any design or instruction furnished or given by you shall not be such as will cause us to infringe any Letters Patent, Registered Design, Trade Mark or Copyright in the execution of your order.

13. **ARBITRATION.**—If at any time any question, dispute or difference whatsoever shall arise between yourselves and ourselves upon, in relation to or in connection with the contract, either of us may give to the other notice in writing of the existence of such question, dispute, or difference, and the same shall be referred to the arbitration of a person to be mutually agreed upon, or failing agreement within 14 days of receipt of such notice, of some person appointed by the President for the time being of the Institution of Electrical Engineers.

14. **LEGAL CONSTRUCTION.**—Unless otherwise agreed in writing the contract shall in all respects be construed and operate as an English contract and in conformity with English law.

15. STATUTORY AND OTHER REGULATIONS.–If the cost to us of performing our obligations under the contract shall be increased or reduced by reason of the making or amendment after the date of tender of any law or of any order, regulation, or bye-law having the force of law that shall affect the performance of our obligations under the contract, the amount of such increase or reduction shall be added to or deducted from the contract price as the case may be.

Published by: The British Electrical and Allied Manufacturers Association Limited, 8 Leicester Street, Leicester Square, London, WC2H 7BN

SAE

Edition December, 1980

Conditions of Sale (SAE) for stock and catalogue articles Export F.O.B.

1. **DEFINITIONS.**–The expression FOB shall bear the meaning assigned to it in **INCOTERMS 1980** save insofar as the same may have been varied by these Conditions.

2. **GENERAL.**–All quotations are made and all orders are accepted subject to the following terms and conditions and no addition thereto or variation therein shall be made unless agreed in writing by the parties.

3. **VALIDITY OF QUOTATIONS.**–We reserve the right to refuse your acceptance of a quotation unless such quotation is stated to be open for a specific period and is not withdrawn within such period.

4. **IMPORT AND EXPORT LICENCES.**–The contract shall be subject to the procurement by you at your own expense of any import licence required for the import of the goods into the country to which the goods are to be despatched from the United Kingdom, and to the procurement by us at our own expense of any export licence required for the export of the goods from the United Kingdom. Provided that where the order is placed from an address in the United Kingdom you shall be responsible for the procurement at your own expense of such export licence.

5. **CATALOGUES.**–Catalogues, price lists and other advertising matter are only an indication of the type of goods offered and no particulars contained therein shall be binding on us.

6. **DESPATCH.**–Any times quoted for despatch are to be treated as estimates only and we shall not be liable for failure to despatch within such time unless you have suffered loss thereby and the amount payable in respect thereof shall have been agreed in writing as liquidated damages, in which case our liability shall be limited to the amount so agreed to be paid. In all cases, whether a time for despatch be quoted or not, the time for despatch shall be extended by a reasonable period if delay in despatch is caused by instructions or lack of instructions from you, or by industrial dispute, or by any cause whatsoever beyond our reasonable control.

7. **STORAGE.**–If by reason of instructions or lack of instructions from you despatch in accordance with the contract is delayed for 21 days after you have been notified that the goods are ready for despatch, the property in the goods shall pass to you who shall take delivery or arrange for storage and for purposes of Clause 11 (Payment) the goods shall thereupon be deemed to have been delivered. If and for so long as our storage facilities permit, we may store the goods and you shall pay a reasonable charge therefor.

8. **DELIVERY.**–Unless otherwise agreed in writing, the price quoted includes delivery f.o.b. at the port stated in our tender. We shall not be required to give you the notice relating to insurance mentioned in Section 32(3) of the Sale of Goods Act, 1979

9. **LOSS OR DAMAGE IN TRANSIT.**–We shall not be liable for loss or damage to the goods beyond the point of shipment by us unless such loss or damage is due to faulty packing.

10. **PACKING.**–Unless otherwise agreed, packing will be charged extra at cost.

11. **PAYMENT.**–Unless otherwise agreed in writing, payment is due on delivery. Subject to the provision of Clause 7 goods shall be deemed to have been delivered when the invoice has been presented in the United Kingdom accompanied by appropriate documents of title.

12. **DEFECTS AFTER DELIVERY.**–We will make good, by repair or at our option by the supply of a replacement, defects which, under proper use, appear in the goods within a period of twelve calendar months after the goods have been delivered and arise solely from faulty design, materials or workmanship: Provided always that defective parts have been returned to us if we shall have so required. We shall refund the cost of carriage on such returned parts and the repaired or new parts will be delivered by us free of charge as provided in Clause 8 (Delivery).

Our liability under this clause shall be in lieu of any warranty or condition implied by law as to the quality or fitness for any particular purpose of the goods, and save as provided in this clause we shall not be under any liability, whether in contract, tort or otherwise, in respect of defects in goods delivered or for any injury, damage or loss resulting from such defects or from any work done in connection therewith.

13. **PATENTS.**–We will indemnify you against any claim of infringement of Letters Patent, Registered Design, Trade Mark or Copyright (published at the date of the Contract) by the use or sale of any article or material supplied by us to you and against all costs and damages which you may incur in any action for such infringement or for which you may become liable in any such action: Provided always that this indemnity shall not apply to any infringement which is due to our having followed a design or instruction furnished or given by you or to the use of such article or material in a manner or for a purpose or in a foreign country not specified by or disclosed to us or to any infringement which is due to the use of such article or material in association or combination with any other article or material not supplied by us. And provided also that this indemnity is conditional on you making no admission in respect of such alleged infringement and giving us the earliest possible notice in writing of any claim being made or action threatened or brought against you and on you permitting us at our own expense to conduct any litigation that may ensue and all negotiations for a settlement of the claim. You on your part warrant that any design or instruction furnished or given by you shall not be such as will cause us to infringe any Letters Patent, Registered Deisgn, Trade Mark or Copyright in the execution of your Order.

14. **ARBITRATION.**–If at any time any question, dispute or difference whatsoever shall arise between yourselves and ourselves upon, in relation to or in connection with the contract, either of us may give to the other notice in writing of the existence of such question, dispute or difference, and the same shall be referred to the arbitration of a person to be mutually agreed upon, or failing agreement within 30 days of receipt of such notice, of some person appointed by the President for the time being of the Institution of Electrical Engineers.

15. **LEGAL CONSTRUCTION.**–Unless otherwise agreed in writing the contract shall in all respects be construed and operate as an English contract and in conformity with English law and the English courts shall have exclusive jurisdiction over any matter arising out of the provisions of Clause 14 (Arbitration).

16. **STATUTORY AND OTHER REGULATIONS.**–If the cost to us of performing our obligations under the contract shall be increased or reduced by reason of the making or amendment after the date of tender of any law or of any order, regulation, or bye-law having the force of law that shall affect the performance of our obligations under the contract, the amount of such increase or reduction shall be added to or deducted from the contract price as the case may be.

Published by: The British Electrical and Allied Manufacturers Assocation Limited, 8 Leicester Street, Leicester Square, London, WC2H 7BN

Edition June, 1979

Conditions of Contract for Commissioning Electronic Equipment United Kingdom

DEFINITION OF TERMS

"The Contract" shall mean such documents as are agreed at the date of the formation of the Contract together with such variations in writing as shall subsequently be agreed in relation to the Commissioning of the Equipment.

"The Contract Price" shall be the sum named in the Contract as the Contract Price.

"Equipment" shall mean all machinery, apparatus, materials and articles which have been provided by us under a separate Contract and which are to be commissioned by us in accordance with the provisions of the Contract.

"Plant" shall mean all machinery, apparatus, materials and articles (other than Equipment) to be provided by you on the site which will be used in association or conjunction with the Equipment.

"Installation" shall mean the fixing in position of the various items of Equipment, their interconnection and the connection to the power supplies and other services.

"Commissioning" shall mean the checking, adjusting, testing and proving of the Equipment, in conjunction with the Plant, and shall include the making of electrical and/or mechanical connections between the Equipment and the Plant all as specified in this Contract.

1. **GENERAL**–The acceptance of our tender includes acceptance of the following terms and conditions:–

2. **VALIDITY**–Unless previously withdrawn, our tender is open for acceptance within the period stated therein, or when no period is so stated, within 30 days only after its date.

3. **ACCEPTANCE**–The acceptance of our tender must be accompanied by all necessary information which we may reasonably require to enable us to proceed with the work on site, without hindrance, from the date or dates specified.

4. **LIMITS OF CONTRACT**–Our tender includes only such work and services as are specified therein.

5. **SITE FACILITIES**–Our tender is based on our estimate of such work and services as will be required to commission the Equipment and assumes the following conditions:–

You will be responsible for ensuring that all Plant and Equipment is correctly installed (except Equipment installed by us), all Plant is sufficient and suitable for its purpose and all Equipment is in accordance with its Specification and that any minor adjustments that may be required by us are carried out expeditiously.

You will provide suitable access to and possession of the site, satisfactory environmental conditions for the Equipment, adequate scaffolding, all unskilled labour, suitable guarding and protection for all Equipment from time of delivery, any electrical power, lighting and heating necessary, and all other necessary facilities and adequate assistance.

You will provide all necessary facilities, including the following, free of charge, as and when required to enable our obligations under the Contract to be expeditiously and properly carried out, and ensure that:–

(a) Associated Plant necessary to the proper functioning of our Equipment is fully operational.

(b) Competent operators are available for the Plant as and when required.

(c) Installation of the Equipment has been completed to the stage where electrical supplies and other services may be safely applied.

(d) The Equipment and Plant is made available to us, and that we have access thereto as necessary to enable the progress of the work to proceed without interruption or hindrance.

6. **HOURS OF WORK**–Our Engineers will work normal hours applicable to the Engineering Industry, Monday to Friday inclusive, local public holidays excepted. Unless stated to the contrary in our tender, or undertaken by us in writing, night work, overtime and holiday working are specifically excluded.

7. **EXTRA COST**–Should we incur extra cost owing to variation or suspension of the work by your instructions, or lack of instructions, or to interruptions, delays, overtime, unusual hours, mistakes or work for which we are not responsible, or to any specified site environmental conditions not being maintained by you, the Contract Price will be adjusted in accordance with our standard rates of charge ruling at the time such extra costs are incurred.

8. **MATERIALS AND APPARATUS USED IN THE WORK OF COMMISSIONING**–Our Engineers will be provided with such tools and test equipment as may be specified in the contract and if it is necessary to keep any such tools or test equipment on site ready for our use you will provide adequate facilities for storage, free of charge, on or adjacent to the location of the Equipment, and you will indemnify us against all loss and damage to such tools and test equipment while in your custody.

You will provide any other tools or equipment, materials, lifting tackle, jacks or any special apparatus or instruments required during Commissioning. Should you fail to provide any such items which it is your responsibility to provide under the terms of the Contract these will be provided by us and you will be charged extra unless such items are available from your stores.

9. **LIABILITY FOR DELAY**–Any times quoted by us in writing for completion of Commissioning are to date from receipt by us of full and final instructions, sufficient and suitable availability of the Equipment and Plant, and full access to the Equipment and Plant to allow us to proceed. The time for completion shall be extended by a reasonable period if delay in completion is caused by instructions or lack of instructions from you or by industrial dispute or by any cause beyond our reasonable control. If a fixed time be quoted for completion, and we fail to complete within that time or within any extension thereof provided by this Clause, and if, as a result, you shall have suffered loss, we undertake to pay for each week or part of a week of delay liquidated damages at the rate of ....per cent. up to a maximum of ....per cent. of the Contract Price which is referable to such portion only of the Commissioning as cannot in consequence of the delay be completed. Such payment shall be in full satisfaction of our liability for delay.

Any time described as an estimate shall not be construed as a fixed time quoted for the purpose of this Clause.

10. **LIABILITY FOR ACCIDENTS AND DAMAGE**–We will indemnify you against direct damage or injury to your property or person or that of others occurring while we are working on site to the extent caused by the negligence of ourselves, our sub-contractors or agents, but not otherwise, by making good such damage to property or compensating personal injury. Provided that:

(a) Our total liability for damage to your property (including damage caused by our breach of contract, tort or breach of statutory duty) shall not exceed £100,000 or the Contract Price, whichever sum is the greater, and

(b) we shall not be liable to you for any loss of profit or of contracts or, save as aforesaid, for any loss or damage of any kind whatsoever and whether caused by our breach of contract, tort, breach of statutory duty or otherwise howsoever.

We shall not be liable for any injury (other than personal injury caused by our negligence as defined in Section 1 of the Unfair Contract Terms Act, 1977), or damage occurring after completion of work on site.

Continued overleaf

11. COMPLETION OF COMMISSIONING–Commissioning, or where the contract provides for Commissioning in Sections, Commissioning of any Section shall be deemed to be complete and a certificate issued accordingly when the Equipment or any Section thereof has satisfactorily passed such tests as may be specified in the contract or as may have been mutually agreed between us in writing, or one calendar month after the Equipment or any Section thereof has been put into use by you whichever is the earlier. Where no such tests are so specified or have been so agreed the Commissioning or any Section thereof shall be deemed to be complete when our Engineer on site advises you that the Equipment or Section thereof has satisfactorily passed such tests as we deem necessary to prove the correct function of the Equipment or any Section thereof, or one calendar month after the Equipment or any Section thereof has been put into use by you whichever is the earlier. Acceptance of completion of Commissioning or any Section thereof shall not be delayed on account of additions, minor omissions or defects which do not materially affect the use of the Equipment.

12. TERMS OF PAYMENT–Unless otherwise agreed, payment in full shall be made by you on completion of Commissioning as defined in Clause 11.

13. EXCEPTIONAL CONDITIONS–Should we incur additional expense by reason that our Engineers are required to work under exceptional conditions which were not made known to us at the time of tender, or by reason of local wage settlements, we shall be entitled to amend the contract price to compensate us for any additional costs that we have reasonably and properly incurred.

14. FAIR WAGES CLAUSE–We undertake to be bound by a fair wages clause in the terms of the House of Commons Resolution of 14th October, 1946. We also undertake that when the installation of machinery and equipment is carried out by men sent from the Contractor's or Sub-Contractor's establishment they shall receive the time rate payable in terms of the Fair Wages Clause to such work-people in such establishment, and in addition shall receive the outworking allowances recognised for outworkers sent from such establishment.

15. ARBITRATION–If at any time any question, dispute, or difference whatsoever shall arise between you and ourselves upon, in relation to, or in connection with the Contract, either of us may give to the other notice in writing of the existence of such question, dispute or difference, and the same shall be referred to the arbitration of a person to be mutually agreed upon, or failing agreement within 30 days of the receipt of such notice, of some person appointed by the President for the time being of the Institution of Electrical Engineers.

16. LEGAL CONSTRUCTION–Unless otherwise agreed in writing, the Contract shall in all respects be construed and operate as an English contract and in conformity with English Law.

17. STATUTORY AND OTHER REGULATIONS–If the cost to us of performing our obligations under the contract shall be increased or reduced by reason of the making or amendment after the date of tender of any law or of any order, regulation, or bye-law having the force of law that shall affect the performance of our obligations under the contract, the amount of such increase or reduction shall be added to or deducted from the contract price as the case may be.

Published by: The British Electrical and Allied Manufacturers' Association (Ltd.), 8 Leicester Street, Leicester Square, London WC2H 7BN

Edition June, 1979

CONDITIONS OF CONTRACT FOR SYSTEMS
INCORPORATING ELECTRONIC EQUIPMENT
(Including Installation)
UNITED KINGDOM

1. DEFINITIONS—In these Conditions the following words shall have the following meanings:

Contract	shall mean such documents as are agreed at the date of formation of the Contract together with such variations in writing as shall subsequently be agreed.
Contract Price	shall mean the sum named in the Contract as the Contract Price.
System	shall mean the Equipment and Programs (if any) and the method by which they are used to perform in association or conjunction with the Plant the specified functions agreed between us.
Equipment	shall mean all machinery apparatus materials and articles to be provided by us under the Contract.
Plant	shall mean all machinery apparatus materials and articles (other than the Equipment) to be provided by you on the site and to be used in association or conjunction with the Equipment.
Programs	shall mean all computer programs to be provided by us under the Contract for use with the Equipment.
Installation	shall mean the fixing in position of the various items of the Equipment, their interconnection and the connection to the electrical power supply all as included in the Contract.
Standard Works Tests	shall mean the tests performed on items of the Equipment in our works to standard test procedures.
Tests on Installation	shall mean the tests to be performed to our standard equipment test specification on the Equipment or any portion thereof after Installation but without any mechanical or electrical connection to the Plant.
Commissioning	shall mean the checking adjusting testing and proving of the Equipment and the System and shall include the making of electrical and/or mechanical connections between the Equipment and the Plant, all as included in the Contract.
System Tests	shall mean the tests to demonstrate that the System is capable of achieving the specified functions.

2. GENERAL—The acceptance of our tender includes the acceptance of the following terms and conditions and of the special conditions (if any) stated in or referred to in the tender.

3. VALIDITY—Unless previously withdrawn, our tender is open for acceptance within the period stated therein or, when no period is so stated, within thirty days only after its date.

4. LIMITS OF CONTRACT—Our tender includes only such equipment and work as are specified therein. Notwithstanding the above we reserve the right as our detailed design proceeds to make minor changes to the equipment and work, provided that such changes shall not affect the Contract Price or the completion date.

5. SUPPLY OF INFORMATION TO US—You will provide us with all necessary information that we may reasonably require from time to time to permit us to proceed uninterruptedly with the design of the System, the design and writing of the Programs (if any), the design and manufacture of the Equipment, the Installation and the Commissioning. In the event that the work is delayed or the extent of the work is increased by reason of delay in the provision by you of the necessary information or to change in such information, we shall be at liberty to amend the Contract Price to compensate us for any additional costs that we have reasonably and properly incurred, and to extend the completion date by a reasonable period.

6. SUPPLY OF INFORMATION BY US—

6.1. Unless otherwise specified in our tender all drawings and particulars of weights and dimensions submitted therewith are approximate only. Descriptions and illustrations contained in our catalogues and other advertisement matter are intended merely to present a general idea of the equipment described therein, and none of these shall form part of the Contract.

6.2. Within such time as may be specified in our tender or if no such time is specified within a reasonable time after acceptance of our tender, we shall submit for your approval two copies of detailed specifications covering the System. Such specifications and any drawings submitted for approval shall be approved or otherwise within the period stated in our tender or where no period is stated within 14 days after such submission. If you disapprove such specifications or drawings we shall, subject to Clause 5 (Supply Of Information To Us) modify them to meet your reasonable requirements, and resubmit for approval. We shall also supply to you two certified copies of the necessary drawings to enable you to make provision for Installation.

6.3. Following completion of System Tests or at such other time as may be agreed we shall furnish you with such information and drawings as may be necessary for the routine operation and maintenance of the Equipment by you, together with documentation describing the Programs (if any) supplied, always provided that we shall be under no obligation to supply copies of manufacturing drawings.

7. COPYRIGHT AND CONFIDENTIALITY—

7.1. All specifications drawings and technical descriptions submitted with or in connection with our tender or the Contract are copyright.

7.2. You shall keep confidential and not without our prior consent in writing disclose to any third party any drawings designs or information (whether of a commercial or technical nature) acquired from us pursuant to our tender or the Contract but you shall use the same only for the purpose of:

(a) adjudicating the tender, or
(b) the Contract, or
(c) the operation of the Equipment.

7.3. We undertake to keep confidential and not to disclose without your prior consent in writing to any third party any trade or business secrets or similar confidential information supplied by you to us relating to your plant or processes except as may be necessary for the proper performance of the Contract.

7.4. Sub-Clauses 7.2. and 7.3. are each subject to the proviso that nothing therein contained shall apply to prevent either of us as the case may be from disclosing information:

(a) in its possession (with full right to disclose) prior to receiving it from the other, or
(b) which is or later becomes public knowledge other than by breach of this clause, or
(c) which it may independently receive from a third party (with full right to disclose).

8. PACKING—Unless otherwise specified in our tender, all packing cases, skids, drums and other packing materials must be returned to our works at your expense and in good condition. If not so returned they will be charged for.

9. TRANSPORT—

9.1. Unless otherwise specified in our tender, the price quoted includes delivery to your site by any method of transport at our option.
9.2. Unless otherwise specified, we shall not be responsible for offloading or transhipment to the place of installation.

10. SITE FACILITIES—

10.1. To enable our obligations under the Contract to be expeditiously and properly carried out you will provide the following facilities free of charge as and when required:

(a) suitable access to and possession of the site, proper foundations and satisfactory environmental conditions for the Equipment, adequate lifting facilities and scaffolding, all unskilled labour, masons', joiners', and builders' work, suitable guarding and protection for the Equipment from time of delivery, any electric power, lighting and heating necessary and all other necessary facilities and adequate assistance.
(b) permanent and suitable electrical supplies for the Equipment.
(c) suitable access to the Plant at reasonable times and for reasonable periods.
(d) competent operators and attendants for the Plant.

10.2. You will be responsible for ensuring that the Plant is correctly installed and is sufficient and suitable for its purpose and that any minor adjustments that may be required by us to be made to the Plant are carried out expeditiously.

11. TESTS—

Works Tests

11.1 The Equipment will be submitted to our standard Works Tests before despatch. Copies of standard test procedures and certification that the Equipment has passed the Standard Works Tests will be supplied on request.

11.2. If you require any additional tests to be conducted in our works these must be agreed and you will be charged for them.

11.3. If any works tests are to be held in the presence of you or your representatives you will be charged therefor. In the event of any delay by you or your representatives attending such tests after 7 days' notice that we are ready, the tests will proceed in your absence and will be deemed to have been made in your presence.

Installation Tests

11.4 Tests on Installation shall be conducted to our standard equipment test specification in the presence of you or your representatives. The Tests on Installation shall be commenced within 7 days of our having given written notice that we are ready to proceed. If we are unable to proceed with the Tests on Installation for reasons within your control or that of other contractors then the Equipment or the relevant portion thereof shall be deemed to have been taken over. Subsequently the Tests on Installation shall be made at a time to be agreed. If any portion of the Equipment fails to pass the Tests on Installation, such tests on the said portion shall, if required, be repeated within a reasonable time and upon the same terms and conditions.

System Tests

11.5. System Tests shall be performed in accordance with the agreed system test specification and commenced within 7 days after we have given written notice that the System is ready for System Tests. It will be your responsibility to ensure that during the period of test the Plant is working normally in accordance with the operating procedure agreed during the system design and within the limits laid down (if any), and that members of your staff whose acts or omissions may affect the operation of the System exercise all appropriate skill and care. For the purpose of Clause 16 (Terms of Payment) the System Tests shall nevertheless be deemed to have been successfully carried out if, due to circumstances within your control or that of other contractors:

(a) System Tests are delayed beyond the time stated in the Contract or when no time is stated for an unreasonable time.
(b) the System fails to pass the System Tests.

In either case System Tests shall subsequently be carried out at a time to be agreed.

11.6. If the results of the System Tests show that we have failed to achieve the requirements of the system test specification then the System Tests shall, if required by either of us, be repeated at a time to be agreed and in the event that:

(a) you or other contractors are responsible for such failure then all our reasonable expenses incurred for the purpose of repeating the System Tests shall be paid by you, or
(b) we are responsible for such failure then all your reasonable expenses incurred for the purpose of witnessing the repeated System Tests shall be paid by us.

12. PERFORMANCE—

12.1 You assume responsibility that any performance requirements or equipment stipulated by you are sufficient and suitable for your purpose, save insofar as your stipulations are in accordance with our advice.

12.2. If the results of the System Tests show that we have failed to achieve performance figures specifically guaranteed by us (subject to any tolerance) in an agreed sum as liquidated damages and the failure is due to factors wholly within our control, then you shall be entitled to payment by us of the liquidated damages provided for the failure at the rates and subject to the terms set out in the said specific guarantee. Before you become entitled to payment of liquidated damages we shall be given reasonable time and opportunity to make good the said failure. The payment by us of liquidated damages shall be in full satisfaction of our liability under this clause. We shall not be liable for failure to meet any performance requirements not so guaranteed.

13. TIME OF TAKING OVER—

13.1. The Equipment or any portion thereof shall be taken over at the earlier of the following times and a take-over certificate issued accordingly:

(a) when Installation has been completed and the Equipment or such portion has been passed or is deemed to have passed the Tests on Installation.
(b) when the Equipment or such portion shall have been put into use.

13.2. The time of taking over shall not be delayed on account of additions, minor omissions or defects which do not materially affect the use of the Equipment.

14. DEFECTS AFTER TAKING OVER—

14.1. We will make good, by repair or at our option by the supply of a replacement, defects which under proper use care and maintenance appear in the Equipment or System within a period of 18 calendar months after the Equipment has been taken over or 12 calendar months after the System has passed the System Tests (whichever period shall expire the earlier) and arise solely from faulty design, materials or workmanship excluding any defects caused by design, materials or workmanship furnished by you; provided that in respect of the items listed in Appendix A there shall be substituted for a period of 12 or 18 months as the case may be the periods se against such items respectively, and provided always that defective parts have been returned to us if we shall have so required. We shall refund the cost of carriage on such returned parts and the repaired or new parts will be delivered by us free of charge to your site.

14.2. If during the period stated in Sub-Clause 14.1. a fault in the Equipment or System occurs and you are unable to locate it using proper skill and following the fault finding procedures (if any) laid down, we will at your request locate and rectify the fault for you. This location and rectification work will be entirely at our cost if the fault was due to our faulty design, materials or workmanship and you had used proper skill in attempting to locate the fault. If however although the fault was due to our faulty design, materials or workmanship you had failed to use such proper skill in locating it, you will reimburse us for our reasonable costs in visiting your site and locating the fault.

14.3. Our liability under this clause shall be in lieu of any warranty or condition implied by law as to the quality or fitness for any particular purpose of the Equipment or System, and save as provided in this clause and in Clause 12 we shall not be under any liability, whether in contract, tort or otherwise, in respect of defects after taking over or for any injury (other than personal injury caused by our negligence as defined in Section 1 of the Unfair Contract Terms Act, 1977), damage or loss resulting from such defects or from any work done in connection therewith.

15. EXTRA COST—Should we incur extra cost owing to variation or suspension of the work by your instructions or lack of instructions, or to interruptions, delays, overtime, unusual hours, mistakes, or work, for which we are not responsible, or to any specified site environmental conditions not being maintained by you, a reasonable sum in respect of such extra cost shall be added to the Contract Price and be paid accordingly.

16. TERMS OF PAYMENT—

16.1. Payments shall be made by you as follows:

(a) per cent of the Contract Price at the time of placing the order or at the time when we receive your written instructions to put the work in hand, whichever is the earlier.
(b) a further per cent of the Contract Price as progress payments at the times and percentages specified in our tender. When no such times and percentages are specified the payments shall be made on delivery of the Equipment or portions thereof from time to time on site.
(c) a further per cent of the Contract Price (or *pro rata* for a portion) when the Equipment or any portion thereof has passed or is deemed to have passed the Tests on Installation.
(d) a further per cent of the Contract Price when the System has passed or is deemed to have passed the System Tests.
(e) the balance of the Contract Price 30 days after the payment under *(d)* falls due.
(f) payments for any work undertaken at an agreed rate shall be made monthly in arrears.

16.2. If we are unable, due to causes within your control or that of other contractors, to deliver all or any of the Equipment when ready or to proceed with the Installation of such Equipment as we have already delivered, we shall be entitled to arrange storage either at our own works or elsewhere on your behalf; all charges for packing and storage, for insurance, for demurrage, for additional carriage and for any retesting and necessary refurbishing shall be payable by you. In any case you will make any payment due to us on delivery under 16.1.*(b)* as though delivery has been made and the payments under 16.1.*(c)* within three calendar months from the date of notification that the Equipment is ready for delivery.

16.3. Payments under 16.1.*(c)*, 16.1.*(d)* and 16.1.*(e)* shall not be withheld on account of minor defects or omissions in the Equipment or the System which do not materially affect its use. You will, however, be entitled to retain from the payments under 16.1.*(e)* such sums as represent the value of such minor defects or omissions until the defects or omissions shall be remedied, which will be done by us at the earliest opportunity.

17. **STATUTORY AND OTHER REGULATIONS.**–If the cost to us of performing our obligations under the Contract shall be increased or reduced by reason of the making or amendment after the date of tender of any law or of any order, regulation, or bye-law having the force of law that shall affect the performance of our obligations under the Contract, the amount of such increase or reduction shall be added to or deducted from the Contract Price as the case may be.

18. VARIATION IN COSTS–If the cost to us of performing our obligations under the Contract shall be increased or reduced by reason of any rise or fall in labour costs or in the cost of material or transport above or below such rates and costs ruling at the date of tender the amount of such increase or reduction shall be added to or deducted from the Contract Price as the case may be provided that no account shall be taken of any amount by which any cost incurred by us has been increased by our default or negligence. For the purposes of this clause "the cost of material" shall be construed as including any duty or tax by whomsoever payable which is payable under or by virtue of any Act of Parliament on the import, purchase, sale, appropriation, processing or use of such material.

19. LIABILITY FOR DELAY—

19.1. Any times quoted for completion of Tests on Installation (including where applicable supply of Programs unless otherwise agreed) are to date from receipt by us of a written order to proceed, and of all necessary information and drawings to enable us to put the work in hand. The time for such completion shall be extended by a reasonable period if delay in such completion is caused by industrial dispute or by any cause beyond our reasonable control.

19.2. If a fixed time for such completion of Tests on Installation be quoted and we fail to complete within that time or any extension thereof, and if as a result you shall have suffered loss, we undertake to pay for each week or part of a week of delay, liquidated damages at the rate of   per cent up to a maximum of   per cent of that portion of the Contract Price which is referable to such portion only of the Equipment as cannot in consequence of the delay be commissioned effectively. Such payment shall be in full satisfaction of our liability for delay.

20. PASSING OF TITLE—Property and risk in the Equipment shall pass on delivery. Unless specifically agreed by us to the contrary you will be responsible for insuring the Equipment thereafter.

21. LIABILITY FOR ACCIDENTS AND DAMAGE–We will indemnify you against direct damage or injury to your property or person or that of others to the extent caused by the negligent acts or omissions of ourselves, our sub-contractors or agents while working on site pursuant to the Contract, but not otherwise, by making good such damage to property or compensating personal injury.
Provided that:

(a) our total liability for damage to your property (including damage caused by our breach of contract, tort or breach of statutory duty) shall not exceed £100,000 or the Contract Price, whichever is the greater, and

(b) we shall not be liable to you for any loss of profits or of contracts or, save as aforesaid, for any loss or damage of any kind whatsoever and whether caused by our breach of contract, tort, breach of statutory duty or otherwise howsoever.

22. PATENTS—We will indemnify you against any claim of infringement of Letters Patent, Registered Design, Trade Mark or Copyright (published at the date of the Contract) by the use or sale of the Equipment and Programs (if any) and against all costs and damages which you may incur in any action for such infringement or for which you may become liable in any such action. Provided always that this indemnity shall not apply to any infringement which is due to our having followed a design or instruction furnished or given by you or to the use of the Equipment and Programs (if any) in a manner or for a purpose or in a foreign country not specified by or disclosed to us, or to any infringement which is due to the use of the Equipment and Programs (if any) in association or combination with the Plant or any other equipment not supplied by us. And provided also that this indemnity is conditional on your giving to us the earliest possible notice in writing of any claim being made or action threatened or brought against you and on your permitting us at our own expense to conduct any litigation that may ensue and all negotiations for a settlement of the claim. You on your part warrant that any design or instructions furnished or given by you shall not be such as will cause us to infringe any Letters Patent, Registered Design, Trade Mark or Copyright in the execution of the Contract.

23. FAIR WAGES CLAUSE—We undertake to be bound by a fair wages clause in the terms of the House of Commons Resolution of 14th October, 1946. We also undertake that when the installation and commissioning of equipment is carried out by men sent from our or our sub-contractor's establishment they shall receive the time rate payable in terms of the Fair Wages Clause to such work people in such establishment, and in addition shall receive the outworking allowances recognised for outworkers sent from such establishment.

24. ARBITRATION–If at any time any question, dispute or difference whatsoever shall arise between you and ourselves upon, in relation to, or in connection with the Contract, either of us may give to the other notice in writing of the existence of such question, dispute or difference, and the same shall be referred to the arbitration of a person to be mutually agreed upon, or failing agreement within 30 days of receipt of such notice, of some person appointed by the President for the time being of the Institution of Electrical Engineers.

25. LEGAL CONSTRUCTION—

25.1. Unless otherwise agreed in writing the Contract shall in all respects be construed and operate as an English contract and in conformity with English law.

25.2. Clause headings are for information only and shall not affect the construction or interpretation hereof.

APPENDIX A
(Guarantees on bought-out components)

Published by:
The British Electrical & Allied Manufacturers' Association Ltd.
8 Leicester Street, London, WC2H 7BN

Edition January 1979
Corrected edition

CONTRACT PRICE
ADJUSTMENT CLAUSE AND FORMULAE FOR USE WITH

HOME CONTRACTS

ELECTRICAL MACHINERY: (for which there is no other specific Formulae)

If the cost to the Contractor of performing his obligations under the Contract shall be increased or reduced by reason of any rise or fall in labour costs or in the cost of material or transport above or below such rates and costs ruling at the date of tender, or by reason of the making or amendment after the date of tender of any law or of any order, regulation, or bye-law having the force of law in the United Kingdom that shall affect the Contractor in the performance of his obligations under the Contract, the amount of such increase or reduction shall be added to or deducted from the Contract Price as the case may be provided that no account shall be taken of any amount by which any cost incurred by the Contractor has been increased by the default or negligence of the Contractor. For the purposes of this clause "the cost of material" shall be construed as including any duty or tax by whomsoever payable which is payable under or by virtue of any Act of Parliament on the import, purchase, sale, appropriation, processing or use of such material.

The operation of this Clause is without prejudice to the effect if any which the imposition of Value Added Tax or any tax of a like nature may have upon the supply of goods or services under the Contract.

Variations in the cost of materials and labour shall be calculated in accordance with the following Formulae:-

(a) Labour

The Contract Price shall be adjusted at the rate of 0.475 per cent of the Contract Price per 1.0 per cent difference between the BEAMA Labour Cost Index published for the month in which the tender date falls and the average of the Index figures published for the last two-thirds of the contract period, this difference being expressed as a percentage of the former Index Figure.

(b) Materials

The Contract Price shall be adjusted at the rate of 0.475 per cent of the Contract Price per 1.0 per cent difference between the Price Index figure of Materials used in the Electrical Machinery Industry last published in the Trade and Industry Journal before the date of tender and the average of the Index Figures commencing with the Index last published before the two-fifths point of the Contract Period and ending with the Index last published before the four-fifths point of the Contract Period, this difference being expressed as a percentage of the former Index figure.

(The following footnotes must be added as appropriate according as to whether the Contract excludes or includes erection).

FOR CONTRACTS EXCLUDING ERECTION

For the purpose of these Formulae:-

(i) Where separate portions of the plant are ready for despatch at different times and are invoiced separately, the Contract Price shall, in relation to each such portion, be an appropriate proportion of the total Contract Price.

(ii) The contract period in respect of any portion of plant shall be deemed to be that period between the date of order and the date when such portion is ready for despatch, or such shorter period corresponding to the manufacturing cycle of the equipment as may be stipulated in the tender or agreed in the Contract.

(iii) Where any index figure is stated to be provisional or is subsequently amended, the figure shall apply as ultimately confirmed, or amended.

FOR CONTRACTS INCLUDING ERECTION:

For the purpose of these Formulae:-

(i) The contract period shall be deemed to be that period between the date of order and the date when the plant is taken over or is ready for commercial use, whichever is the earlier, or such shorter period corresponding to the manufacturing cycle of the equipment as may be stipulated in the tender or agreed in the Contract.

-2-

(ii) The plant shall be deemed to be ready for commercial use even though certain minor matters which do not affect the use for which the plant is intended remain to be completed.

(iii) Where any index figure is stated to be provisional or is subsequently amended, the figure shall apply as ultimately confirmed, or amended.

INTERIM CPA PROCEDURES

Payments on account of CPA shall apply where the Contract is subject to progress or interim terms of payment.

All such claims will be calculated in an identical manner to the final CPA claim except that the contract completion date for this purpose will be the date to which the progress and/or interim payment is calculated and the contract value will be the cumulative total of progress and/or interim payments claimable to that date. Each CPA claim will have deducted from it the cumulative value of all previous CPA claims.

Published by: the British Electrical and Allied Manufacturers Association Limited, 8 Leicester Street, London, WC2H 7BN.

Edition January 1979
Corrected edition

CONTRACT PRICE
ADJUSTMENT CLAUSE AND FORMULA FOR USE WITH
EXPORT CONTRACTS

ELECTRICAL MACHINERY: (for which there is no other specific formula)

If the cost to the Contractor of performing his obligations under the Contract shall be increased or reduced by reason of any rise or fall in labour costs or in the cost of material or transport above or below such rates and costs ruling at the date of tender, or by reason of the making or amendment after the date of tender of any law or of any order, regulation, or bye-law having the force of law in the United Kingdom that shall affect the Contractor in the performance of of his obligations under the Contract, the amount of such increase or reduction shall be added to or deducted from the Contract Price as the case may be provided that no account shall be taken of any amount by which any cost incurred by the Contractor has been increased by the default or negligence of the Contractor. For the purposes of this clause "the cost of material" shall be construed as including any duty or tax by whomsoever payable which is payable under or by virtue of any Act of Parliament on the import, purchase, sale, appropriation, processing or use of such material.

Variations in the cost of materials and labour shall be calculated in accordance with the following Formula

(a) Labour

The FOB Price shall be adjusted at the rate of 0.475 per cent of the FOB Price per 1.0 per cent difference between the BEAMA Labour Cost Index published for the month in which the tender date falls and the average of the Index figures published for the last two-thirds of the delivery period, this difference being expressed as a percentage of the former Index Figure.

(b) Materials

The FOB Price shall be adjusted at the rate of 0.475 per cent of the FOB Price per 1.0 per cent difference between the Price Index figure of Materials used in the Electrical Machinery Industry last published in the Trade and Industry Journal before the date of tender and the average of the Index Figures commencing with the Index last published before the two-fifths point of the Delivery Period and ending with the Index last published before the four-fifths point of the Delivery Period, this difference being expressed as a percentage of the former Index figure.

For the purpose of this Formula:-

(i) Where separate portions of the plant are ready for despatch at different times and are invoiced separately, the FOB Price shall in relation to each such portion be an appropriate proportion of the total FOB Price of the Contract.

(ii) The delivery period of the Contract in respect of any portion of plant shall be deemed to be that period between the date of order and the date when such portion is ready for despatch from the manufacturer's works to Port of Shipment or such shorter period corresponding to the manufacturing cycle of the equipment as may be stipulated in the tender or agreed in the Contract.

(iii) Where any index figure is stated to be provisional or is subsequently amended, the figure shall apply as ultimately confirmed, or amended.

INTERIM CPA PROCEDURES

Payments on account of CPA shall apply where the Contract is subject to progress or interim terms of Payment.

All such claims will be calculated in an identical manner to the final claim except that the contract completion date for this purpose will be the date to which the progress and/or interim payment is calculated and the contract value will be the cumulative total of progress and/or interim payments claimable to that date. Each CPA claim will have deducted from it the cumulative value of all previous CPA claims.

Published by: The British Electrical and Allied Manufacturers Association Limited, 8 Leicester Street, London WC2H 7BN.

Appendix XXVI: Addresses of Institutions for Obtaining Published Conditions, etc.

Institution of Electrical Engineers ('IEE')
Savoy Place, London WC2R 0BL
(personal callers only)
Publications Sales Department,
Station House, 70 Nightingale Road,
Hitchin, Hertfordshire SG5 1RJ
(postal applications)

Institution of Mechanical Engineers ('IMechE')
1 Birdcage Walk, Westminster, London SW1H 9JJ
(personal callers only)
Publications Sales Department,
PO Box 24, Northgate Avenue,
Bury St Edmunds, Suffolk IP32 6BW
(postal applications)

Association of Consulting Engineers ('ACE')
Alliance House, 12 Caxton Street, London SW1H 0QL

Fédération Internationale des Ingénieurs – Conseils ('FIDIC')
PO Box 17334, 2502 CH, The Hague, Netherlands

The British Electrical and Allied Manufacturers' Association Limited ('BEAMA')
8 Leicester Street, Leicester Square, London WC2H 7BN

United Nations Economic Commission for Europe ('UNECE')
– apply to BEAMA

Organisme de Liaison des Industries Métalliques Européennes ('ORGALIME')
13 Rue des Drapiers, Bruxelles 5

Federation of Civil Engineering Contractors ('FCEC')
Cowdray House, 6 Portugal Street, London WC2A 2HH

The Institution of Chemical Engineers ('IChemE')
165–171 Railway Terrace, Rugby CV21 3HQ

International Chamber of Commerce ('ICC')
British National Committee, Centre Point,
103 New Oxford Street, London WC1A 1QB

Bibliography

Hudson's Building and Engineering Contracts, I.N. Duncan Wallace, 10th edn 1970, First Supplement 1979, Sweet & Maxwell Limited, London.
Schmitthoff's Export Trade, Clive M. Schmitthoff, 7th edn 1980, Stevens & Sons Limited, London.
Further Building and Engineering Standard Forms, I.N. Duncan Wallace, 1973, Sweet & Maxwell Limited, London.
Electrical and Mechanical Engineering Contracts, K.F.A. Johnston, 1971, Gower Press Limited, Farnborough, Hampshire.
Croner's Reference Book for Exporters, Croner Publications Limited, New Malden, Surrey.
ECGD Services, 1979, ECGD, London.
Cutting Your Losses, November 1980, ECGD, London.
Contract Bonds and Guarantees, September 1981, Confederation of British Industry, London.
Check List; Establishing a Joint Venture, The Export Group for the Constructional Industries, London.
Guide for Drawing up an International Consortium Agreement, 2nd edn May 1976, ORGALIME, Brussels.
Double Taxation Relief, Deloitte, Haskins & Sells, 1979, Tolley Publishing Company Limited, Croydon.

International Chamber of Commerce (ICC) Publications:
The Problem of Clean Bills of Lading, ICC Publication No 283.
Uniform Customs and Practice for Documentary Credits, ICC Publication No 290.
Uniform Rules for a Combined Transport Document, ICC Publication No 298.
Guide to Documentary Credit Operations, ICC Publication No 305.
Uniform Rules for Collections, ICC Publication No 322.
Standard Forms for Issuing Documentary Credits, ICC Publication No 323.
Uniform Rules for Contract Guarantees, ICC Publication No 325.
Guide to Incoterms, ICC Publication No 354 (1980 edn).

(*Note:* Besides the above publications, which are mentioned in the text, BEAMA members only are entitled to obtain from BEAMA its Conditions of Contract *Brown Book*, which contains a wealth of information on many of the published forms mentioned in this book.)